Abdallah Afoukal, Brahim Es-sebbar, Khalil Ezzinbi, Gaston Mandata N'Gu

Almost Periodicity and Almost Automorphy

Also of Interest

Topics in Complex Analysis
Dan Romik, 2023
ISBN 978-3-11-079678-0, e-ISBN (PDF) 978-3-11-079681-0

Complex Analysis and Special Functions
Cauchy Formula, Elliptic Functions and Laplace's Method
Valery Serov, Markus Harju, 2025
ISBN 978-3-11-163211-7, e-ISBN (PDF) 978-3-11-163227-8

Applied Nonlinear Functional Analysis
An Introduction
Nikolaos S. Papageorgiou, Patrick Winkert, 2024
ISBN 978-3-11-128421-7, e-ISBN (PDF) 978-3-11-128695-2

Applications of Complex Variables
Asymptotics and Integral Transforms
Foluso Ladeinde, 2024
ISBN 978-3-11-135090-5, e-ISBN (PDF) 978-3-11-135117-9s

Differential Geometry
Frenet Equations and Differentiable Maps
Muhittin E. Aydin, Svetlin G. Georgiev, 2024
ISBN 978-3-11-150089-8, e-ISBN (PDF) 978-3-11-150185-7

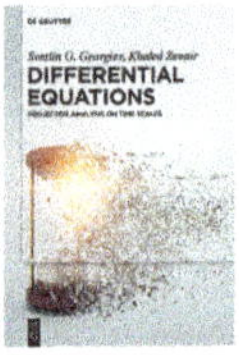

Differential Equations
Projector Analysis on Time Scales
Svetlin G. Georgiev, Khaled Zennir, 2024
ISBN 978-3-11-137509-0, e-ISBN (PDF) 978-3-11-137715-5

Abdallah Afoukal, Brahim Es-sebbar, Khalil Ezzinbi,
Gaston Mandata N'Guérékata

Almost Periodicity and Almost Automorphy

for Evolution Equations and Partial Functional Differential Equations

DE GRUYTER

Mathematics Subject Classification 2020
Primary: 34K14, 47D06, 34K30; Secondary: 35P05, 34K14

Authors

Abdallah Afoukal
Université Cadi Ayyad
Faculté des Sciences Semlalia
Département de Mathématiques
Marrakesh
Morocco
afoukal.abdallah@gmail.com

Khalil Ezzinbi
Université Cadi Ayyad
Faculté des Sciences Semlalia
Département de Mathématiques
Marrakesh
Morocco
ezzinbi@uca.ac.ma

Brahim Es-sebbar
Département de Mathématiques
Faculté des Sciences et Techniques Guéliz
Laboratoire de Mathématiques Appliquées et Informatique
Université Cadi Ayyad
Marrakesh
Morocco
b.es-sebbar@uca.ma

Gaston Mandata N'Guérékata
NEERLab
Mathematics Department
Morgan State University
Baltimore
USA
gaston.nguerekata@morgan.edu

ISBN 978-3-11-168361-4
e-ISBN (PDF) 978-3-11-168471-0
e-ISBN (EPUB) 978-3-11-168514-4

Library of Congress Control Number: 2024950669

Bibliographic information published by the Deutsche Nationalbibliothek
The Deutsche Nationalbibliothek lists this publication in the Deutsche Nationalbibliografie; detailed bibliographic data are available on the Internet at http://dnb.dnb.de.

Cover image: Transfuchsian / iStock / Getty Images Plus
Typesetting: VTeX UAB, Lithuania

www.degruyter.com
Questions about General Product Safety Regulation:
productsafety@degruyterbrill.com

Preface

The main purpose of this book is to present different methods used in the literature to investigate the response of various finite- and infinite-dimensional dynamical systems to different kinds of forced oscillations, especially the almost periodic and almost automorphic oscillations.

Most of the subject covered in this book deals with differential equations in Banach spaces whose coefficients are linear unbounded operators. When we try to study and understand such equations, we feel that we are working with ordinary differential equations; however, the fact that the operator coefficients are unbounded makes things look quite different from what is known in the classical case.

Examples and applications for such equations are naturally found in the theory of partial differential equations. More specifically, if we give importance to the time variable at the expense of the spatial variables, we obtain an "ordinary differential equation" with respect to the variable which was put in evidence. Thus, for example, the heat and wave equations give rise to ordinary differential equations of this kind. Adding boundary conditions can often be translated in terms of considering solutions in some convenient functional Banach space. The theory of semigroups of operators provides an elegant approach to study such systems.

Therefore, we can frequently guess or even prove theorems for differential equations in Banach spaces looking at a corresponding pattern in finite-dimensional ordinary differential equations. The results on the oscillatory aspect of bounded solutions in Sections 4.2, 5.1, 5.2, and 5.4 constitute an example of such a strategy.

Now we will briefly sketch the contents of this work:

Chapter 1 is devoted to exploring the history of almost periodic and almost automorphic motions, their significance in mathematical theory and modeling, and their importance in the qualitative theory of dynamical systems.

Chapter 2 provides the mathematical background necessary to understand the more advanced discussions in later chapters. We introduce key concepts from functional analysis, semigroup theory, and spectral theory, along with the formal definitions of almost periodic and almost automorphic functions. This chapter is meant to serve as a reference for readers who may not be familiar with some of the more technical material used throughout the book.

Chapter 3 explores dynamical systems in finite-dimensional spaces, focusing on both autonomous and nonautonomous ordinary differential equations, functional differential equations of neutral type, and linear renewal equations. We discuss important results, such as the Massera- and Bohr–Neugebauer-type results, and introduce a method based on Lyapunov functionals to obtain sufficient conditions for the existence and uniqueness of a positive, compact, almost automorphic solution to delay differential equations arising in population dynamics.

In Chapter 4, we shift our focus to studying the oscillatory dynamics of various classes of evolution equations in infinite-dimensional Banach spaces, utilizing diverse

https://doi.org/10.1515/9783111684710-201

methods and techniques under a range of hypotheses. These include analyses of nonlinear differential equations under dissipativity and monotonicity conditions; nonhomogeneous linear evolution equations under quasicompactness hypotheses and Favard conditions, with additional geometric assumptions when compactness is absent; the subvariant functional method for semilinear evolution equations; spectral countability conditions for nonautonomous cases; fixed-point approaches for semilinear fractional differential equations; differential inclusions governed by maximal monotone operators; and elliptic equations via the method of sub- and supersolutions.

Chapter 5 investigates methods for establishing the existence and uniqueness of almost periodic and almost automorphic solutions of partial functional differential equations. We extend Massera- and Bohr–Neugebauer-type results for equations with finite and infinite delays, neutral types, and with nonautonomous perturbations. For the autonomous case, we employ the variation of constants formula and spectral decomposition under compactness assumptions to demonstrate that the dynamics of these complex systems can be analyzed through ordinary differential equations, a process known as the reduction principle. Additionally, we derive uniqueness results under exponential dichotomy. In the nonautonomous case, we illustrate how the dynamics of bounded solutions can be reduced to a linear discrete dynamical system.

We would like to thank all our collaborators for their contributions to research on the topics covered in this book.

10 January 2025

Abdallah Afoukal
Brahim Es-sebbar
Khalil Ezzinbi
Gaston M. N'Guérékata

Contents

1 Introduction

1.1 Almost periodic oscillations

Many physical, chemical, biological, economic, and ecological phenomena may exhibit a more or less periodic behavior. This usually happens when the involved systems are strongly influenced by periodic environmental variations or external factors. Since the periods involved in such phenomena are not the same in general, the study of the behavior of such dynamics requires concepts that go beyond the concept of periodicity.

Let us consider the following differential equation in $\mathbb{R}^n$:

$$x'(t) = A(t)x(t) + f(t) \quad \text{for } t \in \mathbb{R}, \tag{1.1}$$

where the matrix $A(t)$ and the vector $f(t)$ are both continuous and ω-periodic for some $\omega > 0$. In [213], Massera proved that the existence of a bounded solution of equation (1.1) on the positive real line is enough to get the existence of an ω-periodic solution. This result is known in the literature as the Massera theorem. Fixed point theory plays an important role in such results.

In general, if $x'(t) = A(t)x(t) + b(t)$ has an ω-periodic solution, the other bounded solutions on $\mathbb{R}$ are not necessarily ω-periodic. Consider, for instance, the following second order differential equation:

$$\ddot{x} + x = \sin(2t).$$

In this case $f(t) := \sin(2t)$ is π-periodic. A simple computation shows that the general solution of the equation is of the form

$$x(t) = C_1 \cos t + C_2 \sin t - \frac{1}{3}\sin(2t).$$

While all the solutions are bounded on $\mathbb{R}$, the only π-periodic solution is $\overline{x}(t) = -\frac{1}{3}\sin(2t)$. However, all the solutions are still 2π-periodic. Can we have solutions that are not periodic for any period? The answer is yes, and this can happen even in the autonomous case. For instance, consider the following scalar system which can describe the dynamic of physical or ecological phenomena:

$$\begin{cases} \dot{x}(t) = -3y(t) + z(t) - 3w(t), \\ \dot{y}(t) = x(t) + z(t), \\ \dot{z}(t) = 3y(t), \\ \dot{w}(t) = x(t) + y(t) + \frac{11}{9}z(t). \end{cases} \tag{1.2}$$

The orbit with initial value $(x(0), y(0), z(0), w(0)) = (1, 0, 0, 0)$ is given by

https://doi.org/10.1515/9783111684710-001

$$\begin{cases} x(t) = -4\cos t + 5\cos(\sqrt{2}t), \\ y(t) = -\sin t + \sqrt{2}\sin(\sqrt{2}t), \\ z(t) = 3(\cos t - \cos(\sqrt{2}t)), \\ w(t) = \frac{1}{3}(3\cos t - 3\cos(\sqrt{2}t) - \sin t + 2\sqrt{2}\sin(\sqrt{2}t)). \end{cases}$$

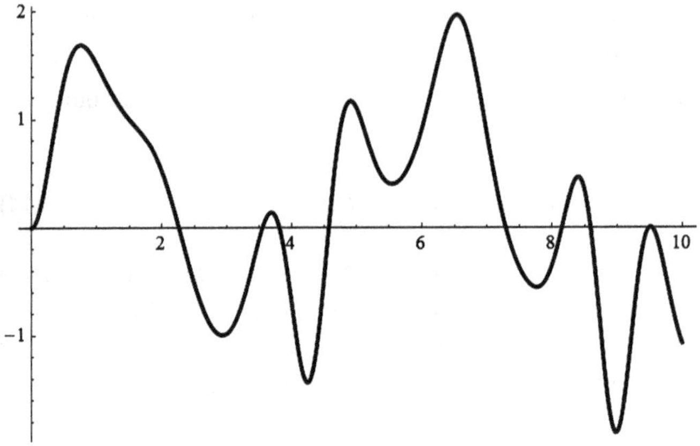

Figure 1.1: Almost periodic self-oscillations.

Those oscillations are bounded in time but are not periodic. They are almost periodic oscillations. Figure 1.1 shows the behavior of the fourth component $w(t)$.

The capability of ordinary and functional differential equations to mimic the dynamics in a periodically fluctuating environment has been the driving force for researchers to make valuable contributions to such problems. There has also been great interest in finding conditions under which periodic and almost periodic functional differential equations admit periodic and almost periodic solutions. Such problems have wide applications in biology and physics.

In general, the occurrence of almost periodic oscillations in nature is actually much more common than that of the periodic. This shows the need for a mathematical theory which addresses such oscillations.

The theory of almost periodic functions was initiated between 1924 and 1926 by Danish mathematician Harald Bohr [78]. Bohr's work was preceded by the important investigations of P. Bohl and E. Esclangon [77, 140]. Subsequently, important contributions were made by A. Besicovitch, S. Bochner, J. von Neumann, V. Stepanov, and B. Levitan [59, 71, 75, 198, 263]. In 1933, Bochner [71] published an important article devoted to the extension of the theory of almost periodic functions to vector-valued (abstract) functions with values in a Banach space.

From the earliest days on, the theory of almost periodic functions has been developed in connection with problems of differential equations, stability theory, dynamical systems. In his fundamental paper [147], J. Favard studied almost periodic differential equations and connected the problem of existence of almost periodic solutions with some separation properties of the bounded solutions. This work of Favard is the starting point of many further investigations.

The range of applications of the theory has been largely extended, and includes not only ordinary differential equations [152], but wider classes of partial differential equations and integrodifferential equation in Banach spaces. In this process an important role has been played starting in 1950s by the investigations of the Italian school, Amerio, Biroli, Prouse, and others [28–32, 35, 64–66, 249, 250], which are directed at extending certain classical results of J. Favard, S. Bochner, J. von Neumann, and S. L. Sobolev to differential equations in Banach spaces. Then in the spirit of the Italian school, Dafermos, Haraux, Ishii, and others have given important contributions to the question of almost periodic solutions [95, 97, 100, 114, 171, 172, 174, 184].

Recently, the existence of almost periodic solutions for differential equations has been extensively studied [8, 42, 45, 46, 61, 62, 69, 80, 81, 87, 89, 95, 97, 100, 123, 124, 145, 146, 158, 180, 182, 201, 204–206, 210, 234, 235, 241, 244, 259, 264, 266, 267, 279]. In the literature, several books are devoted to almost periodic differential equations. For example, let us indicate the books of Amerio and Prouse [35], Corduneanu [105], Fink [152], Levitan and Zhikov [199], and Zaidman [285].

1.2 Almost automorphic oscillations

Let us consider again the differential equation

$$x'(t) = A(t)x(t) + f(t) \quad \text{for } t \in \mathbb{R},$$

where this time A and f are almost periodic. It is well known that this system does not necessarily carry almost periodic dynamics [185, 244, 259]. Although it may have bounded oscillating solutions, these oscillations belongs to a class larger than the class of almost periodic functions – we are talking about *almost automorphic* functions.

Bochner introduced the concept of almost automorphy [72–74, 76] as a generalization of almost periodicity. This concept was then deeply investigated by Veech [268, 269] and many other authors. The name "almost automorphic" was given by Bochner himself because he encountered such functions first in his work on differential geometry. He also observed that almost automorphic functions can sometimes be used in obtaining simpler proofs of certain results concerning almost periodic functions by first proving these results for almost automorphic functions.

Other important contributions to the theory of almost automorphic functions include those by Zaki [287, 288], N'Guérékata [232, 233], Shen and Yi [259].

More recently, the existence of almost automorphic solutions to ordinary, as well as abstract, differential equations has been intensively studied [124, 158, 180, 204, 235, 241]. We refer the reader to N'Guérékata and Diagana's books [119, 234].

1.3 Ergodic perturbations of oscillatory functions

In [290–292], C. Y. Zhang introduced an extension of the almost periodic functions, the so-called pseudo almost periodic functions. A pseudo almost periodic function is an almost periodic function which is perturbed by an ergodic term. This class of functions and some of its generalizations attracted the attention of many researchers due to its applications in the theory of differential equations; we refer to [17–19, 22–24, 111, 116, 122, 200, 203, 275, 276, 278].

The concept of weighted pseudo almost periodic functions was introduced by T. Diagana in [117] as a generalization of pseudo almost periodicity. Since then, there has been an extensive study on the properties of this class of functions in the literature [68, 202, 295], as well as using these properties to establish existence results for differential equations, partial differential equations, and functional differential equations [15, 20, 36, 70, 118, 129, 144, 206].

2 Preliminaries

In this chapter, we provide the foundational concepts and background necessary for developing the main results in this book.

2.1 Spectral theory of linear operators

In this section, we will review various concepts and some fundamental results that will be necessary for this book on the spectral theory of linear operators in a Banach space. We will also demonstrate a spectral decomposition of the space based on the conditions of the essential spectrum of a bounded operator.

Let $B : D(B) \subset X \to X$ be a linear operator with domain $D(B)$. Then $(B, D(B))$ is said to be densely defined if $\overline{D(B)} = X$. The set

$$\mathcal{G}(B) := \{(x, Bx) : x \in D(B)\}$$

is called the graph of $(B, D(B))$.

Definition 2.1.1 ([136]). We say that $(B, D(B))$ is closed if $\mathcal{G}(B)$ is a closed subspace of $X \times X$ endowed with the usual product norm.

Definition 2.1.2 ([136]). Let $(B, D(B))$ be a linear operator on X. Then:

(1) The resolvent set of B, denoted by $\rho(B)$, is defined by

$$\rho(B) := \{\lambda \in \mathbb{C} : \lambda I_X - B : D(B) \to X \text{ has a bounded inverse}\}.$$

For $\lambda \in \rho(B)$, the operator $\mathcal{R}(\lambda, B) := (\lambda I_X - B)^{-1} : X \to D(B)$ is (by definition) bounded and will be called the resolvent of B at λ.

(2) The spectrum of B, denoted by $\sigma(B)$, is defined by

$$\begin{aligned}\sigma(B) &:= \mathbb{C} \setminus \rho(B) \\ &= \{\lambda \in \mathbb{C} : \lambda I_X - B \text{ is not bijective}\}.\end{aligned}$$

(3) The point spectrum of B is the set

$$\begin{aligned}\sigma_p(B) &:= \{\lambda \in \sigma(B) : \lambda I_X - B \text{ is not injective}\}. \\ &= \{\lambda \in \mathbb{C} : \ker(\lambda I_X - B) \neq \{0_X\}\}.\end{aligned}$$

For $\lambda \in \sigma_p(B)$, λ is called an eigenvalue of B and the elements of $\ker(\lambda I_X - B) \setminus \{0_X\}$ are known as eigenvectors of B corresponding to λ. The subspace $\ker(\lambda I_X - B)$ is called the eigenspace of B with respect to λ and $\dim \ker(\lambda I_X - B)$ is said to be the multiplicity of λ.

https://doi.org/10.1515/9783111684710-002

Proposition 2.1.3 ([136]). *Let $(B, D(B))$ be a linear operator on X. Then, the following statements are true:*
(1) *The resolvent $\rho(B)$ is an open set of $\mathbb{C}$.*
(2) *If $\rho(B) \neq \emptyset$, then $(B, D(B))$ is closed.*
(3) *If $\rho(B) \neq \emptyset$, then the map $\mathcal{R}(\cdot, B) : \rho(B) \to \mathcal{L}(X)$ such that $\mathcal{R}(\lambda, B) = (\lambda I_X - B)^{-1}$ is holomorphic.*

Proposition 2.1.4 ([136]). *Let $B \in \mathcal{L}(X)$. Then, $\sigma(B)$ is a nonempty compact set of $\mathbb{C}$ and*

$$\sigma(B) \subset B_{\mathbb{C}}(0, |B|_{\mathcal{L}(X)}) := \{\lambda \in \mathbb{C} : |\lambda|_{\mathbb{C}} \leq |B|_{\mathcal{L}(X)}\}.$$

In this case,

$$r(B) := \sup\{|\lambda|_{\mathbb{C}} : \lambda \in \sigma(B)\} < \infty.$$

The number $r(B)$ is called the spectral radius of B.

Theorem 2.1.5 ([86]). *Let $B \in \mathcal{L}(X)$. Then, the following three statements are equivalent:*
(1) $\lim_{n\to+\infty} B^n = 0$ *in the norm of $\mathcal{L}(X)$.*
(2) *For each real number μ and $x \in X$, the solution of the following difference equation:*

$$\begin{cases} y(n+1) = By(n) + e^{i\mu n}x & \text{for } n \in \mathbb{N}, \\ y(0) = 0, \end{cases}$$

is bounded.
(3) $r(B) < 1$.

Now, we turn to the notion of the essential spectrum.

Definition 2.1.6 ([270]). Let $(B, D(B))$ be a closed operator on X. The essential spectrum, $\sigma_{ess}(B)$, of B is the set of all $\lambda \in \sigma(B)$, when at least one of the following conditions is satisfied:
(1) $\mathrm{range}(\lambda I_X - B)$, the range of $\lambda I_X - B$, is not closed;
(2) $\bigcup_{k\geq 0} \ker((\lambda I_X - B)^k)$ is infinite dimensional;
(3) $\lambda \in \overline{\sigma(B) \setminus \{\lambda\}}$.

Remark 2.1.7 ([270]).
(1) If $B \in \mathcal{L}(X)$, then

$$r_{ess}(B) := \sup\{|\lambda|_{\mathbb{C}} : \lambda \in \sigma_{ess}(B)\} < \infty.$$

The number $r_{ess}(B)$ is said to be the essential spectral radius of B.
(2) If B is a compact operator in X, then $r_{ess}(B) = 0$.

Lemma 2.1.8 ([270]). *Let B_1 and B_2 two bounded linear operators in X. Assume that B_2 is compact. Then,*

$$r_{\text{ess}}(B_1 + B_2) = r_{\text{ess}}(B_1).$$

Theorem 2.1.9. *Let $B \in \mathcal{L}(X)$ and $\lambda \in \sigma(B) \setminus \sigma_{\text{ess}}(B)$. Then:*

(1) *λ is an isolated eigenvalue point of B.*
(2) *There exists $k_\lambda \in \mathbb{N}^*$ such that*

$$\bigcup_{k\geq 0} \ker((\lambda I_X - B)^k) = \ker((\lambda I_X - B)^{k_\lambda}) \quad \textit{and} \quad \dim \ker((\lambda I_X - B)^{k_\lambda}) < +\infty.$$

(3) *λ is a pole of $\mu \mapsto \mathcal{R}(\mu, B)$.*
(4) *The operator*

$$\mathcal{J}_\lambda = \frac{1}{2\pi i} \int_{\gamma_\lambda} \mathcal{R}(\mu, B) \mathrm{d}\mu,$$

where γ_λ is a positively oriented curve in $\mathbb{C}$ enclosing the isolated singularity λ but no other points of $\sigma(B)$, is a projection on X and $\operatorname{range}(\mathcal{J}_\lambda) = \ker((\lambda I_X - B)^{k_\lambda})$.
(5) *$B(\operatorname{range}(\mathcal{J}_\lambda)) \subset \operatorname{range}(\mathcal{J}_\lambda)$ and $B(\operatorname{range}(I_X - \mathcal{J}_\lambda)) \subset \operatorname{range}(I_X - \mathcal{J}_\lambda)$.*
(6) *$\sigma(B|_{\operatorname{range}(\mathcal{J}_\lambda)}) = \{\lambda\}$ and $\sigma(B|_{\operatorname{range}(I_X - \mathcal{J}_\lambda)}) = \sigma(B) \setminus \{\lambda\}$.*

Proof. For (1)–(5), the proofs can be found in [270, Proposition 4.11, p. 166].
See [115, Theorem 1.5.4, p. 30], for the proof of (6). □

2.2 C_0-semigroups and Hille–Yosida operator

This section is devoted to some preliminary results about semigroup theory and the Hille–Yosida operator, along with some examples.

Definition 2.2.1 ([247]). A family $(T(t))_{t\geq 0} \subset \mathcal{L}(X)$ is said to be a *C_0-semigroup* (or a *strongly continuous semigroup*) if the following properties hold:

(i) $T(0) = I_X$,
(ii) $T(t + s) = T(t)T(s)$ for $t, s \geq 0$ (the semigroup property),
(iii) For each $x \in X$,

$$\lim_{t\to 0^+} T(t)x = x.$$

That is, the mapping $t \mapsto T(t)x$ is right-continuous at 0.

Theorem 2.2.2 ([136]). *Let $(T(t))_{t\geq 0}$ be a C_0-semigroup on X. Then, for every $x \in X$, the orbit map $t \mapsto T(t)x$ is continuous at any $t \in \mathbb{R}^+$. In addition, there exist $\omega_0 \in \mathbb{R}$ and $M_0 \geq 1$ such that*

$$|T(t)|_{\mathcal{L}(X)} \le M_0 e^{\omega_0 t} \quad \textit{for all } t \ge 0.$$

Note that, when $\omega_0 = 0$ *and* $M_0 = 1$, $(T(t))_{t\ge0}$ *is said to be a contraction semigroup. If* $\omega_0 < 0$, *we say that* $(T(t))_{t\ge0}$ *is exponentially stable.*

Definition 2.2.3 ([136]). The operator $(A, D(A))$ defined by

$$\begin{cases} D(A) = \{x \in X : \lim_{h\to0} \frac{1}{h}(T(t)x - x) \text{ exists}\}, \\ Ax = \lim_{h\to0} \frac{1}{h}(T(t)x - x) \quad \text{for } x \in D(A), \end{cases}$$

is called the *infinitesimal generator* of $(T(t))_{t\ge0}$.

Example 2.2.4 (Semigroup of transport equation, [247]). Let $X = L^2(\mathbb{R};\mathbb{R})$. For $t \ge 0$, we define the operator $T(t) : X \to X$ by

$$(T(t)f)(s) = f(s-t) \quad \text{for } s \in \mathbb{R} \text{ and } f \in X.$$

Then, $(T(t))_{t\ge0}$ is a strongly continuous semigroup. Its infinitesimal generator A is given by

$$\begin{cases} D(A) = H^1(\mathbb{R}), \\ Af = -f'. \end{cases}$$

Proposition 2.2.5 ([136]). *Let* $(T(t))_{t\ge0}$ *be a* C_0*-semigroup on* X *and* $(A, D(A))$ *its generator. Then:*

(1) $(A, D(A))$ *is a closed linear operator in* X.
(2) $\overline{D(A)} = X$.
(3) *If* $(S(t))_{t\ge0}$ *is another* C_0*-semigroup generated by* $(A, D(A))$, *then* $S(t) = T(t)$ *for all* $t \ge 0$.

The following theorem, due to Hille and Yosida, is a characterization of the generators of strongly continuous semigroups.

Theorem 2.2.6 (Hille–Yosida theorem, [136]). *Let* $(A, D(A))$ *be a linear operator on* X, $\omega_0 \in \mathbb{R}$, *and* $M_0 \ge 1$. *Then,* $(A, D(A))$ *is the infinitesimal generator of a is strongly continuous semigroup* $(T(t))_{t\ge0}$ *satisfying*

$$\|T(t)\| \le Me^{\omega_0 t} \quad \textit{for all } t \ge 0$$

if and only if

(a) A *is closed and* $\overline{D(A)} = X$,
(b) *the resolvent set* $\rho(A)$ *of* A *contains* $(\omega_0, +\infty)$ *and, for every* $\lambda > \omega_0$,

$$|\mathcal{R}(\lambda, A)^n|_{\mathcal{L}(X)} \le \frac{M}{(\lambda - \omega_0)^n} \quad \text{for all } n \in \mathbb{N}^*.$$

Example 2.2.7 (Semigroup of heat equation, [97]). Let Ω is a bounded open subset of $\mathbb{R}^p$ with smooth boundary $\partial\Omega$ and $X = C_0(\Omega;\mathbb{R})$ be the space of all continuous functions from $\overline{\Omega}$ (the closure of Ω) to $\mathbb{R}$ vanishing on $\partial\Omega$, endowed with the uniform norm topology. Define the operator $A : D(A) \subset X \to X$ by

$$\begin{cases} D(A) = \{f \in X \cap H_0^1(\Omega) : \Delta f \in X\}, \\ Af = \Delta f, \end{cases}$$

where Δ is the Laplacian operator. Then $(A, D(A))$ generates an exponentially stable C_0-semigroup $(T(t))_{t\geq 0}$ on X such that

$$|T(t)|_{\mathcal{L}(X)} \leq M_0 e^{-\lambda_0 t} \quad \text{for all } t \geq 0,$$

where λ_0 is the smallest eigenvalue of $-\Delta$ in $H_0^1(\Omega)$ ($\lambda_0 > 0$ since Ω is bounded) and

$$M_0 := \exp\left(\lambda_0 |\Omega|^{\frac{2}{p}} (4\pi)^{-1}\right).$$

Theorem 2.2.8 ([247]). *Let $(T(t))_{t\geq 0}$ be a C_0-semigroup on X and K be a compact set of X. Then,*

$$\lim_{t\to 0^+} \sup_{x\in K} |T(t)x - x| = 0.$$

Definition 2.2.9 ([136]). A C_0-semigroup $(T(t))_{t\geq 0}$ is called *compact* if $T(t)$ is compact for all $t > 0$.

Definition 2.2.1 ([136]). The growth bound $\omega_0(T)$ of the C_0-semigroup $(T(t))_{t\geq 0}$ is defined by

$$\omega_0(T) := \inf\left\{\omega \in \mathbb{R} : \sup_{t\geq 0} e^{-\omega t}|T(t)| < \infty\right\}.$$

For a bounded subset B of X, the Kuratowski measure of noncompactness, $\alpha(B)$, is defined by

$$\alpha(B) := \inf\{d > 0 : \text{there exist finitely many sets of diameter at most } d \text{ which cover } B\}.$$

Moreover, for a bounded linear operator K on X, we define $\alpha(K)$ by

$$\alpha(K) := \inf\{k > 0 : \alpha(K(B)) \leq k\alpha(B) \text{ for any bounded set } B \text{ of } X\}.$$

Definition 2.2.2 ([181]). The essential growth bound $\omega_{\text{ess}}(T)$ of the C_0-semigroup $(T(t))_{t\geq 0}$ is defined by

$$\omega_{\text{ess}}(T) := \lim_{t\to\infty} \frac{\log \alpha(T(t))}{t} = \inf_{t>0} \frac{\log \alpha(T(t))}{t}. \tag{2.1}$$

Theorem 2.2.3 ([181]). *The relation between $r_{\mathrm{ess}}(T(t))$ and $\omega_{\mathrm{ess}}(T)$ is given by the following formula:*

$$r_{\mathrm{ess}}(T(t)) = e^{t\omega_{\mathrm{ess}}(T)}. \tag{2.2}$$

From the spectral mapping inclusion $e^{t\sigma_{\mathrm{ess}}(A_T)} \subset \sigma_{\mathrm{ess}}(T(t))$ and the formula (2.2), one can see that

$$\sigma_{\mathrm{ess}}(A_T) \subset \{\lambda \in \sigma(A_T) : \operatorname{Re}\lambda \leq \omega_{\mathrm{ess}}(T)\}. \tag{2.3}$$

This means that if $\lambda \in \sigma(A_T)$ and $\operatorname{Re}\lambda > \omega_{\mathrm{ess}}(T)$, then λ does not belong to $\sigma_{\mathrm{ess}}(A_T)$. Therefore, by Theorem 2.1.9, λ is an isolated eigenvalue of A_T.

The spectral bound $s(A_T)$ of the infinitesimal generator A_T is defined by

$$s(A_T) := \sup\{\operatorname{Re}\lambda : \lambda \in \sigma(A_T)\}.$$

Theorem 2.2.4 ([136]). *We have the following formula:*

$$\omega_0(T) = \max\{\omega_{\mathrm{ess}}(T), s(A_T)\}.$$

Definition 2.2.10 (Hille–Yosida operator, [113]). A linear operator $(A, D(A))$ is called *a Hille–Yosida operator* if it satisfies the following Hille–Yosida condition: There exist $\omega_0 \in \mathbb{R}$ and $M_0 \geq 1$ such that $]\omega_0, \infty[\subset \rho(A)$ and

$$|\mathcal{R}(\lambda, A)^n|_{\mathcal{L}(X)} \leq \frac{M_0}{(\lambda - \omega_0)^n} \quad \text{for all } n \geq 0 \text{ and } \lambda > \omega_0.$$

Definition 2.2.11 ([39]). Let $(A, D(A))$ be a Hille–Yosida operator on X. The part of $(A, D(A))$ on $\overline{D(A)}$, denoted by $(A_0, D(A_0))$, is defined by

$$\begin{cases} D(A_0) = \{x \in D(A) : Ax \in \overline{D(A)}\}, \\ A_0 x = Ax \quad \text{for } x \in D(A_0). \end{cases}$$

Lemma 2.2.12 ([39]). *Let $(A, D(A))$ be a Hille–Yosida operator on X. Then, the following statements are true:*

(a) *$\rho(A) \subset \rho(A_0)$ and $\mathcal{R}(\lambda, A)|_{\overline{D(A)}} = \mathcal{R}(\lambda, A_0)$ for all $\lambda \in \rho(A)$.*

(b) *$(A_0, D(A_0))$ generates a strongly continuous semigroup $(T(t))_{t\geq 0}$ on $\overline{D(A)}$, which satisfies*

$$|T(t)|_{\mathcal{L}(\overline{D(A)})} \leq M_0 e^{\omega_0 t} \quad \text{for } t \geq 0.$$

In the following, we shall present some examples of linear operators with nondense domain satisfying the Hille–Yosida condition.

Example 2.2.13 ([113]). Let $X = C([0,1];\mathbb{R})$ and the linear operator $A : D(A) \subset X \to X$ be defined by $Af = f'$, where $D(A) = \{f \in C^1([0,1];\mathbb{R}) : f(0) = 0\}$. Then

$$\overline{D(A)} := \{f \in X : f(0) = 0\} \neq X.$$

Example 2.2.14 ([113]). Let $X = C([0,\pi];\mathbb{R})$ with the uniform norm topology and $(A, D(A))$ be the linear operator defined by

$$\begin{cases} D(A) = \{y \in C^2([0,\pi];\mathbb{R}) : y(\pi) = y(0) = 0\}, \\ Ay = y'' \quad \text{for } y \in D(A). \end{cases}$$

From [113, Example 14.5 and Proposition 14.6, pp. 319–320], we have that

$$]0, +\infty[\subset \rho(A)$$

and

$$|\mathcal{R}(\lambda, A)^n|_{\mathcal{L}(X)} \leq \frac{1}{\lambda^n} \quad \text{for } n \geq 1 \text{ and } \lambda > 0.$$

Then, $(A, D(A))$ is a Hille–Yosida operator on X.

Example 2.2.15 ([139]). Let Ω be a bounded open subset of $\mathbb{R}^p$ with a smooth boundary $\partial\Omega$. Consider $X = C(\overline{\Omega};\mathbb{R})$ which is endowed with supremum norm. Let $(A, D(A))$ be the linear operator defined by

$$\begin{cases} Ay = \Delta y, \\ D(A) = \{y \in X \cap H_0^1(\Omega) : \Delta y \in X\}. \end{cases}$$

Let $\lambda_0 > 0$ be the smallest eigenvalue of $(-A, D(A))$ in $H_0^1(\Omega)$. Then, form [139], we have that $]-\lambda_0, +\infty[\subset \rho(A)$ and

$$|\mathcal{R}(\lambda, A)^m| \leq \frac{\exp\left(\lambda_0 |\Omega|^{\frac{2}{p}} (2\pi)^{-1}\right)}{\lambda + \lambda_0} \quad \text{for } m \in \mathbb{N} \text{ and } \lambda > -\lambda_0.$$

Consequently, $(A, D(A))$ is a Hille–Yosida operator on X and $\overline{D(A)} = \{y \in X : y|_{\partial\Omega} = 0\} \neq X$.

Let $(A, D(A))$ be a Hille–Yosida operator on X. Consider the following nonhomogenous linear Cauchy problem:

$$\begin{cases} \frac{\mathrm{d}}{\mathrm{d}t} u(t) = Au(t) + f(t) \quad \text{for } t \geq \sigma, \\ u(\sigma) = x \in X, \end{cases} \tag{2.4}$$

where $f \in L^1_{\mathrm{loc}}([\sigma, +\infty[; X)$.

Definition 2.2.16 ([113]). We say that a continuous function $u : [\sigma, +\infty[\to X$ is a *strict solution* of equation (2.4) in $[\sigma, +\infty[$ if:
(i) $u \in C^1([\sigma, +\infty[; X) \cap C([\sigma, +\infty[; D(A))$.
(ii) u satisfies equation (2.4) on $[\sigma, +\infty[$.

Remark 2.2.17. Assume that u is a strict solution of equation (2.4) in $[\sigma, +\infty[$. Then,

$$\frac{\mathrm{d}u(\sigma)}{\mathrm{d}t} = Au(\sigma) + f(\sigma).$$

Since $u(t) \in D(A)$ for all $t \geq \sigma$, $\frac{\mathrm{d}u(\sigma)}{\mathrm{d}t} \in \overline{D(A)}$. Then, $Au(\sigma) + f(\sigma) \in \overline{D(A)}$. In conclusion,

$$u \text{ is a strict solution of equation (2.4) in } [\sigma, +\infty[\implies Au(\sigma) + f(\sigma) \in \overline{D(A)}.$$

The following result discusses general assumptions for equation (2.4) to have a strict solution in $[\sigma, +\infty[$.

Theorem 2.2.18 ([113]). *Assume that f is absolutely continuous, that is, there is a locally Bochner integrable function g such that*

$$f(t) = f(\sigma) + \int_\sigma^t g(s)\mathrm{d}s.$$

For $x \in D(A)$ such that $Ax + f(\sigma) \in \overline{D(A)}$, equation (2.4) *has a unique strict solution in* $[\sigma, +\infty[$.

In the case where $x \notin D(A)$, or if the function f is not sufficiently regular, equation (2.4) may not have a strict solution. However, it may still have a solution in a weak sense, referred to as a mild or integral solution (see [113, 209, 265]). This motivates the following definition.

Definition 2.2.19 ([265]). An integral solution with initial data $x \in X$ of equation (2.4) is a continuous function $u : [\sigma, +\infty[\to X$ having the following properties:
(i) $\int_\sigma^t u(s)\mathrm{d}s \in D(A)$ for $t \geq \sigma$.
(ii) $u(t) = x + A(\int_\sigma^t u(s)\mathrm{d}s) + \int_\sigma^t f(s)\mathrm{d}s$ for $t \geq \sigma$.
(iii) $u(\sigma) = x$.

Remark 2.2.20 ([265]).
(1) Note that every strict solution of equation (2.4) is an integral solution. In fact, let u be a strict solution of (2.4) on $[\sigma, +\infty[$. For $t \geq \sigma$, $u(t) \in D(A)$ for all $t \geq \sigma$. Then,

$$\int_\sigma^t u(s)\mathrm{d}s \in \overline{D(A)}.$$

We observe that

$$\lim_{h\to0^+}\frac{T(h)\int_\sigma^t u(s)\mathrm{d}s-\int_\sigma^t u(s)\mathrm{d}s}{h}=\lim_{h\to0^+}\int_\sigma^t\frac{(T(h)u(s)-u(s))}{h}\mathrm{d}s$$
$$=\int_\sigma^t\lim_{h\to0^+}\frac{(T(h)u(s)-u(s))}{h}\mathrm{d}s$$
$$=\int_\sigma^t Au(s)\mathrm{d}s.$$

Thus,

$$\int_\sigma^t u(s)\mathrm{d}s\in D(A)\quad\text{and}\quad A\left(\int_\sigma^t u(s)\mathrm{d}s\right)=\int_\sigma^t Au(s)\mathrm{d}s.$$

In addition, by integrating equation (2.4), we obtain

$$u(t)=x+A\left(\int_\sigma^t u(s)\mathrm{d}s\right)+\int_\sigma^t f(s)\mathrm{d}s\quad\text{for } t\geq\sigma.$$

In other words, u is an integral solution of equation (2.4).

(2) Generally, an integral solution u of equation (2.4) may not necessarily be strict, for instance, when $f(t)=x_0$ for all $t\geq\sigma$, where $x_0\in X\setminus\overline{D(A)}$. If $u(\sigma)=0$, then

$$Au(\sigma)+f(\sigma)=0+x_0=x_0\notin\overline{D(A)}.$$

In the light of Remark 2.2.17, u is not strict.

(3) Let u be an integral solution of equation (2.4) on $[\sigma,+\infty[$. Then, by (i), we obtain that $u(t)\in\overline{D(A)}$, for all $t\geq\sigma$. In particular, $u(\sigma)=x\in\overline{D(A)}$. Consequently, the phase space associated with equation (2.4) is $\overline{D(A)}$.

The next theorem gives a variation of constants formula for integral solutions.

Theorem 2.2.21 ([265]). *If $f\in L^1_{\mathrm{loc}}([\sigma,+\infty[;X)$, then equation (2.4) has a unique integral solution u on $[\sigma,+\infty[$ satisfying the following formula:*

$$u(t)=T(t-\sigma)x+\lim_{\lambda\to+\infty}\int_\sigma^t T(t-s)B_\lambda f(s)\mathrm{d}s\quad\textit{for } t\geq\sigma,\tag{2.5}$$

where $B_\lambda=\lambda\mathcal{R}(\lambda,A)$ for $\lambda>\omega_0$. In addition, the following estimate holds for u:

$$|u(t)| \leq M_0 e^{\omega_0(t-\sigma)}|x| + \int_\sigma^t M_0 e^{\omega_0(t-s)}|f(t)|\mathrm{d}s, \quad \forall t \geq \sigma.$$

Remark 2.2.22 ([247]). In the case where $\overline{D(A)} = X$, an integral solution is called a mild solution and satisfies the following formula:

$$u(t) = T(t-\sigma)x + \int_\sigma^t T(t-s)f(s)\,\mathrm{d}s \quad \text{for } t \geq \sigma. \tag{2.6}$$

2.3 Almost periodic and almost automorphic functions

In this section, we will review the basic properties of almost periodicity and almost automorphy, including some recent results on the integration of Stepanov almost periodic and Stepanov almost automorphic functions.

Let $(X, |\cdot|)$ be a Banach space and $\mathrm{BC}(\mathbb{R}, X)$ be the space of bounded continuous functions from $\mathbb{R}$ to X equipped with the supremum norm.

Definition 2.3.1 ([152]). Let $S \subset \mathbb{R}$. Then S is said to be relatively dense on $\mathbb{R}$ if there exists $l > 0$ such that

$$[a, a+l] \cap S \neq \emptyset \quad \text{for all } a \in \mathbb{R}.$$

The number l is known as the inclusion length.

Definition 2.3.2 ([152]). Let $f : \mathbb{R} \to X$ be a bounded continuous function and $\varepsilon > 0$. The ε-translation (or ε-period) set of f is given by

$$\mathbb{T}(f, \varepsilon) := \left\{a \in \mathbb{R} : \sup_{t\in\mathbb{R}} |f(t+a) - f(t)| < \varepsilon\right\}.$$

Definition 2.3.3 ([152]). We say that $f \in C(\mathbb{R}; X)$ is Bohr almost periodic (or simply almost periodic) if $\mathbb{T}(f, \varepsilon)$ is relatively dense for all $\varepsilon > 0$. That is, there exists a positive number l such that every interval of length l contains a number τ such that

$$|f(t+\tau) - f(t)| < \varepsilon \quad \text{for } t \in \mathbb{R}.$$

The following results describe some of the properties of almost periodic functions.

Theorem 2.3.1 ([152]). *Each almost periodic function is uniformly continuous.*

A useful characterization of almost periodic functions was given by Bochner [73].

Theorem 2.3.2 ([73]). *A continuous function $f : \mathbb{R} \to X$ is almost periodic if and only if for every sequence of real numbers $(s_n)_n$ there exist a subsequence $(s_n')_n \subset (s_n)_n$ and a function $\tilde{f}$ such that*

$$f(t + s_n') \to \tilde{f}(t)$$

uniformly on $\mathbb{R}$ *as* $n \to \infty$.

Another characterization of almost periodic functions due to Bochner [73], which uses pointwise convergence, is given as follows.

Theorem 2.3.3 ([73]). *A continuous function* $f : \mathbb{R} \to X$ *is almost periodic if and only if for every pair of sequences* $(s_n)_n$ *and* $(t_n)_n$ *there are common subsequences* $(s_n')_n \subset (s_n)_n$ *and* $(t_n')_n \subset (t_n)_n$ *such that for each* $t \in \mathbb{R}$

$$\lim_{n\to\infty} f(t + s_n' + t_n') = \lim_{n\to\infty} \lim_{m\to\infty} f(t + s_n' + t_m').$$

Let $\mathrm{AP}(\mathbb{R}, X)$ denote the space of almost periodic X-valued functions on $\mathbb{R}$.

Proposition 2.3.4 ([103]). *If* $f_1, f_2 \in \mathrm{AP}(\mathbb{R}, X)$, *then*
(i) $f_1 + f_2 \in \mathrm{AP}(\mathbb{R}, X)$,
(ii) $\lambda f \in \mathrm{AP}(\mathbb{R}, X)$, *for any scalar* λ,
(iii) $f_a \in \mathrm{AP}(\mathbb{R}, X)$, *where* f_a *is the right translation function of* f,
(iv) *the range* $\mathcal{R}_f := \{f(t) : t \in \mathbb{R}\}$ *is relatively compact in* X, *thus* f *is bounded in norm,*
(v) *if* $f_n \to f$ *uniformly on* $\mathbb{R}$, *where each* $f_n \in \mathrm{AP}(\mathbb{R}, X)$, *then* $f \in \mathrm{AP}(\mathbb{R}, X)$.

Let us now present some examples of almost periodic functions.

Example 2.3.4 ([152]). Every continuous periodic function is almost periodic. The following functions:

$$f_1(t) = \sin(\pi t) + \sin(t) \quad \text{for } t \in \mathbb{R},$$
$$f_2(t) = \sum_{n=0}^{+\infty} \frac{1}{2^n} \cos((\sqrt{2})^n t) \quad \text{for } t \in \mathbb{R}$$

are almost periodic, but not periodic As shown in Figure 2.1.

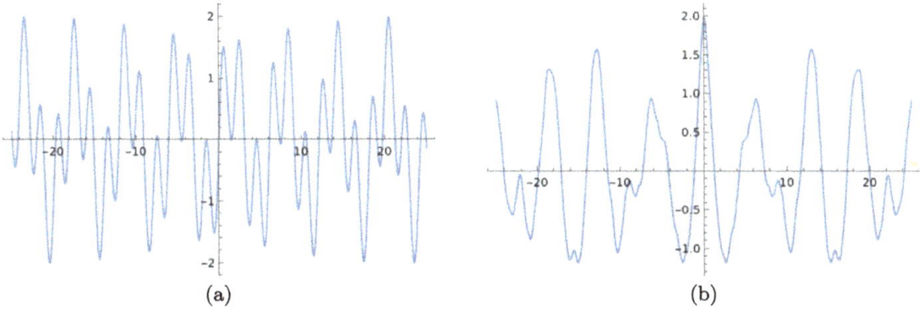

Figure 2.1: Graphs of the functions: (a) f_1; (b) f_2.

Example 2.3.5 ([78, 153]). Let $\zeta : \mathbb{C} \to \mathbb{C}$ be the Riemann zeta function, which is given, for $z = x + iy \in \mathbb{C}$, by

$$\begin{aligned}\zeta(z) &= \sum_{n=1}^{\infty} \frac{1}{n^z} \\ &= \sum_{n=1}^{\infty} e^{-x\ln(n)} e^{-iy\ln(n)}.\end{aligned}$$

For fixed $x > 1$, the function $y \mapsto \zeta(x + iy)$ is almost periodic.

Now, we turn to the notion of an almost periodic sequence.

Definition 2.3.6 ([102]). We say that a sequence $(x_n)_{n\in\mathbb{Z}}$ in X is almost periodic if for all $\varepsilon > 0$ there exists $l(\varepsilon) \in \mathbb{N}^*$ such that for all $m \in \mathbb{Z}$ there exists $p \in \{m, \ldots, m + l(\varepsilon)\}$ satisfying

$$\sup_{n\in\mathbb{Z}} |x_{n+p} - x_n| < \varepsilon.$$

The following theorem provides a Bochner characterization of almost periodic sequences.

Theorem 2.3.7 ([102]). *A sequence $(x_n)_{n\in\mathbb{Z}}$ is almost periodic if and only if for every $\{m'_k\}_{k\in\mathbb{N}} \subset \mathbb{Z}$ there exist a subsequence $\{m_k\}_{k\in\mathbb{N}}$ of $\{m'_k\}_{k\in\mathbb{N}}$ and a sequence $(y_n)_{n\in\mathbb{Z}}$ such that*

$$\lim_{k\to+\infty} \sup_{n\in\mathbb{Z}} |x_{n+m_k} - y_n| = 0.$$

The following propositions discuss the link between almost periodic functions and almost periodic sequences.

Proposition 2.3.8 ([102, 152]). *If $f \in AP(\mathbb{R}, X)$ is an almost periodic function, then $f|_{\mathbb{Z}}$ is an almost periodic sequence. Moreover, $\mathbb{T}(f, \varepsilon) \cap \mathbb{Z}$ is relatively dense on $\mathbb{R}$ for all $\varepsilon > 0$.*

Proposition 2.3.9 ([152]). *Let $(x_n)_{n\in\mathbb{Z}}$ is an almost periodic sequence and $\tilde{x} : \mathbb{R} \to X$ be the canonical extension of $(x_n)_{n\in\mathbb{Z}}$ which is defined by*

$$\tilde{x}(t) = x_{[t]} + (t - [t])(x_{[t]+1} - x_{[t]}) \quad \text{for } t \in \mathbb{R}, \tag{2.7}$$

where $[t]$ is the integer part of t. Then, $\tilde{x} \in AP(\mathbb{R}, X)$.

The following result asserts the completeness of the space of almost periodic sequences.

Theorem 2.3.10 ([105]). *The space of all almost periodic sequences is a closed subspace of*

$$l^\infty(\mathbb{Z}, X) := \Big\{(x_n)_{n\in\mathbb{Z}} \subset X : \sup_{n\in\mathbb{Z}} |x_n| < \infty\Big\},$$

endowed with the following norm:

$$|(x_n)_{n\in\mathbb{Z}}|_{l^\infty(\mathbb{Z},X)} = \sup_{n\in\mathbb{Z}}|x_n|.$$

In [74], Bochner introduced the concept of almost automorphy which is a generalization of almost periodicity.

Definition 2.3.5 ([74]). A continuous function $f : \mathbb{R} \to \mathcal{Y}$ is said to be almost automorphic if for any sequence of real numbers $(t'_n)_{n\in\mathbb{N}}$ there exists a subsequence $(t_n)_{n\in\mathbb{N}}$ of $(t'_n)_{n\in\mathbb{N}}$ such that

$$\lim_{m\to+\infty}\lim_{n\to+\infty} f(t + s_n - s_m) = f(t)$$

for any $t \in \mathbb{R}$. This is equivalent to the following:

$$g(t) = \lim_{n\to+\infty} f(t + t_n) \tag{2.8}$$

is well defined for any $t \in \mathbb{R}$ and

$$\lim_{n\to+\infty} g(t - t_n) = f(t) \quad \text{for all } t \in \mathbb{R}. \tag{2.9}$$

Moreover, if the limits in (2.8) and (2.9) are uniform on any compact subset $K \subset \mathbb{R}$, we say that f is compact almost automorphic.

In the sequel, $\mathrm{AA}(\mathbb{R}, X)$ (resp. $\mathrm{AA}_c(\mathbb{R}, X)$) denotes the space of (resp. compact) almost automorphic X-valued functions.

Remark. By choosing $s_n = -t_n$ in Theorem 2.3.3, one can see that almost periodic functions are almost automorphic. Due to pointwise convergence, the function $\tilde{f}$ in Definition 2.3.5 is only measurable and not necessarily continuous. If either convergence in Definition 2.3.5 is uniform, then $f \in \mathrm{AP}(\mathbb{R}, X)$. For more details about this topic, we refer the reader to the books [234, 284].

Example 2.3.11 ([240, Theorem 2.8]). Every almost periodic function is almost automorphic. Let $f_3, f_4 : \mathbb{R} \to \mathbb{R}$ be given by

$$f_3(t) = \sin\left(\frac{1}{\cos t + \cos(\sqrt{2}t) + 2}\right), \quad t \in \mathbb{R},$$

$$f_4(t) = \frac{2 + e^{it} + e^{i\sqrt{2}t}}{|2 + e^{it} + e^{i\sqrt{2}t}|}, \quad t \in \mathbb{R}.$$

The functions f_3, f_4 are almost automorphic, but not almost periodic (see Figure 2.2).

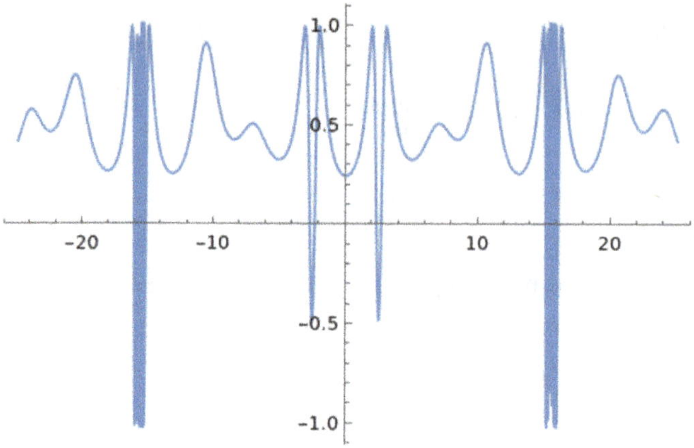

Figure 2.2: Graph of the function f_3.

Proposition 2.3.6. *A function f is compact almost automorphic if and only if it is almost automorphic and uniformly continuous.*

Proof. Let $f : \mathbb{R} \to X$ be an almost automorphic function which is uniformly continuous. Let $(s_n)_n$ be a sequence of real numbers. Then there exist a subsequence $(s'_n)_n \subset (s_n)_n$ and a function $\tilde{f}$ such that, for each $t \in \mathbb{R}$,

$$f(t + s'_n) \to \tilde{f}(t) \tag{2.10}$$

and

$$\tilde{f}(t - s'_n) \to f(t) \tag{2.11}$$

as $n \to \infty$. Consider the sequence of functions defined for each $n \in \mathbb{N}$ by

$$f_n(t) = f(t + s'_n) \quad \text{for } t \in \mathbb{R}.$$

Since f is uniformly continuous, the family $(f_n)_n$ is equicontinuous. It follows that the convergence (2.10) holds uniformly on compact subsets of $\mathbb{R}$.

On the other hand, from (2.10), one can see that $\tilde{f}$ is also uniformly continuous. Using the same arguments, the convergence (2.11) also holds uniformly on compact subsets of $\mathbb{R}$.

Now if f is compact almost automorphic, then it is almost automorphic. To show that f is uniformly continuous, we take two sequences $(t_n)_n$ and $(s_n)_n$ such that $|t_n - s_n| \to 0$ as $n \to \infty$ and show that $a_n = |f(t_n) - f(s_n)| \to 0$ as $n \to \infty$. Let $(a'_n)_n = (|f(t'_n) - f(s'_n)|)_n$ be a subsequence of $(a_n)_n$. Since f is compact almost automorphic, there exist a subsequence $(s''_n)_n \subset (s'_n)_n$ and a function $g : \mathbb{R} \to X$ such that $f(t + s''_n) \to g(t)$ uniformly on compact subsets of $\mathbb{R}$. In addition, the function g is continuous. Let $a, b \in \mathbb{R}$ be such that $a \leq t_n - s_n \leq b$ for all $n \in \mathbb{N}$. Then, from the compact almost automorphy of f and the continuity of g, we have

$$\begin{aligned}\alpha''_n &= |f(t''_n) - f(s''_n)| \\ &\le |f(t''_n - s''_n + s''_n) - g(t''_n - s''_n)| + |g(t''_n - s''_n) - g(0)| + |g(0) - f(s''_n)| \\ &\le \sup_{a\le t\le b} |f(t + s''_n) - g(t)| + |g(t''_n - s''_n) - g(0)| + |g(0) - f(s''_n)| \to 0\end{aligned}$$

as $n \to \infty$. Thus we showed that every subsequence of $(\alpha_n)_n$ has a subsequence which converges to 0. We conclude that the whole sequence $(\alpha_n)_n$ converges to 0. Therefore f is uniformly continuous. □

Theorem 2.3.12 ([234, 240]). *Assume that a sequence* $(f_n)_{n\in\mathbb{N}} \subset \mathrm{AA}(\mathbb{R}, X)$ *(resp.* $\mathrm{AA}_C(\mathbb{R}, X)$*) converges uniformly to a function* f*. Then,* $f \in \mathrm{AA}(\mathbb{R}, X)$ *(resp.* $\mathrm{AA}_C(\mathbb{R}; X)$*). Moreover, the spaces* $\mathrm{AA}(\mathbb{R}, X)$ *and* $\mathrm{AA}_c(\mathbb{R}, X)$ *are closed subspaces of* $\mathrm{BC}(\mathbb{R}, X)$ *under the supremum norm.*

Theorem 2.3.13 ([240, Theorem 2.8]). *Let* $f \in \mathrm{AA}(\mathbb{R}, X)$ *and* $g : \{f(t) : t \in \mathbb{R}\} \to X$ *be a continuous and bounded function. Then,* $g \circ f \in \mathrm{AA}(\mathbb{R}, X)$.

The following discusses the concept of an almost automorphic sequence and its connection to compact almost automorphic functions.

Definition 2.3.14 ([37, 124, 266]). A sequence $(x_n)_{n\in\mathbb{Z}} \subset X$ is called almost automorphic if for every sequence $\{m'_k\}_{k\in\mathbb{N}} \subset \mathbb{Z}$ there exist a subsequence $\{m_k\}_{k\in\mathbb{N}}$ of $\{m'_k\}_{k\in\mathbb{N}}$ and a sequence $(y_n)_{n\in\mathbb{Z}} \subset X$ such that, for all $n \in \mathbb{Z}$,

$$\lim_{k\to+\infty} |x_{n+m_k} - y_n| = 0 \quad \text{and} \quad \lim_{k\to+\infty} |y_{n-m_k} - x_n| = 0.$$

Remark 2.3.15 ([37]). If one of the limits in Definition 2.3.14 is uniform on $\mathbb{Z}$, then $(x_n)_{n\in\mathbb{Z}}$ is almost periodic.

Theorem 2.3.16 ([37]). *The set of all almost automorphic sequences in* X *is a closed subspace of* $l^\infty(\mathbb{Z}, X)$.

Theorem 2.3.17 ([151]). *The following are true:*
(1) *The restriction of any almost automorphic function on* $\mathbb{Z}$ *is an almost automorphic sequence.*
(2) *If* $(x_n)_{n\in\mathbb{Z}}$ *is an almost automorphic sequence, then the canonical extension* $\tilde{x}$ *of* $(x_n)_{n\in\mathbb{Z}}$ *defined by* (2.7) *is compact almost automorphic.*

Example 2.3.18. By the above theorem, the restriction of the function f_3 on $\mathbb{Z}$, defined in Example 2.3.11, is an almost automorphic sequence. In addition, the following function:

$$\tilde{f}_3(t) = f_3([t]) + (t - [t])(f_3([t] + 1) - f_3([t])) \quad \text{for } t \in \mathbb{R}$$

is compact almost automorphic (see Figure 2.3).

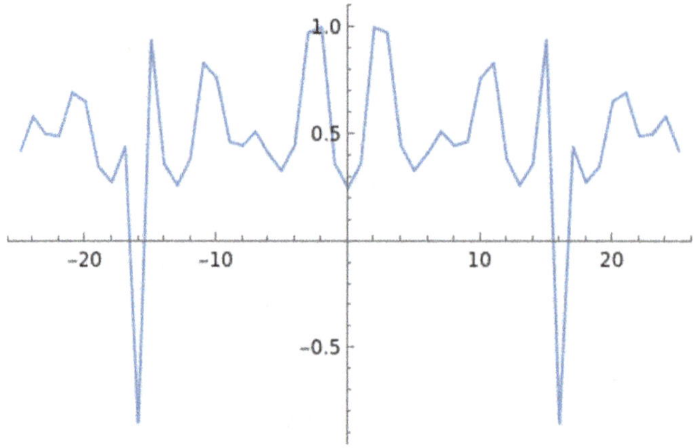

Figure 2.3: Graph of the function $\tilde{f}_3$.

Example 2.3.19 ([151, 240]).

(1) Every almost periodic sequence is almost automorphic, but the converse is not always true. Indeed, the following sequence:

$$x(n) = \begin{cases} 1 & \text{if } \cos(2n\pi\tau_0) > 0, \\ -1 & \text{if } \cos(2n\pi\tau_0) < 0, \end{cases}$$

where $\tau_0 \in \mathbb{R} \setminus \mathbb{Q}$, is almost automorphic but not almost periodic.

(3) Each almost periodic function is a compact almost automorphic function. However, the converse is not always true. For example, the linear extension of $(x(n))_{n\in\mathbb{Z}}$ on $\mathbb{R}$, which is given by

$$\tilde{x}(t) = x_{[t]} + (t - [t])(x_{[t]+1} - x_{[t]}) \quad \text{for } t \in \mathbb{R},$$

is compact almost automorphic but not almost periodic.

We use the Bochner characterization to define the notion of weak almost periodicity.

Definition 2.3.7. A weakly continuous function $f : \mathbb{R} \mapsto X$ is said to be weakly almost periodic if for every sequence of real numbers $(s_n)_n$ there exist a subsequence $(s'_n)_n \subset (s_n)_n$ and a function $\tilde{f}$ such that, for each $\varphi \in X^*$,

$$\langle \varphi, f(t + s'_n) \rangle \to \langle \varphi, \tilde{f}(t) \rangle,$$

uniformly on $\mathbb{R}$ as $n \to \infty$.

Remark 2.3.8. By applying the Bochner characterization in Theorem 2.3.2 to the scalar function $t \mapsto \langle \varphi, f(t) \rangle$, one can see that the notion of weak almost periodicity in Definition 2.3.7 is stronger than that used in [35, p. 39], [105, Definition 5.1, p. 137] and [199,

Definition 1, p. 65], where the authors called a function f weakly almost periodic if the scalar function $t \mapsto \langle \varphi, f(t) \rangle$ is almost periodic for each $\varphi \in X^*$. However, if the space X is weakly complete (for instance, X is reflexive), then both notions are equivalent, see [35, Theorem IX, p. 45].

Definition 2.3.9 ([234]). A weakly continuous function $f : \mathbb{R} \mapsto X$ is said to be weakly almost automorphic if for every sequence of real numbers $(s_n)_n$ there exist a subsequence $(s'_n)_n \subset (s_n)_n$ and a function $\tilde{f}$ such that, for each $t \in \mathbb{R}$,

$$f(t + s'_n) \to \tilde{f}(t)$$

and

$$\tilde{f}(t - s'_n) \to f(t),$$

as $n \to \infty$, where both convergences hold in the weak sense.

Remark 2.3.10. It is clear that if f is weakly almost automorphic, then the scalar function $t \mapsto \langle \varphi, f(t) \rangle$ is almost automorphic for each $\varphi \in X^*$.

An almost periodic (resp. almost automorphic) function is also weakly almost periodic (resp. weakly almost automorphic). The following two propositions establish a necessary and sufficient condition for the converse to hold.

Proposition 2.3.11 ([35, Theorem X, p. 45]). *Let $f : \mathbb{R} \mapsto X$ be a weakly almost periodic function. Then f is almost periodic if and only if its range is relatively compact.*

Proposition 2.3.12 ([234, Theorem 2.3.7, p. 25]). *Let $f : \mathbb{R} \mapsto X$ be a weakly almost automorphic function. Then f is almost automorphic if and only if its range is relatively compact.*

Let us now turn to the notion of almost periodicity in the sense of Stepanov.

Definition 2.3.13 ([105]). A function $f \in L^p_{\mathrm{loc}}(\mathbb{R}, X)$ is said to be Stepanov almost periodic for some $p \geq 1$ (or S^p-almost periodic) if for every $\varepsilon > 0$ there exists a positive number l such that every interval of length l contains a number τ such that

$$\left(\int_t^{t+1} |f(s+\tau) - f(s)|^p \mathrm{d}s \right)^{\frac{1}{p}} < \varepsilon \quad \text{for } t \in \mathbb{R}.$$

Bochner pointed out that the notion of Stepanov almost periodicity can be reduced to Bohr almost periodicity using the following construction.

Definition 2.3.14 ([35, 71, 246]). The Bochner transform f^b of a function $f \in L^p_{\mathrm{loc}}(\mathbb{R}, X)$ is the function $f^b : \mathbb{R} \to L^p([0,1], X)$ defined for each $t \in \mathbb{R}$ by

$$(f^b(t))(s) = f(t+s) \quad \text{for } s \in [0,1].$$

Proposition 2.3.15 ([35, 71, 246]). *A function $f \in L^p_{loc}(\mathbb{R}, X)$ is S^p-almost periodic if and only if the function $f^b : \mathbb{R} \to L^p([0,1], X)$ is almost periodic.*

If f is an S^p-almost periodic function, then from the Bochner characterization in Theorem 2.3.2, one can see that its Bochner transform $f^b : \mathbb{R} \to L^p([0,1], X)$ is bounded on $\mathbb{R}$, that is, f satisfies

$$\sup_{t\in\mathbb{R}} \left(\int_t^{t+1} \|f(s)\|^p \mathrm{d}s \right)^{\frac{1}{p}} < \infty.$$

This leads to the following definition.

Definition 2.3.16 ([35, 71, 246]). Let $p \geq 1$. The space $\mathrm{BS}^p(\mathbb{R}, X)$ consists of all functions $f \in L^p_{loc}(\mathbb{R}, X)$ such that $f^b : \mathbb{R} \to L^p([0,1], X)$ is bounded, that is, $\sup_{t\in\mathbb{R}}(\int_t^{t+1} \|f(s)\|^p \mathrm{d}s)^{\frac{1}{p}} < \infty$. It is a normed space when equipped with the following norm:

$$\|f\|_{\mathrm{BS}^p} = \sup_{t\in\mathbb{R}} \left(\int_t^{t+1} \|f(s)\|^p \mathrm{d}s \right)^{\frac{1}{p}}.$$

Remark. Functions of $\mathrm{BS}^p(\mathbb{R}, X)$ may not be bounded. For example, let $p \geq 1$ and f be the function defined by

$$f(t) = \begin{cases} k & \text{for } k \leq t \leq k + \frac{1}{k^p} \text{ with } k \in \mathbb{N}^*, \\ 0 & \text{otherwise.} \end{cases}$$

Then f is not bounded; however, $f \in \mathrm{BS}^p(\mathbb{R}, X)$. In fact,

$$\begin{aligned} \int_t^{t+1} \|f(s)\|^p \mathrm{d}s &\leq \int_{[t]}^{[t]+2} \|f(s)\|^p \mathrm{d}s \\ &= \sum_{k=[t]}^{[t]+1} \int_k^{k+\frac{1}{k^p}} \|f(s)\|^p \mathrm{d}s \\ &= 2. \end{aligned}$$

Proposition 2.3.17. *$(\mathrm{BS}^p(\mathbb{R}, X), \|\cdot\|_{\mathrm{BS}^p})$ is a Banach space.*

Proof. Let $(f_n)_n$ be a Cauchy sequence in $\mathrm{BS}^p(\mathbb{R}, X)$ and consider a subsequence $(f_{p_n})_n$ of $(f_n)_n$ such that

$$\sup_{t\in\mathbb{R}}\left(\int_t^{t+1}|f_{p_{n+1}}(s)-f_{p_n}(s)|^p\,ds\right)^{\frac{1}{p}}=\|f_{p_{n+1}}-f_{p_n}\|_{BS^p}\leq\frac{1}{2^n}\quad\text{for all } n\geq 1.$$

Let $g := \sum_{k=1}^{\infty}|f_{p_{k+1}}-f_{p_k}|$. Then using the monotone convergence theorem and the triangle inequality, we have, for each $m \in \mathbb{Z}$,

$$\begin{aligned}\left(\int_m^{m+1}g(s)^p\,ds\right)^{\frac{1}{p}}&=\lim_{N\to\infty}\left(\int_m^{m+1}\left(\sum_{k=1}^{N}|f_{p_{k+1}}(s)-f_{p_k}(s)|\right)^p ds\right)^{\frac{1}{p}}\\&\leq\sum_{k=1}^{\infty}\left(\int_m^{m+1}|f_{p_{k+1}}(s)-f_{p_k}(s)|^p\,ds\right)^{\frac{1}{p}}\\&\leq 1.\end{aligned}\tag{2.12}$$

It follows that the series defining g converges a. e. in each interval $[m, m+1]$. Therefore, the series

$$\sum_{k=1}^{\infty}(f_{p_{k+1}}-f_{p_k})$$

converges a. e. on $\mathbb{R}$ and

$$\left|\sum_{k=1}^{\infty}(f_{p_{k+1}}-f_{p_k})\right|\leq\sum_{k=1}^{\infty}|f_{p_{k+1}}-f_{p_k}|=g.\tag{2.13}$$

Let $f = f_{p_1}+\sum_{k=1}^{\infty}(f_{p_{k+1}}-f_{p_k})$. Then the sequence $f_{p_n} = f_{p_1}+\sum_{k=1}^{n-1}(f_{p_{k+1}}-f_{p_k})$ converges a. e. to f. In addition, from (2.12) and (2.13), we deduce that $f \in BS^p(\mathbb{R}, X)$. It remains to prove that $f_{p_n} \to f$ in $BS^p(\mathbb{R}, X)$. For this, let $\varepsilon > 0$ and $N_\varepsilon \in \mathbb{N}$ be such that

$$\sup_{t\in\mathbb{R}}\left(\int_t^{t+1}|f_n(s)-f_m(s)|^p\,ds\right)^{\frac{1}{p}}=\|f_n-f_m\|_{BS^p}\leq\varepsilon,$$

for all $n, m \geq N_\varepsilon$. Let $m \geq N_\varepsilon$, then for $n \geq N_\varepsilon$ we have

$$\left(\int_t^{t+1}|f_{p_n}(s)-f_m(s)|^p\,ds\right)^{\frac{1}{p}}\leq\varepsilon,$$

for all $t \in \mathbb{R}$. Using Fatou's lemma, we get for each $m \geq N_\varepsilon$ and $t \in \mathbb{R}$,

$$\left(\int_t^{t+1}|f(s)-f_m(s)|^p\,ds\right)^{\frac{1}{p}}\leq\liminf_{n\to\infty}\left(\int_t^{t+1}|f_{p_n}(s)-f_m(s)|^p\,ds\right)^{\frac{1}{p}}\leq\varepsilon.$$

Therefore $(f_n)_n$ converges to f in $BS^p(\mathbb{R}, X)$. □

The following Bochner-type characterization for the almost periodicity in the sense of Stepanov is essential for the rest of this study.

Proposition 2.3.18. *A function $f \in L^p_{loc}(\mathbb{R}, X)$ is S^p-almost periodic if and only if for every sequence of real numbers $(s_n)_n$ there exist a subsequence $(s'_n)_n \subset (s_n)_n$ and a function $g \in L^p_{loc}(\mathbb{R}, X)$ such that*

$$\sup_{t\in\mathbb{R}}\left(\int_t^{t+1} \|f(s+s'_n) - g(s)\|^p \mathrm{d}s\right)^{\frac{1}{p}} \to 0, \tag{2.14}$$

as $n \to \infty$.

Proof. If f satisfies (2.14), then from the fact that

$$\begin{aligned}\sup_{t\in\mathbb{R}}\|f^b(t+s'_n) - g^b(t)\|_{L^p([0,1],X)} &= \sup_{t\in\mathbb{R}}\left(\int_0^1 \|f^b(t+s'_n)(s) - g^b(t)(s)\|^p \mathrm{d}s\right)^{\frac{1}{p}} \\ &= \sup_{t\in\mathbb{R}}\left(\int_t^{t+1} \|f(s+s'_n) - g(s)\|^p \mathrm{d}s\right)^{\frac{1}{p}},\end{aligned}$$

it follows using Theorem 2.3.2 that f^b is almost periodic, and hence f is S^p-almost periodic. Now assume that f is S^p-almost periodic, that is, f^b is almost periodic. Let $(s_n)_n$ be a real sequence. Then there exist a subsequence $(s'_n)_n \subset (s_n)_n$ and a function $G : \mathbb{R} \to L^p([0,1], X)$ such that

$$\sup_{t\in\mathbb{R}}\|f^b(t+s'_n) - G(t)\|_{L^p([0,1],X)} \to 0,$$

as $n \to \infty$. It follows that

$$\sup_{t\in\mathbb{R}}\left(\int_t^{t+1} \|f(s+s'_n) - f(s+s'_m)\|^p \mathrm{d}s\right)^{\frac{1}{p}} = \sup_{t\in\mathbb{R}}\|f^b(t+s'_n) - f^b(t+s'_m)\|_{L^p([0,1],X)} \to 0,$$

as $n, m \to \infty$. That is, the sequence $(f(\cdot + s'_n))_n$ is a Cauchy sequence in the Banach space $BS^p(\mathbb{R}, X)$. Therefore, there exists a function $g \in BS^p(\mathbb{R}, X)$ such that

$$\sup_{t\in\mathbb{R}}\left(\int_t^{t+1} \|f(s+s'_n) - g(s)\|^p \mathrm{d}s\right)^{\frac{1}{p}} = \|f(\cdot + s'_n) - g\|_{BS^p} \to 0,$$

as $n \to \infty$. □

The link between the almost periodicity in Bohr's sense and in Stepanov's sense is given by the following result.

Proposition 2.3.19 ([60]). *A function f is almost periodic in Bohr's sense if and only if it is S^p-almost periodic and uniformly continuous.*

Example 2.3.20 ([35]).
- Let $H \in C_0^\infty(\mathbb{R},\mathbb{R})$ with support in $(-\frac{1}{2},\frac{1}{2})$ be such that $H \geq 0$, $H(0) = 1$, and $\int_{-\frac{1}{2}}^{\frac{1}{2}} H(t)dt = 1$. Let β_n be defined on $\mathbb{R}$ by $\beta_n(t) = \sum_{k\in P_n} H(n^2(t-k))$ with $P_n = 3^n(2\mathbb{Z}+1)$. Let f be given by

$$f(t) = \sum_{n\geq 1} \beta_n(t) \quad \text{for } t \in \mathbb{R}. \tag{2.15}$$

  Then, f is S^1-almost periodic, but not bounded. Hence, it is not almost periodic (see [26]).
- The function

$$f(t) = \sin\left(\frac{1}{2+\cos(t)+\cos(\sqrt{2}t)}\right)$$

  is S^p-almost periodic for all $p \geq 1$, but not almost periodic.

Now, let us shift our focus to the concept of almost automorphy in the Stepanov sense.

Definition 2.3.20 ([123]). A function $f \in L^p_{\mathrm{loc}}(\mathbb{R},X)$ is said to be Stepanov almost automorphic for some $p \geq 1$ (or S^p-almost automorphic) if the function $f^b : \mathbb{R} \to L^p([0,1],X)$ is almost automorphic.

Proposition 2.3.21. *A function $f \in L^p_{\mathrm{loc}}(\mathbb{R},X)$ is S^p-almost automorphic if and only if for every sequence of real numbers $(s_n)_n$ there exist a subsequence $(s'_n)_n \subset (s_n)_n$ and a function $g \in L^p_{\mathrm{loc}}(\mathbb{R},X)$ such that, for each $t \in \mathbb{R}$,*

$$\left(\int_t^{t+1} |f(s+s'_n) - g(s)|^p ds\right)^{\frac{1}{p}} \to 0 \tag{2.16}$$

and

$$\left(\int_t^{t+1} |g(s-s'_n) - f(s)|^p ds\right)^{\frac{1}{p}} \to 0, \tag{2.17}$$

as $n \to \infty$.

Proof. We have

$$|f^b(t+s'_n) - g^b(t)|_{L^p([0,1],X)} = \left(\int_t^{t+1} |f(s+s'_n) - g(s)|^p ds\right)^{\frac{1}{p}}$$

and

$$\|g^b(t-s'_n)-f^b(t)\|_{L^p([0,1],X)}=\left(\int_t^{t+1}\|g(s-s'_n)-f(s)\|^p\mathrm{d}s\right)^{\frac{1}{p}}.$$

It is clear that (2.16) and (2.17) imply that f^b is almost automorphic, and hence f is S^p-almost automorphic. Now assume that f is S^p-almost automorphic. Let $(s_n)_n$ be a real sequence. Then there exists a subsequence $(s'_n)_n \subset (s_n)_n$ and a function $G:\mathbb{R}\to L^p([0,1],X)$ such that, for each $t\in\mathbb{R}$,

$$\|f^b(t+s'_n)-G(t)\|_{L^p([0,1],X)}\to 0 \tag{2.18}$$

and

$$\|G(t-s'_n)-f^b(t)\|_{L^p([0,1],X)}\to 0, \tag{2.19}$$

as $n\to\infty$. It follows that for each $t\in\mathbb{R}$,

$$\left(\int_t^{t+1}\|f(s+s'_n)-f(s+s'_m)\|^p\mathrm{d}s\right)^{\frac{1}{p}}=\|f^b(t+s'_n)-f^b(t+s'_m)\|_{L^p([0,1],X)}\to 0,$$

as $n,m\to\infty$. Let $N\in\mathbb{N}$, then we have

$$\left(\int_{-N}^{N}\|f(s+s'_n)-f(s+s'_m)\|^p\mathrm{d}s\right)^{\frac{1}{p}}\to 0,$$

as $n,m\to\infty$. That is, $(f(\cdot+s'_n))_n$ is a Cauchy sequence in $L^p([-N,N],X)$. Hence there exists a function $h_N\in L^p([-N,N],X)$ such that

$$\left(\int_{-N}^{N}\|f(s+s'_n)-h_N(s)\|^p\mathrm{d}s\right)^{\frac{1}{p}}\to 0,$$

as $n\to\infty$. Consider the function $g:\mathbb{R}\to X$ defined by

$$g(t)=h_N(t)\quad \text{if } t\in[-N,N].$$

It is clear that $g\in L^p_{\mathrm{loc}}(\mathbb{R},X)$ and, for each $t\in\mathbb{R}$,

$$\|f^b(t+s'_n)-g^b(t)\|_{L^p([0,1],X)}=\left(\int_t^{t+1}\|f(s+s'_n)-g(s)\|^p\mathrm{d}s\right)^{\frac{1}{p}}\to 0, \tag{2.20}$$

as $n \to \infty$. It follows from (2.18) and (2.20) that $G = g^b$. Hence from (2.19), we deduce that

$$\left(\int_t^{t+1} |g(s - s_n') - f(s)|^p \mathrm{d}s \right)^{\frac{1}{p}} = |g^b(t - s_n') - f^b(t)|_{L^p([0,1],X)} \to 0,$$

as $n \to \infty$. □

Let $\mathrm{SAP}^p(\mathbb{R}, X)$ (resp. $\mathrm{SAA}^p(\mathbb{R}, X)$) denote the space of S^p-almost periodic (resp. S^p-almost automorphic) X-valued functions on $\mathbb{R}$. It is clear that for all $p \geq 1$, $\mathrm{AP}(\mathbb{R}, X) \subset \mathrm{SAP}^p(\mathbb{R}, X)$ and $\mathrm{AP}(\mathbb{R}, X) \subset \mathrm{AA}(\mathbb{R}, X) \subset \mathrm{SAA}^p(\mathbb{R}, X)$. Moreover, if $p \geq q$, then $\mathrm{SAP}^p(\mathbb{R}, X) \subset \mathrm{SAP}^q(\mathbb{R}, X)$ and $\mathrm{SAA}^p(\mathbb{R}, X) \subset \mathrm{SAA}^q(\mathbb{R}, X)$. If $h \in \mathrm{AP}(\mathbb{R}, \mathbb{C})$ (resp. $h \in \mathrm{AA}(\mathbb{R}, \mathbb{C})$) and $f \in \mathrm{SAP}^p(\mathbb{R}, \mathbb{C})$ (resp. $f \in \mathrm{SAA}^p(\mathbb{R}, \mathbb{C})$), then $h.f \in \mathrm{SAP}^p(\mathbb{R}, \mathbb{C})$ (resp. $h.f \in \mathrm{SAA}^p(\mathbb{R}, \mathbb{C})$).

Example 2.3.21 ([241]).

1. Let $(a_n)_{n\in\mathbb{Z}}$ be an almost automorphic sequence and $\varepsilon_0 \in]0, \frac{1}{2}[$. Consider a function $f : \mathbb{R} \to \mathbb{R}$ defined by

$$f(t) = \begin{cases} a_n & \text{if } t \in]n - \varepsilon_0, n + \varepsilon_0[, \\ 0 & \text{otherwise.} \end{cases}$$

Then, f is S^p-almost automorphic for all $p \geq 1$ but not almost automorphic.

In what follows, we denote by $\mathcal{B}$ the Lebesgue σ-field of $\mathbb{R}$ and by $\mathcal{M}$ the set of all positive measures μ on $\mathcal{B}$ satisfying $\mu(\mathbb{R}) = \infty$ and $\mu([a, b]) < \infty$, for all $a, b \in \mathbb{R}$ with $a < b$. For $\mu \in \mathcal{M}$, we formulate the following hypothesis:

(H4) For all $\tau \in \mathbb{R}$, there exist $\beta > 0$ and a bounded interval I such that

$$\mu(\{a + \tau : a \in A\}) \leq \beta\mu(A) \quad \text{when } A \in \mathcal{B} \text{ satisfies } A \cap I = \emptyset.$$

The hypothesis **(H4)** was formulated in [68].

Definition 2.3.22 ([68]). Let $\mu \in \mathcal{M}$. A bounded continuous function $f : \mathbb{R} \to X$ is said to be μ-ergodic if

$$\lim_{r\to\infty} \frac{1}{\mu([-r, r])} \int_{[-r,r]} |f(s)| \mathrm{d}\mu(s) = 0.$$

We denote the space of all such functions by $\mathcal{E}(\mathbb{R}, X, \mu)$.

Proposition 2.3.23 ([68]). *If a sequence $(f_n)_n$ of μ-ergodic functions converges uniformly to a function f, then f is also μ-ergodic.*

We introduce the following notion of ergodicity:

Definition 2.3.24. Let $\mu \in \mathcal{M}$. A function $f \in BS^p(\mathbb{R}, X)$ is said to be μ-ergodic in Stepanov's sense for some $p \geq 1$ (or S^p-μ-ergodic) if

$$\lim_{r\to\infty} \frac{1}{\mu([-r,r])} \int_{[-r,r]} \left(\int_t^{t+1} |f(s)|^p \mathrm{d}s \right)^{\frac{1}{p}} \mathrm{d}\mu(t) = 0.$$

We denote the space of all such functions by $\mathcal{E}^p(\mathbb{R}, X, \mu)$.

Proposition 2.3.25. *If μ satisfies* **(H4)**, *then $\mathcal{E}^p(\mathbb{R}, X, \mu)$ is invariant with respect to translation.*

Proof. Let $f \in \mathcal{E}^p(\mathbb{R}, X, \mu)$ and $F(t) = (\int_t^{t+1} |f(s)|^p \mathrm{d}s)^{\frac{1}{p}}$. Then $F \in \mathcal{E}(\mathbb{R}, \mathbb{R}, \mu)$ and, since $\mathcal{E}(\mathbb{R}, \mathbb{R}, \mu)$ is invariant with respect to translation [68],

$$\frac{1}{\mu([-r,r])} \int_{[-r,r]} \left(\int_t^{t+1} |f(s+a)|^p \mathrm{d}s \right)^{\frac{1}{p}} \mathrm{d}\mu(t) = \frac{1}{\mu([-r,r])} \int_{[-r,r]} F(t+a) \mathrm{d}\mu(t) \to 0,$$

as $r \to \infty$. We deduce that $f(\cdot + a) \in \mathcal{E}^p(\mathbb{R}, X, \mu)$. □

Theorem 2.3.26. *Let $\mu \in \mathcal{M}$ satisfy* **(H4)** *and $f \in \mathcal{E}(\mathbb{R}, X, \mu)$. Then $f \in \mathcal{E}^p(\mathbb{R}, X, \mu)$ for all $p \geq 1$.*

Proof. Let $f \in \mathcal{E}(\mathbb{R}, X, \mu)$, then by Hölder's inequality and Fubini's theorem,

$$\begin{aligned}
\int_{[-r,r]} \left(\int_t^{t+1} |f(s)|^p \mathrm{d}s \right)^{\frac{1}{p}} \mathrm{d}\mu(t) &= \int_{[-r,r]} \left(\int_0^1 |f(s+t)|^p \mathrm{d}s \right)^{\frac{1}{p}} \mathrm{d}\mu(t) \\
&\leq (\mu[-r,r])^{\frac{1}{q}} \left[\int_{[-r,r]} \left(\int_0^1 |f(s+t)|^p \mathrm{d}s \right) \mathrm{d}\mu(t) \right]^{\frac{1}{p}} \\
&\leq |f|_\infty^{\frac{1}{q}} (\mu[-r,r])^{\frac{1}{q}} \left[\int_{[-r,r]} \left(\int_0^1 |f(s+t)| \mathrm{d}s \right) \mathrm{d}\mu(t) \right]^{\frac{1}{p}} \\
&= |f|_\infty^{\frac{1}{q}} (\mu[-r,r])^{\frac{1}{q}} \left[\int_0^1 \left(\int_{[-r,r]} |f(s+t)| \mathrm{d}\mu(t) \right) \mathrm{d}s \right]^{\frac{1}{p}} \\
&= |f|_\infty^{\frac{1}{q}} \mu[-r,r] \left[\int_0^1 \frac{1}{\mu[-r,r]} \left(\int_{[-r,r]} |f(s+t)| \mathrm{d}\mu(t) \right) \mathrm{d}s \right]^{\frac{1}{p}} \\
&= |f|_\infty^{\frac{1}{q}} \mu[-r,r] \left[\int_0^1 \left(\frac{1}{\mu[-r,r]} \int_{[-r,r]} |f(s+t)| \mathrm{d}\mu(t) \right) \mathrm{d}s \right]^{\frac{1}{p}},
\end{aligned}$$

where $q = \infty$ if $p = 1$. Therefore,

$$\frac{1}{\mu[-r,r]} \int_{[-r,r]} \left(\int_t^{t+1} |f(s)|^p \mathrm{d}s \right)^{\frac{1}{p}} \mathrm{d}\mu(t) \le |f|_\infty^{\frac{1}{q}} \left[\int_0^1 \left(\frac{1}{\mu[-r,r]} \int_{[-r,r]} |f(s+t)| \mathrm{d}\mu(t) \right) \mathrm{d}s \right]^{\frac{1}{p}}.$$

Since $\mathcal{E}(\mathbb{R}, X, \mu)$ is translation invariant, then for all $s \in [0,1]$,

$$\frac{1}{\mu[-r,r]} \int_{[-r,r]} |f(s+t)| \mathrm{d}\mu(t) \to 0 \quad \text{when } r \to \infty.$$

From the Lebesgue dominated convergence theorem, it follows that

$$\lim_{r\to\infty} \frac{1}{\mu[-r,r]} \int_{[-r,r]} \left(\int_t^{t+1} |f(s)|^p \mathrm{d}s \right)^{\frac{1}{p}} \mathrm{d}\mu(t) = 0.$$

This ends the proof. □

Now, we give a sufficient condition for an S^p-μ-ergodic function to be μ-ergodic.

Proposition 2.3.27. *Let μ satisfy* **(H4)**. *If a bounded function f is S^p-μ-ergodic and uniformly continuous, then it is μ-ergodic.*

Remark. Unlike in Proposition 2.3.19, the condition of uniform continuity given in Proposition 2.3.27 is only sufficient, but not necessary. For example, we consider the following function:

$$f(t) = \begin{cases} 2k^2 t - 2k^3 & \text{for } k \le t \le k + \frac{1}{2k^2} \text{ with } k \in \mathbb{N}^*, \\ -2k^2 t + 2 + 2k^3 & \text{for } k + \frac{1}{2k^2} \le t \le k + \frac{1}{k^2} \text{ with } k \in \mathbb{N}^*, \\ 0 & \text{otherwise.} \end{cases}$$

The function f looks like this:

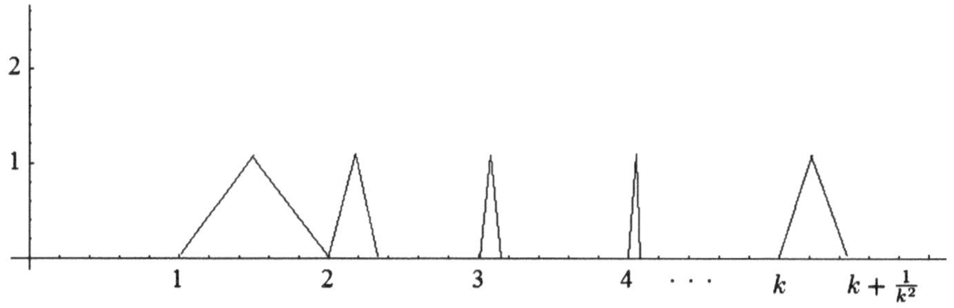

Such a function f is continuous, since it is continuous at the points k, $k + \frac{1}{2k^2}$, and $k + \frac{1}{k^2}$. This f is also λ-ergodic where λ is the Lebesgue measure, in fact,

$$\frac{1}{2r}\int_{-r}^{r}|f(s)|\mathrm{d}s \le \frac{1}{2r}\int_{0}^{\infty}|f(s)|\mathrm{d}s$$
$$\le \frac{1}{2r}\sum_{k=1}^{\infty}\frac{1}{2k^2}$$
$$= \frac{\pi^2}{24r} \to 0, \quad \text{when } r \to \infty.$$

Let $(x_k)_{k\ge1}$ and $(y_k)_{k\ge1}$ be two sequences defined by

$$x_k = k \quad \text{and} \quad y_k = k + \frac{1}{2k^2}.$$

Then

$$|x_k - y_k| = \frac{1}{2k^2} \to 0, \quad \text{when } k \to \infty,$$

but

$$|f(x_k) - f(y_k)| = 1.$$

This implies that f is not uniformly continuous.

Proof of Proposition 2.3.27. Let f be a bounded S^p-μ-ergodic function which is uniformly continuous. Let $(\rho_k)_k$ be a family of mollifier functions, i. e., for each $k \in \mathbb{N}$, $\rho_k > 0$ is smooth with compact support and satisfying

$$\int_{-\infty}^{\infty}\rho_k(s)\mathrm{d}s = 1,$$

and the support of ρ_k goes to $\{0\}$ when k goes to ∞. We consider the following convolution product:

$$(f \star \rho_k)(t) = \int_{-\infty}^{\infty}\rho_k(s)f(t-s)\mathrm{d}s.$$

Then

$$\frac{1}{\mu[-r,r]}\int_{-r}^{r}|(f \star \rho_k)(t)|\mathrm{d}\mu(t) \le \frac{1}{\mu[-r,r]}\int_{-r}^{r}\left(\int_{-M}^{M}|\rho_k(s)f(t-s)|\mathrm{d}s\right)\mathrm{d}\mu(t)$$
$$\le \sum_{n=-[M]-1}^{[M]}\left(\frac{1}{\mu[-r,r]}\int_{-r}^{r}\left(\int_{n}^{n+1}\rho_k(s)|f(t-s)|\mathrm{d}s\right)\mathrm{d}\mu(t)\right)$$

$$\leq |\rho_k|_q \sum_{n=-[M]-1}^{[M]} \left(\frac{1}{\mu[-r,r]} \int_{-r}^{r} \left(\int_{n}^{n+1} \|f(t-s)\|^p \mathrm{d}s \right)^{\frac{1}{p}} \mathrm{d}\mu(t) \right)$$

$$\leq |\rho_k|_q \sum_{n=-[M]-1}^{[M]} \left(\frac{1}{\mu[-r,r]} \int_{-r}^{r} \left(\int_{t}^{t+1} \|f(s-n-1)\|^p \mathrm{d}s \right)^{\frac{1}{p}} \mathrm{d}\mu(t) \right),$$

with $\frac{1}{p} + \frac{1}{q} = 1$. Since $\mathcal{E}^p(\mathbb{R}, X, \mu)$ is invariant with respect to translation (Proposition 2.3.25), $f \star \rho_k \in \mathcal{E}(\mathbb{R}, X, \mu)$. Moreover, as f is bounded and uniformly continuous, by [133] then $f \star \rho_k \to f$ uniformly on $\mathbb{R}$, and it follows from Proposition 2.3.23 that $f \in \mathcal{E}(\mathbb{R}, X, \mu)$. □

Definition 2.3.28 ([68]). Let $\mu \in \mathcal{M}$. A function $f \in BC(\mathbb{R}, X)$ is said to be μ-pseudo almost periodic if f is written in the form

$$f = g + g_0,$$

where $g \in AP(\mathbb{R}, X)$ and $g_0 \in \mathcal{E}(\mathbb{R}, X, \mu)$. We denote this class of functions by $PAP(\mathbb{R}, X, \mu)$.

Definition 2.3.29. Let $\mu \in \mathcal{M}$. A function $f \in BS^p(\mathbb{R}, X)$ is said to be μ-pseudo almost periodic in Stepanov's sense if f is written in the form

$$f = g + g_0,$$

where $g \in SAP^p(\mathbb{R}, X)$ and $g_0 \in \mathcal{E}^p(\mathbb{R}, X, \mu)$. We denote this class of functions by $PAP^p(\mathbb{R}, X, \mu)$.

2.3.1 A Result on Integration of Stepanov Almost Periodic Functions

In this section, we give sufficient conditions for the almost periodicity of the indefinite integral of Stepanov almost periodic functions. Let us first recall the following results.

Theorem 2.3.22 ([35, Theorem II, p. 55]). *Let X be a uniformly convex Banach space and $f : \mathbb{R} \to X$ be an almost periodic function. If $F(t) := \int_0^t f(s)\mathrm{d}s$ is bounded on $\mathbb{R}$, then it is almost periodic.*

The following result relaxes the assumption of almost periodicity and boundedness in Theorem 2.3.22.

Theorem 2.3.23 ([35, Theorem II', p. 82]). *Let X be a uniformly convex Banach space and $f : \mathbb{R} \to X$ be an S^p-almost periodic function with $p > 1$. Let $F(t) := \int_0^t f(s)\mathrm{d}s$. If $F \in BS^p(\mathbb{R}, X)$, then F is almost periodic.*

The proof of Theorem 2.3.23 relies on the fact that the space $L^p([0,1], X)$ is uniformly convex for $p > 1$, when X is uniformly convex. For $p = 1$, the space $L^p([0,1], X)$ is not uniformly convex, thus one cannot use the approach in Theorem 2.3.23 to give a similar result. To solve this problem, we make use of the Bochner-type characterization in Proposition 2.3.18. We have the following result.

Theorem 2.3.24. *Let X be a uniformly convex Banach space and $f : \mathbb{R} \to X$ be an S^1-almost periodic function. Let $F(t) := \int_0^t f(s)\mathrm{d}s$. If F is bounded on $\mathbb{R}$, then it is almost periodic.*

Before proving Theorem 2.3.24, let us first study the scalar case.

Lemma 2.3.25. *Let $f : \mathbb{R} \to \mathbb{C}$ be an S^1-almost periodic function. If the integral $F(t) := \int_0^t f(s)\mathrm{d}s$ is bounded on $\mathbb{R}$, then it is almost periodic.*

Proof of Lemma 2.3.25. We only have to prove the result for $\mathbb{R}$-valued functions. In fact, for a function $f : \mathbb{R} \to \mathbb{C}$, we have

$$\int_0^t f(s)\mathrm{d}s = \int_0^t \mathrm{Re}(f(s))\mathrm{d}s + i\int_0^t \mathrm{Im}(f(s))\mathrm{d}s.$$

Thus, the integral $\int_0^t f(s)\mathrm{d}s$ is bounded if and only if $\int_0^t \mathrm{Re}(f(s))\mathrm{d}s$ and $\int_0^t \mathrm{Im}(f(s))\mathrm{d}s$ are bounded.

Let $f : \mathbb{R} \to \mathbb{R}$ be an S^1-almost periodic function and $F(t) := \int_0^t f(s)\mathrm{d}s$ for each $t \in \mathbb{R}$. Let $m(F) := \inf_{s\in\mathbb{R}} F(s)$, $M(F) := \sup_{s\in\mathbb{R}} F(s)$, and $(s_n)_{n\in\mathbb{N}}$ be an arbitrary sequence. We have for all $n \in \mathbb{N}$ and $t \in \mathbb{R}$,

$$F(t + s_n) = F(s_n) + \int_0^t f(s + s_n)\mathrm{d}s.$$

Since F is bounded and f is S^1-almost periodic, by the Bochner-type characterization in Proposition 2.3.18, there exist a subsequence $(s_n')_n \subset (s_n)_n$, a function $\tilde{f} \in L^1_{\mathrm{loc}}(\mathbb{R}, \mathbb{R})$, and a constant $c \in \mathbb{R}$ such that, for each $t \in \mathbb{R}$,

$$\sup_{t\in\mathbb{R}} \int_t^{t+1} |f(s + s_n') - \tilde{f}(s)|\mathrm{d}s \to 0 \tag{2.21}$$

and

$$F(s_n') \to c, \tag{2.22}$$

as $n \to +\infty$. Consider the function

$$\tilde{F}(t) := c + \int_0^t \tilde{f}(s)\mathrm{d}s \quad \text{for all } t \in \mathbb{R}.$$

We can see that, for each $t \in \mathbb{R}$,

$$\int_0^t f(s+s_n')ds \to \int_0^t \tilde{f}(s)ds \quad \text{as } n \to +\infty. \tag{2.23}$$

In fact, using (2.21), for $t \geq 0$ we obtain

$$\begin{aligned}\left|\int_0^t f(s+s_n')ds - \int_0^t \tilde{f}(s)ds\right| &\leq \int_0^t |f(s+s_n') - \tilde{f}(s)|ds \\ &\leq \sum_{k=0}^{[t]} \int_k^{k+1} |f(s+s_n') - \tilde{f}(s)|ds \to 0,\end{aligned}$$

as $n \to +\infty$. Using the same arguments, for $t \leq 0$ we have

$$\begin{aligned}\left|\int_0^t f(s+s_n')ds - \int_0^t \tilde{f}(s)ds\right| &\leq \int_t^0 |f(s+s_n') - \tilde{f}(s)|ds \\ &\leq \sum_{k=[t]}^{-1} \int_k^{k+1} |f(s+s_n') - \tilde{f}(s)|ds \to 0,\end{aligned}$$

as $n \to +\infty$. From (2.22) and (2.23), we deduce that, for each $t \in \mathbb{R}$,

$$F(t+s_n') \to \tilde{F}(t) \quad \text{as } n \to +\infty.$$

We claim that this convergence is uniform in $t \in \mathbb{R}$. In fact, if the convergence were not uniform, then there would exist $\rho > 0$ and three sequences $(p_n)_n$, $(q_n)_n$, $(t_n)_n$ with $p_n, q_n \geq n$ such that

$$|F(t_n + s_{p_n}') - F(t_n + s_{q_n}')| > \rho. \tag{2.24}$$

Let $a_n := t_n + s_{p_n}'$ and $b_n := t_n + s_{q_n}'$. Since f is S^1-almost periodic and F is bounded, there exist two subsequences, which for simplicity will be also denoted by $(a_n)_n$ and $(b_n)_n$, and two functions $\tilde{f}_1, \tilde{f}_2 \in L^1_{\text{loc}}(\mathbb{R}, \mathbb{R})$ such that

$$\begin{cases} \sup_{t\in\mathbb{R}} \int_t^{t+1} |f(s+a_n) - \tilde{f}_1(s)|ds \to 0, \\ \sup_{t\in\mathbb{R}} \int_t^{t+1} |f(s+b_n) - \tilde{f}_2(s)|ds \to 0, \end{cases} \tag{2.25}$$

and

$$F(a_n) \to c_1 \quad \text{and} \quad F(b_n) \to c_2, \tag{2.26}$$

as $n \to +\infty$, where $c_1, c_2 \in \mathbb{R}$. We have

$$\begin{aligned}\int_t^{t+1} |f(s+a_n) - f(s+b_n)|ds &= \int_t^{t+1} |f(s+t_n+s'_{p_n}) - f(s+t_n+s'_{q_n})|ds \\ &= \int_{t+t_n}^{t+t_n+1} |f(s+s'_{p_n}) - f(s+s'_{q_n})|ds \\ &\le \sup_{t\in\mathbb{R}} \int_t^{t+1} |f(s+s'_{p_n}) - f(s+s'_{q_n})|ds \\ &\le \sup_{t\in\mathbb{R}} \int_t^{t+1} |f(s+s'_{p_n}) - \tilde{f}(s)|ds + \sup_{t\in\mathbb{R}} \int_t^{t+1} |\tilde{f}(s) - f(s+s'_{q_n})|ds.\end{aligned}$$

It follows from (2.21) that

$$\int_t^{t+1} |f(s+a_n) - f(s+b_n)|ds \to 0 \quad \text{as } n \to +\infty. \tag{2.27}$$

Now, since

$$\begin{aligned}\int_t^{t+1} |\tilde{f}_1(s) - \tilde{f}_2(s)|ds \le{}& \int_t^{t+1} |\tilde{f}_1(s) - f(s+a_n)|ds + \int_t^{t+1} |f(s+a_n) - f(s+b_n)|ds \\ &+ \int_t^{t+1} |f(s+b_n) - \tilde{f}_2(s)|ds,\end{aligned}$$

letting $n \to +\infty$ and using (2.25) and (2.27), we conclude that, for all $t \in \mathbb{R}$,

$$\int_t^{t+1} |\tilde{f}_1(s) - \tilde{f}_2(s)|ds = 0.$$

That is, $\tilde{f}_1 = \tilde{f}_2$ a. e. Using (2.25) and (2.26), it is clear that, for all $t \in \mathbb{R}$,

$$F(t+a_n) \to c_1 + \int_0^t \tilde{f}_1(s)ds := \tilde{F}_1(t) \tag{2.28}$$

and

$$F(t+b_n) \to c_2 + \int_0^t \tilde{f}_1(s)ds := \tilde{F}_2(t), \tag{2.29}$$

as $n \to +\infty$. This implies that

$$\begin{cases} M(\widetilde{F}_1) \le M(F), \\ M(\widetilde{F}_2) \le M(F), \\ m(\widetilde{F}_1) \ge m(F), \\ m(\widetilde{F}_2) \ge m(F). \end{cases} \tag{2.30}$$

On the other hand, for all $n \in \mathbb{R}$ and $t \in \mathbb{N}$,

$$\begin{cases} \widetilde{F}_1(t - a_n) = \widetilde{F}_1(-a_n) + \int_0^t \widetilde{f}_1(s - a_n)\mathrm{d}s, \\ \widetilde{F}_2(t - b_n) = \widetilde{F}_2(-b_n) + \int_0^t \widetilde{f}_1(s - b_n)\mathrm{d}s. \end{cases} \tag{2.31}$$

From (2.28) and (2.29), one can see that the functions $\widetilde{F}_1$ and $\widetilde{F}_2$ are also bounded on $\mathbb{R}$. Then, there exist two subsequence $(a'_n)_n \subset (a_n)_n$ and $(b'_n)_n \subset (b_n)_n$ and two constants $d_1, d_2 \in \mathbb{R}$ such that

$$\begin{cases} \widetilde{F}_1(-a'_n) \to d_1, \\ \widetilde{F}_2(-b'_n) \to d_2. \end{cases} \tag{2.32}$$

It is clear from (2.25) that

$$\begin{cases} \sup_{t \in \mathbb{R}} \int_t^{t+1} |f(s) - \widetilde{f}_1(s - a_n)|\mathrm{d}s \to 0, \\ \sup_{t \in \mathbb{R}} \int_t^{t+1} |f(s) - \widetilde{f}_1(s - b_n)|\mathrm{d}s \to 0. \end{cases}$$

Using the same arguments as above, we have

$$\begin{cases} \int_0^t \widetilde{f}_1(s - a'_n)\mathrm{d}s \to \int_0^t f(s)\mathrm{d}s = F(t), \\ \int_0^t \widetilde{f}_1(s - b'_n)\mathrm{d}s \to \int_0^t f(s)\mathrm{d}s = F(t), \end{cases} \tag{2.33}$$

as $n \to +\infty$. From (2.31)–(2.33), we deduce that, for each $t \in \mathbb{R}$,

$$\begin{cases} \widetilde{F}_1(t - a'_n) \to d_1 + F(t), \\ \widetilde{F}_2(t - b'_n) \to d_2 + F(t), \end{cases} \tag{2.34}$$

as $n \to +\infty$. In addition, we observe from (2.34) that

$$\begin{cases} M(d_1 + F) \le M(\widetilde{F}_1), \\ M(d_2 + F) \le M(\widetilde{F}_2), \\ m(d_1 + F) \ge m(\widetilde{F}_1), \\ m(d_2 + F) \ge m(\widetilde{F}_2). \end{cases} \tag{2.35}$$

It follows from (2.30) and (2.35) that

$$d_1 = d_2 = 0.$$

Therefore, $M(\widetilde{F}_1) = M(\widetilde{F}_2)$. However, we know from (2.28) and (2.29) that

$$\widetilde{F}_1(t) = \widetilde{F}_2(t) + c_1 - c_2 \quad \text{for } t \in \mathbb{R}.$$

Hence we deduce that $\widetilde{F}_1 = \widetilde{F}_2$. It follows from (2.28) and (2.29) that, for all $t \in \mathbb{R}$,

$$|F(t + a_n) - F(t + b_n)| = |F(t + t_n + s'_{p_n}) - F(t + t_n + s'_{q_n})| \to 0,$$

which contradicts (2.24). □

Proof of Theorem 2.3.24. Let $F(t) := \int_0^t f(s)\mathrm{d}s$ for $t \in \mathbb{R}$. Then, for all $\phi \in X^*$ we have

$$\langle \phi, F(t)\rangle = \int_0^t \langle \phi, f(s)\rangle \mathrm{d}s.$$

Since $f \in \mathrm{SAP}^1(\mathbb{R}, X)$, one has $t \mapsto \langle \phi, f(t)\rangle \in \mathrm{SAP}^1(\mathbb{R}, \mathbb{R})$ for all $\phi \in X^*$. The boundedness of F implies the boundedness of the function $t \mapsto \langle \phi, F(t)\rangle$. It follows by Lemma 2.3.25 that $t \mapsto \langle \phi, F(t)\rangle$ is almost periodic for each $\phi \in X^*$. That is, F is weakly almost periodic. If $F = 0$, then F is almost periodic. Assume that $F \neq 0$. We claim that F has a relatively compact range. In fact, if the range $R_F = \{F(t) : t \in \mathbb{R}\}$ were not relatively compact, then there would exist a constant $c > 0$ and a sequence $(t_n)_n$ such that

$$|F(t_p) - F(t_q)| > c \quad \text{for all } p \neq q. \tag{2.36}$$

We have, for all $t \in \mathbb{R}$ and $n \in \mathbb{N}$,

$$F(t + t_n) = F(t_n) + \int_0^t f(s + t_n)\mathrm{d}s.$$

It follows that, for all $p, q \in \mathbb{N}$,

$$F(t + t_p) - F(t + t_q) = F(t_p) - F(t_q) + \int_0^t (f(s + t_p) - f(s + t_q))\mathrm{d}s.$$

Hence,

$$F(t_p) - F(t_q) = F(t + t_p) - F(t + t_q) - \int_0^t (f(s + t_p) - f(s + t_q))\mathrm{d}s.$$

It follows from (2.36) that

$$|F(t+t_p)-F(t+t_q)|+\int_0^t |f(s+t_p)-f(s+t_q)|ds \geq |F(t_p)-F(t_q)|>c. \tag{2.37}$$

One can see that

$$\int_0^t |f(s+t_p)-f(s+t_q)|ds \leq \sum_{k=0}^{[t]}\int_k^{k+1} |f(s+t_p)-f(s+t_q)|ds.$$

From the S^1-almost periodicity of f, there exists a subsequence $(t'_n)_n \subset (t_n)_n$ such that

$$\int_k^{k+1} |f(s+t'_p)-f(s+t'_q)|ds \to 0,$$

as $p,q \to +\infty$. Hence for p,q large enough, we get

$$\int_0^t |f(s+t'_p)-f(s+t'_q)|ds \leq \frac{c}{2}. \tag{2.38}$$

Thus, using (2.38) and (2.37), we obtain

$$|F(t+t'_p)-F(t+t'_q)|>\frac{c}{2}. \tag{2.39}$$

Since X is uniformly convex, there exists $\delta>0$ such that for any $u,v\in X$ with $|u|\leq 1$ and $|v|\leq 1$, if $|u-v|\geq \frac{c}{2|F|_\infty}$, then $|\frac{u+v}{2}|\leq 1-\delta$. Thus, (2.39) implies that

$$\left|\frac{F(t+t'_p)+F(t+t'_q)}{2}\right| \leq |F|_\infty - \delta|F|_\infty.$$

For all $\phi\in X^*$ with $|\phi|\leq 1$, we have

$$\left|\left\langle \phi, \frac{F(t+t'_p)+F(t+t'_q)}{2}\right\rangle\right| \leq \left|\frac{F(t+t'_p)+F(t+t'_q)}{2}\right| \leq |F|_\infty - \delta|F|_\infty. \tag{2.40}$$

Since F is weakly almost periodic and X is weakly sequentially complete (being reflexive), by [35, Theorem IX, p. 45], there exist a subsequence $(t''_n)_n \subset (t'_n)_n$ and a function $\widetilde{F}$ such that, for all $t\in\mathbb{R}$,

$$F(t+t''_n)\to \widetilde{F}(t) \tag{2.41}$$

uniformly in the weak sense. This implies that we also have

$$\widetilde{F}(t-t''_n)\to F(t) \tag{2.42}$$

uniformly in the weak sense. It is clear that (2.41) and (2.42) imply

$$|F|_\infty = |\widetilde{F}|_\infty.$$

By taking $p, q \to +\infty$ in (2.40), we get, for all $\phi \in X^*$ with $|\phi| \le 1$,

$$|\langle \phi, \widetilde{F}(t)\rangle| \le |F|_\infty - \delta|F|_\infty = |\widetilde{F}|_\infty - \delta|\widetilde{F}|_\infty,$$

yielding

$$|\widetilde{F}|_\infty = \sup_{t\in\mathbb{R}}|\widetilde{F}(t)| = \sup_{t\in\mathbb{R}} \sup_{|\phi|\le 1, \phi\in X^*} |\langle \phi, \widetilde{F}(t)\rangle| \le |\widetilde{F}|_\infty - \delta|\widetilde{F}|_\infty,$$

which is a contradiction since $|\widetilde{F}|_\infty = |F|_\infty \neq 0$. We conclude that $R_F = \{F(t) : t \in \mathbb{R}\}$ is relatively compact. We deduce from Theorem 2.3.11 that F is almost periodic. □

2.3.2 On the integration of Stepanov almost automorphic functions

Let us find the conditions under which the integral of a Stepanov almost automorphic function is almost automorphic. We first recall the following result.

Theorem 2.3.26 ([240]). *Let us define $F : \mathbb{R} \mapsto \mathbb{X}$ by $F(t) = \int_0^t f(s)ds$ where $f \in AA(\mathbb{X})$. Then $F \in AA(\mathbb{X})$ if and only if $R_F = \{F(t) \mid t \in \mathbb{R}\}$ is precompact.*

Theorem 2.3.30 ([234, Theorem 2.4.6, p. 29]). *Let X be a uniformly convex Banach space and $f : \mathbb{R} \to X$ be an almost automorphic function. If $F(t) := \int_0^t f(s)ds$ is bounded on $\mathbb{R}$, then it is almost automorphic.*

The following result gives the same conclusion as in Theorem 2.3.30 but with weaker assumptions.

Theorem 2.3.31. *Let X be a uniformly convex Banach space and $f : \mathbb{R} \to X$ be an S^p-almost automorphic function with $p > 1$. Let $F(t) := \int_0^t f(s)ds$. If $F \in BS^p(\mathbb{R}, X)$, then F is almost automorphic.*

The following lemmas are needed in the proof of Theorem 2.3.31.

Lemma 2.3.32 ([210, Lemma 3.2]). *Let X be a Banach space and $f : \mathbb{R} \to X$ be an S^p-almost automorphic function with $p \ge 1$. Then the function F defined on $\mathbb{R}$ by $F(t) := \int_0^t f(s)ds$ is uniformly continuous.*

Lemma 2.3.33 ([210, Lemma 3.1]). *Let X be a Banach space and $f : \mathbb{R} \to X$ be an S^p-almost automorphic function with $p \ge 1$. If f is uniformly continuous then f is almost automorphic.*

Proof of Theorem 2.3.31. We first observe that, for each $t \in \mathbb{R}$,

$$\int_0^t f^b(s)ds = F^b(t) - F^b(0),$$

where f^b denotes the Bochner transform of f (see Definition 2.3.14). Since $f \in SAA^p(\mathbb{R}, X)$, we get $f^b \in AA(\mathbb{R}, L^p([0,1], X))$. Note that F^b is bounded on $\mathbb{R}$ since $F \in BS^p(\mathbb{R}, X)$. The uniform convexity of $L^p([0,1], X)$ follows from the uniform convexity of X and the fact that $p > 1$. Now from Theorem 2.3.30, we deduce that F^b is almost automorphic, that is, $F \in SAA^p(\mathbb{R}, X)$. From Lemma 2.3.32, F is uniformly continuous. It follows by Lemma 2.3.33 that F is almost automorphic. □

The proof of Theorem 2.3.31 relies on the fact that the space $L^p[(0,1), X]$ is uniformly convex for $p > 1$, when X is uniformly convex. For $p = 1$, the space $L^p((0,1), X)$ is not uniformly convex, thus one cannot use the approach in Theorem 2.3.31 to give a similar result when $p = 1$.

Theorem 2.3.34. *Let X be a uniformly convex Banach space and $f : \mathbb{R} \to X$ be an S^1-almost automorphic function. Let $F(t) := \int_0^t f(s)ds$. If F is bounded on $\mathbb{R}$, then it is almost automorphic.*

Proof. The proof is divided in two steps:

Step 1. In this step, we will prove that F is almost automorphic in the weak topology. Let $(s_n)_{n\in\mathbb{N}}$ be an arbitrary sequence. We have, for all $n \in \mathbb{N}$ and $t \in \mathbb{R}$,

$$F(t + s_n) = F(s_n) + \int_0^t f(s + s_n)ds.$$

Since the Banach space X is reflexive, F is bounded, and f is S^1-almost automorphic, there exist a subsequence $(s'_n)_n \subset (s_n)_n$, a function $\tilde{f} \in L^1_{loc}(\mathbb{R}, X)$, and a constant $c \in X$ such that, for each $t \in \mathbb{R}$,

$$\begin{cases} \int_t^{t+1} \|f(s + s'_n) - \tilde{f}(s)\|ds \to 0, \\ \int_t^{t+1} \|\tilde{f}(s - s'_n) - f(s)\|ds \to 0, \end{cases} \tag{2.43}$$

and

$$F(s'_n) \to c, \tag{2.44}$$

in the weak topology as $n \to \infty$. Consider the function

$$\tilde{F}(t) := c + \int_0^t \tilde{f}(s)ds \quad \text{for all } t \in \mathbb{R}.$$

We can see that, for each $t \in \mathbb{R}$,

$$\int_0^t f(s+s'_n)\mathrm{d}s \to \int_0^t \tilde{f}(s)\mathrm{d}s \quad \text{as } n \to \infty. \tag{2.45}$$

In fact, using (2.43), we obtain that, for $t \geq 0$,

$$\begin{aligned}\left|\int_0^t f(s+s'_n)\mathrm{d}s - \int_0^t \tilde{f}(s)\mathrm{d}s\right| &\leq \int_0^t |f(s+s'_n) - \tilde{f}(s)|\mathrm{d}s \\ &\leq \sum_{k=0}^{[t]} \int_k^{k+1} |f(s+s'_n) - \tilde{f}(s)|\mathrm{d}s \to 0,\end{aligned}$$

as $n \to \infty$. Using the same arguments, for $t \leq 0$ we have

$$\begin{aligned}\left|\int_0^t f(s+s'_n)\mathrm{d}s - \int_0^t \tilde{f}(s)\mathrm{d}s\right| &\leq \int_t^0 |f(s+s'_n) - \tilde{f}(s)|\mathrm{d}s \\ &\leq \sum_{k=[t]}^{-1} \int_k^{k+1} |f(s+s'_n) - \tilde{f}(s)|\mathrm{d}s \to 0,\end{aligned}$$

as $n \to \infty$.

From (2.44) and (2.45), we deduce that, for each $t \in \mathbb{R}$,

$$F(t+s'_n) \to \tilde{F}(t) \tag{2.46}$$

in the weak topology as $n \to \infty$. Using the fact that $|\tilde{F}(t)| = \sup_{\|\varphi\|\leq 1, \varphi\in X^*} |\langle \varphi, \tilde{F}(t)\rangle|$, one can observe from (2.46) that

$$|\tilde{F}|_\infty \leq |F|_\infty, \tag{2.47}$$

where $|\cdot|_\infty$ denotes the supremum norm. On the other hand, for all $n \in \mathbb{N}$ and $t \in \mathbb{R}$, we have

$$\tilde{F}(t-s'_n) = \tilde{F}(-s'_n) + \int_0^t \tilde{f}(s-s'_n)\mathrm{d}s.$$

Since $\tilde{F}$ is also bounded, there exist a subsequence $(s''_n)_n \subset (s'_n)_n$ and a constant $d \in \mathbb{R}$ such that

$$\tilde{F}(-s''_n) \to d \tag{2.48}$$

in the weak topology as $n \to \infty$. Using the same arguments as above, we have

$$\int_0^t \widetilde{f}(s - s_n'')ds \to \int_0^t f(s)ds = F(t) \quad \text{as } n \to \infty. \tag{2.49}$$

From (2.48) and (2.49), we deduce that, for each $t \in \mathbb{R}$,

$$\widetilde{F}(t - s_n'') \to d + F(t) \tag{2.50}$$

in the weak topology as $n \to \infty$.

Let us now prove that $d = 0$. Using the same notation as in [234], we can write (2.46) and (2.50) as

$$T_{s'}F = \widetilde{F} \tag{2.51}$$

and

$$T_{-s''}\widetilde{F} = d + F, \tag{2.52}$$

where T_s is the operator defined for some sequence $s = (s_n)_n$ by

$$(T_s f)(t) = \lim_{n\to\infty} f(t + s_n)$$

if this limit exists. We note that this notation was used for the first time in the famous paper of Bochner [73]. Since $(s_n')_n$ is a subsequence of $(s_n)_n$, $T_{s''}F$ exists and is equal to $T_{s'}F$. Define the operator A_s by $A_s := T_{-s}T_s$. Then from (2.51) and (2.52), we have

$$A_{s''}F = d + F.$$

We can again apply the operator A and obtain

$$A_{s''}^2 F = A_{s''}(A_{s''}F) = 2d + F.$$

In general, for each $n \in \mathbb{N}$, we have

$$A_{s''}^n F = nd + F. \tag{2.53}$$

From (2.47), we have

$$|A_{s''}F|_\infty = |T_{-s}T_sF|_\infty \leq |T_sF|_\infty \leq |F|_\infty.$$

Thus (2.53) implies that

$$\sup_{t\in\mathbb{R}}|nd + F(t)| \leq \sup_{t\in\mathbb{R}}|F(t)| \quad \text{for all } n \in \mathbb{N},$$

which holds only if $d = 0$. Therefore, we deduce that, for each $t \in \mathbb{R}$,

$$\widetilde{F}(t - s_n'') \to F(t) \tag{2.54}$$

in the weak topology as $n \to \infty$. As a conclusion, using (2.46) and (2.54), we proved that for an arbitrary sequence $(s_n)_n$, there exist a subsequence $(s_n'')_n$ and a measurable function $\widetilde{F}$ such that, for each $t \in \mathbb{R}$,

$$\lim_{n\to\infty} F(t + s_n'') = \widetilde{F}(t)$$

and

$$\lim_{n\to\infty} \widetilde{F}(t - s_n'') = F(t),$$

where both limits hold in the weak topology. This implies that F is weakly almost automorphic. We stress that we did not use the uniform convexity assumption, yet. Thus the result obtained in this step is valid for an arbitrary reflexive Banach space.

Step 2. In this step, we prove that F is almost automorphic in the strong sense. If $F = 0$, then F is almost automorphic. Assume that $F \neq 0$, then all we need to prove is that the range of F is relatively compact. In fact, if the range $R_F = \{F(t) : t \in \mathbb{R}\}$ were not relatively compact, then there would exist a constant $c > 0$ and a sequence $(t_n)_n$ such that

$$|F(t_p) - F(t_q)| > c \quad \text{for all } p \neq q. \tag{2.55}$$

We have, for all $t \in \mathbb{R}$ and $n \in \mathbb{N}$,

$$F(t + t_n) = F(t_n) + \int_0^t f(s + t_n)\mathrm{d}s.$$

It follows that, for all $p, q \in \mathbb{N}$,

$$F(t + t_p) - F(t + t_q) = F(t_p) - F(t_q) + \int_0^t (f(s + t_p) - f(s + t_q))\mathrm{d}s.$$

Hence

$$F(t_p) - F(t_q) = F(t + t_p) - F(t + t_q) - \int_0^t (f(s + t_p) - f(s + t_q))\mathrm{d}s.$$

It follows from (2.55) that

$$|F(t + t_p) - F(t + t_q)| + \int_0^t |f(s + t_p) - f(s + t_q)|\mathrm{d}s \geq |F(t_p) - F(t_q)| > c. \tag{2.56}$$

One can see that, for $t \geq 0$,

$$\int_0^t |f(s+t_p) - f(s+t_q)|ds \leq \sum_{k=0}^{[t]} \int_k^{k+1} |f(s+t_p) - f(s+t_q)|ds.$$

From the S^1-almost automorphy of f, there exists a subsequence $(t'_n)_n \subset (t_n)_n$ such that

$$\int_k^{k+1} |f(s+t'_p) - f(s+t'_q)|ds \to 0,$$

as $p, q \to \infty$. Hence for p, q large enough, we get

$$\int_0^t |f(s+t'_p) - f(s+t'_q)|ds \leq \frac{c}{2}. \tag{2.57}$$

Thus using (2.57) and (2.56), we obtain

$$|F(t+t'_p) - F(t+t'_q)| > \frac{c}{2}. \tag{2.58}$$

Since X is uniformly convex, there exists $\delta > 0$ such that for any $u, v \in X$ with $|u| \leq 1$ and $|v| \leq 1$, if $|u - v| \geq \frac{c}{2|F|_\infty}$, then $|\frac{u+v}{2}| \leq 1 - \delta$. Thus (2.58) implies

$$\left|\frac{F(t+t'_p) + F(t+t'_q)}{2}\right| \leq |F|_\infty - \delta |F|_\infty.$$

For all $\varphi \in X^*$ with $|\varphi| \leq 1$, we have

$$\left|\left\langle \varphi, \frac{F(t+t'_p) + F(t+t'_q)}{2}\right\rangle\right| \leq \left|\frac{F(t+t'_p) + F(t+t'_q)}{2}\right| \leq |F|_\infty - \delta |F|_\infty. \tag{2.59}$$

Since F is weakly almost automorphic, there exist a subsequence $(t''_n)_n \subset (t'_n)_n$ and a function $\widetilde{F}$ such that, for all $t \in \mathbb{R}$,

$$F(t+t''_n) \to \widetilde{F}(t) \tag{2.60}$$

and

$$\widetilde{F}(t-t''_n) \to F(t) \tag{2.61}$$

in the weak topology as $n \to \infty$. It is clear that (2.60) and (2.61) imply

$$|F|_\infty = |\widetilde{F}|_\infty.$$

By taking $p, q \to \infty$ in (2.59), we get, for all $\varphi \in X^*$ with $\|\varphi\| \le 1$,

$$|\langle \varphi, \widetilde{F}(t) \rangle| \le |F|_\infty - \delta |F|_\infty = |\widetilde{F}|_\infty - \delta |\widetilde{F}|_\infty.$$

Therefore,

$$|\widetilde{F}|_\infty = \sup_{t \in \mathbb{R}} |\widetilde{F}(t)| = \sup_{t \in \mathbb{R}} \Big(\sup_{|\varphi| \le 1, \varphi \in X^*} |\langle \varphi, \widetilde{F}(t) \rangle| \Big) \le |\widetilde{F}|_\infty - \delta |\widetilde{F}|_\infty,$$

which is a contradiction since $|\widetilde{F}|_\infty = |F|_\infty \neq 0$. We conclude that $R_F = \{F(t) : t \in \mathbb{R}\}$ is relatively compact. We deduce from Proposition 2.3.12 that F is almost automorphic. □

Proposition 2.3.35. *Let $f : \mathbb{R} \to \mathbb{R}$ be an almost automorphic function such that the improper integral $\int_0^\infty f(s)\mathrm{d}s$ converges. Then $f(t) = 0$ for all $t \in \mathbb{R}$.*

Proof. Consider a sequence $(t_n)_n$ such that $t_n \to \infty$. Then we have a subsequence denoted also by $(t_n)_n$ such that, for all $t \in \mathbb{R}$,

$$\begin{cases} f(t + t_n) \to g(t), \\ g(t - t_n) \to f(t). \end{cases} \tag{2.62}$$

Since the improper integral $\int_0^\infty f(s)\mathrm{d}s$ converges, then also $\int_0^t f(s+t_n)\mathrm{d}s = \int_{t_n}^{t+t_n} f(s)\mathrm{d}s \to 0$. On the other hand, from (2.62) and Lebesgue's dominated convergence theorem, we have $\int_0^t f(s + t_n)\mathrm{d}s \to \int_0^t g(s)\mathrm{d}s$. This implies that, for all $t \in \mathbb{R}$,

$$\int_0^t g(s)\mathrm{d}s = 0. \tag{2.63}$$

From (2.62) and (2.63), we have

$$\int_0^t f(s)\mathrm{d}s = \lim_{n\to\infty} \int_0^t g(s - t_n)\mathrm{d}s = \lim_{n\to\infty} \int_{-t_n}^{t-t_n} g(s)\mathrm{d}s = 0$$

for all $t \in \mathbb{R}$. Since f is continuous then $f(t) = 0$ for all $t \in \mathbb{R}$. □

3 Dynamical systems: finite-dimensional case

This chapter focuses on evolution equations in finite-dimensional spaces, examining both autonomous and nonautonomous ordinary differential equations, functional differential equations of neutral type, and linear renewal equations. We discuss classical results, such as the Massera- and Bohr–Neugebauer-type results, which are crucial for understanding the behavior of bounded solutions. We conclude this chapter by studying sufficient conditions for the existence and uniqueness of a positive, compact, almost automorphic solution to a logistic equation with discrete and continuous delays.

3.1 Linear autonomous nonhomogeneous ordinary differential equations

In this section, we study the behavior of bounded integral solutions of the following finite dimensional differential equation:

$$x'(t) = Bx(t) + g(t) \quad \text{for } t \in \mathbb{R}, \tag{3.1}$$

where $g : \mathbb{R} \to \mathbb{C}^n$ is S^1-almost periodic or S^1-almost automorphic and $B : \mathbb{C}^n \to \mathbb{C}^n$ a matrix.

In the literature, there are two (best) classical results that are fundamental in characterizing the existence of almost periodic solutions for equation (3.1), and they serve as the basis for much of the research in this area. The first result, known as the Bohr–Neugebauer theorem, was established by Bohr and Neugebauer in [79]. It states: Every bounded solution of equation (3.1) on $\mathbb{R}$ is almost periodic. The second result (see [182]) shows that the existence of a bounded solution of equation (3.1) on $\mathbb{R}^+$ implies the existence of an almost periodic solution on $\mathbb{R}$, although not necessarily the same solution. This second result is called Massera theorem, named after the famous work of Massera [213]. In his pioneering research, Massera explored the implications of this theorem for studying the existence of periodic solutions to nonlinear equations. The main point of these results is that the boundedness alone serves as both a necessary and sufficient condition. In [204, Theorem 3.1 and Lemma 3.3], the authors show versions of Bohr–Neugebauer and Massera theorems for equation (3.1) when g is almost automorphic.

The main aim of this section is to extend the Bohr–Neugebauer and Massera theorems for equation (3.1) when g is almost periodic or almost automorphic in the Stepanov sense.

Since g is only locally integrable, by an integral solution of equation (3.1) we mean a locally integrable function $x : \mathbb{R} \to \mathbb{C}^n$ which satisfies the following integral equation:

$$x(t) = x(0) + \int_0^t Bx(s)ds + \int_0^t g(s)ds \quad \text{for } t \in \mathbb{R}.$$

https://doi.org/10.1515/9783111684710-003

Using this convention, an integral solution of equation (3.1) is locally absolutely continuous and is given by the following formula:

$$x(t) = e^{tB}x(0) + \int_0^t e^{(t-s)B}g(s)ds \quad \text{for all } t \in \mathbb{R}.$$

Moreover, an integral solution of equation (3.1) satisfies (3.1) almost everywhere.

We begin by considering the one-dimensional case. Let us examine the following scalar differential equation:

$$x'(t) = \lambda x(t) + g(t) \quad \text{for } t \in \mathbb{R}, \tag{3.2}$$

where $g : \mathbb{R} \to \mathbb{C}$ is a locally integrable function and $\lambda \in \mathbb{C}$.

Theorem 3.1.1. *If g is S^1-almost periodic (resp. S^1-almost automorphic), then every bounded integral solution of equation* (3.2) *on $\mathbb{R}$ is almost periodic (resp. almost automorphic).*

Proof. Let x be a bounded integral solution of (3.2) on $\mathbb{R}$. Then, we have

$$x(t) = e^{\lambda t}\left(c + \int_0^t e^{-\lambda s}g(s)ds\right) \quad \text{for } t \in \mathbb{R},$$

with some constant $c \in \mathbb{C}$. We distinguish three cases:

Case 1: $\operatorname{Re}\lambda = 0$, that is, $\lambda = i\theta$, for some $\theta \in \mathbb{R}$. We have

$$x(t) = e^{i\theta t}c + e^{i\theta t}\int_0^t e^{-i\theta s}g(s)ds \quad \text{for } t \in \mathbb{R}.$$

Since x is bounded, the function $t \mapsto \int_0^t e^{-i\theta s}g(s)ds$ is also bounded. Moreover, the function $s \mapsto e^{-i\theta s}g(s)$ is S^1-almost periodic (resp. S^1-almost automorphic) as a product of a periodic and an S^1-almost periodic (resp. S^1-almost automorphic) function. We deduce from Lemma 2.3.25 (resp. Theorem 2.3.34) that the function $t \mapsto \int_0^t e^{-i\theta s}g(s)ds$ is almost periodic (resp. almost automorphic). This implies that $x \in AP(\mathbb{R}, \mathbb{C})$ (resp. $AA(\mathbb{R}, \mathbb{C})$).

Case 2: $\operatorname{Re}\lambda < 0$. Now the solution x can be written in the following form:

$$x(t) = \int_{-\infty}^t e^{\lambda(t-s)}g(s)ds \quad \text{for } t \in \mathbb{R}.$$

By writing

$$x(t) = \sum_{k=1}^{\infty} x_k(t),$$

where

$$x_k(t) := \int_{t-k}^{t-k+1} e^{\lambda(t-s)} g(s)\mathrm{d}s$$
$$= \int_0^1 e^{\lambda(k-s)} g(s+t-k)\mathrm{d}s,$$

since g is S^1-almost periodic (resp. S^1-almost automorphic), we then obtain

$$\sup_{t\in\mathbb{R}} \int_t^{t+1} |g(s)|\mathrm{d}s < \infty.$$

For $t \in \mathbb{R}$, we observe that

$$\begin{aligned}
|x_k(t)| &\le \left| \int_{t-k}^{t-k+1} e^{\lambda(t-s)} g(s)\mathrm{d}s \right| \\
&\le \int_{t-k}^{t-k+1} e^{\operatorname{Re}\lambda(t-s)} |g(s)|\mathrm{d}s \\
&\le e^{\operatorname{Re}\lambda(k-1)} \int_{t-k}^{t-k+1} |g(s)|\mathrm{d}s \\
&\le e^{-\operatorname{Re}\lambda} e^{\operatorname{Re}\lambda k} \sup_{t\in\mathbb{R}} \int_t^{t+1} |g(s)|\mathrm{d}s.
\end{aligned}$$

Then, using Weierstrass M-test, we deduce that the series $\sum_{k=1}^{\infty} x_k(t)$ is uniformly convergent on $\mathbb{R}$. For $t \in \mathbb{R}$ and $p \ge 1$, we have

$$\begin{aligned}
\left| \sum_{k=1}^{p} x_k(t) - x(t) \right| &\le \left| \int_{t-p}^{t} e^{\lambda(t-s)} g(s)\mathrm{d}s - \int_{-\infty}^{t} e^{\lambda(t-s)} g(s)\mathrm{d}s \right| \\
&\le \left| \int_{-\infty}^{t-p} e^{\lambda(t-s)} g(s)\mathrm{d}s \right| \le \int_{-\infty}^{t-p} e^{\operatorname{Re}\lambda(t-s)} |g(s)|\mathrm{d}s \\
&\le \sum_{k=p+1}^{\infty} \int_{t-k}^{t-k+1} e^{\operatorname{Re}\lambda(t-s)} |g(s)|\mathrm{d}s \\
&\le e^{-\operatorname{Re}\lambda} \sup_{t\in\mathbb{R}} \left(\int_t^{t+1} |g(s)|\mathrm{d}s \right) \sum_{k=p+1}^{\infty} e^{\operatorname{Re}\lambda k}.
\end{aligned}$$

Consequently, taking the supremum over $t \in \mathbb{R}$,

$$\lim_{p\to\infty} \sup_{t\in\mathbb{R}} \left| \sum_{k=1}^{p} x_k(t) - x(t) \right| = 0,$$

which means that $\sum_{k=1}^{\infty} x_k(t)$ converges uniformly to $x(t)$. Then, to prove that x is almost periodic (resp. almost automorphic), it remains to show that x_k is almost periodic (resp. almost automorphic). To that purpose, we start by the case when g is S^1-almost periodic. Let $k \geq 1$ and $\varepsilon > 0$. Then, there exists $l(\varepsilon) > 0$ such that for all $a \in \mathbb{R}$, there exists $\tau_\varepsilon \in [a, a + l(\varepsilon)]$ such that

$$\sup_{t\in\mathbb{R}} \int_t^{t+1} |g(s + \tau_\varepsilon) - g(s)| ds < \varepsilon.$$

Therefore, for $t \in \mathbb{R}$, we have

$$\begin{aligned}
|x_k(t + \tau_\varepsilon) - x_k(t)| &\leq \left| \int_{t+\tau_\varepsilon-k}^{t+\tau_\varepsilon-k+1} e^{\lambda(t+\tau_\varepsilon-s)} g(s) ds - \int_{t-k}^{t-k+1} e^{\lambda(t-s)} g(s)) ds \right| \\
&\leq \left| \int_{t-k}^{t-k+1} e^{\lambda(t-s)} ((g(s + \tau_\varepsilon) - g(s)) ds \right| \\
&\leq \int_{t-k}^{t-k+1} |g(s + \tau_\varepsilon) - g(s)| ds \\
&\leq \varepsilon.
\end{aligned}$$

Thus, we deduce that x_k is almost periodic.

Now assume that g is S^1-almost automorphic. Let $(s'_j)_{j\in\mathbb{N}}$ be any sequence in $\mathbb{R}$. Since g is S^1-almost automorphic, there exists a subsequence $(s_j)_{j\in\mathbb{N}}$ of $(s'_j)_{j\in\mathbb{N}}$ such that, for any $t \in \mathbb{R}$,

$$\lim_{m\to+\infty} \lim_{j\to+\infty} \int_0^1 |g(t + s_j - s_m + s) - g(t + s)| ds = 0.$$

Then, for $t \in \mathbb{R}$, one has

$$\begin{aligned}
|x_k(t + s_j - s_m) - x_k(t)| &\leq \left| \int_0^1 e^{\lambda(k-s)} g(t + s_j - s_m - k + s) - g(t - k + s)) ds \right| \\
&\leq \int_0^1 |g(t + s_j - s_m - k + s) - g(t - k + s)| ds.
\end{aligned}$$

Thus,

$$\lim_{m\to+\infty}\lim_{j\to+\infty}|x_k(t+s_j-s_m)-x_k(t)|=0.$$

That is, x_k is almost automorphic.

Case 3: $\operatorname{Re}\lambda>0$. The solution x can now be expressed in the following form:

$$x(t)=-\int_t^{+\infty} e^{\lambda(t-s)}g(s)ds \quad \text{for } t\in\mathbb{R}.$$

Using the same argument as in Case 2, we can show that x is almost periodic (resp. almost automorphic). □

We have the following Bohr–Neugebauer-type theorem for equation (3.1). This result is similar to [152, Theorem 5.8, p. 86] and [237, Theorem 2.4]; however, we do not require the function g to be almost periodic or almost automorphic in the strong sense.

Theorem 3.1.1. *If g is S^1-almost periodic (resp. S^1-almost automorphic), then every bounded integral solution of equation* (3.1) *on $\mathbb{R}$ is almost automorphic.*

Proof. We use the same strategy as in [237, Theorem 2.4]. Let x be a bounded integral solution of equation (3.1), that is,

$$x(t)=x(0)+\int_0^t Bx(s)ds+\int_0^t g(s)ds \quad \text{for } t\in\mathbb{R}.$$

Using the linear transformation $x=Ty$ where $T:\mathbb{C}^n\to\mathbb{C}^n$ is an invertible matrix, we get

$$y(t)=y(0)+\int_0^t Cy(s)ds+\int_0^t h(s)ds \quad \text{for } t\in\mathbb{R}, \tag{3.3}$$

where C is an upper-triangular matrixgiven by

$$C=T^{-1}BT=\begin{pmatrix}\lambda_1 & c_{12} & c_{13} & \dots & c_{1n}\\ 0 & \lambda_2 & c_{23} & \dots & c_{2n}\\ \vdots & \vdots & \vdots & \ddots & \vdots\\ 0 & 0 & 0 & \dots & \lambda_n\end{pmatrix}.$$

The function h is given by

$$h(t)=T^{-1}g(t) \quad \text{for } t\in\mathbb{R}.$$

We observe that equation (3.3) can be written as

$$\begin{cases} y_1(t) = y_1(0) + \int_0^t \lambda_1 y_1(s)ds + \int_0^t [c_{12}y_2(s) + \cdots + c_{1n}y_n(s) + h_1(s)]ds, \\ y_2(t) = y_2(0) + \int_0^t \lambda_2 y_2(s)ds + \int_0^t [c_{23}y_2(s) + \cdots + c_{2,n}y_n(s) + h_2(s)]ds, \\ \quad \vdots \\ y_{n-1}(t) = y_{n-1}(0) + \int_0^t \lambda_{n-1}y_{n-1}(s)ds + \int_0^t [c_{n-1,n}y_n(s) + h_{n-1}(s)]ds, \\ y_n(t) = y_n(0) + \int_0^t \lambda_n y_n(s)ds + \int_0^t h_n(s)ds, \end{cases}$$

where h_i, $i = 1, \ldots, n$ are the S^1-almost periodic (resp. S^1-almost automorphic) components of h. The result in the scalar case (3.2) ensures that the bounded function y_n is almost periodic (resp. almost automorphic). It follows that the forcing term $c_{n-1,n}y_n + h_{n-1}$ in the second to last equation is S^1-almost periodic (resp. S^1-almost automorphic). This process can be continued using the result in the scalar case (3.2) until we obtain the almost periodicity (resp. almost automorphy) of all the components of y. Therefore $x = Ty$ is almost periodic (resp. almost automorphic). □

The next theorem shows that the existence of a bounded integral solution of equation (3.1) on $\mathbb{R}$ is equivalent to the existence of a bounded integral solution on $\mathbb{R}^+$.

Theorem 3.1.2. *Assume that g is S^1-almost automorphic. If equation* (3.1) *has a bounded integral solution on $\mathbb{R}^+$, then it has a bounded integral solution on $\mathbb{R}$.*

Proof. Let x be a bounded integral solution of equation (3.1) on $\mathbb{R}^+$. Let $(s_n)_n$ be a sequence of positive real numbers such that $s_n \to +\infty$ as $n \to +\infty$. Then, there exist a subsequence $(s'_n)_n \subset (s_n)_n$, a function $\tilde{g} \in L^1_{\text{loc}}(\mathbb{R}, \mathbb{C}^n)$, and a constant $c \in \mathbb{C}^n$ such that

$$\int_t^{t+1} |g(s + s'_n) - \tilde{g}(s)| ds \to 0, \tag{3.4}$$

$$\int_t^{t+1} |\tilde{g}(s - s'_n) - g(s)| ds \to 0, \tag{3.5}$$

for any $t \in \mathbb{R}$, and

$$x(s'_n) \to c \tag{3.6}$$

as $n \to +\infty$. Consider the function

$$y(t) := e^{tB}c + \int_0^t e^{(t-s)B}\tilde{g}(s)ds \quad \text{for all } t \in \mathbb{R}.$$

For each $t \in \mathbb{R}$, there exists $N(t) \in \mathbb{N}$ such that for all $n \geq N(t)$, $t + s'_n \geq 0$, and

$$x(t+s_n') = e^{tB}x(s_n') + \int_0^t e^{(t-s)B}g(s+s_n')\mathrm{d}s,$$

we have, for each $t \in \mathbb{R}$,

$$\int_0^t e^{(t-s)B}g(s+s_n')\mathrm{d}s \to \int_0^t e^{(t-s)B}\tilde{g}(s)\mathrm{d}s \quad \text{as } n \to +\infty. \tag{3.7}$$

In fact, using (3.4), we obtain, for $t \geq 0$,

$$\begin{aligned}\left|\int_0^t e^{(t-s)B}g(s+s_n')\mathrm{d}s - \int_0^t e^{(t-s)B}\tilde{g}(s)\mathrm{d}s\right| &\leq \int_0^t e^{|(t-s)B|}|g(s+s_n')-\tilde{g}(s)|\mathrm{d}s \\ &\leq \sum_{k=0}^{[t]}\int_k^{k+1} e^{|(t-s)B|}|g(s+s_n')-\tilde{g}(s)|\mathrm{d}s \\ &\leq \sum_{k=0}^{[t]}\left(C(k,t,B)\int_k^{k+1}|g(s+s_n')-\tilde{g}(s)|\mathrm{d}s\right) \to 0\end{aligned}$$

as $n \to +\infty$, where $C(k,t,B) := \sup_{k\leq s\leq k+1} e^{|t-s||B|}$. Using the same arguments for $t \leq 0$, we get

$$\left|\int_0^t e^{(t-s)B}g(s+s_n')\mathrm{d}s - \int_0^t e^{(t-s)B}\tilde{g}(s)\mathrm{d}s\right| \to 0 \quad \text{as } n \to +\infty.$$

It follows from (3.6) and (3.7) that, for each fixed $t \in \mathbb{R}$,

$$x(t+s_n') \to y(t) \quad \text{as } n \to +\infty.$$

This implies that the function y is bounded on the whole real line. Now, for all $n \in \mathbb{N}$ and $t \in \mathbb{R}$, we have

$$y(t-s_n') = e^{tB}y(-s_n') + \int_0^t e^{(t-s)B}\tilde{g}(s-s_n')\mathrm{d}s. \tag{3.8}$$

We can extract a subsequence $(s_n'')_n \subset (s_n')_n$ such that

$$y(-s_n'') \to d. \tag{3.9}$$

Consider the function

$$z(t) := e^{tB}d + \int_0^t e^{(t-s)B}g(s)ds \quad \text{for all } t \in \mathbb{R}.$$

It is clear that the function z is an integral solution of equation (3.1) on $\mathbb{R}$. Using (3.5), we have, for each $t \in \mathbb{R}$,

$$\int_0^t e^{(t-s)B}\tilde{g}(s - s'_n)ds \to \int_0^t e^{(t-s)B}g(s)ds \quad \text{as } n \to +\infty. \tag{3.10}$$

From (3.8)–(3.10), we deduce that, for each fixed $t \in \mathbb{R}$,

$$y(t - s''_n) \to z(t) \quad \text{as } n \to +\infty.$$

Therefore, z is a bounded integral solution of equation (3.1) on $\mathbb{R}$. □

The following Massera-type result is a direct consequence of Theorems 3.1.1 and 3.1.2.

Corollary 3.1.3. *Assume that g is S^1-almost periodic (resp. S^1-almost automorphic). If equation* (3.1) *has a bounded integral solution on $\mathbb{R}^+$, then it has an almost periodic integral solution.*

3.2 Nonautonomous ordinary differential equations

Let us consider the following differential equation in $\mathbb{C}^N$:

$$x'(t) = A(t)x(t) + f(t), \tag{3.11}$$

where the matrix $t \mapsto A(t) \in \mathcal{L}(\mathbb{C}^N)$ and $f(t)$ are both almost automorphic.

When $(A(t))_{t\in\mathbb{R}}$ is periodic in time (i. e., $A(t+q) = A(t)$ for some $q > 0$), equation (3.11) can be reduced to an autonomous equation through a Floquet transformation. More precisely, according to the Floquet theorem (see [58, Theorem 6.30, p. 205]), there exists a q-periodic continuous function $Q : \mathbb{R} \to GL(\mathbb{C}^n)$, where $GL(\mathbb{C}^N)$ is the set of all invertible matrices, such that x is a solution of equation (3.11) if and only if the function $t \mapsto y(t) := Q(t)x(t)$ is a solution of the following autonomous equation:

$$y'(t) = Py(t) + Q(t)g(t),$$

where P is a matrix on X, given by

$$\mathbb{U}(q, 0) = e^{qP}, \tag{3.12}$$

and $\mathbb{U}(q,0)$ is the monodromy (Poincaré) operatorassociated with equation (3.11). In this case, we can deduce, by applying the results obtained in Section 3.1, the following Bohr–Neugebauer-type theorem.

Theorem 3.2.1 ([152]). *Assume that $A(t)$ is a periodic matrix-valued function and $f(t)$ an S^1-almost periodic (resp. S^1-almost automorphic) function. Then, every bounded solution of $x'(t) = A(t)x(t) + f(t)$ on $\mathbb{R}$ is almost periodic (resp. almost automorphic). In addition, the existence of a bounded solution on $\mathbb{R}^+$, implies the existence of an almost periodic (resp. almost automorphic) solution.*

One of the greatest impediments to a thorough understanding of the nature of solutions of the almost periodic linear differential equation (3.11) is the lack of a Floquet theory. As soon as we allow $A(t)$ to depend on t, even almost periodically, the Bohr–Neugebauer property fails for equation (3.11) even if $A(t)$ is Hermitian (self-adjoint) or skew-Hermitian and $f = 0$, see Example 3.2.11 and Remark 3.2.8. In the present section, we will provide some sufficient conditions that ensure the Bohr–Neugebauer property for equation (3.11) when $A(t)$ is almost automorphic. This will be established by investigating the nature of bounded solutions of the equation $x'(t) = A(t)x(t)$.

In [147], Favard proved a Massera-type theorem for equation (3.11), when the matrix $A(t)$ and the vector $f(t)$ depend almost periodically on t, under the assumption that each bounded nontrivial solution of the linear equation

$$x'(t) = B(t)x(t) \quad \text{for } t \in \mathbb{R}$$

satisfies the separation condition

$$\inf_{t\in\mathbb{R}} |x(t)| > 0, \tag{3.13}$$

for every matrix $B(t)$ in the hull of $A(t)$. An important tool in the Favard's work was the characterization of almost periodicity which had been obtained by Bochner. It is well known that without the Favard's separation condition (3.13), the existence of a bounded solution of equation (3.11) does not imply the existence of an almost periodic solution [185]. If Favard's property holds only for $A(t)$ (not necessarily for other matrices in its hull), Fink [152] conjectures that we may not have almost periodic solutions. In this work, we show that one can still have compactly almost automorphic solutions, when $A(t)$ and $f(t)$ are only almost automorphic.

3.2.1 Preliminaries

Unlike almost periodicity, the almost automorphy requires us to deal with measurable functions which are not necessarily continuous. For this reason, we introduce the following notion of weak solutions.

Definition 3.2.2. Let $B(t)$ and $g(t)$ be locally integrable functions. A continuous function $u(t)$ is said to be a weak solution (or Carathéodory solution) of $x' = B(t)x + g(t)$ on $\mathbb{R}$ if, for all $t, \sigma \in \mathbb{R}$,

$$u(t) = u(\sigma) + \int_{\sigma}^{t} B(s)u(s)ds + \int_{\sigma}^{t} g(s)ds.$$

In this case u is locally absolutely continuous and thus satisfies the equation $x'(t) = B(t)x(t) + g(t)$ almost everywhere.

We have the following backward–forward uniqueness property.

Proposition 3.2.3. *Let $B(t)$ and $g(t)$ be bounded measurable functions and let u, v be two weak solutions of $x' = B(t)x + g(t)$ on $\mathbb{R}$. If $u(t_0) = v(t_0)$ for some $t_0 \in \mathbb{R}$, then $u(t) = v(t)$ for all $t \in \mathbb{R}$.*

Proof. For all $t \in \mathbb{R}$, we have

$$u(t) - v(t) = u(t_0) - v(t_0) + \int_{t_0}^{t} B(s)(u(s) - v(s))ds.$$

If $t \geq t_0$, we get

$$\|u(t) - v(t)\| \leq \|u(t_0) - v(t_0)\| + \|B\|_{\infty} \int_{t_0}^{t} \|u(s) - v(s)\|ds.$$

Thus, by Grönwall's inequality,

$$\|u(t) - v(t)\| \leq \|u(t_0) - v(t_0)\| e^{(t-t_0)\|B\|_{\infty}} \tag{3.14}$$

for all $t \geq t_0$. Using the same arguments, we obtain

$$\|u(t) - v(t)\| \leq \|u(t_0) - v(t_0)\| e^{(t_0-t)\|B\|_{\infty}} \tag{3.15}$$

for all $t \leq t_0$. If $u(t_0) = v(t_0)$, we deduce from (3.14) and (3.15) that $u(t) = v(t)$ for all $t \in \mathbb{R}$. □

The following fundamental lemma will be used several times during this work.

Lemma 3.2.4. *Let $A(t)$ and $f(t)$ be bounded measurable functions and let u be a weak solution of $x' = A(t)x + f(t)$ which is bounded on $\mathbb{R}$. If there exist a sequence $(t'_n)_n$, a function $g(t)$, and a matrix-valued function $B(t)$ such that, for all $t \in \mathbb{R}$, $f(t + t'_n) \to g(t)$ and $A(t + t'_n) \to B(t)$ as $n \to \infty$, then there exists a subsequence $(t_n)_n \subset (t'_n)_n$ such that $u(t + t_n) \to v(t)$ uniformly on each compact subset of $\mathbb{R}$, where v is a weak solution of $x' = B(t)x + g(t)$.*

Proof. Let $K = \overline{\{u(t) : t \in \mathbb{R}\}}$ and let $(t'_n)_n$ be a sequence such that $f(t + t'_n) \to g(t)$ and $A(t + t'_n) \to B(t)$ as $n \to \infty$. Define u_n by $u_n(t) = u(t + t'_n)$. Then for each $n \in \mathbb{N}$, $u_n \in C(\mathbb{R}, \mathbb{C}^N)$ and $u_n(t) \in K$ for all $t \in \mathbb{R}$. Therefore the family of functions $(u_n)_n$ is equibounded. For all $t, \sigma \in \mathbb{R}$, we have

$$\|u(t) - u(\sigma)\| \leq (\|A\|_\infty \|u\|_\infty + \|f\|_\infty)\|t - \sigma\|.$$

The weak solution u is then uniformly continuous on $\mathbb{R}$. This implies that $(u_n)_n$ is also equiuniformly continuous on $\mathbb{R}$. It follows by the Arzelà–Ascoli theorem that there exist a function v and a subsequence $(t_n^1)_n \subset (t'_n)_n$ such that $u(t + t_n^1) \to v(t)$ as $n \to \infty$ uniformly on $[-1, 1]$. Using the same arguments, we deduce inductively that for each $p \in \mathbb{N}^*$ there exists a subsequence $(t_n^p)_n \subset (t_n^{p-1})_n \subset \cdots \subset (t_n^1)_n \subset (t'_n)_n$ such that $u(t + t_n^p) \to v(t)$ as $n \to \infty$ uniformly on $[-p, p]$. Now consider the diagonal sequence $(t_n)_n := (t_n^n)_n$. It is clear that $u(t + t_n) \to v(t)$ as $n \to \infty$ uniformly on each compact subset of $\mathbb{R}$. Since for each $t, \sigma \in \mathbb{R}$ and $n \in \mathbb{N}$,

$$u(t + t_n) = u(\sigma + t_n) + \int_\sigma^t A(s + t_n)u(s + t_n)ds + \int_\sigma^t f(s + t_n)ds,$$

by letting $n \to \infty$, we obtain

$$v(t) = v(\sigma) + \int_\sigma^t B(s)v(s)ds + \int_\sigma^t g(s)ds$$

for all $t, \sigma \in \mathbb{R}$. This shows that v is a weak solution of $x' = B(t)x + g(t)$. □

3.2.2 The equation $x' = A(t)x$

In this section, $\mathbb{K}$ denotes either $\mathbb{R}$ or $\mathbb{C}$, and $M_N(\mathbb{K})$ is the space of $\mathbb{K}$-valued $N \times N$ matrices.

Definition 3.2.5. Let M be a Hermitian matrix. We say that M is positive semidefinite, and write $M \succeq 0$, if all its eigenvalues are nonnegative, or equivalently $\langle Mx, x\rangle \geq 0$ for all $x \in \mathbb{R}^N$. We say that M is positive definite, and write $M \succ 0$, if all its eigenvalues are positive, or equivalently $\langle Mx, x\rangle > 0$ for all $x \in \mathbb{R}^N$ with $x \neq 0$. Similarly, M being negative semidefinite and M negative definite is denoted respectively by $M \preceq 0$ and $M \prec 0$.

Let A^* be the conjugate transpose of A. In what follows, we denote respectively by A_H and A_S the Hermitian and skew-Hermitian part of A, that is,

$$A_H = \frac{A + A^*}{2} \quad \text{and} \quad A_S = \frac{A - A^*}{2}.$$

We have the following splitting:

$$A = A_H + A_S.$$

We denote by $\sigma(A)$ the set of the eigenvalues of A.

Proposition 3.2.6. *Assume that $A_H(t) \succeq 0$ for all $t \in \mathbb{R}$ and let x be a solution of $x' = A(t)x$. Then x has a constant norm if and only if $x(t) \in \ker A_H(t)$ for all $t \in \mathbb{R}$.*

Proof. Let $y(t) = \|x(t)\|^2$, then $y'(t) = 2\langle A_H(t)x(t), x(t)\rangle \geq 0$. Since $A_H(t)$ is Hermitian and its eigenvalues are nonnegative, by the spectral theorem, $A_H(t)x(t) = 0$ if and only if $y'(t) = 2\langle A_H(t)x(t), x(t)\rangle = 0$. □

3.2.2.1 Behavior of the bounded solutions of $x' = A(t)x$

Proposition 3.2.7. *Assume that $A(t)$ is continuous. Then all solutions of $x' = A(t)x$ have a constant norm if and only if $A^*(t) = -A(t)$ for all $t \in \mathbb{R}$.*

Proof. If $A^*(t) = -A(t)$ then $A_H(t) = 0$, and this implies that $\frac{\mathrm{d}}{\mathrm{d}t}\|x(t)\|^2 = 2\langle A_H(t)x(t), x(t)\rangle = 0$, which yields that $x(t)$ has a constant norm. Conversely, if all the solutions have a constant norm then, for all $c \in \mathbb{C}^n$,

$$\frac{\mathrm{d}}{\mathrm{d}t}\|X(t)c\|^2 = 2\langle A_H(t)X(t)c, X(t)c\rangle = 0.$$

Hence $\langle X(t)^* A_H(t)X(t)c, c\rangle = 0$ for all $c \in \mathbb{C}^n$. It follows that $X(t)^* A_H(t)X(t) = 0$, which implies that $A_H(t) = 0$, that is, $A^*(t) = -A(t)$. □

Remark 3.2.8. If we assume that $A^*(t) = -A(t)$ for all $t \in \mathbb{R}$ and $A(t)$ is almost automorphic, then the solutions have constant norm. However, the solutions are not necessarily almost automorphic. For example, consider the following differential equation taken from [173]:

$$x' = A(t)x, \tag{3.16}$$

with $A(t) = ab(t)\left(\begin{smallmatrix} 0 & 1 \\ -1 & 0 \end{smallmatrix}\right)$, where $a > 0$ and $b(t) = \sum_{n=0}^{\infty} \frac{1}{2^n} \sin(\frac{t}{2^n})\cos(\frac{t}{2^n})$. Then $A^*(t) = -A(t)$ and a solution of (3.16) is given by

$$x(t) = \begin{pmatrix} \cos(\frac{a}{2}g(t)) \\ \sin(\frac{a}{2}g(t)) \end{pmatrix},$$

where $g(t) = \sum_{n=0}^{\infty} \sin^2(\frac{t}{2^n})$. In [173], the author showed that the function $v(t) = \sin(\frac{a}{2}g(t))$ is not almost periodic. We claim that this function is not even almost automorphic. Let us first prove the following lemma.

Lemma 3.2.9. *Consider the function φ defined on $\mathbb{R}$ by*

$$\varphi(t) = \sum_{k=1}^{\infty} \sin^2\left(\frac{\pi}{2^k}t\right).$$

Then φ is continuous on $\mathbb{R}$, $\varphi(\mathbb{R}) = \mathbb{R}^+$, and for all $t \in \mathbb{R}$ we have

$$\lim_{n\to\infty} \varphi(t + 2^n) = \lim_{n\to\infty} \varphi(t - 2^n) = \varphi(t) + \varphi(1).$$

Proof. Let $\varphi_k(t) = \sin^2(\frac{\pi}{2^k}t)$ for each positive integer k and $t \in \mathbb{R}$. The continuity of $\varphi(t)$ on $\mathbb{R}$ follows from the uniform convergence of the series $\sum_{k\geq 1} \varphi_k(t)$ on the compact subsets of $\mathbb{R}$. Recall that the mean value of a T-periodic function is given by

$$\mathcal{M}(f) = \lim_{r\to\infty} \frac{1}{2r} \int_{-r}^{r} f(t)\mathrm{d}t = \frac{1}{T}\int_0^T f(t)\mathrm{d}t.$$

We claim that $\varphi(t)$ is not bounded. In fact, if $\varphi(t) \leq c$ for some constant $c > 0$ then for all $n \geq 1$ and $t \in \mathbb{R}$, we have

$$S_n(t) = \sum_{k=1}^{n} \varphi_k(t) \leq \varphi(t) \leq c. \tag{3.17}$$

Since φ_k is 2^k-periodic, by computing the mean value of (3.17), we get, for all $n \geq 1$,

$$\mathcal{M}(S_n) = \sum_{k=1}^{n} \mathcal{M}(\varphi_k) = \sum_{k=1}^{n} \frac{1}{2}\mathrm{d}t \leq c,$$

which is a contradiction. Thus $\varphi(t)$ cannot be bounded. Moreover, as $\varphi(t)$ is continuous, nonnegative, and $\varphi(0) = 0$, we have $\varphi(\mathbb{R}) = \mathbb{R}^+$. In the other hand, for all $n \geq 2$, we have

$$\begin{aligned}
\varphi(t + 2^n) &= \sum_{k=1}^{n} \sin^2\left(\frac{\pi}{2^k}(t + 2^n)\right) + \sum_{k=n+1}^{\infty} \sin^2\left(\frac{\pi}{2^k}(t + 2^n)\right) \\
&= \sum_{k=1}^{n} \sin^2\left(\frac{\pi}{2^k}t\right) + \sum_{k=1}^{\infty} \sin^2\left(\frac{\pi}{2^{k+n}}t + \frac{\pi}{2^k}\right).
\end{aligned}$$

By letting $n \to \infty$ and using the uniform convergence of the series $\sum_{k\geq 1} \sin^2(\frac{\pi t}{2^{k+n}} + \frac{\pi}{2^k})$ with respect to $n \in \mathbb{N}^*$, we get $\lim_{n\to\infty} \varphi(t + 2^n) = \varphi(t) + \varphi(1)$. Since φ is even, we also have $\lim_{n\to\infty} \varphi(t - 2^n) = \varphi(t) + \varphi(1)$. □

Proposition 3.2.10. *Let $a > 0$ be such that $a \notin \frac{2\pi}{\varphi(1)}\mathbb{N}$. Then the function*

$$v(t) = \sin\left(\frac{a}{2}g(t)\right) = \sin\left(\frac{a}{2}\sum_{k=0}^{\infty} \sin^2\left(\frac{t}{2^k}\right)\right)$$

is not almost automorphic.

Proof. Letting $w(t) = v(\frac{\pi}{2}t)$, it is sufficient to prove that w is not almost automorphic. We have $\varphi(t) = \sum_{k=1}^{\infty} \sin^2(\frac{\pi}{2^k}t)$, hence $w(t) = \sin(\frac{\alpha}{2}\varphi(t))$. From Lemma 3.2.9, we have

$$\begin{cases} \lim_{n\to\infty} w(t+2^n) = \sin(\frac{\alpha}{2}[\varphi(t)+\varphi(1)]) := s(t), \\ \lim_{n\to\infty} s(t-2^n) = \sin(\frac{\alpha}{2}[\varphi(t)+2\varphi(1)]). \end{cases}$$

If w is almost automorphic then, for all $t \in \mathbb{R}$,

$$\sin\left(\frac{\alpha}{2}\varphi(t) + \alpha\varphi(1)]\right) = \sin\left(\frac{\alpha}{2}\varphi(t)\right).$$

But since $\varphi(\mathbb{R}) = \mathbb{R}^+$, this would imply that $\sin(x + \alpha\varphi(1)) = \sin(x)$ for all $x \geq 0$ and thus, by the periodicity of $\sin x$, we would have $\sin(x + \alpha\varphi(1)) = \sin(x)$ for all $x \in \mathbb{R}$. This implies that $\alpha\varphi(1) \in 2\pi\mathbb{N}$, that is, $\alpha \in \frac{2\pi}{\varphi(1)}\mathbb{N}$, which contradicts our assumption. Therefore v cannot be almost automorphic. □

Remark 3.2.11. An example of an α that satisfies the condition in Proposition 3.2.10 is $\alpha = 1$. In fact,

$$0 < \varphi(1) = \sum_{k=1}^{\infty} \sin^2\left(\frac{\pi}{2^k}\right) \leq \sum_{k=1}^{\infty} \frac{\pi^2}{4^k} = \frac{\pi^2}{3} < 2\pi.$$

Thus $\varphi(1) \notin 2\pi\mathbb{N}$, that is, $1 \notin \frac{2\pi}{\varphi(1)}\mathbb{N}$.

We now present the first main result of this subsection.

Theorem 3.2.12. *Assume that $A(t)$ is almost automorphic and $A(t) \succeq 0$ for all $t \in \mathbb{R}$. Then, every bounded solution $u(t)$ of $x' = A(t)x$ on $\mathbb{R}$ is constant. More specifically, $u(t) = c \in \bigcap_{t\in\mathbb{R}} \ker A(t)$.*

In order to prove Theorem 3.2.12, we need the following lemma.

Lemma 3.2.13. *Assume that $A(t)$ is almost automorphic and $A(t) \succeq 0$ for all $t \in \mathbb{R}$. If $u(t)$ is a bounded solution of $x' = A(t)x$ on $\mathbb{R}$, then for every sequence $(t_n)_n$ with $t_n \to \infty$ there exist a subsequence $(t_n)_n$ and a constant $c \in \bigcap_{t\in\mathbb{R}} \ker A(t)$ such that $\|c\| = \|u\|_\infty$ and $u(t+t_n) \to c$ for all $t \in \mathbb{R}$.*

Proof. Let $u(t)$ be a bounded solution of $x' = A(t)x$ on $\mathbb{R}$. Since

$$\frac{\mathrm{d}}{\mathrm{d}t}\|u(t)\|^2 = 2\langle A(t)u(t), u(t)\rangle \geq 0, \tag{3.18}$$

it follows that $t \mapsto \|u(t)\|$ is nondecreasing and

$$\lim_{t\to\infty} \|u(t)\| = \|u\|_\infty. \tag{3.19}$$

Let $t_n \to \infty$, then from Lemma 3.2.4 and the almost automorphy of $A(t)$ we get a subsequence, also denoted by $(t_n)_n$, and functions v, w such that

$$\begin{cases} u(t+t_n) \to v(t), \\ v(t-t_n) \to w(t) \end{cases} \quad \text{and} \quad \begin{cases} A(t+t_n) \to B(t), \\ B(t-t_n) \to A(t) \end{cases} \tag{3.20}$$

when $n \to \infty$, where v is a weak solution of $x' = B(t)x$ and w is a strong solution of $x' = A(t)x$ (since $A(t)$ is continuous). It follows from (3.19) and (3.20) that

$$\|v(t)\| = \|w(t)\| = \|u\|_\infty. \tag{3.21}$$

On the other hand, from (3.18) we have

$$|u(t+t_n)|^2 - |u(\sigma+t_n)|^2 = \int_t^\sigma 2\langle A(s+t_n)u(s+t_n), u(s+t_n)\rangle \mathrm{d}s.$$

It follows, by letting $n \to \infty$ and using the dominated convergence theorem, that $\int_t^\sigma \langle B(s)v(s), v(s)\rangle \mathrm{d}s = 0$, that is, $\langle B(s)v(s), v(s)\rangle = 0$ almost everywhere. But since $B(s) \geq 0$, it follows that $v'(s) = B(s)v(s) = 0$ almost everywhere. Now since v is absolutely continuous, it follows that $v(t) = c$ for all $t \in \mathbb{R}$, which also implies by (3.20) that $w(t) = c$ for all $t \in \mathbb{R}$. Now w being a strong solution of $x' = A(t)x$, it follows that $c \in F = \bigcap_{t\in\mathbb{R}} \ker A(t)$. From (3.21) we have $\|c\| = \|u\|_\infty$. □

Proof of Theorem 3.2.12. Let $u(t)$ be a bounded solution of $x' = A(t)x$ on $\mathbb{R}$ and $t_n \to \infty$. Then, by Lemma 3.2.13, there exist a subsequence $(t_n)_n$ and a constant $c \in \bigcap_{t\in\mathbb{R}} \ker A(t)$ such that $\|c\| = \|u\|_\infty$ and, for all $t \in \mathbb{R}$,

$$u(t+t_n) \to c. \tag{3.22}$$

Let P be the orthogonal projection on the subspace $F = \bigcap_{t\in\mathbb{R}} \ker A(t)$ and consider the following orthogonal decomposition:

$$u(t) = \underbrace{Pu(t)}_{\in F} + \underbrace{z(t)}_{\in F^\perp}.$$

Since $A(t)$ is Hermitian,

$$A(t)u(t) \in \operatorname{Im} A(t) = (\ker A(t))^\perp \subset \Big(\bigcap_{s\in\mathbb{R}} \ker A(s)\Big)^\perp = F^\perp,$$

which implies that

$$\frac{\mathrm{d}}{\mathrm{d}t} Pu(t) = Pu'(t) = PA(t)u(t) = 0. \tag{3.23}$$

It follows from (3.22) and (3.23) that

$$Pu(t) = \text{constant} = \lim_{n\to\infty} Pu(t+t_n) = Pc = c.$$

That is,

$$u(t) = \underbrace{c}_{\in F} + \underbrace{z(t)}_{\in F^\perp}.$$

The function $z(t)$ is also a solution of $x' = A(t)x$ which is bounded on $\mathbb{R}$. It follows again by Lemma 3.2.13 that there exist a subsequence $(t_n)_n$ and a constant $d \in F$ such that $z(t+t_n) \to d$ for all $t \in \mathbb{R}$ and $\|d\| = \|z\|_\infty$. However, since $z(t) \in F^\perp$, we get that $d \in F \cap F^\perp = \{0\}$, that is, $z = 0$. We conclude that $u(t) = c$ for all $t \in \mathbb{R}$. □

Remark 3.2.14. Theorem 3.2.12 does not hold if we drop the assumption $A(t) \geq 0$, even if $A(t)$ is Hermitian. Consider $A(t) = a(t) = \sum_{k=1}^{\infty} a_k(t)$, where $a_k(t) = \frac{-1}{k^2}\sin(\frac{\pi t}{k^3})$. The function $a(t)$ is a sign-changing odd function. It is also almost periodic as a uniform limit of periodic functions. As $\int_0^t a_k(s)ds = \frac{k(\cos(\frac{\pi t}{k^3})-1)}{\pi} \leq 0$ for all $t \in \mathbb{R}$, we have $\int_0^t a(s)ds \leq 0$ for all $t \in \mathbb{R}$. This implies that all the solutions of $x' = a(t)x$ are bounded on $\mathbb{R}$ because these solutions have the form $x(t) = \exp(\int_0^t a(s)ds)x_0$. It is clear that these solutions are not constant (except the trivial solution).

Corollary 3.2.15. *Assume that $A(t)$ is almost automorphic, $A(t) \geq 0$ for all $t \in \mathbb{R}$, and $\bigcap_{t\in\mathbb{R}} \ker A(t) = \{0\}$. Then, the trivial solution is the only bounded solution of $x' = A(t)x$ on $\mathbb{R}$.*

Example 3.2.1. Consider $A(t) = \begin{pmatrix} 1-\sin t & \cos t \\ \cos t & 1+\sin t \end{pmatrix}$, $\sigma(A(t)) = \{0, 2\} \subset \mathbb{R}^+$, and $\ker A(0) = \operatorname{span}\{(1,-1)\}$, $\ker A(\pi) = \operatorname{span}\{(1,1)\}$. Here $\ker A(0) \cap \ker A(\pi) = \{0\}$.

Remark 3.2.16. In Corollary 3.2.15, one can replace $\bigcap_{t\in\mathbb{R}} \ker A(t) = \{0\}$ by $\bigcap_{i=1}^N \ker A(t_i) = \{0\}$ for some real numbers $t_1, t_2, \ldots, t_N$. In fact, since $\dim M_N(\mathbb{K}) = N^2$, there exists $r \in [1, N^2]$ such that $\operatorname{span}(A(t))_{t\in\mathbb{R}} = \operatorname{span}(A(\tau_i))_{1\leq i\leq r}$. Thus

$$\bigcap_{i=1}^{r} \ker A(\tau_i) = \{0\}. \tag{3.24}$$

On the other hand, we have that (3.24) is equivalent to $\sum_{i=1}^r [\ker A(\tau_i)]^\perp = \mathbb{K}^N$, thus $\mathbb{K}^N$ has a basis ε extracted from $\sum_{i=1}^r [\ker A(\tau_i)]^\perp$. This implies that for each $i \in [1, N]$, there exists $t_i \in \{\tau_j : 0 \leq j \leq r\}$ such that $\varepsilon_i \in [\ker A(t_i)]^\perp$. Therefore $\sum_{i=1}^N [\ker A(t_i)]^\perp = \mathbb{K}^N$ and thus $\bigcap_{i=1}^N \ker A(t_i) = \{0\}$.

We present the second main result of this subsection.

Theorem 3.2.17. *Assume that $A(t)$ is almost automorphic and $A_H(t) \geq 0$ for all $t \in \mathbb{R}$. Then, every bounded solution of $x' = A(t)x$ on $\mathbb{R}$ has a constant norm.*

Lemma 3.2.18. *Assume that $A(t)$ is almost automorphic and $A_H(t) \geq 0$ for all $t \in \mathbb{R}$. If u is a bounded solution of $x' = A(t)x$, then there exists a solution w such that $\|w(t)\| = \|u\|_\infty$ for all $t \in \mathbb{R}$.*

Proof. Let $u(t)$ be a bounded solution of $x' = A(t)x$ on $\mathbb{R}$. Then

$$\frac{\mathrm{d}}{\mathrm{d}t}\|u(t)\|^2 = 2\langle A_H(t)u(t), u(t)\rangle \geq 0, \tag{3.25}$$

the function $t \mapsto \|u(t)\|$ is nondecreasing, and one has

$$\lim_{t\to\infty}\|u(t)\| = \|u\|_\infty. \tag{3.26}$$

Letting $t_n \to \infty$, due to Lemma 3.2.4 and the almost automorphy of $A(t)$, we get a subsequence, also denoted by $(t_n)_n$, and functions v, w such that

$$\begin{cases} u(t+t_n) \to v(t), \\ v(t-t_n) \to w(t) \end{cases} \quad \text{and} \quad \begin{cases} A(t+t_n) \to B(t), \\ B(t-t_n) \to A(t) \end{cases} \tag{3.27}$$

when $n \to \infty$, where v is a weak solution of $x' = B(t)x$ and w is a strong solution of $x' = A(t)x$. It follows from (3.26) and (3.27) that $\|w(t)\| = \|v(t)\| = \|u\|_\infty$. □

Let S_b be the subspace of bounded solutions of $x' = A(t)x$ on $\mathbb{R}$ and $S_c \subset S_b$ be the subspace of solutions having a constant norm.

Lemma 3.2.19. *Let x, y be two solutions of $x' = A(t)x$ such that $y \in S_c$. Then $t \mapsto \langle x(t), y(t)\rangle$ is constant.*

Proof. Using Proposition 3.2.6, we have

$$\frac{\mathrm{d}}{\mathrm{d}t}\langle x(t), y(t)\rangle = \langle x'(t), y(t)\rangle + \langle x(t), y'(t)\rangle = \langle x(t), 2A_H y(t)\rangle = \langle x(t), 0\rangle = 0.$$

It follows that $t \mapsto \langle x(t), y(t)\rangle$ is constant. □

Lemma 3.2.20. *Let $(f_i)_{1\leq i\leq r}$ be a linearly independent family of S_c. Then there exists a family $(\varphi_i)_{1\leq i\leq r}$ of S_c such that $\operatorname{span}(f_i)_{1\leq i\leq r} = \operatorname{span}(\varphi_i)_{1\leq i\leq r}$ and $(\varphi_i(t))_{1\leq i\leq r}$ is orthonormal in $\mathbb{C}^N$ for each $t \in \mathbb{R}$.*

Proof. Since $(f_i)_{1\leq i\leq r}$ are linearly independent, this family can be extended to a basis of the space of solutions of $x' = A(t)x$, and thus the Wronskian does not vanish. It follows that $(f_i(t))_{1\leq i\leq r}$ is a linearly independent family of $\mathbb{C}^N$ for all $t \in \mathbb{R}$ as a subfamily of a basis of $\mathbb{C}^N$. One can then apply the Gram–Schmidt process and obtain an orthonormal family $(\varphi_i(t))_{1\leq i\leq r}$ of $\mathbb{C}^N$. Let $\varphi_1(t) = \frac{f_1(t)}{\lambda_1}$ where $\lambda_1 = \|f_1(t)\|$ is constant, hence $\varphi_1 \in S_c$. Assume that we constructed $\varphi_1, \dots, \varphi_{k-1} \in S_c$. Let $W_k(t) = f_k(t) - \sum_{i=1}^{k-1}\langle\varphi_i(t), f_k(t)\rangle\varphi_i(t)$. Then, by Lemma 3.2.19, the function $t \mapsto \langle\varphi_i(t), f_k(t)\rangle = c_{i,k}$ is constant. Hence $W_k =$

$f_k - \sum_{i=1}^{k-1} c_{i,k}\varphi_i \in S_c$, and it is then sufficient to take $\varphi_k(t) = \frac{W_k(t)}{\mu_1}$ where $\mu_1 = \|W_k(t)\|$ is constant. □

Proof of Theorem 3.2.17. Let $x(t)$ be a bounded solution of $x' = A(t)x$ on $\mathbb{R}$, that is, $x \in S_b$. If $x = 0$, then $x \in S_c$. If $x \neq 0$, then, by Lemma 3.2.18, there exists $v_x \in S_c$ such that $\|v_x(t)\| = \|x\|_\infty$ for all $t \in \mathbb{R}$. This implies that $S_c \neq \{0\}$. Let $(f_i)_{1\le i\le r}$ be a basis of S_c. Then, by Lemma 3.2.20, there exists a basis $(\varphi_i)_{1\le i\le r}$ of S_c such that $(\varphi_i(t))_{1\le i\le r}$ is orthonormal for all $t \in \mathbb{R}$. If $x \notin S_c$, then consider

$$\varphi_{r+1}(t) = x(t) - \sum_{i=1}^{r} c_i\varphi_i(t),$$

where $c_i = \langle x(t), \varphi_i(t)\rangle$ is constant by Lemma 3.2.19. It follows that $\varphi_{r+1} \in S_b$ and $(\varphi_i(t))_{1\le i\le r+1}$ is orthogonal for all $t \in \mathbb{R}$. For each $i = 1, \dots, r+1$, there exists $v_i \in S_c$ such that $\|v_i(t)\| = \|\varphi_i\|_\infty \neq 0$ for all $t \in \mathbb{R}$. Moreover for all $1 \le i \neq j \le r+1$, $\langle \varphi_i(t), \varphi_j(t)\rangle = 0$. Hence one can see from the construction of v_i in the proof of Lemma 3.2.18 that $\langle v_i(t), v_j(t)\rangle = 0$. This implies that $(v_i(t))_{1\le i\le r+1}$ is a family of linearly independent nonzero vectors of S_c, which is in contradiction with $\dim S_c = r$, thus one must have $x \in S_c$. □

The following result is a direct consequence of Proposition 3.2.7 and Theorem 3.2.17.

Corollary 3.2.21. *Assume that $A(t)$ is almost automorphic, $A_H(t) \succeq 0$ for all $t \in \mathbb{R}$. Then all the solutions of $x' = A(t)x$ are bounded on $\mathbb{R}$ if and only if $A^*(t) = -A(t)$ for all $t \in \mathbb{R}$.*

Corollary 3.2.22. *Assume that $A(t)$ is almost automorphic, $A_H(t) \succeq 0$ for all $t \in \mathbb{R}$ and* $\ker A_H(t_0) = \{0\}$ *for some $t_0 \in \mathbb{R}$. Then, the trivial solution is the only bounded solution of $x' = A(t)x$ on $\mathbb{R}$.*

Proof. Let $u(t)$ be a bounded solution of $x' = A(t)x$ on $\mathbb{R}$. Then, by Theorem 3.2.17, u has a constant norm, which implies, by Proposition 3.2.6, that $A_H(t_0)u(t_0) = 0$. Since $A_H(t_0)$ is invertible, it follows that $u(t_0) = 0$, which then implies, by the uniqueness of solutions, that $u(t) = 0$ for all $t \in \mathbb{R}$. □

Corollary 3.2.23. *Assume that $A_H(t)$ is almost automorphic, $A_H(t) \succeq 0$ for all $t \in \mathbb{R}$, and $A_S(t)$ is periodic. Then, every bounded solution of $x' = A(t)x$ on $\mathbb{R}$ is almost periodic.*

Proof. Let $u(t)$ be a bounded solution of $x' = A(t)x$ on $\mathbb{R}$. Then, by Theorem 3.2.17, u has a constant norm, which implies, by Proposition 3.2.6, that $A_H(t)u(t) = 0$. It follows that

$$u'(t) = A(t)u(t) = A_S(t)u(t).$$

Therefore, by Theorem 3.2.1, $u(t)$ is almost periodic. □

Remark 3.2.24. Using the change of variable $\tilde{A}(t) = -A(-t) \succeq 0$ and $\tilde{u}(t) = u(-t)$, one can see that the results of this section hold if we replace $A(t) \succeq 0$ by $A(t) \preceq 0$.

3.2.2.2 Favard's condition

Proposition 3.2.25. *Let $A(t)$ be almost automorphic. Then the nontrivial almost automorphic solutions of $x' = A(t)x$ satisfy $\inf_{t\in\mathbb{R}} \|x(t)\| > 0$.*

Proof. Notice first that an almost automorphic solution of $x' = A(t)x$ has a bounded derivative and thus is automatically compactly almost automorphic (Proposition 2.3.6). Assume that there exits a nontrivial almost automorphic solution u of $x' = A(t)x$ such that $\inf_{t\in\mathbb{R}} \|u(t)\| = 0$. Then there exists a sequence $(t_n)_n$ such that

$$\|u(t_n)\| \to 0. \tag{3.28}$$

From the almost automorphy of A and u, there exist a subsequence, also denoted (for the sake of simplicity) by $(t_n)_n$, a continuous function $v(t)$, and a measurable matrix-valued function $B(t)$ such that $u(t + t_n) \to v(t)$, $v(t - t_n) \to u(t)$, and $A(t + t_n) \to B(t)$ as $n \to \infty$, where the first two convergences hold uniformly on compact subsets of $\mathbb{R}$. Using the same arguments as at the end of the proof of Lemma 3.2.4, we deduce that v is a weak solution of $x' = B(t)x$. But, due to (3.28), then $v(0) = 0$. It follows by Proposition 3.2.3 that $v(t) = 0$ for all $t \in \mathbb{R}$, and thus $u(t) = 0$ for all $t \in \mathbb{R}$, which is a contradiction. □

Definition 3.2.26 ([152]). We say that the equation $x' = A(t)x$ (or simply $A(t)$) satisfies Favard's property if all bounded nontrivial solutions of $x' = A(t)x$ satisfy $\inf_{t\in\mathbb{R}} \|x(t)\| > 0$.

Example 3.2.2. If $A(t)$ is constant (or even periodic), then it satisfies Favard's property. In fact, by Theorem 3.2.1 all bounded nontrivial solutions of $x' = A(t)x$ are almost periodic and thus satisfy $\inf_{t\in\mathbb{R}} \|x(t)\| > 0$ by Proposition 3.2.25.

Example 3.2.3. If $A(t)$ is exponentially dichotomic, then it satisfies Favard's property. In fact, in this case the only bounded solution of $x' = A(t)x$ on $\mathbb{R}$ is the trivial solution.

Example 3.2.4. If $A(t)$ is anti-Hermitian for all $t \in \mathbb{R}$, then it satisfies Favard's property. In fact, if x is a nontrivial solution of $x' = A(t)x$ then $\frac{\mathrm{d}}{\mathrm{d}t}\|x(t)\|^2 = \langle A(t)x, x\rangle + \langle x, A(t)x\rangle = 0$. Thus $\inf_{t\in\mathbb{R}} \|x(t)\| = \|x(0)\| > 0$.

Proposition 3.2.27. *If $M(t) \succeq 0$ for all $t \in \mathbb{R}$ (respectively $M(t) \preceq 0$ for all $t \in \mathbb{R}$) where*

$$M(t) := A_H'(t) + A_H(t)A(t) + A(t)^* A_H(t),$$

then $A(t)$ satisfies Favard's property.

Proof. Let x be a bounded nontrivial solution of $x' = A(t)x$. Set $y(t) = \|x(t)\|^2 = \langle x(t), x(t)\rangle$, thus $y' = 2\langle A_H(t)x, x\rangle$ and

$$y'' = 2\langle (A_H'(t) + A_H(t)A(t) + A(t)^* A_H(t))x, x\rangle = 2\langle M(t)x, x\rangle \geq 0 \quad (\text{respectively} \leq 0).$$

Thus the function y is convex (respectively concave). Since it is bounded on $\mathbb{R}$, it is constant. Therefore $\inf_{t\in\mathbb{R}} \|x(t)\| = \|x(0)\| > 0$. □

Example 3.2.5. If $A(t)$ is Hermitian, then $M(t) = A'(t) + 2A(t)^2$. If $A(t)$ is normal ($AA^* = A^*A$), then $M(t) = A'_H(t) + 2A_H(t)^2$. This implies that if $A(t)$ is Hermitian and $A'(t) \succeq 0$ or, more generally, if $A(t)$ is normal and $A'_H(t) \succeq 0$, then it satisfies Favard's condition.

Example 3.2.6. If $A(t)$ is almost automorphic and $A_H(t) \succeq 0$ for all $t \in \mathbb{R}$, then it satisfies Favard's condition by Theorem 3.2.17.

Proposition 3.2.28. *Let $X(t)$ be the fundamental matrix of $x' = A(t)x$. If $X(t)$ is bounded on $\mathbb{R}$ and there exists $c \in \mathbb{R}$ such that $\int_0^t \mathrm{Re}(\mathrm{tr}(A(s)))ds \geq c$ for all $t \in \mathbb{R}$, then $A(t)$ satisfies Favard's property.*

Proof. Let u be a nontrivial solution of $x' = A(t)x$. We complete u to get a basis $u, u_2, \ldots, u_N$ of the space of solutions of $x' = A(t)x$. It is known that the Wronskian $W(t) = \det(u(t), u_2(t), \ldots, u_N(t)) \neq 0$ for all $t \in \mathbb{R}$ satisfies the Liouville's formula

$$W(t) = W(0)\exp\left(\int_0^t \mathrm{tr}(A(s))ds\right).$$

It then follows that

$$\|W(t)\| = \|W(0)\|\exp\left(\int_0^t \mathrm{Re}(\mathrm{tr}(A(s)))ds\right) \geq \|W(0)\|e^c.$$

This implies that $\inf_{t\in\mathbb{R}} \|W(t)\| > 0$. If $\inf_{t\in\mathbb{R}} \|u(t)\| = 0$ then there exists a sequence $(t_n)_n$ such that $u(t_n) \to 0$ and, since $X(t)$ is bounded on $\mathbb{R}$, we have $W(t_n) = \det(u(t_n), u_2(t), \ldots, u_N(t_n)) \to 0$, which contradicts the fact that $\inf_{t\in\mathbb{R}} \|W(t)\| > 0$. Therefore we must have $\inf_{t\in\mathbb{R}} \|u(t)\| > 0$. □

Example 3.2.7. If $X(t)$ is bounded on $\mathbb{R}$, $A_H(t) \succeq 0$ for all $t \geq 0$, and $A_H(t) \preceq 0$ for all $t \leq 0$ then, since $\mathrm{tr}(A_H(t)) = \frac{\mathrm{tr}(A(t))+\overline{\mathrm{tr}(A(t))}}{2} = \mathrm{Re}\,\mathrm{tr}(A(t))$, it follows that $\int_0^t \mathrm{Re}(\mathrm{tr}(A(s)))ds = \int_0^t \mathrm{tr}(A_H(t))ds \geq 0$ and thus $A(t)$ satisfies Favard's condition.

Example 3.2.8. If $X(t)$ is bounded on $\mathbb{R}$ and $A(t)$ is nilpotent or anti-Hermitian, then $\mathrm{tr}(A(s)) = 0$ and thus $A(t)$ satisfies Favard's condition.

Example 3.2.9. Assume that $A(t)$ is almost automorphic, nilpotent, and commutative, for example $A(t) = \begin{pmatrix} a(t) & -a(t) \\ a(t) & -a(t) \end{pmatrix}$. We will see later in Proposition 3.2.48 that all bounded solutions of $x' = A(t)x$ on $\mathbb{R}$ are almost automorphic, and thus, by Proposition 3.2.25, $A(t)$ satisfies Favard's condition.

Now we move to some examples where Favard's condition is not satisfied.

Example 3.2.10. In the scalar case, $A(t) = a(t) = -2t$ does not satisfy Favard's condition, since all the solutions are bounded on $\mathbb{R}$ and $\inf_{t\in\mathbb{R}} \|x(t)\| = \inf_{t\in\mathbb{R}} e^{-t^2} \|x_0\| = 0$.

Example 3.2.11. An example of $A(t)$ which does not satisfy Favard's condition and which is almost periodic is given in Remark 3.2.14. That is, $A(t) = a(t) = \sum_{k=1}^{\infty} a_k(t)$, where $a_k(t) = \frac{-1}{k^2}\sin(\frac{\pi t}{k^3})$. We have seen in Remark 3.2.14 that all the solutions of $x' = a(t)x$ are bounded on $\mathbb{R}$ and have the form $x(t) = \exp(\int_0^t a(s)ds)x_0$. However,

$$\int_0^{n^3} a(s)ds = \sum_{k=1}^{\infty} \int_0^{n^3} a_k(s)ds \le \int_0^{n^3} a_n(s)ds = \frac{-2n}{\pi},$$

which implies that all these solutions satisfy $\lim_{n\to\infty} \|x(n^3)\| = 0$. Hence the equation $x' = a(t)x$ does not satisfy Favard's property.

3.2.3 The equation $x' = A(t)x + f(t)$

Let $F = \bigcap_{t\in\mathbb{R}} \ker A(t)$ and consider the orthogonal decomposition

$$\mathbb{C}^N = F \oplus F^{\perp}. \tag{3.29}$$

For each $x \in \mathbb{C}^N$, x_F and $x_{F^\perp}$ denote the components of x corresponding to the orthogonal decomposition (3.29).

Proposition 3.2.29. *Assume that $A(t), f(t)$ are almost automorphic and $A(t) \ge 0$ for all $t \in \mathbb{R}$. Then the equation $x' = A(t)x + f(t)$ has an almost automorphic solution $u(t)$ if and only if $t \mapsto \int_0^t f_F(s)ds$ is bounded. In this case, all almost automorphic solutions are equal modulo a constant $c \in F$.*

Proof. Assume that $x' = A(t)x+f(t)$ has an almost automorphic solution $u(t)$. By projecting this equation along F, we get $u_F'(t) = f_F(t)$. Thus $\int_0^t f_F(s)ds = u_F(t) + cte$ is bounded. Now, conversely, assume that $\int_0^t f_F(s)ds$ is bounded. The subspace $F^\perp$ is invariant under $A(t)$. Let $A_{F^\perp}(t)$ be the restriction of $A(t)$ on $F^\perp$. Then we have $\bigcap_{t\in\mathbb{R}} \ker A_{F^\perp}(t) = \{0\}$ and $A_{F^\perp}(t) \ge 0$. It follows by Corollary 3.2.15 that $x' = A_{F^\perp}(t)x$ has an exponential dichotomy in the subspace $F^\perp$. Hence by [25, Proposition 31], the equation $x' = A_{F^\perp}(t)x + f_{F^\perp}(t)$ has a unique almost automorphic solution $v(t)$ in the subspace $F^\perp$. Now consider the function

$$u(t) = \underbrace{\int_0^t f_F(s)ds}_{\in F} + \underbrace{v(t)}_{\in F^\perp}. \tag{3.30}$$

Since $A_F(t) = 0$, u is a solution of $x' = A(t)x + f(t)$. The function $\int_0^t f_F(s)ds$ is almost automorphic because it is bounded and, from (3.30), we deduce that $u(t)$ is also almost automorphic. If $w(t)$ is another almost automorphic solution of $x' = A(t)x + f(t)$ then, by Theorem 3.2.12, one can see that there exists a constant $c \in F$ such that $w(t) = u(t) + c$. □

3.2.3.1 A Massera-type theorem

Definition 3.2.30. Let $A(t)$ be a vector- or a matrix-valued function. Then the set

$$H(A) = \left\{B(t) = \lim_n A(t + t_n) \text{ for some sequence } (t_n)_n\right\}$$

is called the hull of A.

Consider the equation

$$x' = A(t)x + f(t), \tag{3.31}$$

where $A(t)$ and $f(t)$ are both almost periodic. If all the matrices $B \in H(A)$ satisfy Favard's property, then it is known [152] that the existence of a solution bounded on $\mathbb{R}^+$ implies the existence of an almost periodic function. However, if Favard's property holds only for $A(t)$ (not necessarily for other matrices in its hull), Fink [152] conjectured that we may not have almost periodic solutions. In the following, we show that one can still have compactly almost automorphic solutions. And that is true even if $A(t)$ and $f(t)$ are only almost automorphic.

Set the following hypothesis:

(H) $A(t)$ and $f(t)$ are both almost automorphic and equation (3.31) has a solution u_0 which is bounded on $\mathbb{R}^+$.

Theorem 3.2.31. *Assume that* (H) *holds and* $A(t)$ *satisfies Favard's condition. Then equation* (3.31) *has at least one compactly almost automorphic solution.*

We divide the proof of Theorem 3.2.31 into several lemmas. Let K be the closed convex hull (or convex envelope) of $\overline{\{u_0(t) : t \geq 0\}}$.

Lemma 3.2.32. *Assume that* (H) *holds. Then there exists at least one solution u of equation* (3.31) *on $\mathbb{R}$ such that* $\{u(t) : t \in \mathbb{R}\} \subset K$.

Proof. Let $(t_n)_n$ be a sequence of real numbers such that $t_n \to \infty$. Since f and A are almost automorphic, there exist g, B, and a subsequence $(t'_n)_n \subset (t_n)_n$ such that

$$\begin{cases} f(t + t'_n) \to g(t), \\ g(t - t'_n) \to f(t) \end{cases} \quad \text{and} \quad \begin{cases} A(t + t'_n) \to B(t), \\ B(t - t'_n) \to A(t) \end{cases}$$

as $n \to \infty$. The solution u_0 can be naturally defined on $\mathbb{R}$ (but can be unbounded on $(-\infty, 0)$). Consider the sequence of functions $u_n : t \mapsto u_0(t + t'_n)$. Then, by applying an argument similar to that in Lemma 3.2.4, there exists a subsequence $(t''_n)_n \subset (t'_n)_n$ such that $u_0(t + t''_n) \to v(t)$ uniformly on each compact subset of $\mathbb{R}$, where v is a weak solution of $x' = B(t)x + g(t)$ which is bounded on $\mathbb{R}$. By applying the above argument to the returning sequence $(-t_n')_n$, we obtain a subsequence $(t_n''')_n \subset (t_n'')_n$ such that $v(t-t_n''') \to u(t)$ uniformly on each compact subset of $\mathbb{R}$, where u is a weak solution of $x' = A(t)x+f(t)$ on $\mathbb{R}$ with the property $\{u(t) : t \in \mathbb{R}\} \subset \overline{\{v(t) : t \in \mathbb{R}\}} \subset \overline{\{u_0(t) : t \geq 0\}} \subset K$. Now since $A(t)$ and $f(t)$ are continuous, u is a strong solution. □

Let Ω_K be the set of all solutions of equation (3.31) on $\mathbb{R}$ that have a range in K. Notice, due to Lemma 3.2.32, that $\Omega_K \neq \emptyset$. Since K is convex, it is clear that Ω_K is a closed convex subset of $C_b(\mathbb{R}, \mathbb{C}^n)$.

Lemma 3.2.33. *Assume that* (**H**) *holds. Then there exists at least one solution* $u_* \in \Omega_K$ *such that*

$$\|u_*\|_\infty = \inf_{u\in\Omega_K} \|u\|_\infty.$$

Proof. Consider a minimizing sequence $(u_n)_n$, that is, a sequence of functions $u_n \in \Omega_K$ such that

$$\lim_{n\to\infty} \|u_n\|_\infty = \inf_{u\in\Omega_K} \|u\|_\infty.$$

The family of solutions $(u_n)_n$ is equibounded, that is, $\|u_n(t)\| \leq \|K\|$ for all $t \in \mathbb{R}$ and $n \in \mathbb{N}$, where $\|K\| := \sup\{\|x\| : x \in K\}$. Moreover, for all $t, \sigma \in \mathbb{R}$, we have

$$u_n(t) = u_n(\sigma) + \int_\sigma^t A(s)u_n(s)ds + \int_\sigma^t f(s)ds. \tag{3.32}$$

This implies that, for all $t, \sigma \in \mathbb{R}$,

$$\left\|u_n(t) - u_n(\sigma)\right\| \leq M\|t - \sigma\|,$$

where $M = \|A\|_\infty\|K\| + \|f\|_\infty$. The constant M, being independent of n, implies the equiuniform continuity on $\mathbb{R}$. Using Arzelà–Ascoli theorem and a diagonal argument similar to that used in the proof of Lemma 3.2.4, one can extract a subsequence, denoted also by $(u_n)_n$, such that $u_n(t) \to u_*(t)$ uniformly on compact subsets of $\mathbb{R}$. It is clear, by letting $n \to \infty$ in (3.32), that u_* is also a solution of $x' = A(t)x + f(t)$ and $u_* \in \Omega_K$. This implies that $\|u_*\|_\infty \geq \inf_{u\in\Omega_K} \|u\|_\infty$. On the other hand, by taking $n \to \infty$ in the following inequality:

$$\left\|u_*(t)\right\| \leq \left\|u_*(t) - u_n(t)\right\| + \|u_n\|_\infty,$$

we get $\|u_*\|_\infty \leq \inf_{u\in\Omega_K} \|u\|_\infty$. We conclude that $\|u_*\|_\infty = \inf_{u\in\Omega_K} \|u\|_\infty$. □

Let $\Lambda_K \subset \Omega_K$ be the set of all minimal solutions, i. e.,

$$\Lambda_K := \left\{x \in \Omega_K : \|x\|_\infty = \inf_{u\in\Omega_K} \|u\|_\infty\right\}.$$

Notice, by Lemma 3.2.33, that $\Lambda_K \neq \emptyset$. In the following, we give some properties enjoyed by the elements of Λ_K.

Proposition 3.2.34. *Assume that* (H) *holds. Then* Λ_K *is a closed convex subset of* $C_b(\mathbb{R}, \mathbb{C}^n)$.

Proof. The set Λ_K is closed since it is the intersection of the closed set Ω_K and the centered sphere of radius $r = \inf_{u\in\Omega_K} \|u\|_\infty$ in $C_b(\mathbb{R}, \mathbb{C}^n)$. For $u, v \in \Lambda_K$ and $\theta \in [0, 1]$, we have

$$\|\theta u + (1-\theta)v\|_\infty \le \theta\|u\|_\infty + (1-\theta)\|v\|_\infty = \inf_{u\in\Omega_K} \|u\|_\infty.$$

Since Ω_K is convex, $\theta u+(1-\theta)v \in \Omega_K$, which implies that $\|\theta u + (1-\theta)v\|_\infty = \inf_{u\in\Omega_K} \|u\|_\infty$ and thus $\theta u + (1-\theta)v \in \Lambda_K$. □

Lemma 3.2.35. *Assume that* (H) *holds and let* $u, v \in \Lambda_K$. *Then* $\inf_{t\in\mathbb{R}} \|u(t) - v(t)\| = 0$.

Proof. Let $u, v \in \Lambda_K$, then $\frac{u+v}{2} \in \Lambda_K$. Consider

$$\begin{cases} \lambda := \|u\|_\infty = \|v\|_\infty = \|\frac{u+v}{2}\|_\infty = \inf_{u\in\Omega_K} \|u\|_\infty, \\ \delta := \inf_{t\in\mathbb{R}} \|u(t) - v(t)\|. \end{cases}$$

Using the parallelogram law, we get

$$\frac{\|u(t)\|^2 + \|v(t)\|^2}{2} = \left\|\frac{u(t) + v(t)}{2}\right\|^2 + \left\|\frac{u(t) - v(t)}{2}\right\|^2 \ge \left\|\frac{u(t) + v(t)}{2}\right\|^2 + \frac{\delta^2}{4}.$$

Now applying the supremum norm, we have $\lambda^2 \ge \lambda^2 + \frac{\delta^2}{4}$, yielding $\delta = 0$. □

Proposition 3.2.36. *Assume that* (H) *holds. Then* Λ_K *contains at most one almost automorphic solution.*

Proof. Let u, v be two different elements of Λ_K such that both are almost automorphic. Then $u - v$ is an almost automorphic solution of $x' = A(t)x$ and thus satisfies $\inf_{t\in\mathbb{R}} \|u(t) - v(t)\| > 0$ by Proposition 3.2.25, which contradicts Lemma 3.2.35. □

Favard's condition has not been used so far. We will use it now to prove the following lemma.

Lemma 3.2.37. *Assume that* (H) *holds and* $A(t)$ *satisfies Favard's condition. Then* Λ_K *contains a unique element* u_*.

Proof. Assume that Λ_K contains at least two different elements $u, v \in \Omega_K$. On the one hand, we have, by Lemma 3.2.35, that $\inf_{t\in\mathbb{R}} \|u(t) - v(t)\| = 0$. On the other hand, $u - v$ is a bounded nontrivial solution of $x' = A(t)x$. Hence, from Favard's condition, we have $\inf_{t\in\mathbb{R}} \|u(t) - v(t)\| > 0$, which is a contradiction. □

Remark 3.2.38. Since $d(0, u_*) = \|u_*\|_\infty = \inf_{u\in\Omega_K} \|u\|_\infty = d(0, \Omega_K)$, Lemma 3.2.37 deals with the existence of a unique minimizer just like in the famous Hilbert projection theorem (also valid for uniformly convex Banach spaces). While Ω_K is a closed convex subset of $C_b(\mathbb{R}, X)$, one cannot use the Hilbert projection theorem here because $C_b(\mathbb{R}, X)$ is not uniformly convex.

We are now ready to give the proof of Theorem 3.2.31.

Proof of Theorem 3.2.31. We will prove that the solution u_*, i. e., the unique element in Λ_K is, in fact, the almost automorphic solution we are looking for. Let $(t_n')_n$ be a sequence of real numbers. From Lemma 3.2.4 and the fact that A and f are almost automorphic, there exists a subsequence $(t_n)_n \subset (t_n')_n$ such that

$$\begin{cases} f(t + t_n) \to g(t), \\ g(t - t_n) \to f(t) \end{cases} \quad \text{and} \quad \begin{cases} A(t + t_n) \to B(t), \\ B(t - t_n) \to A(t), \end{cases}$$

as well as

$$u_*(t + t_n) \to v(t) \tag{3.33}$$

uniformly on each compact subset of $\mathbb{R}$, where v is a weak solution of $x' = B(t)x + g(t)$. As a consequence of (3.33), we get $\|v\|_\infty \le \|u_*\|_\infty$. Using the same arguments this time on v and the returning sequence $(-t_n)_n$, we get a subsequence, also denoted by $(t_n)_n$, such that

$$v(t - t_n) \to w(t) \tag{3.34}$$

uniformly on each compact subset of $\mathbb{R}$, where w is a strong solution of $x' = A(t)x + f(t)$ with $w \in \Omega_K$. As a consequence, we get $\|w\|_\infty \le \|v\| \le \|u_*\|_\infty$, and thus $w \in \Lambda_K$. But since Lemma 3.2.37 tells us that u_* is the unique element in Λ_K, we get $w = u_*$ and (3.33) and (3.34) give the compact almost automorphy of u_*. □

We have the following Massera-type result.

Corollary 3.2.39. *Assume that $A_H(t) \ge 0$ and $A(t)$ and $f(t)$ are both almost automorphic. If equation* (3.31) *has a solution u_0 which is bounded on $\mathbb{R}^+$, then it has at least one compactly almost automorphic solution.*

Proof. From Theorem 3.2.17, $A(t)$ almost automorphic and $A_H(t) \ge 0$ implies Favard's condition. Hence Corollary 3.2.39 is just a consequence of Theorem 3.2.31. □

3.2.3.2 Several Bohr–Neugebauer-type results

We have the following Bohr–Neugebauer-type result.

Proposition 3.2.40. *Assume that $A(t)$ and $f(t)$ are both almost automorphic and $A(t) \geq 0$ for all $t \in \mathbb{R}$. Then, every bounded solution of $x' = A(t)x + f(t)$ on $\mathbb{R}$ is compactly almost automorphic.*

Proof. Let u be a bounded solution of $x' = A(t)x + f(t)$ on $\mathbb{R}$. Then, by Corollary 3.2.39, there exists u_* which is a compactly almost automorphic solution of $x' = A(t)x+f(t)$. As the function $u-u_*$ is a bounded solution of $x' = A(t)x$, it follows by Theorem 3.2.12 that $u-u_* = c$, where c is a constant. Therefore $u = u_* + c$ is compactly almost automorphic. □

Remark 3.2.41. Proposition 3.2.40 does not hold if we drop the assumption $A(t) \geq 0$ for all $t \in \mathbb{R}$. For example, consider the sign-changing almost periodic function $a(t) = \sum_{k=1}^{\infty} \frac{-1}{k^2} \sin(\frac{\pi t}{k^3})$. We saw in Example 3.2.11 that all solutions $x(t)$ of $x' = a(t)x$ are bounded on $\mathbb{R}$ and satisfy $\lim_{n\to\infty} \|x(n^3)\| = 0$. Hence Proposition 3.2.25 tells us that none of these solutions is almost automorphic (of course, except the trivial solution).

Remark 3.2.42. Proposition 3.2.40 does not hold if we replace the assumption $A(t) \geq 0$ for all $t \in \mathbb{R}$ by $A_H(t) \geq 0$ for all $t \in \mathbb{R}$. We still have an almost automorphic solution if a bounded solution exists (Corollary 3.2.39), but the other bounded solutions may not be almost automorphic (see Remark 3.2.8).

Remark 3.2.43. Proposition 3.2.40 also holds in the almost periodic case. In fact, it is sufficient to notice that if $A(t) \geq 0$, then $B(t) \geq 0$ for every matrix in the hull of A. This implies that the Favard condition is satisfied in all the hull of A. Hence we can apply Favard's theorem [152, Theorem 6.3] together with Theorem 3.2.12.

Proposition 3.2.44. *Assume that $A_H(t)$ is almost automorphic, $A_H(t) \geq 0$ for all $t \in \mathbb{R}$, and $A_S(t)$ is periodic. Then, every bounded solution of $x' = A(t)x + f(t)$ on $\mathbb{R}$ is compactly almost automorphic.*

Proof. Let u be a bounded solution of $x' = A(t)x + f(t)$ on $\mathbb{R}$. Then, by Corollary 3.2.39, there exists u_* which is a compactly almost automorphic solution of $x' = A(t)x+f(t)$. As the function $u - u_*$ is a bounded solution of $x' = A(t)x$, it follows by Corollary 3.2.23 that $u-u_*$ is almost periodic. Therefore $u = (u-u_*)+u_*$ is compactly almost automorphic. □

In what follows, we investigate the case where $A(t)$ is almost automorphic and the matrices $A(t)$ are commutative, that is, $A(t)A(s) = A(s)A(t)$ for all $t, s \in \mathbb{R}$. For example, in two dimensions $A(t) = \begin{pmatrix} a(t) & b(t)-a(t) \\ b(t)-a(t) & a(t) \end{pmatrix}$ where $a(t)$ and $b(t)$ are scalar almost automorphic functions. Then, in this case, there exists a simultaneous triangularization of the form

$$\begin{cases} T(t) = P^{-1}A(t)P, \\ y(t) = P^{-1}x(t), \\ g(t) = P^{-1}f(t), \end{cases}$$

where $T(t)$ is upper triangular,

$$T(t) = \begin{pmatrix} \lambda_1(t) & b_{1,2}(t) & b_{1,3}(t) & \dots & b_{1,n}(t) \\ 0 & \lambda_2(t) & b_{2,3}(t) & \dots & b_{2,n}(t) \\ \vdots & \vdots & \vdots & \ddots & \vdots \\ 0 & 0 & 0 & \lambda_{n-1}(t) & b_{n-1,n}(t) \\ 0 & 0 & 0 & \dots & \lambda_n(t) \end{pmatrix}. \tag{3.35}$$

Hence the equation $x' = A(t)x + f(t)$ is equivalent to $y' = T(t)y + g(t)$. Let $y(t) = (y_1(t) \ \dots \ y_n(t))^T$ and $g(t) = (g_1(t) \ \dots \ g_n(t))^T$. Then the equation $y' = T(t)y + g(t)$ becomes

$$\begin{cases} y_1' = \lambda_1(t)y_1 + b_{1,2}(t)y_2 + \dots + b_{1,n}(t)y_n + g_1(t), \\ y_2' = \lambda_2(t)y_2 + b_{2,3}(t)y_3 + \dots + b_{2,n}(t)y_n + g_2(t), \\ \vdots \\ y_{n-1}' = \lambda_{n-1}(t)y_{n-1} + b_{n-1,n}(t)y_n + g_{n-1}(t), \\ y_n' = \lambda_n(t)y_n + g_n(t). \end{cases}$$

The problem $x' = A(t)x + f(t)$ can then be reduced to the scalar problems $y_i' = \lambda_i(t)y_i + b_i(t)$ where λ_i are almost automorphic, $b_n(t) = g_n(t)$, and $b_i(t) = \sum_{j=i+1}^n b_{i,j}(t)y_j + g_i(t)$ for $i < n$. This shows that it is sufficient to study the scalar equation

$$x' = a(t)x + g(t),$$

where $a(t)$ and $g(t)$ are $\mathbb{K}$-valued almost automorphic functions. Let us first investigate the equation $x' = a(t)x$. The solutions of $x' = a(t)x$ have the form $x(t) = \exp(\int_0^t a(s)ds)c$ for some $c \in \mathbb{C}^N$.

Proposition 3.2.45. *The function $\varphi(t) = \exp(\int_0^t a(s)ds)$ is almost automorphic if and only if $t \mapsto \int_0^t \operatorname{Re} a(s)ds$ is bounded on $\mathbb{R}$ and $t \mapsto \exp(i\int_0^t \operatorname{Im} a(s)ds)$ is almost automorphic.*

Proof. If $\varphi(t) = \exp(\int_0^t a(s)ds)$ is almost automorphic then the same can be said about

$$|\varphi(t)| = \exp\left(\int_0^t \operatorname{Re} a(s)ds\right),$$

which implies that $t \mapsto \int_0^t \operatorname{Re} a(s)ds$ is bounded above. In the other hand, $|\varphi(t)|$ is an almost automorphic solution of solution of $x' = \operatorname{Re} a(t)x$, which implies by Proposition 3.2.25 that $\inf_{t\in\mathbb{R}} |\varphi(t)| > 0$, hence $t \mapsto \int_0^t \operatorname{Re} a(s)ds$ is bounded below and thus bounded, and even almost automorphic by Theorem 2.3.30. It follows by the continuity of $t \mapsto e^{-t}$ that

$$\exp\left(i\int_0^t \operatorname{Im} a(s)ds\right) = \exp\left(\int_0^t a(s)ds\right)\exp\left(-\int_0^t \operatorname{Re} a(s)ds\right)$$

is almost automorphic.

Now if $t \mapsto \int_0^t \operatorname{Re} a(s)ds$ is bounded on $\mathbb{R}$ and $t \mapsto \exp(i\int_0^t \operatorname{Im} a(s)ds)$ is almost automorphic, then $t \mapsto \int_0^t \operatorname{Re} a(s)ds$ is almost automorphic and thus

$$\exp\left(\int_0^t a(s)ds\right) = \exp\left(\int_0^t \operatorname{Re} a(s)ds\right)\exp\left(i\int_0^t \operatorname{Im} a(s)ds\right)$$

is almost automorphic. □

The problem is then reduced to the almost automorphy of functions of the form $t \mapsto \exp(i\int_0^t a(s)ds)$ where $a(t)$ is a real-valued almost automorphic function. If $t \mapsto \int_0^t a(s)ds$ is bounded then we know that $t \mapsto \exp(i\int_0^t a(s)ds)$ is almost automorphic. But the boundedness of $t \mapsto \int_0^t a(s)ds$ is not necessary since for constant functions $a(t) = k$, $t \mapsto \exp(i\int_0^t a(s)ds)$ is periodic and thus almost automorphic. We have

Proposition 3.2.46. *If there exists a (necessarily unique) number $k \in \mathbb{R}$ such that $t \mapsto \int_0^t a(s)ds - kt$ is bounded on $\mathbb{R}$, then $t \mapsto \exp(i\int_0^t a(s)ds)$ is almost automorphic.*

Proof. Since $a(s)-k$ is almost atomorphic, $\int_0^t a(s)ds - kt = \int_0^t (a(s) - k)ds$ being bounded implies its almost automorphy. It follows that $\exp(i\int_0^t a(s)ds) = \exp(i(\int_0^t a(s)ds - kt))e^{ikt}$ is almost automorphic. In this case, k in unique and $k = \lim_{|t|\to\infty} \frac{1}{t}\int_0^t a(s)ds$. □

For a general study of the almost automorphy of the function $\exp(i\int_0^t a(s)ds)$, assume that $a(t)$ is almost automorphic and consider a sequence $(t_n')_n$. Then there exist a measurable function $\beta(t)$ and a subsequence $(t_n)_n$ such that for all $t \in \mathbb{R}$, $a(t+t_n) \to \beta(t)$ and $\beta(t - t_n) \to a(t)$. We have

$$\exp\left(i\int_0^{t+t_n} a(s)ds\right) = \exp\left(i\int_0^{t_n} a(s)ds\right)\exp\left(i\int_0^t a(s + t_n)ds\right).$$

Since $\exp(i\int_0^{t_n} a(s)ds)$ is bounded, we get a subsequence, also denoted by $(t_n)_n$, such that

$$\exp\left(i\int_0^{t_n} a(s)ds\right) \to \lambda \in \mathbb{C}.$$

It follows that

$$\varphi(t) = \exp\left(i\int_0^{t+t_n} a(s)ds\right) \underset{n\to\infty}{\to} \psi(t) = \lambda\exp\left(i\int_0^t \beta(s)ds\right).$$

On the other hand, we have

$$\psi(t-t_n)=\lambda\exp\left(i\int_0^{t-t_n}\beta(s)\mathrm{d}s\right)=\lambda\exp\left(i\int_0^{-t_n}\beta(s)\mathrm{d}s\right)\exp\left(i\int_0^{t}\beta(s-t_n)\mathrm{d}s\right).$$

We get a subsequence, also denoted by $(t_n)_n$, such that $\exp(i\int_0^{-t_n}\beta(s)\mathrm{d}s)\to\mu\in\mathbb{C}$. This implies that $\psi(t-t_n)\to\lambda\mu\varphi(t)$ as $n\to\infty$. One can see then that φ is almost automorphic if and only if $\lambda\mu=1$, which is also equivalent to

$$\lim_{n\to\infty}\exp\left(i\int_0^{t_n}(\alpha(s)-\beta(-s))\mathrm{d}s\right)=1. \tag{3.36}$$

(From the above extraction process, we note that (3.36) holds for the whole sequence $(t_n)_n$ and not only for some subsequence, since 1 is the only adherent point.) Equation (3.36) is also equivalent to

$$\lim_{n\to\infty}\int_0^{t_n}(\alpha(s)-\beta(-s))\mathrm{d}s-2k_n\pi=0\quad\text{for some sequence } k_n\in\mathbb{Z}.$$

In fact, this is a consequence of the following proposition:

Proposition 3.2.47. *Let $(u_n)_n$ be a sequence of real numbers. Then $\lim_{n\to\infty}e^{iu_n}=1$ if and only if there exists a sequence $(k_n)_n\in\mathbb{Z}$ such that $\lim_{n\to\infty}(u_n-2k_n\pi)=0$.*

Proof. First notice that $\lim_{n\to\infty}e^{iu_n}=1$ if and only if $\lim_{n\to\infty}|\sin(\frac{u_n}{2})|=0$. Let $k_n=\lfloor\frac{\pi+u_n}{2\pi}\rfloor$, where $\lfloor\cdot\rfloor$ is the floor function. Then

$$k_n\pi\le\frac{\pi+u_n}{2}<(1+k_n)\pi.$$

Thus $-\frac{\pi}{2}\le\frac{u_n}{2}-k_n\pi<\frac{\pi}{2}$ and $|\sin(\frac{u_n}{2})|=|\sin(\frac{u_n}{2}-k_n\pi)|\to0$ if and only if $\lim_{n\to\infty}(u_n-2k_n\pi)=0$. One can also write $\lim_{n\to\infty}d(u_n,2\pi\mathbb{Z})=0$. □

We return now to the scalar equation

$$x'=a(t)x+g(t).$$

The solutions have the form

$$x(t)=\exp\left(\int_0^t a(s)\mathrm{d}s\right)\left[c+\int_0^t g(s)\exp\left(-\int_0^s a(u)\mathrm{d}u\right)\mathrm{d}s\right].$$

From Proposition 3.2.45, $\exp(\int_0^t a(s)\mathrm{d}s)$ is almost automorphic if and only if $\exp(-\int_0^t a(s)\mathrm{d}s)$ is almost automorphic. Hence if $g(t)$ is almost automorphic, then all solutions of $x' = a(t)x + g(t)$ are almost automorphic if and only if $t \mapsto \int_0^t \mathrm{Re}(a(s))ds$, $t \mapsto \int_0^t (g(s)\exp(-\int_0^s a(u)du))ds$ are bounded on $\mathbb{R}$ and $t \mapsto \exp(i\int_0^t \mathrm{Im}(a(s))ds)$ is almost automorphic.

We have another Bohr–Neugebauer-type result.

Proposition 3.2.48. *Assume that $A(t)$ and $f(t)$ are almost automorphic, $A(t)$ are nilpotent and simultaneously triangularizable. Then all bounded solutions of $x' = A(t)x + f(t)$ on $\mathbb{R}$ are almost automorphic.*

Proof. Consider the following triangularization:

$$\begin{cases} T(t) = PA(t)P^{-1}, \\ y(t) = Px(t), \\ g(t) = Pf(t), \end{cases}$$

where $T(t)$ is the upper-triangular nilpotent matrix (3.35) with $\lambda_i(t) = 0$ for all $i = 1, \dots, N$ and $t \in \mathbb{R}$. The equation $x' = A(t)x + f(t)$ is then equivalent to $y' = T(t)y + g(t)$, and we have

$$\begin{cases} y_1' = b_{1,2}(t)y_2 + \cdots + b_{1,n}(t)y_n + g_1(t), \\ y_2' = b_{2,3}(t)y_3 + \cdots + b_{2,n}(t)y_n + g_2(t), \\ \quad\vdots \\ y_{n-1}' = b_{n-1,n}(t)y_n + g_{n-1}(t), \\ y_n' = g_n(t). \end{cases}$$

Starting from the last equation, y_n is almost automorphic by Theorem 2.3.30. This implies that $b_{n-1,n}(t)y_n + g_{n-1}(t)$ is also almost automorphic, and then the same can be said about y_{n-1}. By repeating this argument, one can show that all the components y_i, $i = 1, \dots, n$ are almost automorphic. Thus $x(t) = P^{-1}y(t)$ is almost automorphic. □

3.3 Functional differential equations of neutral type

Let us consider the following functional differential equation of neutral type:

$$\begin{cases} \frac{d(Dx_t)}{dt} = L(x_t) + f(t) \quad \text{for } t \geq 0, \\ x_0 = \varphi \in C = C([-r, 0], \mathbb{R}^n), \end{cases} \tag{3.37}$$

where C is the space of continuous functions on $[-r, 0]$ with values in $\mathbb{R}^n$ provided with the supremum norm, D and L are two bounded linear operators from C into $\mathbb{R}^n$, while f is a continuous function on $[0, \infty[$ with values in $\mathbb{R}^n$. For every t, the history function $x_t \in C$ is defined by

$$x_t(\theta) = x(t+\theta) \quad \text{for } \theta \in [-r, 0].$$

The initial value problem associated with equation (3.37) is: given $\varphi \in C$, find a continuous function $x : [-r, \infty) \to \mathbb{R}^n$ such that $x_0 = \varphi$ while $t \to Dx_t$ is continuously differentiable and satisfies equation (3.37) for $t \geq 0$. For the well-posedness of equation (3.37), we assume that D is given by

$$D(\varphi) = \varphi(0) - D_0(\varphi),$$

where D_0 is a bounded linear operator from C to $\mathbb{R}^n$. By the Riesz representation theorem, it is well known that there exist two matrix-valued functions μ and η of bounded variation on $[-r, 0]$ such that

$$D_0(\varphi) = \int_{-r}^{0} \mathrm{d}\mu(\theta)\varphi(\theta) \quad \text{and} \quad L(\varphi) = \int_{-r}^{0} \mathrm{d}\eta(\theta)\varphi(\theta), \quad \varphi \in C. \tag{3.38}$$

In this work, we assume that μ is nonatomic at zero, which means that

$$\int_{-\varepsilon}^{0} |\mathrm{d}\mu(\theta)| \to 0 \quad \text{as } \varepsilon \to 0.$$

This condition implies the well-posedness of equation (3.37) in the sense that, for all $\varphi \in C$, equation (3.37) has a unique solution which is defined for all $t \geq 0$. The theory of functional differential equations of neutral type has been extensively developed; we refer the reader to [163].

In this work we prove that Massera-type theorem holds when f is almost automorphic. We show the equivalence between the existence of an almost automorphic solution and the existence of a bounded solution on $\mathbb{R}^+$. To achieve this goal, we use the variation of constants formula for equation (3.37) obtained in [4] and develop some fundamental results about the spectral decomposition of the solutions of equation (3.37).

3.3.1 Variation of constants formula

We consider the following linear functional differential equation of neutral type:

$$\begin{cases} \frac{\mathrm{d}(Dx_t)}{\mathrm{d}t} = Lx_t & \text{for } t > 0, \\ x_0 = \varphi \in C. \end{cases} \tag{3.39}$$

Let $(T(t))_{t\geq 0}$ be defined by

$$T(t)\varphi = x_t(\cdot, \varphi), \quad t \geq 0 \text{ and } \varphi \in C.$$

Then, $(T(t))_{t\geq 0}$ is a strongly continuous semigroup on C. Its generator is given by the following theorem.

Theorem 3.3.1 ([166]). *Let $(A, D(A))$ be defined by*

$$\begin{cases} D(A) = \{\varphi \in C^1([-r,0];\mathbb{R}^n) : D(\varphi') = L(\varphi)\}, \\ A\varphi = \varphi'. \end{cases}$$

Then, A is the infinitesimal generator of $(T(t))_{t\geq 0}$.

In order to give a variation of constants formula, we need to recall some notations and results which are taken from [4]. Let $\langle X_0 \rangle$ be the space defined by

$$\langle X_0 \rangle = \{X_0 c : c \in \mathbb{R}^n\},$$

where the function $X_0 c$ is defined by

$$(X_0 c)(\theta) = \begin{cases} 0 & \text{if } \theta \in [-r, 0), \\ c & \text{if } \theta = 0. \end{cases}$$

The space $C \oplus \langle X_0 \rangle$ is equipped with the norm $\|\phi + X_0 c\| = |\phi| + |c|$ for $(\phi, c) \in C \times \mathbb{R}^n$ is a Banach space and we consider the extension $\widetilde{A}$ of the operator A defined on $C \oplus \langle X_0 \rangle$ by

$$\begin{cases} D(\widetilde{A}) = C^1([-r,0];\mathbb{R}^n), \\ \widetilde{A}\phi = \varphi' + X_0(D(\varphi') - L(\varphi)). \end{cases}$$

Lemma 3.3.2 ([4]). *Operator $\widetilde{A}$ satisfies the Hille–Yosida condition on $C \oplus \langle X_0 \rangle$: there exist $\widetilde{M} \geq 0$, $\widetilde{\omega} \in \mathbb{R}$ such that $(\widetilde{\omega}, +\infty) \subset \rho(\widetilde{A})$, and*

$$|R(\lambda, \widetilde{A})^n| \leq \frac{\widetilde{M}}{(\lambda - \widetilde{\omega})^n}, \quad n \in \mathbb{N}, \lambda > \widetilde{\omega},$$

where $R(\lambda, \widetilde{A}) = (\lambda - \widetilde{A})^{-1}$.

Theorem 3.3.3 ([4, Theorem 16]). *The solution x of equation (3.37) is given by the following variation of constants formula:*

$$x_t = T(t)\varphi + \lim_{\lambda \to +\infty} \int_0^t T(t-s)\widetilde{B}_\lambda(X_0 f(s))ds, \quad \text{for } t \geq 0,$$

where $\widetilde{B}_\lambda = \lambda R(\lambda, \widetilde{A})$ for $\lambda > \widetilde{\omega}$.

3.3.2 Spectral decomposition and reduction principle

To deal with the qualitative behavior of solutions, we need to make more assumptions on the homogeneous equation

$$\begin{cases} Du_t = 0, & t \geq 0, \\ u_0 = \varphi \in C. \end{cases} \tag{3.40}$$

Definition 3.3.4 ([166, p. 59]). The operator D is said to be stable if there exist positive constants η and μ such that the solution of the difference equation (3.40) with $\varphi \in \{\psi \in C : D\psi = 0\}$ satisfies

$$|u_t(\cdot, \varphi)| \leq \mu \exp(-\eta t)|\varphi| \quad \text{for } t \geq 0.$$

Remark 3.3.5. The operator D defined by

$$D\varphi = \varphi(0) - q\varphi(-r)$$

is stable if and only if $|q| < 1$.

In the following, we assume that:
($\mathbf{H_1}$) The operator D is stable.

We give the following fundamental result on the semigroup $(T(t))_{t\geq 0}$.

Theorem 3.3.6 ([166, Theorem 9]). *Assume that* ($\mathbf{H_1}$) *holds. Then the semigroup* $(T(t))_{t\geq 0}$ *is decomposed on* C *as follows:*

$$T(t) = T_1(t) + T_2(t) \quad \text{for } t \geq 0,$$

where $(T_1(t))_{t\geq 0}$ *is an exponentially stable semigroup on* C, *which means that there are positive constants* a_0 *and* N_0 *such that*

$$|T_1(t)\varphi| \leq N_0 e^{-a_0 t}|\varphi| \quad \text{for } t \geq 0 \text{ and } \varphi \in C,$$

and $T_2(t)$ *is compact for every* $t > 0$.

Using Theorem 3.3.6 and by a sample computation, we get that $\omega_{ess}(T) < 0$. As a consequence, one has the following spectral decomposition result.

Theorem 3.3.7 ([166, Theorem 5.3.7, p. 333]). *Space* C *is decomposed as follows:*

$$C = S \oplus V,$$

where S *is the stable space which is* T*-invariant and there are positive constants* a *and* N *such that*

$$|T(t)\varphi| \leq Ne^{-at}|\varphi| \quad \text{for } t \geq 0 \text{ and } \varphi \in S, \tag{3.41}$$

while V is a finite-dimensional space and the restriction of T to V becomes a group.

In the sequel, $T^s(t)$ and $T^v(t)$ denote the restrictions of $T(t)$ on S and V, respectively, which corresponds to the above decomposition of C.

Let $d = \dim V$ with a basis of vectors $\Phi = \{\phi_1, \ldots, \phi_d\}$. Then there exist d-elements $\{\psi_1, \ldots, \psi_d\}$ in C^* such that

$$\begin{cases} \langle \psi_i, \phi_j \rangle = \delta_{ij}, & \text{where } \delta_{ij} = \begin{cases} 1 & \text{if } i = j, \\ 0 & \text{if } i \neq j, \end{cases} \\ \langle \psi_i, \phi \rangle = 0 & \text{for all } \phi \in S \text{ and } i \in \{1, \ldots, d\}, \end{cases} \tag{3.42}$$

where $\langle \cdot, \cdot \rangle$ denotes the duality pairing between C^* and C. Letting $\Psi = \operatorname{col}\{\psi_1, \ldots, \psi_d\}$, $\langle \Psi, \Phi \rangle$ is a $d \times d$ matrix, with its (i,j)-component being $\langle \psi_i, \phi_j \rangle$. Denote by Π^s and Π^v the projections on S and V, respectively. For each $\varphi \in C$, we have

$$\Pi^v \varphi = \Phi \langle \Psi, \varphi \rangle.$$

In fact, for $\varphi \in C$, we have $\varphi = \Pi^s \varphi + \Pi^v \varphi$ with $\Pi^v \varphi = \sum_{i=1}^d a_i \phi_i$ and $a_i \in \mathbb{R}$. By (3.42), we conclude that

$$a_i = \langle \psi_i, \varphi \rangle.$$

Hence

$$\Pi^v \varphi = \sum_{i=1}^d \langle \psi_i, \varphi \rangle \phi_i = \Phi \langle \Psi, \varphi \rangle.$$

Since $(T^v(t))_{t \geq 0}$ is a group on V, there exists a $d \times d$ matrix G such that

$$T^v(t)\Phi = \Phi e^{tG} \quad \text{for } t \in \mathbb{R}.$$

Moreover, $\sigma(G) = \{\lambda \in \sigma(A) : \operatorname{Re}(\lambda) \geq 0\}$.

For $n, n_0 \in \mathbb{N}$ such that $n \geq n_0 \geq \widetilde{\omega}$ and $i \in \{1, \ldots, d\}$, we define the linear mapping $x^*_{i,n}$ by

$$x^*_{i,n}(a) = \langle \psi_i, \widetilde{B}_n X_0 a \rangle \quad \text{for } a \in \mathbb{R}^n.$$

Since $|\widetilde{B}_n| \leq \frac{n}{n-\widetilde{\omega}} \widetilde{M}$, for any $n > n_0$, then

$$|x^*_{i,n}| \leq \frac{n}{n - n_0} \widetilde{M} |\psi_i| \quad \text{for } n \geq n_0.$$

Define the d-column vector $x^*_n = \operatorname{col}(x^*_{1,n}, \ldots, x^*_{d,n})$ for $n \geq n_0$. We have

$$\langle x_n^*, a\rangle = \langle \Psi, \widetilde{B}_n X_0 a\rangle,$$

with

$$\langle x_n^*, a\rangle_i = \langle \psi_i, \widetilde{B}_n X_0 a\rangle \quad \text{for } i = 1, \dots, d \text{ and } a \in \mathbb{R}^n.$$

Then

$$\sup_{n>n_0} |x_n^*| \le \widetilde{M} \sup_{1\le i\le d} |\psi_i| < \infty,$$

which implies that $(x_n^*)_{n\ge n_0}$ is a bounded sequence in $\mathcal{L}(\mathbb{R}^n, \mathbb{R}^d)$. Consequently, there exists a subsequence $(x_{n_k}^*)_{k\ge 0}$ which is convergent to $x^* \in \mathcal{L}(\mathbb{R}^n, \mathbb{R}^d)$.

Lemma 3.3.8. *The whole sequence $(x_n^*)_{n\ge n_0}$ converges to x^*.*

Proof. In fact, we proceed by contradiction and suppose that there exists a subsequence $(x_{n_p}^*)_{p\in\mathbb{N}}$ of $(x_n^*)_{n\ge n_0}$ which converges weakly to some $\widetilde{x}^*$ with $\widetilde{x}^* \neq x^*$. Let $u_t(\cdot, \sigma, \varphi, f)$ denote the solution of equation (3.37). Then

$$\Pi^\nu u_t(\cdot, \sigma, 0, f) = \lim_{n\to+\infty} \int_\sigma^t T^\nu(t-\xi)\Pi^\nu(\widetilde{B}_n X_0 f(\xi))\mathrm{d}\xi$$

and

$$\Pi^\nu(\widetilde{B}_n X_0 f(\xi)) = \Phi\langle \Psi, \widetilde{B}_n X_0 f(\xi)\rangle = \Phi\langle x_n^*, f(\xi)\rangle.$$

It follows that

$$\begin{aligned}
\Pi^\nu u_t(\cdot, \sigma, 0, f) &= \lim_{n\to+\infty} \Phi \int_\sigma^t e^{(t-\xi)G}\langle \Psi, \widetilde{B}_n X_0 f(\xi)\rangle \mathrm{d}\xi, \\
&= \lim_{n\to+\infty} \Phi \int_\sigma^t e^{(t-\xi)G}\langle x_n^*, f(\xi)\rangle \mathrm{d}\xi.
\end{aligned}$$

For any $a \in \mathbb{R}^n$, set $f(\cdot) = a$. Then

$$\lim_{k\to+\infty} \int_\sigma^t e^{(t-\xi)G}\langle x_{n_k}^*, a\rangle \mathrm{d}\xi = \lim_{p\to+\infty} \int_\sigma^t e^{(t-\xi)G}\langle x_{n_p}^*, a\rangle \mathrm{d}\xi \quad \text{for } a \in \mathbb{R},$$

which implies that

$$\int_\sigma^t e^{(t-\xi)G}\langle x^*, a\rangle \mathrm{d}\xi = \int_\sigma^t e^{(t-\xi)G}\langle \widetilde{x}^*, a\rangle \mathrm{d}\xi \quad \text{for } a \in \mathbb{R}.$$

Consequently, $x^* \equiv \widetilde{x}^*$, which gives a contradiction. Therefore,

$$x_n^* \to x^* \quad \text{as } n \to \infty.$$ □

As a consequence, we conclude that

Corollary 3.3.9. *For any continuous function $h : \mathbb{R} \to \mathbb{R}$, we have*

$$\lim_{n\to+\infty} \int_{\sigma}^{t} T^{\nu}(t-\xi)\Pi^{\nu}(\widetilde{B}_n X_0 h(\xi))\mathrm{d}\xi = \Phi \int_{\sigma}^{t} e^{(t-\xi)G}\langle x^*, h(\xi)\rangle \mathrm{d}\xi \quad \textit{for all } t, \sigma \in \mathbb{R}.$$

Now we are in a position to present the following fundamental reduction principle.

Theorem 3.3.10. *Assume that* $(\mathbf{H}_1)$ *holds. Let u be the solution of equation* (3.37) *on* $\mathbb{R}$. *Then $z(t) = \langle \Psi, u_t\rangle$ is a solution of the following ordinary differential equation:*

$$\frac{\mathrm{d}}{\mathrm{d}t} z(t) = Gz(t) + \langle x^*, f(t)\rangle, \quad t \in \mathbb{R}. \tag{3.43}$$

Conversely, if f is a bounded function on $\mathbb{R}$ and z is a solution of equation (3.43) *on* $\mathbb{R}$, *then the function u given by*

$$u(t) = \left[\Phi z(t) + \lim_{n\to+\infty} \int_{-\infty}^{t} T^{s}(t-\xi)\Pi^{s}(\widetilde{B}_n X_0 f(\xi))\mathrm{d}\xi\right](0) \quad \textit{for } t \in \mathbb{R},$$

is a solution of equation (3.37) *on* $\mathbb{R}$.

Proof. Let u be a solution of equation (3.37) on $\mathbb{R}$. Then

$$u_t = \Pi^{s} u_t + \Pi^{\nu} u_t \quad \text{for all } t \in \mathbb{R}$$

and

$$\Pi^{\nu} u_t = T^{\nu}(t-\sigma)\Pi^{\nu} u_{\sigma} + \lim_{n\to+\infty} \int_{\sigma}^{t} T^{\nu}(t-\xi)\Pi^{\nu}(\widetilde{B}_n X_0 f(\xi))\mathrm{d}\xi \quad \text{for } t, \sigma \in \mathbb{R}.$$

Since $\Pi^{\nu} u_t = \Phi\langle \Psi, u_t\rangle$, by Corollary 3.3.9, we get

$$\begin{aligned}\Phi\langle \Psi, u_t\rangle &= T^{\nu}(t-\sigma)\Phi\langle \Psi, u_{\sigma}\rangle + \Phi \int_{\sigma}^{t} e^{(t-\xi)G}\langle x^*, f(\xi)\rangle \mathrm{d}\xi \\ &= \Phi e^{(t-\sigma)G}\langle \Psi, u_{\sigma}\rangle + \Phi \int_{\sigma}^{t} e^{(t-\xi)G}\langle x^*, f(\xi)\rangle \mathrm{d}\xi \quad \text{for } t, \sigma \in \mathbb{R}.\end{aligned}$$

Let $z(t) = \langle \Psi, u_t \rangle$. Then

$$z(t) = e^{(t-\sigma)G} z(\sigma) + \int_{\sigma}^{t} e^{(t-\xi)G} \langle x^*, f(\xi) \rangle \mathrm{d}\xi \quad \text{for } t, \sigma \in \mathbb{R}.$$

Consequently, z is a solution of the ordinary differential equation (3.43) on $\mathbb{R}$.

Conversely, assume that f is bounded on $\mathbb{R}$. Then $\int_{-\infty}^{t} T^s(t-\xi)\Pi^s(\widetilde{B}_n X_0 f(\xi))\mathrm{d}\xi$ is well defined on $\mathbb{R}$. Let z be a solution of (3.43) on $\mathbb{R}$ and define v by

$$v(t) = \Phi z(t) + \lim_{n\to+\infty} \int_{-\infty}^{t} T^s(t-\xi)\Pi^s(\widetilde{B}_n X_0 f(\xi))\mathrm{d}\xi \quad \text{for } t \in \mathbb{R}.$$

Since

$$z(t) = e^{(t-\sigma)G} z(\sigma) + \int_{\sigma}^{t} e^{(t-\xi)G} \langle x^*, f(\xi) \rangle \mathrm{d}\xi \quad \text{for } t, \sigma \in \mathbb{R},$$

using Corollary 3.3.9, the function v_1 given by

$$v_1(t) = \Phi z(t) \quad \text{for } t \in \mathbb{R}$$

satisfies the integral equation

$$v_1(t) = T^v(t-\sigma)v_1(\sigma) + \lim_{n\to+\infty} \int_{\sigma}^{t} T^v(t-\xi)\Pi^v(\widetilde{B}_n X_0 f(\xi))\mathrm{d}\xi \quad \text{for } t, \sigma \in \mathbb{R}.$$

Moreover, the function v_2 given by

$$v_2(t) = \lim_{n\to+\infty} \int_{-\infty}^{t} T^s(t-\xi)\Pi^s(\widetilde{B}_n X_0 f(\xi))\mathrm{d}\xi \quad \text{for } t \in \mathbb{R}$$

satisfies

$$v_2(t) = T^s(t-\sigma)v_2(\sigma) + \lim_{n\to+\infty} \int_{\sigma}^{t} T^s(t-\xi)\Pi^s(\widetilde{B}_n X_0 f(\xi))\mathrm{d}\xi \quad \text{for all } t \geq \sigma.$$

Then, for all $t \geq \sigma$ with $t, \sigma \in \mathbb{R}$, one has

$$
\begin{aligned}
T(t-\sigma)v(\sigma) &= T^{v}(t-\sigma)v_1(\sigma) + T^{s}(t-\sigma)v_2(\sigma),\\
&= v_1(t) - \lim_{n\to+\infty}\int_{\sigma}^{t} T^{v}(t-\xi)\Pi^{v}(\widetilde{B}_n X_0 f(\xi))d\xi\\
&\quad + v_2(t) - \lim_{n\to+\infty}\int_{\sigma}^{t} T^{s}(t-\xi)\Pi^{s}(\widetilde{B}_n X_0 f(\xi))d\xi\\
&= v(t) - \lim_{n\to+\infty}\int_{\sigma}^{t} T(t-\xi)(\widetilde{B}_n X_0 f(\xi))d\xi.
\end{aligned}
$$

Therefore,

$$
v(t) = T(t-\sigma)v(\sigma) + \lim_{n\to+\infty}\int_{\sigma}^{t} T(t-\xi)(\widetilde{B}_n X_0 f(\xi))\xi \quad \text{for } t \ge \sigma.
$$

By Theorem 3.3.3, we obtain that the function u defined by $u(t) = v(t)(0)$ is a solution of equation (3.37) on $\mathbb{R}$. □

3.3.3 Almost automorphic solutions for equation (3.37)

In the following, we assume
$(\mathbf{H}_2)$ f is an almost automorphic function.

Theorem 3.3.11. *Assume that* $(\mathbf{H}_1)$ *and* $(\mathbf{H}_2)$ *hold. If equation* (3.37) *has a bounded solution on* $\mathbb{R}^+$, *then it has an almost automorphic solution.*

Proof. Let u be a bounded solution of equation (3.37) on $\mathbb{R}^+$. By Theorem 3.3.10, the function $z(t) = \langle \Psi, u_t\rangle$, for $t \ge 0$, is a solution of the ordinary differential equation (3.43) and z is bounded on $\mathbb{R}^+$. Moreover, the function

$$
\rho(t) = \langle x^*, f(t)\rangle \quad \text{for } t \in \mathbb{R}
$$

is almost automorphic from $\mathbb{R}$ to $\mathbb{R}^d$. By Theorem 3.1.3, we get that the reduced system (3.43) has an almost automorphic solution $\tilde{z}$. Consequently, $\Phi\tilde{z}(\cdot)$ is an almost automorphic function on $\mathbb{R}$. By Theorem 3.3.10, the function $u(t) = v(t)(0)$, where

$$
v(t) = \Phi\tilde{z}(t) + \lim_{n\to+\infty}\int_{-\infty}^{t} T^{s}(t-\xi)\Pi^{s}(\widetilde{B}_n X_0 f(\xi))d\xi \quad \text{for } t \in \mathbb{R}
$$

is an integral solution of equation (3.37) on $\mathbb{R}$. So to prove that v is almost automorphic, it is enough to show that the function

$$y(t) = \lim_{n\to+\infty} \int_{-\infty}^{t} T^s(t-\xi)\Pi^s(\widetilde{B}_n X_0 f(\xi))d\xi \quad \text{for } t \in \mathbb{R}$$

is almost automorphic. Since f is almost automorphic, for any sequence of real numbers $(a'_p)_{p\geq 0}$ there exists a subsequence $(a_p)_{p\geq 0}$ of $(a'_p)_{p\geq 0}$ such that

$$\lim_{p\to\infty} f(t+a_p) = h(t) \quad \text{for each } t \in \mathbb{R}$$

and

$$\lim_{p\to\infty} h(t-a_p) = f(t) \quad \text{for each } t \in \mathbb{R}.$$

Now

$$y(t+a_p) = \lim_{n\to+\infty} \int_{-\infty}^{t+a_p} T^s(t+a_p-\xi)\Pi^s(\widetilde{B}_n X_0 f(\xi))d\xi \quad \text{for } t \in \mathbb{R},$$

which gives

$$y(t+a_p) = \lim_{n\to+\infty} \int_{-\infty}^{t} T^s(t-\xi)\Pi^s(\widetilde{B}_n X_0 f(\xi+a_p))d\xi \quad \text{for } t \in \mathbb{R}.$$

By the Lebesgue's dominated convergence theorem, we get

$$y(t+a_p) \to w(t) \quad \text{as } p \to \infty,$$

where w is given by

$$w(t) = \lim_{n\to+\infty} \int_{-\infty}^{t} T^s(t-\xi)\Pi^s(\widetilde{B}_n X_0 h(\xi))d\xi \quad \text{for } t \in \mathbb{R}.$$

Using the same arguments as above, we prove that

$$w(t-a_p) \to \lim_{n\to+\infty} \int_{-\infty}^{t} T^s(t-\xi)\Pi^s(\widetilde{B}_n X_0 f(\xi))d\xi \quad \text{as } p \to \infty,$$

which implies that y is almost automorphic and v is an almost automorphic solution of equation (3.37). □

3.3.4 Applications

Consider the following functional differential equation of neutral type:

$$\begin{cases} \frac{\mathrm{d}}{\mathrm{d}t}[x(t) - qx(t-r)] = -[x(t) - qx(t-r)] + \int_{-r}^{0} g(\theta)x(t+\theta)\mathrm{d}\theta + f(t) & \text{for } t \geq 0, \\ x_0 = \varphi \in C = C([-r,0]; \mathbb{R}^n), \end{cases} \tag{3.44}$$

where $0 < q < 1$ and g is a continuous function from $[-r, 0]$ to $\mathbb{R}$ such that there exists a constant $\beta \in (0, 1)$ satisfying

$$\int_{-r}^{0} |g(\theta)|\mathrm{d}\theta \leq (1-q)\beta. \tag{3.45}$$

We assume that f is almost automorphic from $\mathbb{R}$ to $\mathbb{R}$.

Proposition 3.3.12. *Under the above assumptions, equation* (3.44) *has a bounded solution on* $\mathbb{R}^+$. *Consequently, it has an almost automorphic solution.*

Proof. The first goal is to prove that equation (3.44) has a bounded solution on $\mathbb{R}^+$. Let $\rho = \frac{1}{1+q}(1 + \frac{|f|}{1-\beta})$, where $|f| = \sup_{s\in\mathbb{R}} |f(s)|$. Take $\varphi \in C$ such that $|\varphi| < \rho$. Then

$$|\varphi(0) - q\varphi(-r)| < (1+q)\rho.$$

We claim that

$$|u(t) - qu(t-r)| \leq (1+q)\rho \quad \text{for all } t \geq 0. \tag{3.46}$$

We proceed by contradiction. Let t_0 be the first time for which (3.46) is not true. Then

$$t_0 = \inf\{t > 0 : |u(t) - qu(t-r)| > (1+q)\rho\}.$$

By continuity, one has

$$|u(t_0) - qu(t_0 - r)| = (1+q)\rho$$

and there exists a positive constant $\varepsilon > 0$ such that

$$|u(t_0) - qu(t_0 - r)| > (1+q)\rho \quad \text{for } t \in]t_0, t_0 + \varepsilon[.$$

Without loss of generality, we assume that $u(t) - qu(t-r) > 0$. Then

$$\frac{\mathrm{d}}{\mathrm{d}t}[u(t_0) - qu(t_0 - r)] \leq -(1+q)\rho + \int_{-r}^{0} |g(\theta)||x(t_0 + \theta)|\mathrm{d}\theta + |f|.$$

Since $|u(t) - qu(t-r)| \le (1+q)\rho$, for $t \le t_0$, then

$$\left|u(t)\right| \le (1+q)\rho + q\left|u(t-r)\right| \quad \text{for } t \in [-r, t_0].$$

Moreover, since $|\varphi| < \rho$, we can prove that

$$\left|u(t)\right| \le \frac{1+q}{1-q}\rho \quad \text{for } t \in [-r, t_0].$$

Then

$$\frac{\mathrm{d}}{\mathrm{d}t}[u(t_0) - qu(t_0 - r)] \le -(1+q)\rho + \left[\int_{-r}^{0} |g(\theta)| \mathrm{d}\theta \frac{1+q}{1-q}\rho + |f|\right].$$

Using condition (3.45), we obtain that

$$\frac{\mathrm{d}}{\mathrm{d}t}[u(t_0) - qu(t_0 - r)] \le -(1+q)\rho + ((1+q)\beta\rho + |f|)$$

and, by a simple computation,

$$\frac{\mathrm{d}}{\mathrm{d}t}[u(t_0) - qu(t_0 - r)] \le \beta - 1.$$

By continuity, there exists a positive ε_0 such that

$$\left|u(t) - qu(t-r)\right| < (1+q)\rho \quad \text{for } t \in]t_0, t_0 + \varepsilon_0[,$$

which gives a contradiction. We deduce that

$$\left|u(t) - qu(t-r)\right| \le (1+q)\rho \quad \text{for } t \ge 0.$$

We claim that

$$\left|u(t)\right| \le \frac{1+q}{1-q}\rho \quad \text{for } t \ge 0.$$

Let $t \in [0, r]$. Then

$$\left|u(t)\right| \le (1+q)\rho + q\rho \le (1+q)(1+q)\rho$$

and, for $t \in [r, 2r]$,

$$\left|u(t)\right| \le (1+q)(1+q+q^2)\rho.$$

We proceed in steps and have that, for $t \in [(n-1)r, nr]$,

$$|u(t)| \leq (1+q)(1+q+q^2+\cdots+q^n)\rho.$$

Consequently,

$$|u(t)| \leq (1+q)\rho \sum_{n\geq 0} q^n = \frac{1+q}{1-q}\rho \quad \text{for all } t \geq 0.$$

Thus equation (3.37) has a bounded solution u on $\mathbb{R}^+$ and, by Theorem 3.3.11, we deduce that equation (3.37) has an almost automorphic solution. □

3.4 Linear renewal equations

The main aim of this section is to prove some results concerning the existence and uniqueness of almost periodic (resp. almost automorphic) solutions for the following nonhomogeneous linear renewal equation with infinite delay:

$$x(t) = \int_{-\infty}^{t} K(t-s)x(s)\,\mathrm{d}s + f(t) \quad \text{for } t \in \mathbb{R}, \tag{3.47}$$

where $K : [0,+\infty[\to \mathbb{C}^{m\times m}$ is a measurable function satisfying the following assumption:

$$\|K\|_{1,\rho} := \int_0^{\infty} \|K(\tau)\|e^{\rho\tau}\,\mathrm{d}\tau < \infty, \quad \|K\|_{\infty,\rho} := \operatorname{ess\,sup}\{\|K(\tau)\|e^{\rho\tau}, \tau \geq 0\} < \infty, \tag{H1}$$

for some $\rho > 0$, and f is a Stepanov almost periodic (resp. Stepanov almost automorphic) function.

Renewal equations (REs) are a type of mathematical equations that do not involve derivatives and describe the present value of an unknown function in terms of its past values. This is the difference between delay differential equations and REs: for the former the rule gives the derivative of the function at the current time, while for the latter the rule specifies the value of the function itself. REs can be expressed as follows:

$$x(t) = F(t, x_t) \quad \text{for } t \in \mathbb{R},$$

where $F : \mathbb{R}\times X \to \mathbb{C}^m$ is a function that satisfies certain conditions, X is a linear normed space of mappings from $]-\infty, 0]$ into $\mathbb{C}^m$, and $x_t \in X$ is the history function, given as

$$x_t(\theta) = x(t+\theta) \quad \text{for } \theta \leq 0.$$

Renewal equations arise in the formulation of various phenomena in population dynamics, including birth processes, cannibalism, physiologically structured populations, epidemic models, and so on. In this context, we mention the work of Kermack and

McKendrick [187] and Lotka [208] on problems in mathematical biology. For instance, in a closed population, let $S(t)$ be the density of susceptible individuals, $F(t)$ the force of infection, and $A(a)$ transmission potential. Then, the force of infection can be described by the following RE:

$$F(t) = \int_0^{+\infty} S(t-\tau)F(t-\tau)A(\tau)d\tau \quad \text{for } t \geq 0.$$

The population growth structured by age can be described by the linear renewal equation:

$$B(t) = \int_0^{+\infty} B(t-a)\beta(a)F(\tau)da \quad \text{for } t \geq 0,$$

where $B(t)$ is the birth rate, $F(a)$ is the age dependent survival probability, and $\beta(a)$ is the age dependent fertility. In [47], C. Barril et al. developed a size-structured model to describe the dynamics of tree growth in the forest, associated with the following renewal equation:

$$b(t) = \int_0^{\infty} \beta\left(x_m + \int_0^{a} g\left(e^{-\mu(\tau-a)} \int_a^{\infty} e^{-\mu s} b(t-s)ds \right) d\tau \right) e^{-\mu a} b(t-a)da,$$

where $b(t)$ is the tree population birth rate at time t, $g(x)$ denotes the growth rate of an individual of height x, the positive parameter $\mu > 0$ represents the per capita death rate, and $\beta(x)$ is the per capita reproduction rate (depending only on height x). It is also assumed in [47] that all newborn individuals have the minimal height $x_m \geq 0$ and that mappings $\beta : [x_m, \infty) \to [0, \infty)$, $g : [0, \infty) \to (0, \infty)$ are continuous, β is increasing and g is decreasing. For more examples, see [27, 47, 82, 125, 187, 208, 229–231, 270].

Fundamental progress has been achieved in the treatment of REs using functional-analytic methods involving the semigroup theory. Diekmann and Gyllenberg [125] utilized perturbation theory for adjoint semigroups to demonstrate the principle of linearized stability for nonlinear renewal equations with infinite delay. In [178], the authors established a new variation of constants formula for equation (3.47) in the phase space, along with a decomposition of the phase space corresponding to the solution semigroup associated with equation (3.47) when $f = 0$. Building on this, they developed a fundamental reduction principle which plays a crucial role in the study of bounded solutions of equation (3.47). This approach allows for the study of asymptotic behaviors of solutions from a dynamical systems viewpoint (see [178, 214, 215]).

In this section, we employ the reduction principle obtained in [178], to prove that the existence of a bounded solution of equation (3.47) on $\mathbb{R}^+$ implies the existence of an almost periodic (resp. almost automorphic) solution. Furthermore, we examine the case when f is continuous purely almost periodic (resp. almost automorphic) in Stepanov case.

3.4.1 Variation of constants formula and spectral decomposition

Throughout the rest of this work, we consider $X = L^1_\rho(\mathbb{R}^-, \mathbb{C}^m)$, the phase space of equation (3.47), consists of equivalence classes of measurable functions $\varphi : \mathbb{R}^- \to \mathbb{C}^m$ such that

$$\int_{-\infty}^{0} |\varphi(\theta)| e^{\rho\theta} \mathrm{d}\theta < +\infty.$$

Then X is a Banach space equipped with the following norm:

$$\|\varphi\| = \int_{-\infty}^{0} |\varphi(\theta)|\, e^{\rho\theta}\, \mathrm{d}\theta \quad \text{for } \varphi \in X,$$

where $\rho > 0$ is the constant mentioned in the assumption **(H1)**. To equation (3.47) we associate the following initial value problem:

$$\begin{cases} x(t) = \int_{-\infty}^{0} K(-\theta)x_t(\theta)\mathrm{d}\theta + f(t) & \text{for } t \geq \sigma, \\ x_\sigma = \varphi, \end{cases} \tag{3.48}$$

where $\sigma \in \mathbb{R}$, $K : \mathbb{R}^+ \to \mathbb{C}^{m\times m}$ is a measurable function verifying condition **(H1)**, and $f : \mathbb{R} \to \mathbb{C}^m$ is a continuous function. For $t \geq \sigma$, $x_t : \mathbb{R}^- \to \mathbb{C}^m$ is the history function defined, for $\theta \leq 0$, by

$$x_t(\theta) = x(t + \theta).$$

Definition 3.4.1 ([178]). A function $x : \mathbb{R} \to \mathbb{C}^m$ is said to be a solution of the initial value problem (3.48) on $\mathbb{R}$ if x satisfies the following conditions:
(i) $x_\sigma = \varphi$ on $\mathbb{R}^-$,
(ii) $x \in L^1_{\mathrm{loc}}([\sigma, +\infty[; \mathbb{C}^m)$,
(iii) $x(t)$ satisfies equation (3.48) for $t \geq \sigma$.

The following result ensures the existence and uniqueness of the solution of equation (3.48).

Proposition 3.4.2 ([178]). *Under the above conditions, equation* (3.48) *has a unique solution defined for $t \geq \sigma$.*

For $\sigma \in \mathbb{R}$ and $\varphi \in X$, we denote by $x(\cdot, \sigma, \varphi, f)$ the unique solution of equation (3.48) on $[\sigma, \infty[$. Observe that $x(\cdot, \sigma, \varphi, f)$ is continuous on $]\sigma, +\infty[$ and $\lim_{t\to\sigma^+} x(t, \sigma, \varphi, f)$ exists. Consequently, the function $t \mapsto x_t(\cdot, \sigma, \varphi, f)$ is also continuous on $]\sigma, +\infty[$, with values in X. In addition, if $\varphi(\theta) = \psi(\theta)$ a. e. $\theta \in \mathbb{R}^-$, for some $\varphi, \psi \in X$, then $x(\cdot, \sigma, \varphi, f) = x(\cdot, \sigma, \psi, f)$. We refer to [178] for the basic properties of equation (3.48).

For $t \geq 0$, we define operator $T(t) : X \to X$ by

$$T(t)\varphi = x_t(\cdot, 0, \varphi, 0),$$

where $x_t(\cdot, 0, \varphi, 0)$ is the unique solution of equation (3.48) with $\sigma = 0$ and $f = 0$.

Theorem 3.4.3 ([178]). *The family $(T(t))_{t\geq 0}$ is a linear strongly continuous semigroup on X. Moreover, the operator $(\mathcal{A}, D(\mathcal{A}))$ defined by*

$$\begin{cases} D(\mathcal{A}) = \{\varphi \in X \mid \varphi(\theta) = \tilde{\varphi}(\theta) \text{ a. e. } \theta \in \mathbb{R}^- \text{ for some } \tilde{\varphi} \in \tilde{X}\}, \\ \mathcal{A}\varphi = \frac{d\tilde{\varphi}}{d\theta}, \quad \varphi \in D(\mathcal{A}), \end{cases}$$

where

$$\tilde{X} = \left\{ \tilde{\varphi} \in X \,\middle|\, \tilde{\varphi} \text{ is locally absolutely continuous on } \mathbb{R}^-, \frac{d\tilde{\varphi}}{d\theta} \in X, \text{ and } \tilde{\varphi}(0) = \int_{-\infty}^{0} K(-\theta)\tilde{\varphi}(\theta)d\theta \right\},$$

is its infinitesimal generator.

For $n \geq 0$, we define $\Gamma^n : \mathbb{R}^- \to \mathbb{R}^+$ by

$$\Gamma^n(\theta) = \begin{cases} 2n(n\theta + 1), & -\frac{1}{n} \leq \theta \leq 0, \\ 0, & \theta < -\frac{1}{n}. \end{cases}$$

For each $n \in \mathbb{N}^*$,

$$\int_{-\infty}^{0} \Gamma^n(\theta)\, d\theta = 1.$$

Moreover, for all $v \in \mathbb{C}^m$, we have $\Gamma^n v \in X$ and

$$\|\Gamma^n v\| \leq |v|.$$

The following result provides a representation of the solutions of equation (3.48) in the phase space X in terms of $(T(t))_{t\geq 0}$ and f. It allows us to study the asymptotic behavior of the solutions from the perspective of an infinite-dimensional dynamical system, which plays a crucial role in this study.

Theorem 3.4.4 ([178]). *The unique solution $x(\cdot, \sigma, \varphi, f)$ of equation* (3.48) *on $\mathbb{R}$ satisfies the following variation of constants formula in X:*

$$x_t(\cdot,\sigma,\varphi,f) = T(t-\sigma)\varphi + \lim_{n\to\infty}\int_{\sigma}^{t} T(t-s)(\Gamma^n f(s))ds \quad \text{for } t \geq \sigma. \tag{VCF}$$

In addition, the above limit exists uniformly on $[\sigma, T]$ *for any* $T > \sigma$.

In the following result, we will give the relation between the solutions of equation (3.48) and a continuous function with values in X which satisfies (VCF). For that purpose, let us define

$$\overline{X} := \{\phi \in X : \phi \text{ is continuous on } [-\varepsilon_\phi, 0] \text{ for some } \varepsilon_\phi > 0\}$$

and

$$X_0 = \{\varphi \in X : \varphi = \phi \text{ a. e. on } \mathbb{R}^- \text{ for some } \phi \in \overline{X}\}.$$

The space X_0 is a normed space equipped with the following norm:

$$\|\varphi\|_{X_0} = \|\varphi\| + |\varphi[0]| \quad \text{for } \varphi \in X_0,$$

where $\varphi[0] = \phi(0)$ represents the value of φ at zero.

Theorem 3.4.5 ([178]). *If* $x(\cdot,\sigma,\varphi,f)$ *is a solution of equation* (3.48) *on* $\mathbb{R}$, *then the function* $\xi(t) = x_t(\cdot,\sigma,\varphi,f)$ *is continuous on* $\mathbb{R}$ *with values in* X_0 *and satisfies in* X *the following formula:*

$$\xi(t) = T(t-\sigma)\xi(\sigma) + \lim_{n\to\infty}\int_{\sigma}^{t} T(t-s)(\Gamma^n f(s))ds \quad \text{for } t \geq \sigma. \tag{3.49}$$

Conversely, if a continuous function $\xi : \mathbb{R} \to X$ *satisfies* (3.49), *then* $\xi \in C(\mathbb{R}, X_0)$ *and the function* $x(t,\sigma,\xi(\sigma),f) = \xi(t)[0]$ *is a solution of equation* (3.48) *on* $\mathbb{R}$.

Next, we summarize the spectral properties of $\mathcal{A}$. Before that, we denote by $\sigma(\mathcal{A})$ (respectively, $\sigma_p(\mathcal{A})$) the spectrum (respectively, point spectrum) of $\mathcal{A}$.

Theorem 3.4.6 ([178, Proposition 5 and Theorem 1]). *The following statements are true:*
1. $\{\lambda \in \sigma(A) : \mathrm{Re}(\lambda) > -\rho\} \subset \sigma_p(\mathcal{A})$.
2. *Let* $\lambda \in \mathbb{C}$ *be such that* $\mathrm{Re}(\lambda) > -\rho$. *Then,* $\lambda \in \sigma(A)$ *if and only if* $\det\Delta(\lambda) = 0$, *where* $\Delta(\lambda) = I_m - \int_{-\infty}^{0} K(-\theta)e^{\lambda\theta}d\theta$ *and* I_m *is the identity* $m \times m$ *matrix.*
3. $r_{\mathrm{ess}}(T(t)) \leq e^{-\rho t}$ *for all* $t > 0$.

By [270, Proposition 4.13] and statement (3) in the above theorem, we have

$$\omega_{\mathrm{ess}}(T) \leq -\rho < 0.$$

As a result of [136, Corollary 2.11, Chapter IV and Theorem 3.1, Chapter V], we have the following spectral decomposition result.

Theorem 3.4.7 ([178]). *The space X is decomposed as follows:*

$$X = S \oplus U,$$

such that

(i) *S and U are closed subspaces of X which are $T(\cdot)$-invariant,*

(ii) *$\dim U < \infty$, and the restriction of $(T(t))_{t\geq 0}$ to U, denoted by $(T^U(t))_{t\in\mathbb{R}}$, can be extended to a C_0-group,*

(iii) *$\sigma(\mathcal{A}|_U) = \{\lambda \in \sigma(\mathcal{A}) : \mathrm{Re}(\lambda) \geq 0\}$ is finite and $\sigma(\mathcal{A}|_{S\cap D(\mathcal{A})}) = \sigma(\mathcal{A}) \setminus \sigma(\mathcal{A}|_U)$,*

(iv) *there exist $C \geq 1$ and $a > 0$ such that*

$$\|T^S(t)\varphi\| \leq C e^{-a t}\|\varphi\| \quad \text{for } \varphi \in S \text{ and } t \geq 0,$$

where $(T^S(t))_{t\geq 0}$ is the restriction of $(T(t))_{t\geq 0}$ to S.

In addition, if $(T(t))_{t\geq 0}$ is hyperbolic, that is,

$$\sigma(\mathcal{A}) \cap i\mathbb{R} = \emptyset, \tag{H.P}$$

then

$$\|T^U(t)\varphi\| \leq C e^{a t}\|\varphi\| \quad \text{for } \varphi \in U \text{ and } t \leq 0.$$

Let $\Phi = \{\phi_1, \phi_2, \dots, \phi_d\}$ be the vector basis of U with $d := \dim U$. Then, there exist d-elements $(\psi_1, \psi_2, \dots, \psi_d)$ in X^* (the dual space of X) such that

$$\begin{cases} \langle \psi_i, \phi_j \rangle = \delta_{ij}, \\ \langle \psi_i, \phi \rangle = 0 \quad \text{for all } \phi \in \mathcal{S} \text{ and } i \in \{1, \dots, d\}, \end{cases} \tag{3.50}$$

where $\langle \cdot, \cdot \rangle$ denotes the duality pairing between X and X^* and

$$\delta_{ij} = \begin{cases} 1, & \text{if } i = j, \\ 0, & \text{if } i \neq j. \end{cases}$$

Letting $\Psi = \mathrm{col}\{\psi_1, \psi_2, \dots, \psi_d\}$, $\langle \Psi, \Phi \rangle$ is a $(d \times d)$-matrix, where the (i,j)-component is $\langle \psi_i, \phi_j \rangle$. Denote by Π^U (resp. Π^S) the projection operator from X into U (resp. S). For each $\phi \in X$, we have

$$\Pi^U \phi = \Phi \langle \Psi, \phi \rangle,$$

where $\langle\Psi,\phi\rangle := (\langle\psi_1,\phi\rangle,\dots,\langle\psi_d,\phi\rangle)$. Since $T^U(t)$ is a C_0-group on U, there exists a $(d\times d)$-matrix G such that

$$T^U(t)\Phi = \Phi e^{Gt}, \quad t \ge 0,$$

where the spectrum of the matrix G is identical with the set $\{\lambda \in \sigma(\mathcal{A}) : \mathrm{Re}(\lambda) \ge 0\}$.

Lemma 3.4.8 ([178, Lemma 2]). *Let $\{e_1,\dots,e_m\}$ be the canonical basis of $\mathbb{C}^m$. Then, there exits a $(d\times m)$-matrix H such that, for any $i \in \{1,\dots,m\}$,*

$$\lim_{n\to\infty} \langle\Psi,\Gamma^n e_i\rangle = He_i.$$

Let consider the following reduced ordinary differential equation:

$$z'(t) = Gz(t) + Hf(t) \quad \text{for } t \in \mathbb{R}. \tag{3.51}$$

The following theorem shows that the dynamics of bounded solutions of equation (3.47) can be described by the ordinary differential equation (3.51) posed in the finite-dimensional space U. This result pertains to the reduction principle.

Theorem 3.4.9 ([178, Proposition 6 and Theorem 8]). *Assume that $f : \mathbb{R} \to \mathbb{C}^m$ is continuous and bounded. Then, the following statements are true:*

1. *For any $t \in \mathbb{R}$, the limit*

$$Y(t) := \lim_{n\to+\infty} \int_{-\infty}^{t} T^S(t-s)\Pi^S(\Gamma^n f(s))ds$$

exits in X. Moreover, if $\eta : \mathbb{R} \to S$ is a continuous functions which satisfies the following formula:

$$\eta(t) = T^S(t-\sigma)\eta(\sigma) + \lim_{n\to+\infty} \int_{\sigma}^{t} T^S(t-s)\Pi^S(\Gamma^n f(s))ds \quad \textit{for } t \ge \sigma,$$

then $\eta(t) = Y(t)$ for all $t \in \mathbb{R}$.

2. *If x is a (resp. bounded) solution of equation* (3.47) *on $\mathbb{R}$, then the function $z(t) := \langle\Psi,x_t\rangle$ is a (resp. bounded) solution of equation* (3.51) *on $\mathbb{R}$.*
3. *If z is a (resp. bounded) solution of equation* (3.51) *on $\mathbb{R}$, then the function $\xi : \mathbb{R} \to X$ defined by*

$$\xi(t) = \Phi z(t) + \lim_{n\to\infty} \int_{-\infty}^{t} T^S(t-s)\Pi^S(\Gamma^n f(s))ds \quad \textit{for } t \in \mathbb{R}$$

is continuous from $\mathbb{R}$ to X_0 and the function $t \mapsto \xi(t)[0]$ is a (resp. bounded) solution of equation (3.47) *on $\mathbb{R}$.*

We get the same result in the above theorem by weakening the boundedness assumption of the forcing term f. Namely, we require that f is only bounded by mean values.

Theorem 3.4.10. *Assume that $f : \mathbb{R} \to \mathbb{C}^m$ is continuous and*

$$\sup_{t\in\mathbb{R}} \int_t^{t+1} |f(s)|ds < +\infty.$$

Then, the following statements are true:

1. *For any $t \in \mathbb{R}$, the limit $Y(t) := \lim_{n\to+\infty} \int_{-\infty}^t T^S(t-s)\Pi^S(\Gamma^n f(s))ds$ exists in X. Moreover, if $\eta : \mathbb{R} \to S$ is a continuous function satisfying the following formula:*

$$\eta(t) = T^S(t-\sigma)\eta(\sigma) + \lim_{n\to+\infty} \int_\sigma^t T^S(t-s)\Pi^S(\Gamma^n f(s))ds \quad \text{for } t \geq \sigma \tag{SVCF}$$

   and $\sup_{t\in\mathbb{R}} \|\eta(t)\| < +\infty$, then $\eta(t) = Y(t)$ for all $t \in \mathbb{R}$.
2. *If x is a solution of equation* (3.47) *on $\mathbb{R}$, then the function $z(t) := \langle \Psi, x_t \rangle$ is a solution of equation* (3.51) *on $\mathbb{R}$.*
3. *If z is a solution of equation* (3.51) *on $\mathbb{R}$, then the function $\xi : \mathbb{R} \to X$ defined by*

$$\xi(t) = \Phi z(t) + \lim_{n\to\infty} \int_{-\infty}^t T^S(t-s)\Pi^S(\Gamma^n f(s))ds \quad \text{for } t \in \mathbb{R}$$

   is continuous from $\mathbb{R}$ to X_0 and the function $t \mapsto \xi(t)[0]$ is a solution of equation (3.47) *on $\mathbb{R}$.*

Proof. (1) Let $t \in \mathbb{R}$ and n sufficiently large. Then,

$$\begin{aligned}\left\| \int_{-\infty}^t T^S(t-s)\Pi^S(\Gamma^n f(s))\,ds \right\| &\leq C\|\Pi^S\| \int_\infty^t e^{-a(t-s)}\|\Pi^S\||f(s)|ds \\ &\leq C\|\Pi^S\| \sum_{k=1}^\infty \int_{t-k}^{t-k+1} e^{-a(t-s)}|f(s)|ds \\ &\leq C\|\Pi^S\| \sum_{k=1}^\infty e^{-a(k-1)} \int_{t-k}^{t-k+1} |f(s)|ds \\ &\leq K,\end{aligned}$$

where

$$K = C\|\Pi^S\| \sup_{t\in\mathbb{R}} \left(\int_t^{t+1} |f(s)|\,ds \right) \frac{1}{1-e^{-a}}.$$

Let

$$H(n,s,t) = T^S(t-s)\Pi^S(\Gamma^n f(s)) \quad \text{for } s \le t \text{ and } n \in \mathbb{N}.$$

Then, for n and m large enough and $\sigma \le t$, we have

$$\begin{aligned}\left\| \int_{-\infty}^{t} H(n,s,t)\mathrm{d}s - \int_{-\infty}^{t} H(m,s,t)\mathrm{d}s \right\| &\le \left\| \int_{-\infty}^{\sigma} H(n,s,t)\mathrm{d}s \right\| + \left\| \int_{-\infty}^{\sigma} H(m,s,t)\mathrm{d}s \right\| \\ &\quad + \left\| \int_{\sigma}^{t} H(n,s,t)\mathrm{d}s - \int_{\sigma}^{t} H(m,s,t)\mathrm{d}s \right\| \\ &\le 2Ke^{-\alpha(t-\sigma)} + \left\| \int_{\sigma}^{t} H(n,s,t)\mathrm{d}s - \int_{\sigma}^{t} H(m,s,t)\mathrm{d}s \right\|.\end{aligned}$$

Since $\lim_{n\to\infty} \int_{\sigma}^{t} H(n,s,t)\mathrm{d}s$ exists in X,

$$\limsup_{n,m\to\infty} \left\| \int_{-\infty}^{t} H(n,s,t)\mathrm{d}s - \int_{-\infty}^{t} H(m,s,t)\mathrm{d}s \right\| \le 2Ke^{-\alpha(t-\sigma)}.$$

By letting $\sigma \to -\infty$, we obtain

$$\limsup_{n,m\to\infty} \left\| \int_{-\infty}^{t} H(n,s,t)\mathrm{d}s - \int_{-\infty}^{t} H(m,s,t)\mathrm{d}s \right\| = 0.$$

Thus, by the completeness of X, we deduce that the limit $\lim_{n\to\infty} \int_{-\infty}^{t} H(n,s,t)\mathrm{d}s$ exists in X.

Now, let $\eta : \mathbb{R} \to S$ be a bounded function which satisfies (SVCF). For $t \in \mathbb{R}$, we observe that

$$\begin{aligned}\|\eta(t) - Y(t)\| &= \Bigg\| T^S(t-\sigma)\eta(\sigma) + \lim_{n\to+\infty} \int_{\sigma}^{t} T^S(t-s)\Pi^S(\Gamma^n f(s))\mathrm{d}s \\ &\quad - \lim_{n\to+\infty} \int_{-\infty}^{t} T^S(t-s)\Pi^S(\Gamma^n f(s))\mathrm{d}s \Bigg\| \\ &\le Ce^{-\alpha(t-\sigma)} \sup_{t\in\mathbb{R}}\|\eta(t)\| + \left\| \lim_{n\to+\infty} \int_{-\infty}^{\sigma} T^S(t-s)\Pi^S(\Gamma^n f(s))\mathrm{d}s \right\| \\ &\le Ce^{-\alpha(t-\sigma)} \sup_{t\in\mathbb{R}}\|\eta(t)\| + \left\| Y(0) - \lim_{n\to+\infty} \int_{\sigma}^{0} T^S(t-s)\Pi^S(\Gamma^n f(s))\mathrm{d}s \right\|.\end{aligned}$$

By the first part of the proof, we assert, for any $t \in \mathbb{R}$, that

$$\lim_{\sigma \to -\infty} \lim_{n \to +\infty} \int_{\sigma}^{t} T^S(t-s)\Pi^S(\Gamma^n f(s))ds = Y(t).$$

Then,

$$\lim_{\sigma \to -\infty} \left\| Y(0) - \lim_{n \to +\infty} \int_{\sigma}^{0} T^S(t-s)\Pi^S(\Gamma^n f(s))ds \right\| = 0.$$

Consequently,

$$\|\eta(t) - Y(t)\| \le \lim_{\sigma \to -\infty} \left(Ce^{-a(t-\sigma)} \sup_{t \in \mathbb{R}} \|\eta(t)\| + \left\| Y(0) - \lim_{n \to +\infty} \int_{\sigma}^{0} T^S(t-s)\Pi^S(\Gamma^n f(s))ds \right\| \right) = 0.$$

Hence, for all $t \in \mathbb{R}$,

$$\eta(t) = Y(t).$$

Claims 2 and 3 can be deduced from [178, Theorem 7]. □

3.4.2 Almost automorphic and almost periodic solutions of equation (3.47)

In this section, we study the existence and uniqueness almost periodic (resp. almost automorphic) solution of equation (3.47) when f almost periodic (resp. almost automorphic) using reduction principal.

3.4.2.1 Almost periodic solutions of equation (3.47)

We start by the following lemma.

Lemma 3.4.11. *Assume that f is a continuous S^p-almost periodic for some $p \ge 1$. Then, the function $Y : \mathbb{R} \to X$ defined, for $t \in \mathbb{R}$, by*

$$Y(t) = \lim_{n \to +\infty} \int_{-\infty}^{t} T^S(t-s)(\Pi^S \Gamma^n f(s))ds$$

is almost periodic.

Proof. For $k \ge 1$, we define the following function:

$$Y_k(t) = \lim_{n \to +\infty} \int_{t-k}^{t-k+1} T^S(t-s)\Pi^S(\Gamma^n f(s))ds.$$

By the strongly continuity of $(T(t))_{t\geq 0}$, we can assert that, for all $k \geq 1$, Y_k is continuous on $\mathbb{R}$. Let $\varepsilon > 0$. Since f is S^p-almost periodic, it is S^1-almost periodic. Thus, there exists $l(\varepsilon) > 0$ such that for all $a \in \mathbb{R}$, there exists $\tau_\varepsilon \in [a, a + l(\varepsilon)]$ such that

$$\sup_{t\in\mathbb{R}} \int_t^{t+1} |f(s + \tau_\varepsilon) - f(s)| ds < \frac{\varepsilon}{C\|\Pi^S\|}.$$

Therefore, for $t \in \mathbb{R}$, we have

$$\begin{aligned}
\|Y_k(t + \tau_\varepsilon) - Y_k(t)\| &\leq \Bigg\| \lim_{n\to+\infty} \int_{t+\tau_\varepsilon-k}^{t+\tau_\varepsilon-k+1} T^S(t + \tau_\varepsilon - s)\Pi^S(\Gamma^n f(s)) ds \\
&\quad - \lim_{n\to+\infty} \int_{t-k}^{t-k+1} T^S(t - s)\Pi^S(\Gamma^n f(s)) ds \Bigg\| \\
&\leq \Bigg\| \lim_{n\to+\infty} \int_{t-k}^{t-k+1} T^S(t + \tau_\varepsilon - s)\Pi^S(\Gamma^n (f(s + \tau_\varepsilon) - f(s)) ds \Bigg\| \\
&\leq \int_{t-k}^{t-k+1} Ce^{-\alpha(t-s)} \|\Pi^S\| |f(s + \tau_\varepsilon) - f(s)| ds \\
&\leq C\|\Pi^S\| \sup_{t\in\mathbb{R}} \int_t^{t+1} |f(s + \tau_\varepsilon) - f(s)| ds \\
&\leq C\|\Pi^S\| \frac{\varepsilon}{C\|\Pi^S\|} \\
&= \varepsilon.
\end{aligned}$$

Then, we deduce that Y_k is almost periodic for all $k \geq 1$. On the other hand, we observe that

$$\begin{aligned}
\|Y_k(t)\| &\leq \Bigg\| \lim_{n\to+\infty} \int_{t-k}^{t-k+1} T^S(t - s)\Pi^S(\Gamma^n f(s)) ds \Bigg\| \\
&\leq C\|\Pi^S\| \int_{t-k}^{t-k+1} e^{-\alpha(t-s)} |f(s)| ds \\
&\leq C\|\Pi^S\| e^{-\alpha(k-1)} \int_{t-k}^{t-k+1} |f(s)| ds \\
&\leq C\|\Pi^S\| e^{\alpha} e^{-\alpha k} \sup_{t\in\mathbb{R}} \int_t^{t+1} |f(s)| ds.
\end{aligned}$$

Then, by Weierstrass M-test, we deduce that the series $\sum_{k=1}^{\infty} Y_k(t)$ is uniformly convergent on $\mathbb{R}$. For $t \in \mathbb{R}$ and $p \geq 1$, we have

$$\begin{aligned}
\left\| \sum_{k=1}^{p} Y_k(t) - Y(t) \right\| &\leq \left\| \lim_{n\to+\infty} \int_{t-p}^{t} T^S(t-s)\Pi^S(\Gamma^n f(s))\mathrm{d}s - \lim_{n\to+\infty} \int_{-\infty}^{t} T^S(t-s)\Pi^S(\Gamma^n f(s))\mathrm{d}s \right\| \\
&\leq \left\| \lim_{n\to+\infty} \int_{-\infty}^{t-p} T^S(t-s)\Pi^S(\Gamma^n f(s))\mathrm{d}s \right\| \\
&\leq C\|\Pi^S\| \int_{-\infty}^{t-p} e^{-a(t-s)}|f(s)|\mathrm{d}s \\
&\leq C\|\Pi^S\| \sum_{k=p+1}^{\infty} \int_{t-k}^{t-k+1} e^{-a(t-s)}|f(s)|\mathrm{d}s \\
&\leq C\|\Pi^S\| e^{a} \sup_{t\in\mathbb{R}} \left(\int_{t}^{t+1} |f(s)|\mathrm{d}s \right) \sum_{k=p+1}^{\infty} e^{-ak}.
\end{aligned}$$

Consequently, for any $t \in \mathbb{R}$,

$$\lim_{p\to\infty} \left\| \sum_{k=1}^{p} Y_k(t) - Y(t) \right\| = 0.$$

This means that $\sum_{k=1}^{\infty} Y_k(t)$ converges uniformly to $Y(t)$. Hence, Y is almost periodic. □

Theorem 3.4.12. *Assume that f is almost periodic. Then:*
1. *If equation* (3.47) *has a bounded solution on $\mathbb{R}^+$, then it has an almost periodic solution on $\mathbb{R}$.*
2. *If x is a bounded solution of equation* (3.47) *on $\mathbb{R}$, then the function $t \mapsto x_t$ is almost periodic, hence x is also almost periodic.*

Proof. (1) Let x be a bounded solution of equation (3.47) on $\mathbb{R}^+$. Then, for $t \geq 0$, we have

$$\begin{aligned}
\|x_t\| &\leq \int_{-\infty}^{0} |x(t+\theta)| e^{\rho\theta}\mathrm{d}\theta \\
&\leq \int_{-\infty}^{-t} |x(t+\theta)| e^{\rho\theta}\mathrm{d}\theta + \int_{-t}^{0} |x(t+\theta)| e^{\rho\theta}\mathrm{d}\theta \\
&\leq \int_{-\infty}^{0} |x_0(\theta)| e^{-t\rho} e^{\rho\theta}\mathrm{d}\theta + \frac{1}{\rho}(1 - e^{-\rho t}) \sup_{t\in\mathbb{R}^+} |x(t)| \\
&\leq \|x_0\| + \frac{1}{\rho} \sup_{t\in\mathbb{R}^+} |x(t)|.
\end{aligned}$$

Consequently, the function $t \mapsto x_t$ is also bounded on $\mathbb{R}^+$ with values in X. Therefore, the function $t \mapsto \langle \Psi, x_t \rangle$ is a bounded solution of equation (3.51) on $\mathbb{R}^+$ (see Theorem 3.4.9). Thanks to [152, Theorem 5.8, p. 86], we obtain that equation (3.51) has an almost periodic solution z. Using Lemma 3.4.11, we conclude that the function $\xi : \mathbb{R} \to X$ defined, for $t \in \mathbb{R}$, by

$$\xi(t) = \Phi z(t) + \lim_{n \to +\infty} \int_{-\infty}^{t} T^S(t-s)\Pi^S \Gamma^n f(s) ds$$

is almost periodic. By Theorem 3.4.9, the function $\tilde{x}(t) = \xi(t)[0]$ is a solution of equation (3.47) on $\mathbb{R}$ with $\tilde{x}_t = \xi(t)$ for all $t \in \mathbb{R}$. Then,

$$\tilde{x}(t) = \int_{-\infty}^{0} K(-\theta)\xi(t)(\theta) d\theta + f(t) \quad \text{for } t \in \mathbb{R}.$$

Hence, $\tilde{x}$ is an almost periodic solution of equation (3.47).

(2) Let x be a bounded solution of equation (3.47) on $\mathbb{R}$. Then, the function $t \mapsto x_t$ is bounded, since

$$\|x_t\| = \int_{-\infty}^{0} |x(t+\theta)| e^{\rho\theta} d\theta \leq \frac{1}{\rho} \sup_{t \in \mathbb{R}} |x(t)|.$$

Therefore, using Theorem 3.4.9, the function $z(t) := \langle \Psi, x_t \rangle$ is a bounded solution of equation (3.51) on $\mathbb{R}$. Once again by using [152, Theorem 5.8, p. 86], we deduce that $z(t)$ is almost periodic. By Theorem 3.4.9, we have that

$$x_t = \Phi\langle \Psi, x_t \rangle + \lim_{n \to +\infty} \int_{-\infty}^{t} T^S(t-s)\Pi^S \Gamma^n f(s) ds.$$

Form Lemma 3.4.11, we obtain that the functions x and $t \mapsto x_t$ are almost periodic, ending the proof. □

Theorem 3.4.13. *Assume that f is almost periodic and*

$$\sigma(\mathcal{A}) \cap i\mathbb{R} = \emptyset. \tag{3.52}$$

Then, equation (3.47) *has a unique almost periodic solution. In addition, if*

$$\{\lambda \in \sigma(\mathcal{A}) : \operatorname{Re}(\lambda) \geq 0\} = \emptyset,$$

then the unique almost periodic solution is globally attractive.

Proof. By condition (H.P), we can assert that

$$\sigma(G) = \{\lambda \in \sigma(\mathcal{A}) : \mathrm{Re}(\lambda) > 0\}.$$

Then, equation (3.51) has a unique bounded solution on $\mathbb{R}$ given by

$$z(t) = -\int_t^{+\infty} e^{(t-s)G} Hf(s)\mathrm{d}s \quad \text{for } t \in \mathbb{R},$$

which is almost periodic (see [152, Chapter 5, Section 11, pp. 92–93]). Consequently, in the light of Theorem 3.4.9 and Lemma 3.4.11, the function $\xi : \mathbb{R} \to X$ defined by

$$\xi(t) = \Phi z(t) + \lim_{n\to+\infty} \int_{-\infty}^{t} T^S(t-s)\Pi^S \Gamma^n f(s)\mathrm{d}s \quad \text{for } t \in \mathbb{R},$$

is almost periodic and $x(t) := \xi(t)[0]$ is an almost periodic solution of equation (3.47). Let $\tilde{x}$ be another almost periodic solution of equation (3.47). Then, the function $\tilde{z}(t) := \langle \Psi, \tilde{x}_t \rangle$ is a bounded solution of (3.51) on $\mathbb{R}$. By the uniqueness of z, we deduce that $\tilde{z} = z$. Thus, $\tilde{x}_t = \xi(t)$ and then we conclude the uniqueness of x.

Now, we suppose that

$$\{\lambda \in \sigma(\mathcal{A}) : \mathrm{Re}(\lambda) \geq 0\} = \emptyset.$$

Then, $U = \{0_X\}$ and $X = S$. By the first part of this theorem, we have that equation (3.47) has a unique almost periodic solution denoted by x. Let $\tilde{x}$ be any solution of equation (3.47) on $\mathbb{R}$. Then, by Theorem 3.4.4, for any $t \geq 0$,

$$\|x_t - \tilde{x}_t\| = \|T(t)(x_0 - \tilde{x}_0)\| \leq Ce^{-at}\|x_0 - \tilde{x}_0\|.$$

Thus,

$$\lim_{t\to+\infty} \|x_t - \tilde{x}_t\| = 0.$$

On the other hand, for $t \geq 0$,

$$\begin{aligned} |x(t) - \tilde{x}(t)| &= \left| \int_{-\infty}^{0} K(-\theta)(x(t+\theta) - \tilde{x}(t+\theta))\mathrm{d}\theta \right| \\ &\leq \|K\|_{\infty,p}\|x_t - \tilde{x}_t\|. \end{aligned}$$

Hence,

$$\lim_{t\to+\infty} |x(t) - \tilde{x}(t)| = 0,$$

which completes the proof. □

3.4.2.2 Almost automorphic solution of equation (3.47)

We begin with the following lemma.

Lemma 3.4.14. *Assume that f is a continuous S^p-almost automorphic for some $p \geq 1$. Then, the function $Y : \mathbb{R} \to X$ given by*

$$Y(t) = \lim_{n\to+\infty} \int_{-\infty}^{t} T^S(t-s)(\Pi^S\Gamma^n f(s))ds \quad \text{for } t \in \mathbb{R}$$

is compact almost automorphic.

Proof. For $k \geq 1$, we define the following function:

$$\begin{aligned} Y_k(t) &:= \lim_{n\to\infty} \int_{t-k}^{t-k+1} T^S(t-s)\Pi^S(\Gamma^n f(s))ds \\ &= \lim_{n\to\infty} \int_0^1 T^S(k-s)\Pi^S(\Gamma^n f(t-k+s))ds \quad \text{for } t \in \mathbb{R}. \end{aligned}$$

Obviously, the function $Y_k(\cdot)$ is continuous. Let $(s_j')_{j\in\mathbb{N}}$ be any sequence in $\mathbb{R}$. Since f is S^p-almost automorphic, it is S^1-almost automorphic. Therefore, there exists a subsequence $(s_j)_{j\in\mathbb{N}}$ of $(s_j')_{j\in\mathbb{N}}$ such that for any $t \in \mathbb{R}$,

$$\lim_{m\to+\infty} \lim_{j\to+\infty} \int_0^1 |f(t+s_j-s_m+s) - f(t+s)|ds = 0.$$

Then, for $t \in \mathbb{R}$, one has

$$\begin{aligned} \|Y_k(t+s_j-s_m) - Y_k(t)\| &\leq \left\| \lim_{n\to\infty} \int_0^1 T^S(k-s)\Pi^S(\Gamma^n f(t+s_j-s_m-k+s) - f(t-k+s))ds \right\| \\ &\leq C\|\Pi^S\| \int_0^1 |f(t+s_j-s_m-k+s) - f(t-k+s)|ds. \end{aligned}$$

Thus,

$$\lim_{m\to+\infty} \lim_{j\to+\infty} \|Y_k(t+s_j-s_m) - Y_k(t)\| = 0.$$

On the other hand, by using similar arguments as in the proof of Lemma 3.4.11, we can assert that the series $\sum_{k=1}^{\infty} Y_k(\cdot)$ converges uniformly to $Y(\cdot)$. Hence, $Y(\cdot)$ is almost automorphic.

Now, we claim that Y is uniformly continuous. Let $(t'_j)_{j\in\mathbb{N}}$ and $(s'_j)_{j\in\mathbb{N}} \subset \mathbb{R}$ be any real sequences such that

$$\lim_{j\to+\infty} |t'_j - s'_j| = 0. \tag{3.53}$$

Then, if $t'_j \ge s'_j$ (see Theorem 3.4.10), we have

$$\begin{aligned}
\|Y(t'_j) - Y(s'_j)\| &= \left\| T^S(t'_j - s'_j)Y(s'_j) - Y(s'_j) + \lim_{n\to+\infty} \int_{s'_j}^{t'_j} T^S(t'_j - \sigma)\Pi^S(\Gamma^n f(\sigma))d\sigma \right\| \\
&\le \sup_{\phi\in\overline{Q}} \|T^S(t'_j - s'_j)\phi - \phi\| + \int_0^{t'_j - s'_j} Ce^{-a(t'_j - s'_j - \sigma)}\|\Pi^S\| |f(s'_j + \sigma)| d\sigma \\
&\le \sup_{\phi\in\overline{Q}} \|T^S(t'_j - s'_j)\phi - \phi\| + C\|\Pi^S\| \int_0^{t'_j - s'_j} |f(s'_j + \sigma)| d\sigma,
\end{aligned}$$

where $Q := \{Y(t) : t \in \mathbb{R}\}$. Since f is S^1-almost automorphic, there exists a subsequence $(s_j)_{j\in\mathbb{N}}$ of $(s'_j)_{j\in\mathbb{N}}$ and a function $\tilde{f} \in L^1_{\mathrm{loc}}(\mathbb{R}; \mathbb{C}^m)$ such that, for any $t \in \mathbb{R}$,

$$\lim_{j\to+\infty} \int_0^1 |f(t + s_j + \sigma) - \tilde{f}(t + \sigma)| d\sigma = 0$$

and

$$\lim_{j\to+\infty} \int_0^1 |\tilde{f}(t - s_j + \sigma) - f(t + \sigma)| d\sigma = 0.$$

In particular,

$$\lim_{j\to+\infty} \int_0^1 |f(s_j + \sigma) - \tilde{f}(\sigma)| d\sigma = 0. \tag{2}$$

Therefore,

$$\begin{aligned}
\|Y(t_j) - Y(s_j)\| \le{}& \sup_{\phi\in\overline{Q}} \|T^S(t_j - s_j)\phi - \phi\| + C\|\Pi^S\| \int_0^{t_j - s_j} |f(s_j + \sigma) - \tilde{f}(\sigma)| d\sigma \\
&+ C\|\Pi^S\| \int_0^{t_j - s_j} |\tilde{f}(\sigma)| d\sigma.
\end{aligned}$$

Consequently, for all $n \in \mathbb{N}$ large enough such that $|t_n - s_n| \le 1$, one has

$$\|Y(t_j) - Y(s_j)\| \le \sup_{\phi \in \overline{\mathcal{Q}}} \|T^S(|t_n - s_n|)\phi - \phi\| + C\|\Pi^S\| \int_0^1 |f(s_j + \sigma) - \tilde{f}(\sigma)| d\sigma$$
$$+ C\|\Pi^S\| \int_0^{|t_n - s_n|} |\tilde{f}(\sigma)| d\sigma.$$

Since Y is almost automorphic, $\overline{\mathcal{Q}}$ is compact in S. Therefore,

$$\lim_{n \to +\infty} \sup_{\phi \in \overline{\mathcal{Q}}} \|T^S(|t_n - s_n|)\phi - \phi\| = 0.$$

By using (3.53) and (2), we deduce

$$\lim_{n \to +\infty} \|Y(t) - Y(s)\| = 0.$$

Hence, Y is uniformly continuous, which concludes the proof. □

Theorem 3.4.15. *Suppose that f is almost automorphic. Then, the following assertions hold:*

(i) *The existence of a bounded solution on $\mathbb{R}^+$ of equation* (3.47) *implies the existence of an almost automorphic solution.*

(ii) *If x is a bounded solution on $\mathbb{R}$ of equation* (3.47)*, then the function $t \mapsto x_t$ is compact almost automorphic and x is almost automorphic.*

(iii) *If $\sigma(\mathcal{A}) \cap i\mathbb{R} = \emptyset$, then equation* (3.47) *has a unique almost automorphic solution.*

(iv) *If $\{\lambda \in \sigma(\mathcal{A}) : \mathrm{Re}(\lambda) \ge 0\} = \emptyset$, then equation* (3.47) *has a unique almost automorphic solution which is globally attractive.*

(v) *If, in addition, f is compact almost automorphic, then every almost automorphic solution of equation* (3.47) *is also compact almost automorphic.*

Proof. (i) Suppose that x is a bounded solution on $\mathbb{R}^+$ of equation (3.47). Then, the function $z_0(t) := \langle \Phi, x_t \rangle$ is a bounded solution on $\mathbb{R}^+$ of equation (3.51). According to [204, Theorem 3.1 and Lemma 3.3], equation (3.51) has an almost automorphic solution z. By Lemma 3.4.14, the function $\xi : \mathbb{R} \to X$ given, for $t \in \mathbb{R}$, by

$$\xi(t) = \Phi z(t) + + \lim_{n \to +\infty} \int_{-\infty}^{t} T^S(t - s)\Pi^S \Gamma^n f(s) ds$$

is almost automorphic. Consequently, $\tilde{x}(t) := \xi(t)[0]$ is an almost automorphic solution of equation (3.47).

(ii) Let x be a bounded solution of equation (3.47) on $\mathbb{R}$. For any $t \in \mathbb{R}$, we have

$$x_t = \Phi z(t) + \lim_{n\to\infty} \int_{-\infty}^{t} T^S(t-s)\Pi^S \Gamma^n f(s)\mathrm{d}s,$$

where $z(t) := \langle \Psi, x_t \rangle$. Obviously, z is a bounded solution of equation (3.51) on $\mathbb{R}$, which is almost automorphic (see [204, Theorem 3.1]). Due to [138, Proposition 34], z is uniformly continuous and, consequently, compact almost automorphic. As a result, the map $t \mapsto x_t$ is also compact almost automorphic. Hence, x is almost automorphic.

(iii) Assume that $\{\lambda \in \sigma(\mathcal{A}) : \mathrm{Re}(\lambda) \geq 0\} = \emptyset$. Then,

$$\sigma(G) \subset \{\lambda \in \mathbb{C} : \mathrm{Re}(\lambda) > 0\}.$$

Consequently, as shown in [152, Chapter 5, Section 11, pp. 92–93], equation (3.51) has a unique bounded solution z on $\mathbb{R}$, defined by

$$z(t) = -\int_{t}^{+\infty} e^{(t-s)G} Hf(s)\mathrm{d}s \quad \text{for } t \in \mathbb{R}.$$

According to [204, Theorem 3.1], z is almost automorphic. Therefore, the function $x(t) = \xi(t)[0]$, where

$$\xi(t) = \Phi z(t) + \lim_{n\to+\infty} \int_{-\infty}^{t} T^S(t-s)\Pi^S \Gamma^n f(s)\mathrm{d}s \quad \text{for } t \in \mathbb{R},$$

is an almost automorphic solution of equation (3.47). The uniqueness of x can be shown in a similar manner as in the proof of Theorem 3.4.13.

(iv) Based on the previous statement, we can conclude that the equation has a unique almost automorphic solution x. Using a similar approach as that employed in proving the second part of Theorem 3.4.13, we can assert that x is globally attractive.

(v) Let x be an almost automorphic solution of equation (3.47). As a result of (ii), the map $t \mapsto x_t$ is compact almost automorphic. Since f is compact almost automorphic and, for $t \in \mathbb{R}$,

$$x(t) = \int_{-\infty}^{0} K(-\theta)x_t(\theta)\mathrm{d}\theta + f(t),$$

then also x is compact almost automorphic. □

3.4.3 Stepanov version of Massera- and Bohr–Neugebauer-type results

In this section, we use reduction principal to investigate the nature of mean value bounded solutions of equation (3.47).

3.4.3.1 Almost periodic case

Theorem 3.4.16. *Assume that f is a continuous S^p-almost periodic for some $p \geq 1$. Then:*

1. *If equation (3.47) has a solution $x(\cdot)$ on $\mathbb{R}^+$ such that*
$$\sup_{t\in\mathbb{R}^+} \int_t^{t+1} |x(s)| ds < \infty,$$
then it has an S^p-almost periodic solution on $\mathbb{R}$.
2. *If $x(\cdot)$ is a solution of equation (3.47) on $\mathbb{R}$ such that*
$$\sup_{t\in\mathbb{R}} \int_t^{t+1} |x(s)| ds < \infty,$$
then the function $t \mapsto x_t$ is almost periodic and $x(\cdot)$ is S^p-almost periodic.
3. *If equation (3.47) has an almost periodic in Bohr sense solution, then f is Bohr almost periodic.*

Proof. First of all, we notice that the function $t \mapsto Hf(t)$ is also S^p-almost periodic.

(1) Let x be a solution of equation (3.47) on $\mathbb{R}^+$ with
$$\sup_{t\in\mathbb{R}^+} \int_t^{t+1} |x(s)| ds < \infty.$$
Then, for $t \geq 0$, we have
$$\begin{aligned}
\|x_t\| &= \int_{-\infty}^{0} |x(t+\theta)| e^{\rho\theta} d\theta \\
&= \int_{-\infty}^{-t} |x(t+\theta)| e^{\rho\theta} d\theta + \int_{-t}^{0} |x(t+\theta)| e^{\rho\theta} d\theta \\
&\leq \int_{-\infty}^{0} |x_0(\theta)| e^{-t\rho} e^{\rho\theta} d\theta + \int_0^t e^{-\rho(t-s)} |x(s)| ds \\
&\leq \|x_0\| + e^{-\rho[t]} \int_0^{[t]+1} e^{\rho s} |x(s)| ds \\
&\leq \|x_0\| + e^{-\rho[t]} \sum_{k=0}^{[t]} \int_k^{k+1} e^{\rho s} |x(s)| ds \\
&\leq \|x_0\| + \left(\sup_{t\in\mathbb{R}^+} \int_t^{t+1} |x(s)| ds \right) e^{-\rho[t]} \sum_{k=0}^{[t]} e^{\rho(k+1)}
\end{aligned}$$

$$\leq \|x_0\| + \left(\sup_{t \in \mathbb{R}^+} \int_t^{t+1} |x(s)| \mathrm{d}s \right) e^{\rho} e^{-\rho[t]} \left(\frac{1 - e^{\rho([t]+1)}}{1 - e^{\rho}} \right)$$

$$\leq \|x_0\| + \left(\sup_{t \in \mathbb{R}^+} \int_t^{t+1} |x(s)| \mathrm{d}s \right) e^{\rho} \left(\frac{e^{-\rho[t]} - e^{\rho}}{1 - e^{\rho}} \right).$$

Consequently, the function $t \mapsto x_t$ is bounded on $\mathbb{R}^+$ with values in X. Therefore, the function $t \mapsto \langle \Psi, x_t \rangle$ is a bounded solution of equation (3.51) on $\mathbb{R}^+$ (see Theorem 3.4.9). Thanks to [26, Corollary 6.4], we obtain that equation (3.51) has an almost periodic solution z. Using Lemma 3.4.11, we conclude that the function $\xi : \mathbb{R} \to X$ defined, for $t \in \mathbb{R}$, by

$$\xi(t) = \Phi z(t) + \lim_{n \to +\infty} \int_{-\infty}^{t} T^S(t-s) \Pi^S \Gamma^n f(s) \mathrm{d}s$$

is almost periodic. By Theorem 3.4.9, the function $\tilde{x}(t) = \xi(t)[0]$ is a solution of equation (3.47) on $\mathbb{R}$ with $\tilde{x}_t = \xi(t)$ for all $t \in \mathbb{R}$. Then,

$$\tilde{x}(t) = \int_{-\infty}^{0} K(-\theta) \xi(t)(\theta) \mathrm{d}\theta + f(t) \quad \text{for } t \in \mathbb{R}.$$

Hence, $\tilde{x}$ is an S^p-almost periodic solution of equation (3.47).

(2) Let x be a solution of equation (3.47) on $\mathbb{R}$ such that

$$\sup_{t \in \mathbb{R}} \int_t^{t+1} |x(s)| \mathrm{d}s < \infty.$$

Then, the function $t \mapsto x_t$ is bounded, since

$$\begin{aligned}
\|x_t\| &= \int_{-\infty}^{0} |x(t+\theta)| e^{\rho\theta} \mathrm{d}\theta \\
&= \int_{-\infty}^{t} e^{-\rho(t-\theta)} |x(\theta)| \mathrm{d}\theta \\
&\leq \sum_{k=1}^{\infty} \int_{t-k}^{t-k+1} e^{-\rho(t-\theta)} |x(\theta)| \mathrm{d}\theta \\
&\leq \sum_{k=1}^{\infty} e^{-\rho(k-1)} \int_{t-k}^{t-k+1} |x(\theta)| \mathrm{d}\theta \\
&\leq \frac{e^{\rho}}{e^{\rho} - 1} \sup_{t \in \mathbb{R}} \int_t^{t+1} |x(s)| \mathrm{d}s.
\end{aligned}$$

Therefore, using Theorem 3.4.9, the function $z(t) := \langle \Psi, x_t \rangle$ is a bounded solution of equation (3.51) on $\mathbb{R}$. By [26, Theorem 6.2], we deduce that $z(t)$ is almost periodic. Since $t \mapsto \Pi^S x_t$ is also bounded, then by Theorem 3.4.9(1), for all $t \in \mathbb{R}$,

$$\Pi^S x_t = \lim_{n\to+\infty} \int_{-\infty}^{t} T^S(t-s)\Pi^S \Gamma^n f(s)\mathrm{d}s.$$

Thus, $t \mapsto \Pi^S x_t$ is almost periodic (see Lemma 3.4.11). Using Theorem 3.4.9(3), we have

$$x_t = \Phi\langle \Psi, x_t \rangle + \lim_{n\to+\infty} \int_{-\infty}^{t} T^S(t-s)\Pi^S \Gamma^n f(s)\mathrm{d}s.$$

It follows that $t \mapsto x_t$ is almost periodic. As, for any $t \in \mathbb{R}$,

$$x(t) = \int_{-\infty}^{0} K(\theta)x_t(\theta)\mathrm{d}\theta + f(t),$$

this $x(\cdot)$ is S^p-almost periodic.

(3) Assume that equation (3.47) has an almost periodic solution $x(\cdot)$ in Bohr sense. Then, by (2), the function $t \mapsto x_t$ is also almost periodic. Consequently, the function $t \mapsto \int_{-\infty}^{0} K(\theta)x_t(\theta)\mathrm{d}\theta$ is almost periodic. Since, for $t \in \mathbb{R}$,

$$f(t) = x(t) - \int_{-\infty}^{0} K(\theta)x_t(\theta)\mathrm{d}\theta,$$

this f is Bohr almost periodic, ending the proof. □

Now, we will give a sufficient condition to existence and uniqueness of a Stepanov almost periodic solution.

Theorem 3.4.17. *Let f be a continuous S^p-almost periodic for some $p \geq 1$ and assume that $(T(t))_{t\geq 0}$ is hyperbolic. Then, equation (3.47) has a unique S^p-almost periodic solution $x(\cdot)$. Moreover, for any $t \in \mathbb{R}$,*

$$x_t = \lim_{n\to+\infty} \int_{-\infty}^{t} T^S(t-s)\Pi^S \Gamma^n f(s)\mathrm{d}s - \lim_{n\to+\infty} \int_{t}^{+\infty} T^U(t-s)\Pi^U \Gamma^n f(s)\mathrm{d}s. \tag{3.54}$$

In addition, if $\{\lambda \in \sigma(\mathcal{A}) : \mathrm{Re}(\lambda) \geq 0\} = \emptyset$, then the unique S^p-almost periodic $x(\cdot)$ is globally attractive. Namely, for any solution $\tilde{x}(\cdot)$ of equation (3.47) on $\mathbb{R}^+$, we have

$$\lim_{t\to+\infty} |\tilde{x}(t) - x(t)| = 0.$$

Proof. Under condition (H.P), we can affirm that

$$\sigma(G) = \{\lambda \in \sigma(\mathcal{A}) : \mathrm{Re}(\lambda) > 0\}.$$

Then, equation (3.51) has a unique bounded solution on $\mathbb{R}$ given by

$$z(t) = -\int_t^{+\infty} e^{(t-s)G} Hf(s)\mathrm{d}s \quad \text{for } t \in \mathbb{R},$$

which is almost periodic (see [211, Lemma 3.3]). Consequently, in the light of Theorem 3.4.10 and Lemma 3.4.11, the function $\xi : \mathbb{R} \to X$ defined by

$$\xi(t) = \Phi z(t) + \lim_{n\to+\infty} \int_{-\infty}^{t} T^S(t-s)\Pi^S \Gamma^n f(s)\mathrm{d}s \quad \text{for } t \in \mathbb{R}$$

is almost periodic and $x(t) := \xi(t)[0]$ is an S^p-almost periodic solution of equation (3.47). On the other hand, for any $t \in \mathbb{R}$, we have

$$\begin{aligned} \Phi z(t) &= -\Phi \int_t^{+\infty} e^{(t-s)G} Hf(s)\mathrm{d}s \\ &= -\lim_{n\to+\infty} \int_t^{+\infty} \Phi e^{(t-s)G} \langle \Psi, \Gamma^n f(s)\rangle \mathrm{d}s \\ &= -\lim_{n\to+\infty} \int_t^{+\infty} T^U(t-s)\Phi \langle \Psi, \Gamma^n f(s)\rangle \mathrm{d}s \\ &= -\lim_{n\to+\infty} \int_t^{+\infty} T^U(t-s)\Pi^U \Gamma^n f(s)\mathrm{d}s. \end{aligned}$$

Therefore, for $t \in \mathbb{R}$,

$$x_t = \xi(t) = \lim_{n\to+\infty} \int_{-\infty}^{t} T^S(t-s)\Pi^S\Gamma^n f(s)\mathrm{d}s - \lim_{n\to+\infty} \int_t^{+\infty} T^U(t-s)\Pi^U\Gamma^n f(s)\mathrm{d}s.$$

Let $\tilde{x}$ be another S^p-almost periodic solution of equation (3.47). Then,

$$\sup_{t\in\mathbb{R}} \int_t^{t+1} |\tilde{x}(s)|\mathrm{d}s < \infty.$$

Therefore, by Theorem 3.4.16, the function $t \mapsto \tilde{x}_t$ is almost periodic and hence $\tilde{z}(t) := \langle \Psi, \tilde{x}_t\rangle$ is an almost periodic solution of (3.51) on $\mathbb{R}$. By the uniqueness of z, we deduce that $\tilde{z} = z$. Thus, $\tilde{x}_t = \xi(t)$ and then we conclude the uniqueness of x.

Let us assume that $\{\lambda \in \sigma(\mathcal{A}) : \mathrm{Re}(\lambda) \geq 0\} = \emptyset$. Consequently, we have $U = \{0_X\}$ and $X = S$. As per the initial part of this theorem, it follows that equation (3.47) has a unique S^p-almost periodic solution $x(\cdot)$. Let $\tilde{x}(\cdot)$ be an arbitrary solution of equation (3.47) on $\mathbb{R}^+$. According to Theorem 3.4.4, for any $t \geq 0$, we have

$$\|x_t - \tilde{x}_t\| = \|T(t)(x_0 - \tilde{x}_0)\| \leq Ce^{-at}\|x_0 - \tilde{x}_0\|.$$

Therefore,

$$\lim_{t\to+\infty} \|x_t - \tilde{x}_t\| = 0.$$

Moreover, for $t \geq 0$,

$$\begin{aligned} |x(t) - \tilde{x}(t)| &= \left|\int_{-\infty}^{0} K(-\theta)(x(t+\theta) - \tilde{x}(t+\theta))\mathrm{d}\theta\right| \\ &\leq \|K\|_{\infty,p}\|x_t - \tilde{x}_t\|. \end{aligned}$$

Consequently,

$$\lim_{t\to+\infty} |x(t) - \tilde{x}(t)| = 0,$$

completing the proof. □

Remark 3.4.18. Assuming that f is purely Stepanov almost periodic, according to the claim (v) of the above theorem, equation (3.47) will never have a Bohr almost periodic solution.

3.4.3.2 Almost automorphic case

Now we focus our attention on the almost automorphic case. We have then the following extension of Massera- and Bohr–Neugebauer-type results.

Theorem 3.4.19. *Assume that f is a continuous S^p-almost automorphic for some $p \geq 1$. Then, the following assertions hold:*

(i) *The existence of a solution $x(\cdot)$ on $\mathbb{R}^+$ of equation* (3.47) *such that*

$$\sup_{t\in\mathbb{R}^+} \int_{t}^{t+1} |x(s)|\mathrm{d}s < \infty$$

implies the existence of an S^p-almost automorphic solution.

(ii) *Let x be a solution on $\mathbb{R}$ of equation* (3.47) *such that*

$$\sup_{t\in\mathbb{R}} \int_{t}^{t+1} |x(s)|\mathrm{d}s < \infty.$$

Then the function $t \mapsto x_t$ *is compact almost automorphic and* x *is* S^p*-almost automorphic.*

(iii) *If equation* (3.47) *has an almost automorphic (resp. compact automorphic) solution, then* f *must be almost automorphic (resp. compact almost automorphic).*

Proof. Obviously, we have that the function $t \mapsto Hf(t)$ is also S^p-almost automorphic.

(i) Suppose x is a solution of equation (3.47) on $\mathbb{R}^+$ such that

$$\sup_{t\in\mathbb{R}^+} \int_t^{t+1} |x(s)| ds < \infty.$$

By employing a similar argument to that used in establishing (1) of Theorem 3.4.16, we can affirm that the function $t \mapsto x_t$ is bounded on $\mathbb{R}^+$. Then the function $z_0(t) := \langle \Phi, x_t \rangle$ is a bounded solution of equation (3.51) on $\mathbb{R}^+$. According to [138, Corollary 35], equation (3.51) has a compact almost automorphic solution z. By Lemma 3.4.14, the function $\xi : \mathbb{R} \to X$ given, for $t \in \mathbb{R}$, by

$$\xi(t) = \Phi z(t) + \lim_{n\to+\infty} \int_{-\infty}^{t} T^S(t-s)\Pi^S \Gamma^n f(s) ds$$

is compact almost automorphic. Consequently, $\tilde{x}(t) := \xi(t)[0]$ is an S^p-almost automorphic solution of equation (3.47).

(ii) Let x be a solution of equation (3.47) on $\mathbb{R}$ with

$$\sup_{t\in\mathbb{R}} \int_t^{t+1} |x(s)| ds < \infty.$$

As in the proof of (2) of Theorem 3.4.16, we can show that the function $t \mapsto x_t$ is bounded. Obviously, $z(t) := \langle \Psi, x_t \rangle$ is a bounded solution of equation (3.51) on $\mathbb{R}$, which is almost automorphic (see [57, Theorem 4.14]). Due to [138, Proposition 34], z is uniformly continuous and, consequently, compact almost automorphic. On the other hand, for any $t \in \mathbb{R}$, we have

$$x_t = \Phi z(t) + \lim_{n\to\infty} \int_{-\infty}^{t} T^S(t-s)\Pi^S \Gamma^n f(s) ds.$$

As a result, the map $t \mapsto x_t$ is also compact almost automorphic. Hence, x is S^p-almost automorphic.

(iii) Assume that equation (3.47) has an almost automorphic (resp. compact almost automorphic) solution $x(\cdot)$. Then, by (ii), the function $t \mapsto x_t$ is compact almost automorphic. Consequently, the function $t \mapsto \int_{-\infty}^{0} K(\theta)x_t(\theta)d\theta$ is compact almost automorphic. Since, for $t \in \mathbb{R}$,

$$f(t) = x(t) - \int_{-\infty}^{0} K(\theta)x_t(\theta)d\theta,$$

this f is almost automorphic (resp. compact almost automorphic). The proof is complete. □

Now we will study the uniqueness of almost automorphic solution.

Theorem 3.4.20. *Let f be a continuous S^p-almost automorphic for some $p \geq 1$ and assume that $(T(t))_{t\geq 0}$ is hyperbolic. Then, equation (3.47) has a unique S^p-almost automorphic solution; additionally, for any $t \in \mathbb{R}$,*

$$x_t = \lim_{n\to+\infty} \int_{-\infty}^{t} T^S(t-s)\Gamma^n f(s)ds - \lim_{n\to+\infty} \int_{t}^{+\infty} T^U(t-s)\Gamma^n f(s)ds. \tag{3.55}$$

In addition, if $\{\lambda \in \sigma(\mathcal{A}) : \operatorname{Re}(\lambda) \geq 0\} = \emptyset$, then the unique S^p-almost automorphic $x(\cdot)$ is globally attractive.

Proof. Given condition (H.P), we can conclude from [211, Lemma 3.3] that equation (3.51) has a unique bounded solution on $\mathbb{R}$ given by

$$z(t) = -\int_{t}^{+\infty} e^{(t-s)G} Hf(s)ds \quad \text{for } t \in \mathbb{R}.$$

From [57, Theorem 4.14], we have that z is almost automorphic. Hence, the function $x(t) := \xi(t)[0]$ is an S^p-almost automorphic solution of equation (3.47), where

$$\xi(t) = \Phi z(t) + \lim_{n\to+\infty} \int_{-\infty}^{t} T^S(t-s)\Pi^S\Gamma^n f(s)ds \quad \text{for } t \in \mathbb{R}.$$

The rest of the proof can be demonstrated using a similar approach to that employed in proving Theorem 3.4.17. □

Remark 3.4.21. As a result of (iii) in the above theorem, if f is purely Stepanov almost automorphic, then equation (3.47) does not have an almost automorphic solution.

3.4.4 Application to an epidemic model with waning immunity

As an illustration of the main results obtained in this work, we consider the following system (see [229]):

$$y(t) = \int_0^{+\infty} k(a)y(t-a)da + f(t) \quad \text{for } t \in \mathbb{R}, \tag{3.56}$$

where

- $k(a) := \frac{\beta(a)F(a)}{R_0} - \frac{(R_0-1)}{\int_0^{+\infty}\mathcal{L}(a)da}\mathcal{L}(a)$ for $a \geq 0$;
- $\beta : \mathbb{R}^+ \to \mathbb{R}^+$ is the age-specific transmission coefficient of infected individuals whose infection-age is a;
- $F : \mathbb{R}^+ \to \mathbb{R}^+$ is a probability function, for an infected individual, to be infectious until his or her infection-age becomes a;
- $R_0 := \int_0^{+\infty}\beta(a)F(a)da$ is the basic reproduction number;
- $\mathcal{L}(a)$ is the probability for an individual who was infected not to obtain susceptibility since the last infection such that

  $$\mathcal{L}(a) = 1 - \int_0^a G(s)ds,$$

  where $G : \mathbb{R}^+ \to \mathbb{R}^+$ is a probability per unit of time to obtain susceptibility after infection, satisfying

  $$\int_0^{+\infty} G(a)da = 1;$$

- f is an S^p-almost periodic (resp. S^p-almost automorphic, almost periodic, almost automorphic) function.

Under the above conditions, the study of almost periodicity or almost automorphy dynamics for (3.56) can be investigated using the Massera and Bohr–Neugebauer criteria. This means that it suffices to analyze the boundedness and boundedness of the mean values of the solutions.

On the other hand, if $\lambda \in \{\mu \in \sigma(\mathcal{A}) : \operatorname{Re}(\mu) \geq 0\}$, then, by Theorem 3.4.6, λ satisfies the following characteristic equation:

$$1 = \int_0^{+\infty} \frac{\beta(a)F(a)e^{-\lambda a}}{R_0}da - \int_0^{+\infty} \frac{(R_0-1)}{\int_0^{+\infty}\mathcal{L}(\sigma)d\sigma}\mathcal{L}(a)e^{-\lambda a}da.$$

Hence, by [229, Proposition 3.1], there is $\varepsilon_1 > 0$ such that if

$$1 < R_0 < 1 + \varepsilon_1,$$

we have $\{\mu \in \sigma(\mathcal{A}) : \operatorname{Re}(\mu) \geq 0\} = \emptyset$. Consequently, thanks to Theorem 3.4.17 (resp. Theorem 3.4.20), equation (3.56) has a unique S^p-almost periodic (resp. S^p-almost automorphic, almost periodic, almost automorphic) solution which is globally attractive.

3.5 Delay differential equations arising in populations dynamics

In [130], we investigated the existence and uniqueness of a positive almost automorphic solution for the following delay differential equation:

$$N'(t) = N(t)\left(a(t) - b(t)N(t-\tau_1) - c(t)\int_0^{\tau_2} k(s)N(t-s)\mathrm{d}s \right) \quad \text{for } t \in \mathbb{R}, \tag{3.57}$$

where $k : [0,\infty) \to [0,\infty)$ is piecewise continuous while the functions $a(\cdot)$, $b(\cdot)$, and $c(\cdot)$ are almost automorphic and satisfy the following conditions:

$$0 < a_0 \le a(t) \le a_1, \quad 0 < b_0 \le b(t) \le b_1, \quad \text{and} \quad 0 < c_0 \le c(t) \le c_1 \quad \text{for } t \in \mathbb{R}, \tag{3.58}$$

where $a_0, a_1, b_0, b_1, c_0, c_1$ are positive constants. This equation can model the dynamics of population of a species in a time-fluctuating environment.

The following lemma gives a priori lower and upper bounds for positive solutions of equation (3.57).

Lemma 3.5.1. *Let N be a positive solution of equation* (3.57) *on $\mathbb{R}^+$. Then,*

$$\limsup_{t\to\infty} N(t) \le \frac{a_1}{\tilde{b}} := m_1 \tag{3.59}$$

and

$$\liminf_{t\to\infty} N(t) \ge \frac{a_0}{b_1 e^{-d_0\tau_1} + c_1 \int_0^{\tau_2} k(s)e^{-d_0 s}\mathrm{d}s} := m_0, \tag{3.60}$$

where $\tilde{b} = b_0 e^{-a_1\tau_1} + c_0 \int_0^{\tau_2} k(s)e^{-a_1 s}\mathrm{d}s$ and $d_0 = a_0 - 2m_1 b_1 - 2c_1 m_1 \int_0^{\tau_2} k(s)\mathrm{d}s$.

Proof. We have, for each $t \ge 0$,

$$N'(t) \le a_1 N(t) - b_0 N(t)N(t-\tau_1) - c_0 N(t)\int_0^{\tau_2} k(s)N(t-s)\mathrm{d}s, \tag{3.61}$$

which implies by the positivity of N that $N'(t) < a_1 N(t)$. It follows that $N(t) < e^{a_1\tau_1}N(t-\tau_1)$ and $N(t) < e^{a_1 s}N(t-s)$ for all $s \in [0,\tau_2]$. We deduce from (3.61) that

$$N'(t) < a_1 N(t) - \tilde{b}N^2(t),$$

where $\tilde{b} = b_0 e^{-a_1\tau_1} + c_0 \int_0^{\tau_2} k(s)e^{-a_1 s}\mathrm{d}s$. Putting $y(t) = \frac{1}{N(t)}$, we then have

$$y'(t) > e^{-a_1 t}y(0) + \frac{\tilde{b}}{a_1}(1 - e^{-a_1 t}).$$

Thus

$$N(t) < \frac{1}{\frac{e^{-a_1 t}}{N(0)} + \frac{\widetilde{b}}{a_1}(1 - e^{-a_1 t})},$$

which implies that

$$\limsup_{t\to\infty} N(t) \le \frac{a_1}{\widetilde{b}} := m_1. \tag{3.62}$$

On the other hand, we have

$$N'(t) = d(t)N(t),$$

where $d(t) = a(t) - b(t)N(t - \tau_1) - c(t)\int_0^{\tau_2} k(s)N(t - s)\mathrm{d}s$. It follows that $N(t) = e^{\int_{t-\tau_1}^{t} \mathrm{d}(\theta)\mathrm{d}\theta} N(t - \tau_1)$ and $N(t) = e^{\int_{t-s}^{t} \mathrm{d}(\theta)\mathrm{d}\theta} N(t - s)$ for all $s \in [0, \tau_2]$. For sufficiently large t, we have, by (3.62), $N(\theta) \le 2m_1$ for all $\theta \in [t - \max(\tau_1, \tau_2), t]$ and thus

$$d(\theta) \ge a_0 - 2m_1 b_1 - 2c_1 m_1 \int_0^{\tau_2} k(s)\mathrm{d}s := d_0.$$

This implies that $N(t - \tau_1) \le e^{-d_0\tau_1} N(t)$ and $N(t - s) \le e^{-d_0 s} N(t)$ for all $s \in [0, \tau_2]$. We get the following differential inequality:

$$N'(t) \ge a_0 N(t) - \left(b_1 e^{-d_0\tau_1} + c_1 \int_0^{\tau_2} k(s)e^{-d_0 s}\mathrm{d}s \right) N^2(t).$$

By differentiating again the function $y(t) = \frac{1}{N(t)}$, we deduce that

$$\liminf_{t\to\infty} N(t) \ge \frac{a}{b_1 e^{-d_0\tau_1} + c_1 \int_0^{\tau_2} k(s)e^{-d_0 s}\mathrm{d}s} := m_0.$$

This ends the proof of the lemma. □

Consider the following Cauchy problem:

$$\begin{cases} N'(t) = N(t)(a(t) - b(t)N(t - \tau_1) - c(t)\int_0^{\tau_2} k(s)N(t - s)\mathrm{d}s) & \text{for } t \ge 0, \\ N(t) = \varphi(t) & \text{for } -\tau \le t \le 0, \end{cases} \tag{3.63}$$

where $\tau = \max\{\tau_1, \tau_2\}$ and φ is a continuous function from $[-\tau, 0]$ to $\mathbb{R}$.

Proposition 3.5.2. *For each nonnegative initial data φ with $\varphi(0) > 0$, there exists a unique positive global solution of equation* (3.63).

Proof. The local existence of a solution is guaranteed by [166, Theorem 2.3] where

$$f(t,\phi) = \phi(0)\left(a(t) - b(t)\phi(-\tau_1) - c(t)\int_0^{\tau_2} k(s)\phi(-s)ds \right).$$

The positivity of this solution can be proved using the same arguments as in [149, Lemma 1]. This local solution is global since otherwise the solution must blow up at the maximal time of existence and thus will contradict the a priori estimate in Lemma 3.5.1. □

Theorem 3.5.3. *There exists a positive solution S of equation* (3.57) *on* $\mathbb{R}$ *such that*

$$m_0 \le S(t) \le m_1, \tag{3.64}$$

where m_0 *and* m_1 *are the positive constants defined by* (3.59) *and* (3.60). *Moreover, this solution is unique provided that*

$$\left(b_1\tau_1 + c_1\int_0^{\tau_2} sk(s)ds \right) m_1^2 < \frac{b_0 + c_0\int_0^{\tau_2} k(s)ds}{b_1 + c_1\int_0^{\tau_2} k(s)ds} m_0. \tag{3.65}$$

Proof. Let N be a positive solution of equation (3.57) on $\mathbb{R}^+$. Using (3.5.1), one can see that N and its derivative are bounded, thus N is uniformly continuous. Let $(t_n)_n$ be a sequence of real numbers such that $\lim_{t\to\infty} t_n = \infty$. Then for sufficiently large n, the sequence of functions $N_n : t \mapsto N(t + t_n)$ is well defined on $[-1, 1]$ and is equicontinuous. It follows by Arzelà–Ascoli theorem that there exist a function Q and a subsequence $(t_n^1)_n \subset (t_n)_n$ such that

$$N(t + t_n^1) \to Q(t) \quad \text{as } n \to \infty$$

uniformly on $[-1, 1]$. By applying the same argument to the subsequence $(t_n^1)_n$, we extract a subsequence $(t_n^2)_n \subset (t_n^1)_n \subset (t_n)_n$ such that

$$N(t + t_n^2) \to Q(t) \quad \text{as } n \to \infty$$

uniformly on $[-2, 2]$. By proceeding inductively, we obtain for each $m \in \mathbb{N}^*$ a subsequence $(t_n^m)_n \subset \cdots \subset (t_n^1)_n \subset (t_n)_n$ such that

$$N(t + t_n^m) \to Q(t) \quad \text{as } n \to \infty$$

uniformly on $[-m, m]$. Let $(t_n')_n := (t_n^n)_n$ be the diagonal sequence, then we have

$$N(t + t_n') \to Q(t) \quad \text{as } n \to \infty \tag{3.66}$$

uniformly on each compact subset of $\mathbb{R}$. Note that

$$m_0 \le Q(t) \le m_1.$$

Since $a(\cdot)$, $b(\cdot)$, and $c(\cdot)$ are almost automorphic functions, we can extract a subsequence $(t_n'')_n \subset (t_n')_n$ such that $a(t+t_n'') \to \tilde{a}(t)$, $\tilde{a}(t-t_n'') \to a(t)$, $b(t+t_n'') \to \tilde{b}(t)$, $\tilde{b}(t-t_n'') \to b(t)$, $c(t+t_n'') \to \tilde{c}(t)$, and $\tilde{c}(t-t_n'') \to c(t)$ as $n \to \infty$. For each $t \ge s$ and for $n \in \mathbb{N}$ sufficiently large, we have

$$\begin{aligned} N(t+t_n'') &= N(s+t_n'') \\ &\quad + \int_s^t N(u+t_n'')\Big(a(u+t_n'') - b(u+t_n'')N(u+t_n''-\tau_1) \\ &\quad - c(u+t_n'') \int_0^{\tau_2} k(\theta)N(u+t_n''-\theta)d\theta \Big)du. \end{aligned} \tag{3.67}$$

Letting $n \to \infty$, we get, for each $t \ge s$,

$$Q(t) = Q(s) + \int_s^t Q(u)\Big(\tilde{a}(u) - \tilde{b}(u)Q(u-\tau_1) - \tilde{c}(u) \int_0^{\tau_2} k(\theta)Q(u-\theta)d\theta \Big)du.$$

By applying the above argument to the returning sequence $(-t_n'')_n$, we obtain a subsequence $(t_n''')_n \subset (t_n'')_n$ and a function S such that

$$Q(t-t_n''') \to S(t) \quad \text{as } n \to \infty \tag{3.68}$$

uniformly on each compact subset of $\mathbb{R}$. One can see that S is a solution of (3.57) on $\mathbb{R}$ which satisfies

$$m_0 \le S(t) \le m_1.$$

Now for the uniqueness assume that (3.65) holds. Let $\tilde{S}$ be another solution on $\mathbb{R}$ which satisfies (3.64). Let $x(t) = \log S(t)$, $y(t) = \log \tilde{S}(t)$, and $L(t) = x(t) - y(t)$. For each fixed $t_0 \in \mathbb{R}$, we set $L_0(t) := L(t+t_0)$. One can see that there exists a continuous function $\theta(t)$ such that, for all $t \in \mathbb{R}$,

$$S(t) - \tilde{S}(t) = \theta(t)L(t) \quad \text{and} \quad m_0 \le \theta(t) \le m_1.$$

Then L_0 satisfies the following differential equation:

$$L_0'(t) = -b_0(t)\theta_0(t-\tau_1)L_0(t-\tau_1) - c_0(t)\int_0^{\tau_2} k(s)\theta_0(t-s)L_0(t-s)ds.$$

Consider the following Lyapunov function:

$$V(t) = V_1(t) + V_2(t) + V_3(t),$$

where

$$V_1(t) = \left[L_0(t) - \int_{t-\tau_1}^{t} b_0(s+\tau_1)\theta_0(s)L_0(s)\mathrm{d}s - \int_0^{\tau_2}\int_{t-s}^{t} k(s)c_0(u+s)\theta_0(u)L_0(u)\mathrm{d}u\mathrm{d}s \right]^2,$$

$$V_2(t) = \left(b_1 + c_1 \int_0^{\tau_2} k(s)\mathrm{d}s \right) b_1 m_1^2 \int_{t-\tau_1}^{t}\int_s^t L_0^2(u)\mathrm{d}u\mathrm{d}s,$$

and

$$V_3(t) = \left(b_1 + c_1 \int_0^{\tau_2} k(s)\mathrm{d}s \right) c_1 m_1^2 \int_0^{\tau_2} k(s) \int_{t-s}^{t}\int_u^t L_0^2(\xi)\mathrm{d}\xi\mathrm{d}u\mathrm{d}s.$$

Differentiating V and using $2xy \le x^2 + y^2$, we get the following inequality:

$$V'(t) \le -CL_0^2(t),$$

where

$$C = 2\left(m_0 \left[b_0 + c_0 \int_0^{\tau_2} k(s)\mathrm{d}s \right] - \left(b_1 + c_1 \int_0^{\tau_2} k(s)\mathrm{d}s \right)\left(b_1\tau_1 m_1^2 + c_1 m_1^2 \int_0^{\tau_2} sk(s)\mathrm{d}s \right)\right).$$

Notice that $C > 0$ by (3.65). Integrating the above inequality and using the positivity of $V(t)$, we obtain

$$\int_0^t L_0^2(s)\mathrm{d}s \le \frac{V(0)}{C}. \tag{3.69}$$

Notice also that there exists a constant M independent of t_0 such that $V(0) \le M$. Thus

$$\int_{-\infty}^{\infty} L^2(s)\mathrm{d}s \le \frac{M}{C}.$$

Since the function $s \mapsto L^2(s)$ is uniformly continuous ($\frac{\mathrm{d}}{\mathrm{d}s}L^2(s)$ is bounded), we obtain by Barbalat's lemma [159, Lemma 1.2.2] that $\lim_{t\to\pm\infty} L^2(t) = 0$. Letting $\varepsilon > 0$, there exists $T > 0$ such that for all $t \in \mathbb{R}$ with $\|t\| > T$ we have $\|L(t)\| < \varepsilon$. We fix $t_0 \in \mathbb{R}$ such that $t_0 < -T$. Thus from (3.69) we have

$$\int_{t_0}^{\infty} L^2(s)\mathrm{d}s \le \frac{V(0)}{C}.$$

On the other hand, since $t_0 < -T$, one can see that there exists a constant $M_1 > 0$ independent of t_0 such that $V(0) \le M_1\varepsilon^2$. Therefore we get

$$\int_{-\infty}^{\infty} L^2(s)\mathrm{d}s \le \frac{M_1}{C}\varepsilon^2.$$

Since $\varepsilon > 0$ is arbitrary, $L(t) = 0$ for all $t \in \mathbb{R}$, and thus $S(t) = \widetilde{S}(t)$ for all $t \in \mathbb{R}$. □

We are now in a position to present our main result.

Theorem 3.5.4. *Assume that*

$$\left(b_1\tau_1 + c_1 \int_0^{\tau_2} sk(s)\mathrm{d}s\right)m_1^2 < \frac{b_0 + c_0 \int_0^{\tau_2} k(s)\mathrm{d}s}{b_1 + c_1 \int_0^{\tau_2} k(s)\mathrm{d}s} m_0. \tag{3.70}$$

Then equation (3.57) *has a unique compact almost automorphic solution S on $\mathbb{R}$ such that $m_0 \le S(t) \le m_1$ for $t \in \mathbb{R}$. Furthermore, S attracts all positive solutions on* $(0, \infty)$.

Proof. By Theorem 3.5.3, equation (3.57) has a unique solution S on $\mathbb{R}$ such that

$$m_0 \le S(t) \le m_1 \quad \text{for } t \in \mathbb{R}. \tag{3.71}$$

We claim that S is compact almost automorphic. In fact, S is uniformly continuous as it has a bounded derivative. Let $(t_n)_n$ be a sequence of real numbers. Using the equicontinuity of the family of functions $S_n : t \mapsto S(t + t_n)$ and the Arzelà–Ascoli theorem, there exist a function Q and a subsequence $(t_n')_n \subset (t_n)_n$ such that

$$S(t + t_n') \to Q(t) \quad \text{as } n \to \infty \tag{3.72}$$

uniformly on each compact subset of $\mathbb{R}$. In the other hand, we can extract a subsequence $(t_n'')_n \subset (t_n')_n$ such that $a(t+t_n'') \to \tilde{a}(t)$, $\tilde{a}(t-t_n'') \to a(t)$, $b(t+t_n'') \to \tilde{b}(t)$, $\tilde{b}(t-t_n'') \to b(t)$, $c(t+t_n'') \to \tilde{c}(t)$, and $\tilde{c}(t-t_n'') \to c(t)$ as $n \to \infty$. Thus Q satisfies the following differential equation:

$$Q'(t) = Q(t)\left(\tilde{a}(t) - \tilde{b}(t)Q(t) - \tilde{c}(t)\int_0^{\tau_2} k(s)Q(t-s)\mathrm{d}s\right) \quad \text{for } t \in \mathbb{R}.$$

By applying the above argument to the returning sequence $(-t_n'')_n$, we obtain a subsequence $(t_n''')_n \subset (t_n')_n$ such that

$$Q(t - t_n''') \to R(t) \quad \text{as } n \to \infty \tag{3.73}$$

uniformly on each compact subset of $\mathbb{R}$, where R is a solution of (3.57) on $\mathbb{R}$. In addition, it follows from (3.71)–(3.73) that

$$m_0 \le R(t) \le m_1 \quad \text{for } t \in \mathbb{R}.$$

We deduce using Theorem (3.5.3) that $R(t) = S(t)$ for all $t \in \mathbb{R}$ and thus, by (3.72) and (3.73), S is compact almost automorphic. The attractiveness of the solution S follows from the proof of Theorem (3.5.3). □

4 Dynamical systems: infinite-dimensional case

When the dimension of the phase space is infinite, the situation becomes more complicated. Without additional hypotheses, the Bohr–Neugebauer and Massera theorems from the previous chapter do not hold. For example, consider the following simple differential equation in c_0 (the space of numerical sequences that converge to zero):

$$x'(t) = f(t) \quad \text{for } t \in \mathbb{R},$$

where $f(t) = \{\frac{\cos(\frac{t}{n})}{n}\}_{n>0}$. Note that the function f is almost periodic, and all solutions take the following form:

$$x(t) = c + F(t),$$

where $c \in c_0$ and

$$F(t) = \int_0^t f(s)\,\mathrm{d}s = \left\{\sin\left(\frac{t}{n}\right)\right\}_{n>0}.$$

As a result, all solutions of the above equation are bounded, but not almost periodic (see [35, 226]).

To overcome this difficulty, much work has been done using powerful tools from functional analysis and operator theory, including imposing geometric conditions on the phase space. The main objective of this chapter is to expose some methodologies and explore different situations in the study of the existence and uniqueness of almost periodic and almost automorphic solutions for evolution equations in abstract Banach spaces. These include nonlinear differential equations under dissipativity or monotonicity conditions, nonhomogeneous linear evolution equations under quasicompactness hypotheses and Favard conditions with geometric conditions in the absence of compactness assumptions, the subvariant functional method for semilinear evolution equations, spectral countability conditions for the nonautonomous case, and fixed point arguments for semilinear fractional differential equations and some differential inclusions governed by a maximal monotone operator.

4.1 Dissipative ordinary differential equations in Banach space

In [131], under some dissipativity condition, we investigated the existence and uniqueness of an almost automorphic solution for the following ordinary differential equation:

$$\frac{\mathrm{d}}{\mathrm{d}t}x(t) = f(t, x(t)) \quad \text{for } t \in \mathbb{R}, \tag{4.1}$$

https://doi.org/10.1515/9783111684710-004

where f satisfies some weaker assumptions compared to the works found in literature. We assume that X is a Banach space endowed with the norm $|\cdot|$ and $f : \mathbb{R} \times X \to X$ is Stepanov almost automorphic in t and dissipative with respect to x.

In what follows, we introduce the following notion:

Definition 4.1.1. A function $f : \mathbb{R}\times X \to X$ is said to be Stepanov almost automorphic in t uniformly with respect to x in bounded sets if for any bounded set B of X, $\sup_{x\in B} \|f(\cdot, x)\|$ is locally integrable and if, for all sequences of real numbers $(t_n)_n$, there exist a map $g : \mathbb{R} \times X \to X$ with $\sup_{x\in B} \|g(\cdot, x)\|$ locally integrable and a subsequence $(t'_n)_n \subset (t_n)_n$ such that, for each $t \in \mathbb{R}$,

$$\lim_{n\to\infty} \int_t^{t+1} \sup_{x\in B} \|f(s + t'_n, x) - g(s, x)\| ds = 0,$$

$$\lim_{n\to\infty} \int_t^{t+1} \sup_{x\in B} \|g(s - t'_n, x) - f(s, x)\| ds = 0.$$

We denote by $\mathrm{SAAU}(\mathbb{R} \times X, X)$ the set of all such functions.

Remark 4.1.2. The authors in [96] assumed that f is compact almost automorphic in t uniformly with respect to x in the following sense: For all sequences of real numbers $(t_n)_n$, there exist a map $g : \mathbb{R} \times X \to X$ and a subsequence $(t'_n)_n \subset (t_n)_n$ such that, for all compact subsets I of $\mathbb{R}$ and for all bounded compact subsets K of X,

$$\lim_{n\to\infty} \sup_{t\in I} \sup_{x\in K} \|f(t + t'_n, x) - g(t, x)\| = 0,$$

$$\lim_{n\to\infty} \sup_{t\in I} \sup_{x\in K} \|g(t - t'_n, x) - f(t, x)\| = 0.$$

We denote by $\mathrm{AAU}(\mathbb{R} \times X, X)$ the set of all such functions.

Lemma 4.1.3. *Let $f \in \mathrm{SAAU}(\mathbb{R} \times X, X)$. Then for any bounded subset B of X and for all sequences of real numbers $(t_n)_n$, there exist a function $g : \mathbb{R}\times X \to X$ with $\sup_{x\in B} \|g(\cdot, x)\|$ locally integrable, a subsequence $(t'_n)_n \subset (t_n)_n$, and a subset $N \subset \mathbb{R}$ with null Lebesgue measure such that, for each $t \in \mathbb{R} \setminus N$,*

$$\sup_{x\in B} \|f(t + t'_n, x) - g(t, x)\| \to 0,$$

$$\sup_{x\in B} \|g(t - t'_n, x) - f(t, x)\| \to 0$$

as $n \to \infty$.

Proof. Let $(t_n)_n$ be a sequence of real numbers. Then there exist $g : \mathbb{R} \times X \to X$ with $\sup_{x\in B} \|g(\cdot, x)\|$ locally integrable and a subsequence $(t'_n)_n \subset (t_n)_n$ such that, for each $t \in \mathbb{R}$,

$$\lim_{n\to\infty}\int_t^{t+1}\sup_{x\in B}\|f(s+t'_n,x)-g(s,x)\|ds=0,$$

$$\lim_{n\to\infty}\int_t^{t+1}\sup_{x\in B}\|g(s-t'_n,x)-f(s,x)\|ds=0.$$

Observe that, for each $a, b \in \mathbb{R}$ with $a < b$, we have

$$\lim_{n\to\infty}\int_a^{b}\sup_{x\in B}\|f(s+t'_n,x)-g(s,x)\|ds=0,$$

$$\lim_{n\to\infty}\int_a^{b}\sup_{x\in B}\|g(s-t'_n,x)-f(s,x)\|ds=0.$$

Thus there exist a subset $N_{a,b}$ of $\mathbb{R}$ with null Lebesgue measure and a subsequence $(t_n^{a,b})_n \subset (t'_n)_n$ such that, for all $t \in [a,b] \setminus N_{a;b}$,

$$\sup_{x\in B}\|f(t+t_n^{a,b},x)-g(t,x)\| \to 0,$$

$$\sup_{x\in B}\|g(s-t_n^{a,b},x)-f(s,x)\| \to 0$$

as $n \to \infty$. Note that the subsequence $(t_n^{a,b})_n$ and the null set $N_{a,b}$ both depend on a and b. To eliminate this dependence, one can use the classical diagonal procedure to obtain a null set N and a subsequence $(t''_n)_n$ such that, for all $t \in \mathbb{R} \setminus N$,

$$\sup_{x\in B}\|f(t+t''_n,x)-g(t,x)\| \to 0,$$

$$\sup_{x\in B}\|g(s-t''_n,x)-f(s,x)\| \to 0$$

as $n \to \infty$. □

The functionals $[\cdot,\cdot]_\pm : X \times X \to \mathbb{R}$ are defined for each $x, y \in X$ by the following formulas:

$$[x,y]_+ := \lim_{h\to 0^+}\frac{|x+hy|-|x|}{h}$$

and

$$[x,y]_- := \lim_{h\to 0^-}\frac{|x+hy|-|x|}{h} = \lim_{h\to 0^+}\frac{|x|-|x-hy|}{h}.$$

The above functionals are well defined. For more details, we refer to [212].

Now we give a list of hypotheses which will be used in the sequel:

(M1) There exist $p \in C(\mathbb{R},\mathbb{R})$, $T_0, T_1 \in \mathbb{R}$ such that $T_0 < T_1$ and δ an almost periodic function from $\mathbb{R}$ to $\mathbb{R}$ such that $\mathcal{M}(\delta) < 0$ and $p(t) \le \delta(t)$ for every $t \notin [T_0, T_1]$.

(M2) There exists $\theta \in C(\mathbb{R}^+, \mathbb{R}^+)$ such that θ is nondecreasing and satisfies $\theta(u) > 0$ for each $u > 0$.

(M3) The function $u \mapsto p(t)\theta(u)$ is nonincreasing on $\mathbb{R}^+$, for each $t \in \mathbb{R}$.

(M4) $f : \mathbb{R} \times X \to X$ is continuous and, for all $t \in \mathbb{R}$ and $x, y \in X$,

$$[x - y, f(t,x) - f(t,y)]_- \le p(t)\|x - y\|\theta(\|x - y\|).$$

(B) $\sup_{t\in\mathbb{R}} |f(t,0)| < \infty$.

(B′) $\sup_{t\in\mathbb{R}} \int_t^{t+1} |f(s,0)|\mathrm{d}s < \infty$.

(A) $f \in \mathrm{AAU}(\mathbb{R} \times X, X)$.

(A′) $f \in \mathrm{SAAU}(\mathbb{R} \times X, X)$.

Remark. Note that **(B)** implies **(B′)**, but the converse is not true. In fact, every function $f : \mathbb{R} \times X \to X$ such that $f(\cdot, 0)$ is unbounded but integrable over $\mathbb{R}$ satisfies **(B′)** but not **(B)**.

Remark. We assume that the conditions **(M1)** and **(M2)** are fulfilled. First, there exists $t_0 \in \mathbb{R}$ such that $p(t_0) < 0$; therefore function $u \mapsto p(t)\theta(u)$ is nonincreasing. Second, **(M3)** is equivalent to the following alternative: either $p(t) \le 0$ for each $t \in \mathbb{R}$, or $\theta(u) = \theta_0$ for each $u > 0$.

Lemma 4.1.4 ([20]). *Under condition* **(M1)**, *there exist $k > 0$ and $c > 0$ such that*

$$e^{\int_s^t p(\sigma)\mathrm{d}\sigma} \le k e^{-c(t-s)} \quad \textit{for all } s, t \in \mathbb{R} \textit{ with } s \le t.$$

Lemma 4.1.5 ([20]). *Assume that* **(M1)**–**(M4)** *hold. Let k and c be the positive constants given in Lemma 4.1.4. Let I be an interval of $\mathbb{R}$ and $g, h \in C(\mathbb{R}, X)$. If x and y are respective solutions on I of*

$$x'(t) = f(t, x(t)) + g(t), \tag{4.2}$$

$$y'(t) = f(t, y(t)) + h(t), \tag{4.3}$$

then we have the following inequalities for all $t, s \in I$ with $s \le t$:

(i) $\|x(t)-y(t)\| \le e^{\int_s^t p(\tau)\theta(\|x(\tau)-y(\tau)\|)\mathrm{d}\tau}\|x(s)-y(s)\| + \int_s^t e^{\int_\sigma^t p(\tau)\theta(\|x(\tau)-y(\tau)\|)\mathrm{d}\tau}\|g(\sigma)-h(\sigma)\|\mathrm{d}\sigma.$

(ii) *If $m \in \mathbb{R}$ is such that $0 < m \le \inf_{s\le\tau\le t} \|x(\tau) - y(\tau)\|$, then*

$$\|x(t) - y(t)\| \le k^{\theta(m)} e^{-c\theta(m)(t-s)}\|x(s) - y(s)\| + k^{\theta(m)} \int_s^t e^{-c\theta(m)(t-\sigma)}\|g(\sigma) - h(\sigma)\|\mathrm{d}\sigma.$$

(iii) *In particular, if* $g = h$, *then*

$$\|x(t) - y(t)\| \leq k^{\theta(\|x(t)-y(t)\|)} e^{-c\theta(\|x(t)-y(t)\|)(t-s)} \|x(s) - y(s)\|.$$

4.1.1 Compact almost automorphic solutions for equation (4.1)

Let k and c be the positive constants given in Lemma 4.1.4 and

$$R_0 := \inf_{\varepsilon>0}\left(\varepsilon k^{\theta(\varepsilon)} + \frac{k^{\theta(\varepsilon)}}{c\theta(\varepsilon)} \sup_{t\in\mathbb{R}} |f(t,0)| \right).$$

Recall the following result:

Theorem 4.1.6 ([20]). *Assume that* **(M1)**–**(M4)** *and* **(B)** *hold. Then equation* (4.1) *has a unique bounded solution* x_f *on* $\mathbb{R}$ *satisfying*

$$\sup_{t\in\mathbb{R}} |x_f(t)| \leq R_0. \tag{4.4}$$

Moreover, x_f *is globally attractive.*

The following result relaxes the assumptions of Theorem 4.1.6 in the sense that only **(B′)** is needed instead of **(B)**.

Theorem 4.1.7. *Assume that* **(M1)**–**(M4)** *and* **(B′)** *hold. Then equation* (4.1) *has a unique bounded solution* x_f *on* $\mathbb{R}$ *satisfying*

$$\sup_{t\in\mathbb{R}} |x_f(t)| \leq \widetilde{R_0}, \tag{4.5}$$

where

$$\widetilde{R_0} := \inf_{\varepsilon>0}\left(\varepsilon k^{\theta(\varepsilon)} + \frac{2k^{\theta(\varepsilon)} e^{c\theta(\varepsilon)}}{1 - e^{-c\theta(\varepsilon)}} \sup_{t\in\mathbb{R}} \int_t^{t+1} |f(s,0)| \mathrm{d}s \right).$$

Moreover, x_f *is globally attractive.*

Proof of Theorem 4.1.7. The uniqueness of the bounded solution on $\mathbb{R}$ comes from Lemma 4.1.5. The proof of the existence of the bounded solution is similar to that given in [20, Theorem 2.1]. In fact, consider the following Cauchy problem:

$$\begin{cases} x'(t) = f(t, x(t)), \\ x(-n) = 0. \end{cases} \tag{4.6}$$

Then equation (4.6) has a unique solution u_n on $[-n, n]$. For $\varepsilon > 0$ and $t \in [-n, n]$, we have either $\|u_n(t)\| \leq \varepsilon$ or $\|u_n(t)\| > \varepsilon$. In the second case, there exists $t_0 \in (-n, t]$ such that $\|u_n(t_0)\| = \varepsilon$ and $\|u_n(s)\| > \varepsilon$ for each $s \in (t_0, t]$; therefore $\inf_{t_0 \le s \le t} \|u_n(s)\| = \varepsilon$. Since $y(t) = 0$ is a solution of equation (4.3) with $h(t) = -f(t, 0)$, by applying Lemma 4.1.5 with $x = u_n$, $g = 0$, $y = 0$, and $h = -f(\cdot, 0)$ we have

$$\begin{aligned}
\|u_n(t)\| &\le k^{\theta(\varepsilon)} e^{-c\theta(\varepsilon)(t-t_0)} \|u_n(t_0)\| + k^{\theta(\varepsilon)} \int_{t_0}^{t} e^{-c\theta(\varepsilon)(t-\sigma)} \|f(\sigma, 0)\| d\sigma \\
&\le \varepsilon k^{\theta(\varepsilon)} + k^{\theta(\varepsilon)} \int_{[t_0]}^{[t]+1} e^{-c\theta(\varepsilon)(t-\sigma)} \|f(\sigma, 0)\| d\sigma \\
&\le \varepsilon k^{\theta(\varepsilon)} + k^{\theta(\varepsilon)} \sum_{m=[t_0]}^{[t]} e^{-c\theta(\varepsilon)(t-m-1)} \int_{m}^{m+1} \|f(\sigma, 0)\| d\sigma \\
&\le \varepsilon k^{\theta(\varepsilon)} + k^{\theta(\varepsilon)} \sup_{t \in \mathbb{R}} \int_{t}^{t+1} \|f(\sigma, 0)\| d\sigma \sum_{m=[t_0]}^{[t]} e^{-c\theta(\varepsilon)(t-m-1)} \\
&\le \varepsilon k^{\theta(\varepsilon)} + k^{\theta(\varepsilon)} \sup_{s \in \mathbb{R}} \int_{s}^{s+1} \|f(\sigma, 0)\| d\sigma \frac{2e^{c\theta(\varepsilon)}}{1 - e^{-c\theta(\varepsilon)}}.
\end{aligned}$$

Thus for all $t \in [-n, n]$ and $\varepsilon > 0$,

$$\|u_n(t)\| \le \varepsilon k^{\theta(\varepsilon)} + k^{\theta(\varepsilon)} \sup_{s \in \mathbb{R}} \int_{s}^{s+1} \|f(\sigma, 0)\| d\sigma \frac{2e^{c\theta(\varepsilon)}}{1 - e^{-c\theta(\varepsilon)}}.$$

Therefore for all $t \in [-n, n]$,

$$\|u_n(t)\| \le \widetilde{R_0}. \tag{4.7}$$

The rest of the proof is similar to that of [20, Theorem 4.1]. We only have to prove that $(u_n)_n$ is a uniform Cauchy sequence in every bounded subset of $\mathbb{R}$. So its limit is a bounded solution of equation (4.1) which satisfies (4.7). The global attractivity of this solution can be proved using the same argument as that of [20, Theorem 4.1]. □

Our main goal is to relax the assumptions of the following result:

Theorem 4.1.8 ([96]). *Assume that* **(M1)**–**(M4)** *and* **(A)** *hold. Then equation* (4.1) *has a unique compact almost automorphic solution* x_f. *Moreover,* x_f *is globally attractive.*

And so we have

Theorem 4.1.9. *Assume that* **(M1)**–**(M4)** *and* **(A′)** *hold. Then equation* (4.1) *has a unique compact almost automorphic solution* x_f. *Moreover,* x_f *is globally attractive.*

The following lemmas are needed for the proof of Theorem 4.1.9.

Lemma 4.1.10. *Under the assumptions of Theorem* 4.1.9, *the solution* x_f *has a relatively compact range.*

Proof of Lemma 4.1.10. Let $(s_n)_n$ be a sequence of real numbers and B be a bounded subset of X which contains the range of x_f. From the Stepanov almost automorphy of f, there exists a subsequence $(s'_n)_n \subset (s_n)_n$ such that, for each $t \in \mathbb{R}$,

$$\int_t^{t+1} \sup_{x\in B} \|f(s+s'_p,x) - f(s+s'_q,x)\| ds \to 0 \tag{4.8}$$

as $p, q \to \infty$. Recall that f satisfies the following dissipativity condition:

$$[x-y, f(t,x) - f(t,y)]_- \le p(t)\|x-y\|\theta(\|x-y\|).$$

Consider the following functions:

$$\begin{cases} x_n(t) := x_f(t+s'_n), \\ f_n(t,x) := f(t+s'_n, x), \\ p_n(t) := p(t+s'_n). \end{cases}$$

The function f_n satisfies the following dissipativity condition:

$$[x-y, f_n(t,x) - f_n(t,y)]_- \le p_n(t)\|x-y\|\theta(\|x-y\|).$$

It is clear that the function p_n satisfies the same assumptions as p and also fits the setting of Lemma 4.1.4 with the same constants k and c. We claim that $(x_n(t))_n$ is a Cauchy sequence in X for each $t \in \mathbb{R}$. In fact, **(M3)** is equivalent to the following alternative: either $\theta(u) = \theta_0 > 0$ for all $u > 0$, or $p(t) \le 0$ for each $t \in \mathbb{R}$.

Case 1. $\theta(u) = \theta_0 > 0$. The functions x_p and x_q satisfy the following differential equations:

$$\begin{cases} x'_p(t) = f_p(t, x_p(t)), \\ x'_q(t) = f_q(t, x_q(t)) = f_p(t, x_q(t)) + f_q(t, x_q(t)) - f_p(t, x_q(t)). \end{cases} \tag{4.9}$$

The function f_p is also dissipative and has the same rate of dissipativity as f. Thus, by applying Lemma 4.1.5 to (4.9) with $x = x_p, y = y_p, g = 0$, and $h = f_q(\cdot, x_q(\cdot)) - f_p(\cdot, x_q(\cdot))$, we obtain, for $\sigma \le t$,

$$|x_p(t) - x_q(t)| \le k^{\theta_0} e^{-c\theta_0(t-\sigma)} |x_p(\sigma) - x_q(\sigma)| + \int_\sigma^t e^{-c\theta_0(t-s)} \|f_q(s, x_q(s)) - f_p(s, x_q(s))\| ds.$$

Letting $\sigma \to -\infty$, we deduce that, for each $t \in \mathbb{R}$,

$$
\begin{aligned}
|x_p(t) - x_q(t)| &\le k^{\theta_0} \int_{-\infty}^{t} e^{-c\theta_0(t-s)} \|f_q(s, x_q(s)) - f_p(s, x_q(s))\| ds \\
&\le k^{\theta_0} \int_{-\infty}^{t} e^{-c\theta_0(t-s)} \sup_{x \in B} \|f_q(s, x) - f_p(s, x)\| ds \\
&= \sum_{k=1}^{\infty} \int_{t-k}^{t-k+1} e^{-c\theta_0(t-s)} \sup_{x \in B} \|f_q(s, x) - f_p(s, x)\| ds \\
&\le \sum_{k=1}^{\infty} e^{-c\theta_0(k-1)} \int_{t-k}^{t-k+1} \sup_{x \in B} \|f(s + s'_p, x) - f(s + s'_q, x)\| ds.
\end{aligned}
$$

From the definition of SAAU functions, one can observe that $\sup_{t \in \mathbb{R}} \int_t^{t+1} \sup_{x \in B} \|f(s, x)\| ds < \infty$. Thus by using (4.8) together with Lebesgue's dominated convergence theorem, we conclude that $(x_n(t))_n$ is a Cauchy sequence in X for each $t \in \mathbb{R}$.

Case 2. $p(t) \le 0$ for all $t \in \mathbb{R}$. If $(x_n(t_0))_n$ is not a Cauchy sequence in X for some $t_0 \in \mathbb{R}$, then there exist $\varepsilon > 0$ and two subsequences $(x_{p_n}(t_0))_n$ and $(x_{q_n}(t_0))_n$ such that, for all $n \in \mathbb{N}$,

$$
|x_{p_n}(t_0) - x_{q_n}(t_0)| \ge \varepsilon. \tag{4.10}
$$

Thus x_{p_n} and x_{q_n} satisfy the following differential equations:

$$
\begin{cases}
x'_{p_n}(t) = f_{p_n}(t, x_{p_n}(t)), \\
x'_{q_n}(t) = f_{q_n}(t, x_{q_n}(t)) = f_{p_n}(t, x_{q_n}(t)) + f_{q_n}(t, x_{q_n}(t)) - f_{p_n}(t, x_{q_n}(t)).
\end{cases} \tag{4.11}
$$

Let $s_0 \in \mathbb{R}$ with $s_0 < t_0$. Then by applying Lemma 4.1.5 to (4.11) with $x = x_p, y = x_q, g = 0$, and $h = f_q(\cdot, x_q(\cdot)) - f_p(\cdot, x_q(\cdot))$, we obtain for $s_0 \le t \le t_0$ the estimate

$$
\begin{aligned}
\varepsilon \le |x_{p_n}(t_0) - x_{q_n}(t_0)| &\le |x_{p_n}(t) - x_{q_n}(t)| + \int_t^{t_0} \|f_{q_n}(s, x_{q_n}(s)) - f_{p_n}(s, x_{q_n}(s))\| ds \\
&\le |x_{p_n}(t) - x_{q_n}(t)| + \int_t^{t_0} \sup_{x \in B} \|f(s + s'_{q_n}, x) - f(s + s'_{p_n}, x)\| ds \\
&\le |x_{p_n}(t) - x_{q_n}(t)| + \frac{\varepsilon}{2}.
\end{aligned}
$$

Therefore

$$
0 < \frac{\varepsilon}{2} \le \inf_{s_0 \le t \le t_0} |x_{p_n}(t) - x_{q_n}(t)|.
$$

By applying Lemma 4.1.5(ii) with $m = \frac{\varepsilon}{2}$, we get

$$
\begin{aligned}
|x_{p_n}(t_0) - x_{q_n}(t_0)| &\le k^{\theta(\frac{\varepsilon}{2})} e^{-c\theta(\frac{\varepsilon}{2})(t_0 - s_0)} |x_{p_n}(t_0) - x_{q_n}(t_0)| \\
&\quad + k^{\theta(\frac{\varepsilon}{2})} \int_{s_0}^{t_0} e^{-c\theta(\frac{\varepsilon}{2})(t_0 - s)} \|f_{q_n}(s, x_{q_n}(s)) - f_{p_n}(s, x_{q_n}(s))\| ds \\
&\le 2\widetilde{R_0} k^{\theta(\frac{\varepsilon}{2})} e^{-c\theta(\frac{\varepsilon}{2})(t_0 - s_0)} + k^{\theta(\frac{\varepsilon}{2})} \int_{s_0}^{t_0} \sup_{x \in B} \|f(s + s'_{q_n}, x) - f(s + s'_{p_n}, x)\| ds.
\end{aligned}
$$

It follows that

$$
\limsup_{n \to \infty} |x_{p_n}(t_0) - x_{q_n}(t_0)| \le 2\widetilde{R_0} k^{\theta(\frac{\varepsilon}{2})} e^{-c\theta(\frac{\varepsilon}{2})(t_0 - s_0)}.
$$

Letting $s_0 \to -\infty$, we obtain

$$
\limsup_{n \to \infty} |x_{p_n}(t_0) - x_{q_n}(t_0)| = 0,
$$

which contradicts (4.10). We conclude in both cases that $(x_n(t))_n = (x_f(t+s'_n))_n$ is a Cauchy sequence in X for each $t \in \mathbb{R}$. Therefore x_f has a relatively compact range. □

Lemma 4.1.11. *Under the assumptions of Theorem* 4.1.9, *the solution* x_f *is uniformly continuous on* $\mathbb{R}$.

Proof of Lemma 4.1.11. If x_f is not uniformly continuous, then there exist $\varepsilon > 0$ and two real sequences $(s_n)_n$ and $(h_n)_n$ with $\lim_{n \to \infty} h_n = 0$ such that

$$
|x_f(s_n + h_n) - x_f(s_n)| > \varepsilon \quad \text{for all } n \in \mathbb{N}. \tag{4.12}
$$

Assume without loss of generality that $0 \le h_n \le 1$ for all $n \in \mathbb{N}$. Then we have

$$
\begin{aligned}
x_f(s_n + h_n) &= x_f(s_n) + \int_{s_n}^{s_n + h_n} f(s, x_f(s)) ds \\
&= x_f(s_n) + \int_0^{h_n} f(s + s_n, x_f(s + s_n)) ds.
\end{aligned}
$$

Let $B = \{x_f(t) : t \in \mathbb{R}\}$ and $(s'_n)_n \subset (s_n)_n$ be a subsequence such that, for each $t \in \mathbb{R}$,

$$
\lim_{n \to \infty} \int_t^{t+1} \sup_{x \in B} \|f(s + s'_n, x) - g(s, x)\| ds = 0,
$$

where $g : \mathbb{R} \times X \to X$ is such that $\sup_{x \in B} \|g(\cdot, x)\|$ is locally integrable. Let $(h'_n)_n$ be the corresponding subsequence of $(h_n)_n$. Then

$$
\begin{aligned}
|x_f(s_n' + h_n') - x_f(s_n')| &\le \int_0^{h_n'} |f(s + s_n', x_f(s + s_n'))| \mathrm{d}s \\
&\le \int_0^{h_n'} |f(s + s_n', x_f(s + s_n')) - g(s, x_f(s + s_n'))| \mathrm{d}s + \int_0^{h_n'} |g(s, x_f(s + s_n'))| \mathrm{d}s \\
&\le \int_0^{1} \sup_{x \in B} |f(s + s_n', x) - g(s, x)| \mathrm{d}s + \int_0^{h_n'} \sup_{x \in B} |g(s, x)| \mathrm{d}s \to 0
\end{aligned}
$$

as $n \to \infty$, which contradicts (4.12). We conclude that x_f must be uniformly continuous. □

Proof of Theorem 4.1.9. Let $B = \{x_f(t) : t \in \mathbb{R}\}$ and $(t_n)_n$ be a sequence of real numbers. Then there exist a subsequence $(t_n')_n \subset (t_n)_n$ and a measurable function $g : \mathbb{R} \times X \to X$ such that, for each $t \in \mathbb{R}$,

$$
\sup_{x \in B} \int_t^{t+1} \|f(s + t_n', x) - g(s, x)\| \mathrm{d}s \to 0, \tag{4.13}
$$

$$
\sup_{x \in B} \int_t^{t+1} \|g(s - t_n', x) - f(s, x)\| \mathrm{d}s \to 0.
$$

Let $x_n(t) := x_f(t + t_n')$, then for each $n \in \mathbb{N}$, $x_n \in C(\mathbb{R}, X)$ and $x_n(t) \in B$ for each $t \in \mathbb{R}$. By Lemma 4.1.10, the set $\{x_n(t) : n \in \mathbb{N}\}$ is a relatively compact subset of X for each $t \in \mathbb{R}$. Since x_f is uniformly continuous (Lemma 4.1.11), the sequence $(x_n)_n$ is equicontinuous on $\mathbb{R}$. In view of Arzelà–Ascoli theorem, we can assert that $\{x_n : n \in \mathbb{N}\}$ is a relatively compact subset of $C(\mathbb{R}, X)$ endowed with the topology of compact convergence. Thus from the sequence $(t_n')_n$ one can extract another subsequence $(t_n'')_n \subset (t_n')_n \subset (t_n)_n$ such that

$$
x_f(t + t_n'') \to y(t) \tag{4.14}
$$

uniformly on compact subsets of $\mathbb{R}$, where $y \in C(\mathbb{R}, X)$.

Since x_f is a solution of equation (4.1), we obtain for each $t, s \in \mathbb{R}$ with $t \ge s$ and $n \in \mathbb{N}$,

$$
x_f(t + t_n'') = x_f(s + t_n'') + \int_s^t f(\tau + t_n'', x_f(\tau + t_n'')) \mathrm{d}\tau. \tag{4.15}
$$

Since $f : \mathbb{R} \times X \to X$ is continuous, it follows by Lemma 4.1.3 that $x \mapsto g(\tau, x)$ is continuous on B for almost all $\tau \in \mathbb{R}$. Using the triangle inequality, we obtain

$$\begin{aligned}\int_s^t \|f(\tau + t_n'', x_f(\tau + t_n'')) - g(\tau, y(\tau))\| d\tau &\le \int_s^t \|f(\tau + t_n'', x_f(\tau + t_n'')) - g(\tau, x_f(\tau + t_n''))\| d\tau \\ &\quad + \int_s^t \|g(\tau, x_f(\tau + t_n'')) - g(\tau, y(\tau))\| d\tau \\ &\le \int_s^t \sup_{x \in B} \|f(\tau + t_n'', x) - g(\tau, x)\| d\tau \\ &\quad + \int_s^t \|g(\tau, x_f(\tau + t_n'')) - g(\tau, y(\tau))\| d\tau.\end{aligned}$$

Letting $n \to \infty$ in (4.15), we deduce that

$$y(t) = y(s) + \int_s^t g(\tau, y(\tau)) d\tau.$$

From (4.14), we can see that y is also uniformly continuous and has a relatively compact range. Thus by applying the same procedure to the function y using the returning sequence $(-t_n'')_n$, we obtain another subsequence $(t_n''')_n \subset (t_n'')_n \subset (t_n')_n \subset (t_n)_n$ such that

$$y(t - t_n''') \to z(t) \tag{4.16}$$

uniformly on compact subsets of $\mathbb{R}$, where z is a bounded solution of equation (4.1) on $\mathbb{R}$. From the uniqueness of the bounded solution (Theorem 4.1.6), we conclude that $z = x_f$. We deduce from (4.14) and (4.16) that x_f is compact almost automorphic. □

Remark. The approach in [96, 142] cannot be used to prove Theorem 4.1.9. In fact, under our assumptions, the function $g : \mathbb{R} \times X \to X$ in the proof of Theorem 4.1.9 is only measurable and not necessarily continuous. Thus one cannot say anything about the existence of a solution x_g of the limiting equation

$$x'(t) = g(t, x(t)).$$

4.1.2 Examples

4.1.2.1 Differential equations on Banach spaces

Consider the following differential equation:

$$x'(t) + q(t) g(\|x(t)\|) x(t) = e(t), \tag{4.17}$$

where $q : \mathbb{R} \to \mathbb{R}$ is almost automorphic and there exists an almost periodic function δ with $\mathcal{M}(\delta) > 0$ and $0 \le \delta(t) \le q(t)$, $g : \mathbb{R}^+ \to \mathbb{R}^+$ is continuous nondecreasing, continuously differentiable, $g(u) > 0$ for $u > 0$ and there exists $0 \leq \alpha < 1$ such that $ug'(u) \leq \alpha g(u)$ for each $u > 0$. An example of such a function is $g(u) = u^r$ with $0 \leq r < 1$. The function $e : \mathbb{R} \to X$ is a continuous Stepanov almost automorphic function; see [264] for an example of such a function.

Proposition 4.1.12. *Equation* (4.17) *has a unique bounded solution x that is compact almost automorphic and globally attractive.*

Proof. Consider the function

$$f(t,x) := -q(t)g(\|x\|)x + e(t).$$

Denote by $\langle \cdot,\cdot \rangle_{\pm}$ the semiinner products associated to the duality mapping on X. That is, for each $x, y \in X$,

$$\langle y, x\rangle_- =: \inf_{\varphi \in F(x)} \operatorname{Re}\langle \varphi, y\rangle \quad \text{and} \quad \langle y, x\rangle_+ =: \sup_{\varphi \in F(x)} \operatorname{Re}\langle \varphi, y\rangle,$$

where F is the multivalued duality mapping defined for each $x \in X$ by

$$F(x) = \{\varphi \in X^* : \operatorname{Re}\langle \varphi, x\rangle = \|x\|^2 = \|\varphi\|^2\}.$$

The functionals $[\cdot,\cdot]_{\pm}$ and the semiinner products $\langle \cdot,\cdot \rangle_{\pm}$ are related by the following formula:

$$\|x\|[x,y]_{\pm} = \langle y, x\rangle_{\pm}, \tag{4.18}$$

for all $x, y \in X$; see [212, p. 228] for more details.

It was proved in [40] that there exists $C > 0$ such that, for all $x, y \in X$,

$$\langle g(\|x\|)x - g(\|y\|)y, x - y\rangle_+ \geq C\|x-y\|^2 g\left(\frac{\|x-y\|}{4}\right).$$

Thus by (4.18), we have, for all $x, y \in X$,

$$[x-y, g(\|x\|)x - g(\|y\|)y]_+ \geq C\|x-y\| g\left(\frac{\|x-y\|}{4}\right).$$

Since

$$[x-y, -q(t)g(\|x\|)x + q(t)g(\|y\|)y]_- = -q(t)[x-y, g(\|x\|)x - g(\|y\|)y]_+,$$

we get

$$[x-y, -q(t)g(\|x\|)x + q(t)g(\|y\|)y]_- \leq -Cq(t)\|x-y\| g\left(\frac{\|x-y\|}{4}\right).$$

That is,

$$[x-y, f(t,x)-f(t,y)]_- \leq -Cq(t)\|x-y\|g\left(\frac{\|x-y\|}{4}\right).$$

Therefore **(M1)**–**(M4)** are satisfied with $p(t) := -Cq(t)$ and $\theta(u) = g(\frac{u}{4})$. On the other hand, one can see that $f \in \mathrm{SAAU}(\mathbb{R}, X)$, that is, **(A′)** holds. The result is a consequence of Theorem 4.1.9. □

4.1.2.2 A hematopoiesis model

Consider the following hematopoiesis model:

$$x'(t) = -a(t)x(t) + b(t)\frac{x^m(t)}{1+x^n(t)} + c(t) \quad \text{for } t \in \mathbb{R}, \tag{4.19}$$

where $x(t)$ denotes the density of mature cells in blood circulation at time t. The cells are lost from the circulation at a time dependent rate $a(t)$, the term $b(t)\frac{x^m(t)}{1+x^n(t)}$ represents the flux of the cells into the circulation from the stem cell compartment. One can see that this flux takes into consideration the density of mature cells in blood circulation. On the other hand, the term $c(t)$ represents a flux of blood cells which does not take into consideration the density of blood cells in circulation (blood donation, for example).

In the real-world phenomena, the periodic variations of the environment (e. g., seasonal effects of weather, resource availability, reproduction, food supplies, mating habits, etc.) play an important role in many biological and ecological systems. In particular, the effects of a periodically varying environment for such systems differ from those of a stable environment. Thus, the assumption of periodicity of the parameters are a way of incorporating the periodicity of the environment.

On the other hand, compared with periodic effects, almost periodic effects are more frequent in many real world applications. Almost automorphic effects are, on the other hand, present in most almost periodic dynamics [185]. In view of this, it is realistic to assume that the parameters in the models are almost automorphic functions.

In what follows, we assume that $a(\cdot)$, $b(\cdot)$, and $c(\cdot)$ are nonnegative continuous Stepanov almost automorphic functions. We assume that $m, n \geq 0$ with $m \leq n+1$. Thus $\sup_{x\in\mathbb{R}} g'(x) := \overline{g} < \infty$ where g is the function defined for each $x \in \mathbb{R}$ by

$$g(x) := \frac{x^m}{1+x^n}.$$

We assume that there exists an almost periodic function δ with $\mathcal{M}(\delta) < 0$ such that, for all $t \in \mathbb{R}$,

$$\overline{g}b(t) - a(t) \leq \delta(t).$$

Consider the function $f : \mathbb{R} \times \mathbb{R} \to \mathbb{R}$ defined for $t, x \in \mathbb{R}$ by

$$f(t,x) = -a(t)x + b(t)g(x) + c(t).$$

For $x, y, t \in \mathbb{R}$ and $h > 0$,

$$\|x-y\| - \|x-y-h(f(t,x)-f(t,y))\| = \|x-y\|[1 - \|1 - h(-a(t) + b(t)g'(c_{x,y}))\|],$$

where $c_{x,y} \in [\min(x,y), \max(x,y)]$. For h small enough, we have, for all $x, y \in \mathbb{R}$,

$$\begin{aligned}\|x-y\| - \|x-y-h(f(t,x)-f(t,y))\| &= \|x-y\|h(-a(t) + b(t)g'(c_{x,y})) \\ &\leq h(\overline{g}b(t) - a(t))\|x-y\|.\end{aligned}$$

Thus for all $x, y, t \in \mathbb{R}$,

$$[x-y, f(t,x)-f(t,y)]_- \leq (\overline{g}b(t) - a(t))\|x-y\|.$$

Therefore **(M1)**–**(M4)** are satisfied with $p(t) := \overline{g}b(t) - a(t)$ and $\theta(u) = 1$. In the other hand, since the function g has a bounded derivative (because $m \leq n+1$), it is Lipschitz and then maps bounded sets of $\mathbb{R}$ into bounded sets. Thus one can see that $f \in SAAU(\mathbb{R} \times \mathbb{R}, \mathbb{R})$. It follows from Theorem 4.1.9 that (4.19) has a unique compact almost automorphic solution that is globally attractive.

Remark 4.1.13. If $c(t) = 0$, for all $t \in \mathbb{R}$, then zero is a solution of (4.19), thus it is the unique compact almost automorphic solution and it is globally attractive. That is, all the trajectories vanish as time passes.

Figure 4.1 shows the asymptotic behavior of the trajectories and the attractiveness of the compact almost automorphic solution for $m = 1$, $n = 2$, and

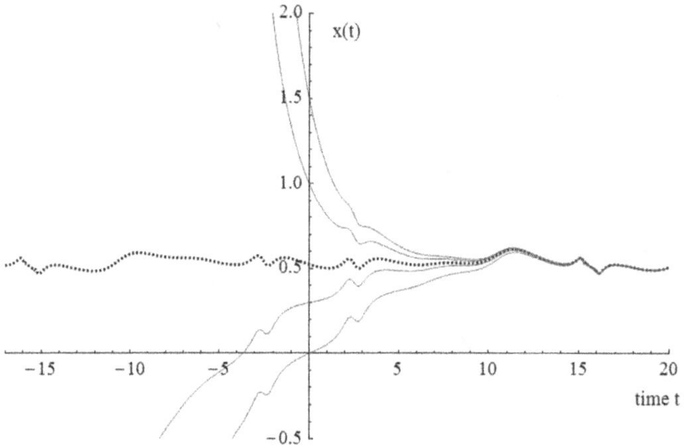

Figure 4.1: Attractiveness of the compact almost automorphic solution (dashed line).

$$\begin{cases} a(t) = 0.2\sin(\frac{1}{2+\cos(t)+\cos(\sqrt{2}t)}) + 0.5, \\ b(t) = 0.2\sin(\frac{1}{2+\cos(t)+\cos(\sqrt{2}t)}) + 0.45, \\ c(t) = 0.2\sin(\frac{1}{2+\cos(t)+\cos(\sqrt{2}t)}). \end{cases}$$

We note that these functions are almost automorphic but not uniformly continuous, and thus are not compact almost automorphic (see [49]). In this case we have $\overline{g} = 1$ and $p(t) = -0.05$.

4.2 Bohr–Neugebauer property for abstract evolution equations

In [101], Cooke proved the Bohr–Neugebauer property for the following differential equation:

$$y^{(n)}(t) + A_1 y^{(n-1)}(t) + \cdots + A_n y(t) = f(t),$$

where $f : \mathbb{R} \to H$ and A_i, $i = 1, \ldots, n$ are compact operators in a separable Hilbert space H. In [280, 281], Zaidman proved that the following equation:

$$\frac{\mathrm{d}}{\mathrm{d}t} x(t) = Ax(t) + f(t) \quad \text{for } t \in \mathbb{R} \tag{4.20}$$

possesses the Bohr–Neugebauer property when A is a self-adjoint operator in a Hilbert space. In another work, Zaidman [282] studied this property for equation (4.20) when A is a finite rank operator. In [157], Goldstein proved the Bohr–Neugebauer property for equation (4.20) under a more general "finite dimensionality assumption" when A is a closed linear operator in a Hilbert space.

While the above works investigated the Bohr–Neugebauer property in Hilbert spaces, it is interesting to know whether this property holds in Banach spaces. This is an important question because many partial differential equations can be formulated as abstract differential equations such as (4.20) with solutions living in purely Banach spaces. Age and size-structured population models are an example of such equations (see Section 4.2.1).

Motivated by this observation, we investigate in Section 4.2 the Bohr–Neugebauer property for equation (4.20) when A is the generator of a C_0-semigroup $(T(t))_{t\geq 0}$ on a Banach space X. We prove that the Bohr–Neugebauer property holds for equation (4.20) when the semigroup $(T(t))_{t\geq 0}$ is quasicompact, that is, when $(T(t))_{t\geq 0}$ has a negative essential growth bound. A special case of this is when the semigroup is eventually compact. An application will be given for some structured-population model.

Semigroup methods have been applied with great success to equations arising from biomathematical models describing the growth (and/or properties like diffusion or convection) of certain populations [218, 252, 253, 270]. In many applications, like population equations, heat equations, transport processes, or some reaction–diffusion equations, solutions with a positive initial value should remain positive. This property is called positivity. With the development of the theory of ordered Banach spaces and positive operators in the 1960s and 1970s, positivity constitutes an elegant way to get access to asymptotic information of semigroups. For more information about this subject, we refer to the monograph of Nagel [227] and the recent monograph of Bátkai, Kramar-Fijavz, and Rhandi [51].

Interest in understanding the dynamics of biological populations is old. Classical, ordinary differential equation models assume homogeneity of individuals within population classes, and involve equations for total population sizes. However, individuals in biological populations differ in their physiological characteristics. Vital rates, such as those of birth, death, and development, vary among individuals. Therefore physiologically structured partial differential equation models are often more useful to understand the dynamics of biological populations. We refer the interested reader to the monographs [183] for basic concepts and results in the theory of structured populations.

Let $(X, |\cdot|)$ be a Banach space. Consider the following evolution equation in X:

$$\frac{\mathrm{d}}{\mathrm{d}t}x(t) = Ax(t) + f(t) \quad \text{for } t \in \mathbb{R}, \tag{4.21}$$

where A is a linear (unbounded) operator which generates a C_0-semigroup $(T(t))_{t\geq 0}$ in the Banach space X.

Definition 4.2.1. A continuous function $x : \mathbb{R} \to X$ is said to be a mild solution of (4.21) on $\mathbb{R}$ if for all $t, \sigma \in \mathbb{R}$ with $t \geq \sigma$,

$$x(t) = T(t-\sigma)x(\sigma) + \int_{\sigma}^{t} T(t-s)f(s)\mathrm{d}s.$$

In what follows, mild solutions of (4.21) will be simply called solutions.

Our goal now is to investigate the nature of solutions of (4.21) which are bounded on $\mathbb{R}$. The following result gives a new sufficient condition under which equation (4.21) has the Bohr–Neugebauer property.

Theorem 4.2.2. *If* $\omega_{\mathrm{ess}}(T) < 0$ *(see Definition* 2.1*) and* $f(\cdot)$ *is almost periodic, then every bounded solution of* (4.21) *on* $\mathbb{R}$ *is almost periodic.*

Remark. Theorem 4.2.2 generalizes the Bohr–Neugebauer theorem in [79, 152] since C_0-semigroups are always compact in finite-dimensional spaces, and thus $\omega_{\mathrm{ess}}(T) = -\infty$ by (2.1). In an abstract Banach space, a special case of $\omega_{\mathrm{ess}}(T) = -\infty$ is when $(T(t))_{t\geq 0}$ is eventually compact, that is, when $T(t_0)$ is compact for some $t_0 > 0$.

Proof of Theorem 4.2.2. The condition $\omega_{\mathrm{ess}}(T) < 0$ ensures that the essential spectrum of the operator A does not touch the imaginary axis $\{\lambda \in \mathbb{C} : \operatorname{Re}\lambda = 0\}$ since, by (2.3), the essential spectrum is in general located at the left half-plane with respect to the vertical line $\{\lambda \in \mathbb{C} : \operatorname{Re}\lambda = \omega_{\mathrm{ess}}(T)\}$ (see Figure 4.2).

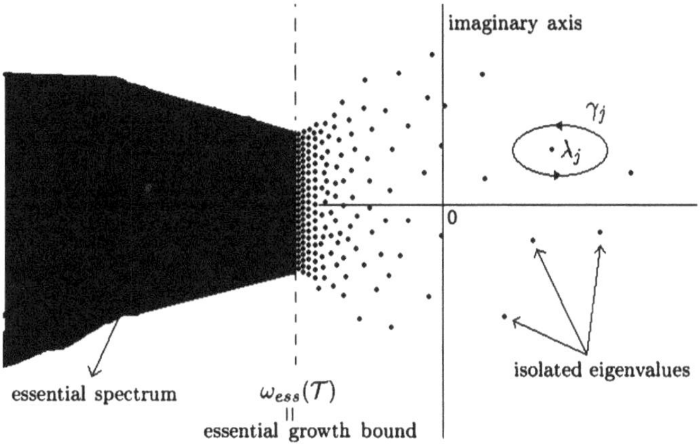

Figure 4.2: Spectrum of the operator A.

Consider the following sets:

$$\sigma_s := \{\lambda \in \sigma(A) : \operatorname{Re}\lambda < 0\}$$

and

$$\sigma_u := \{\lambda \in \sigma(A) : \operatorname{Re}\lambda \geq 0\}.$$

From [136, Corollary IV.2.11 and Theorem V.3.1] and (2.3), the set σ_u is finite and $\sigma_u \cap \sigma_{\mathrm{ess}}(A) = \emptyset$. Thus σ_u contains only isolated eigenvalues of A. Let $\sigma_u = \{\lambda_1, \ldots, \lambda_n\}$ and define the following operators:

$$\Pi_j := \frac{1}{2\pi i} \int_{\gamma_j} R(\lambda, A) d\lambda$$

for each $1 \leq j \leq n$, where γ_j is a positively oriented closed curve in $\mathbb{C}$ enclosing the isolated singularity λ_j, but no other points of $\sigma(A)$ (see Figure 4.2). Then Π_j is a projection in X and $\Pi_j\Pi_h = 0$ for $j \neq h$. Let $U_j := R(\Pi_j)$ be the range of Π_j. Then A restricted to U_j is a bounded operator with spectrum consisting of the single point λ_j. Let $\Pi^U := \sum_{j=1}^{k} \Pi_j$, $\Pi^S = I - \Pi^U$, $S = R(\Pi^S)$, and $U = U_1 \oplus \cdots \oplus U_n$. Then

$$X = U \oplus S,$$

and U and S are closed subspaces of X which are invariant under the semigroup $(T(t))_{t\geq 0}$. The subspace U is finite dimensional. Moreover, for every sufficiently small $\varepsilon > 0$, there exists $C_\varepsilon > 0$ such that

$$\|T(t)x\| \le C_\varepsilon e^{-\varepsilon t}\|x\| \quad \text{for } t \ge 0 \text{ and } x \in S. \tag{4.22}$$

For more details, we refer the reader to [270, Proposition 4.15].

In what follows, $T^U(t)$ and $T^S(t)$ denote the restrictions of $T(t)$ on U and S, respectively. Then $(T^U(t))_{t\in\mathbb{R}}$ is a group of operators and

$$T^U(t) = e^{tA_U} \quad \text{with } A_U \in \mathcal{L}(U).$$

Let x be a bounded solution of equation (4.21) on $\mathbb{R}$. Then we have, for all $t \in \mathbb{R}$,

$$x(t) = \Pi^S x(t) + \Pi^U x(t). \tag{4.23}$$

On the other hand, we have for each $t \ge \sigma$, with $t, \sigma \in \mathbb{R}$,

$$x(t) = T(t-\sigma)x(\sigma) + \int_\sigma^t T(t-\tau)f(\tau)\mathrm{d}\tau. \tag{4.24}$$

It follows that

$$\Pi^S x(t) = T^S(t-\sigma)\Pi^S x(\sigma) + \int_\sigma^t T^S(t-\tau)\Pi^S f(\tau)\mathrm{d}\tau. \tag{4.25}$$

Since $T^S(t)$ is exponentially stable by (4.22), by letting σ tend to $-\infty$ in (4.25), we then obtain

$$\Pi^S x(t) = \int_{-\infty}^t T^S(t-\tau)\Pi^S f(\tau)\mathrm{d}\tau,$$

which is well defined since $f(\cdot)$ is bounded on $\mathbb{R}$.

We claim that $t \mapsto \Pi^S x(t)$ is almost periodic. In fact, let $(t_n)_{n\in\mathbb{N}}$ be sequence of real numbers. Since $f \in AP(\mathbb{R}, X)$, by Bochner's characterization there exist a subsequence $(t_n') \subset (t_n)$ and a function $\tilde{f}(t)$ such that $(f(\cdot + t_n))_{n\in\mathbb{N}}$ converges uniformly in $\mathbb{R}$ to $\tilde{f}(\cdot)$. Let

$$\tilde{x}(t) := \int_{-\infty}^t T^S(t-\tau)\tilde{f}(\tau)\mathrm{d}\tau \quad \text{for } t \in \mathbb{R}.$$

Then

$$\begin{aligned} |\Pi^S x(t+t_n) - \tilde{x}(t)| &\le C_\varepsilon \int_{-\infty}^t e^{-\varepsilon(t-\tau)}|f(\tau+t_n) - \tilde{f}(\tau)|\mathrm{d}\tau \\ &\le \frac{C_\varepsilon}{\varepsilon} \sup_{\tau\in\mathbb{R}}|f(\tau+t_n) - \tilde{f}(\tau)| \to 0 \end{aligned}$$

as $n \to \infty$ uniformly with respect to $t \in \mathbb{R}$.

On the other hand, from (4.24) we have

$$\Pi^U x(t) = e^{(t-\sigma)A_U}\Pi^U x(\sigma) + \int_\sigma^t e^{(t-\tau)A_U}\Pi^U f(\tau)\mathrm{d}\tau.$$

Since A_U is a bounded operator in U, one can see that

$$\frac{\mathrm{d}}{\mathrm{d}t}\Pi^U x(t) = A_U\Pi^U x(t) + \Pi^U f(t). \tag{4.26}$$

We remind that U is a finite-dimensional subspace of X. Let $\{e_1,\dots,e_n\}$ be a basis of U. Let $(x_i(t))_{i=1,\dots,n}$ and $(f_i(t))_{i=1,\dots,n}$ be the coordinates of $\Pi^U x(t)$ and $\Pi^U f(t)$, respectively. For each $i = 1,\dots,n$, $(a_{ij})_{j=1,\dots,n}$ denote the coordinates of $A_U e_i$. Then, we get

$$A_U\Pi^U x(t) = \sum_{i=1}^n\left(\sum_{j=1}^n x_j(t)a_{ji}\right)e_i.$$

Define the matrix $M := (a_{ij})_{i,j}$. Then equation (4.26) is equivalent to

$$\begin{pmatrix}\dot{x}_1(t)\\ \dot{x}_i(t)\\ \dot{x}_n(t)\end{pmatrix} = M^T\begin{pmatrix}x_1(t)\\ x_i(t)\\ x_n(t)\end{pmatrix} + \begin{pmatrix}f_1(t)\\ f_i(t)\\ f_n(t)\end{pmatrix}.$$

Since $x(\cdot)$ is bounded on $\mathbb{R}$, $(x_1(\cdot),\dots,x_n(\cdot))^T$ is also bounded on $\mathbb{R}$. On the other hand, $(f_1(\cdot),\dots,f_n(\cdot))^T$ is almost periodic because $f(\cdot) \in AP(\mathbb{R}, X)$. By using Theorem 3.1.1, we deduce that $(x_1(\cdot),\dots,x_n(\cdot))^T$ is almost periodic. As a result, $x_i(t) \in AP(\mathbb{R},\mathbb{R})$ for each $i = 1,\dots,n$. It follows that $\Pi^U x(t)$ is almost periodic. Now from (4.23) we conclude that $x(\cdot)$ is almost periodic. □

Corollary 4.2.3. *If $\omega_0(T) < 0$ and $f(\cdot)$ is almost periodic, then there exists a unique bounded solution of* (4.21) *on $\mathbb{R}$ which is almost periodic and globally attractive. As a consequence, all other solutions are asymptotically almost periodic.*

Proof. Let $M \geq 1$, $\omega > 0$ be such that $\|T(t)\| \leq Me^{-\omega t}$ for $t \geq 0$. Consider the function defined on $\mathbb{R}$ by

$$z(t) := \int_{-\infty}^t T(t-s)f(s)\mathrm{d}s.$$

We claim that z is a bounded mild solution of equation (4.21) on $\mathbb{R}$. In fact, for each $t \geq \sigma$ with $t,\sigma \in \mathbb{R}$, we have

$$T(t-\sigma)z(\sigma) = \int_{-\infty}^\sigma T(t-s)f(s)\mathrm{d}s.$$

It follows that

$$T(t-\sigma)z(\sigma)+\int_{\sigma}^{t}T(t-s)f(s)\mathrm{d}s=\int_{-\infty}^{\sigma}T(t-s)f(s)\mathrm{d}s+\int_{\sigma}^{t}T(t-s)f(s)\mathrm{d}s$$
$$=\int_{-\infty}^{t}T(t-s)f(s)\mathrm{d}s$$
$$=z(t).$$

Thus z is a solution of equation (4.21) on $\mathbb{R}$. On the other hand, we have, for each $t\in\mathbb{R}$,

$$\begin{aligned}\|z(t)\| &\le \int_{-\infty}^{t}\|T(t-s)\|\|f(s)\|\mathrm{d}s\\ &\le M\left(\int_{-\infty}^{t}e^{-\omega(t-s)}\mathrm{d}s\right)\|f\|_{\infty}\\ &\le \frac{M}{\omega}\|f\|_{\infty}.\end{aligned}$$

Thus z is bounded. But since

$$\omega_{\mathrm{ess}}(T)\le\omega_0(T)<0,$$

by Theorem 4.2.2 we then conclude that the bounded solution z is almost periodic.

Let y be another solution which is bounded on $\mathbb{R}$. Then for each $t\ge\sigma$ with $t,\sigma\in\mathbb{R}$,

$$y(t)=T(t-\sigma)y(\sigma)+\int_{\sigma}^{t}T(t-s)f(s)\mathrm{d}s.$$

But since

$$z(t)=T(t-\sigma)z(\sigma)+\int_{\sigma}^{t}T(t-s)f(s)\mathrm{d}s,$$

it follows that

$$\|z(t)-y(t)\|\le Me^{-\omega(t-\sigma)}\|z(\sigma)-y(\sigma)\|\le\widetilde{M}e^{-\omega(t-\sigma)},\tag{4.27}$$

where $\widetilde{M}:=M\sup_{\sigma\in\mathbb{R}}\|z(\sigma)-y(\sigma)\|$. By letting σ tend to $-\infty$ in (4.27), we obtain

$$\|z(t)-y(t)\|=0\quad\text{for each } t\in\mathbb{R}.$$

Hence the uniqueness of z. Next we show that this solution is globally attractive. For this, consider another solution w. Then we obtain for each $t \geq 0$,

$$\|z(t) - w(t)\| \leq Me^{-\omega t}\|z(0) - w(0)\|.$$

This implies that

$$\|z(t) - w(t)\| \to 0,$$

as $t \to \infty$. Consequently, if we consider the following decomposition of w:

$$w(t) = z(t) + \big(w(t) - z(t)\big),$$

then it is clear that w is asymptotically almost periodic. □

4.2.1 Application to a size-structured population model

In this subsection, we give certain concrete sufficient conditions for the Bohr–Neugebauer property to hold for a model of size-structured population model with distributed delay and infinite states-at-birth. Moreover, we give sufficient conditions for the existence of a globally attractive almost periodic solution. We consider a population of individuals that are distinguished by their individual size. Therefore, we can describe the density of population of size s at time t by the number $u(t,s)$. More precisely, $\int_{s_1}^{s_2} u(t,s)ds$ is the number of individuals that at time t have size s between s_1 and s_2. As time passes, the following processes are supposed to take place in this population:

- Individuals grow linearly in time at constant speed 1.
- Individuals are subject to a size-dependent mortality denoted by μ.
- It is assumed that individuals may have different sizes at birth, and therefore $\beta(\sigma, s, b)$ gives the rate at which an individual of size b produces offspring of the size s. This process is assumed to occur with a time delay $-\sigma$ smaller than 1 (e. g., pregnancy duration).
- The population is subject to two different kinds of migration:
  - A density-dependent migration process with a time lag $r = 1$, represented by the term $v(s)u(t-1,s)$;
  - A time-dependent migration process due to local oscillating environment, which does not take into account the state of the population, represented by the oscillating almost periodic term $h(t)w(s)$.

From these assumptions the following evolution equation can be derived:

$$
\begin{cases}
\frac{\partial}{\partial t}u(t,s) = -\frac{\partial}{\partial s}u(t,s) - \mu(s)u(t,s) + \nu(s)u(t-1,s) \\
\qquad\qquad + \int_0^\infty \int_{-1}^0 \beta(\sigma,s,b)u(t+\sigma,b)d\sigma db + h(t)w(s) & \text{for } t \geq 0, s \in \mathbb{R}^+, \\
u(t,0) = 0 & \text{for } t \geq 0, \\
u(\sigma,s) = \varphi(\sigma,s) & \text{for } (\sigma,s) \in [-1,0] \times \mathbb{R}^+.
\end{cases}
\tag{4.28}
$$

The model (4.28) is a size-structured population model with infinite states-at-birth. This model is similar to that given in [43] when we take $\nu = 0$ and $h = 0$. However, here we do not impose a maximal size. Thus the semigroup corresponding to (4.28) will not be eventually compact, unlike in [43]. The compactness of semigroups plays an essential role in analyzing the spectral properties and thus understanding the asymptotic behavior. To overcome the lack of compactness, we will use tools of the theory of positive semigroups [51, 161, 227].

In the sequel, μ, ν, and β are supposed to satisfy the following conditions: $\mu, \nu \in L^\infty(\mathbb{R}^+)$ are positive functions such that the limits

$$
\mu_\infty = \lim_{s\to\infty} \mu(s) \quad \text{and} \quad \nu_\infty = \lim_{s\to\infty} \nu(s)
$$

exist. The birth function $\beta : [-1,0] \times \mathbb{R}^+ \times \mathbb{R}^+ \to \mathbb{R}^+$ satisfies

$$
\sup_{\substack{-1\leq\sigma\leq 0 \\ b\geq 0}} \int_0^\infty \beta(\sigma,s,b)ds < \infty.
\tag{4.29}
$$

The function $h : \mathbb{R} \to \mathbb{R}$ is almost periodic and $w \in L^1(\mathbb{R}^+)$.

Note that our assumptions are tailored toward the mathematical analysis of the model. In case of a specific population, one can make additional assumptions on the model ingredients, such as $\beta(\sigma,s,b) = 0$ whenever $s > b$, that is, individuals can only produce offspring of smaller size.

To write this equation in an abstract form, we introduce the Banach lattice $X = L^1(\mathbb{R}^+)$ and the operator A defined on X by

$$
\begin{cases}
D(A) = \{z \in W^{1,1}(\mathbb{R}^+) : z(0) = 0\}, \\
(Az)(s) = -z'(s) - \mu(s)z(s) \quad \text{for each } s \in \mathbb{R}^+.
\end{cases}
$$

The operator A generates on X the semigroup given for each $z \in X$ by

$$
(T(t)z)(s) = \begin{cases}
0 & \text{for } s < t, \\
e^{-\int_{s-t}^s \mu(b)db} z(s-t) & \text{for } s > t.
\end{cases}
\tag{4.30}
$$

We introduce also the delay operator $\Phi : W^{1,1}([-1,0],X) \to X$ defined for each $\varphi \in W^{1,1}([-1,0],X)$ and $s \geq 0$ by

$$(\Phi\varphi)(s) := \nu(s)\varphi(-1)(s) + \int_0^\infty \int_{-1}^0 \beta(\sigma, s, b)\varphi(\sigma)(b)\mathrm{d}\sigma\mathrm{d}b. \tag{4.31}$$

System (4.28) can be written on the Banach lattice $X = L^1(\mathbb{R}^+)$ in the following abstract form:

$$\begin{cases} \dot{u}(t) = Au(t) + \Phi(u_t) + H(t) & \text{for } t \geq 0, \\ u(0) = y \in X, \\ u_0 = \varphi \in L^1([-1,0],X), \end{cases} \tag{4.32}$$

where $H(t) = h(t)w(\cdot)$. One can hide the delay term in equation (4.32) by formulating this equation in a more abstract form. To do this, we introduce the product space $\mathcal{X} = X \times L^1([-1,0],X)$ and the function

$$\mathcal{U}(t) := \begin{pmatrix} u(t) \\ u_t \end{pmatrix} \in \mathcal{X}.$$

Further, on this product space we define the operator

$$\begin{cases} D(\mathcal{A}) := \{ \left(\begin{smallmatrix} z \\ \varphi \end{smallmatrix}\right) \in D(A) \times W^{1,1}([-1,0],X) : \varphi(0) = z \}, \\ \mathcal{A} := \left(\begin{smallmatrix} A & \Phi \\ 0 & \frac{\mathrm{d}}{\mathrm{d}\sigma} \end{smallmatrix}\right), \end{cases}$$

where $\frac{\mathrm{d}}{\mathrm{d}\sigma}$ denotes the derivative.

Proposition 4.2.4. *Equation* (4.32) *is equivalent to the abstract inhomogeneous Cauchy problem*

$$\begin{cases} \dot{\mathcal{U}}(t) = \mathcal{A}\mathcal{U}(t) + \mathcal{H}(t), & t \geq 0, \\ \mathcal{U}(0) = \left(\begin{smallmatrix} y \\ \varphi \end{smallmatrix}\right) \end{cases}$$

on $\mathcal{X}$, where $\mathcal{H}(t) := \left(\begin{smallmatrix} H(t) \\ 0 \end{smallmatrix}\right)$.

Proof. The proof is similar to the one in [52] with $H = 0$. □

To show that $\mathcal{A}$ generates a C_0-semigroup on $\mathcal{X}$, we split it as

$$\mathcal{A} := \begin{pmatrix} A & \Phi \\ 0 & \frac{\mathrm{d}}{\mathrm{d}\sigma} \end{pmatrix} = \begin{pmatrix} A & 0 \\ 0 & \frac{\mathrm{d}}{\mathrm{d}\sigma} \end{pmatrix} + \begin{pmatrix} 0 & \Phi \\ 0 & 0 \end{pmatrix} =: \mathcal{A}_0 + \mathcal{A}_\Phi, \tag{4.33}$$

where

$$\begin{cases} D(\mathcal{A}_0) := D(\mathcal{A}), \\ \mathcal{A}_0 := \left(\begin{smallmatrix} A & 0 \\ 0 & \frac{\mathrm{d}}{\mathrm{d}\sigma} \end{smallmatrix}\right) \end{cases} \quad \text{and} \quad \begin{cases} D(\mathcal{A}_\Phi) := D(\mathcal{A}), \\ \mathcal{A}_\Phi := \left(\begin{smallmatrix} 0 & \Phi \\ 0 & 0 \end{smallmatrix}\right). \end{cases}$$

Proposition 4.2.5 ([53, Theorem 3.25, p. 64]). *The operator $\mathcal{A}_0$ generates a C_0-semigroup given explicitly by the formula*

$$\mathcal{T}_0(t) := \begin{pmatrix} T(t) & 0 \\ T_t & T_l(t) \end{pmatrix}, \tag{4.34}$$

where $(T_l(t))_{t\geq 0}$ is the nilpotent left-shift semigroup on $L^1([-1,0],X)$ and $T_t : X \to L^1([-1,0],X)$ is defined by

$$(T_t f)(\tau) := \begin{cases} T(t+\tau)f & \text{if } -t < \tau \leq 0, \\ 0 & \text{if } -1 \leq \tau \leq -t. \end{cases}$$

One can see that the perturbation operator $\mathcal{A}_\Phi$ satisfies the Miyadera–Voigt condition [53, Theorem 1.37]. Moreover, we observe that the semigroup $(T(t))_{t\geq 0}$ (see (4.30)) and the delay operator Φ (see (4.31)) are positive. Thus from [53, Theorem 6.10], we have the following result.

Proposition 4.2.6. *$\mathcal{A}$ generates a positive C_0-semigroup $(\mathcal{T}(t))_{t\geq 0}$ on $\mathcal{X}$.*

We recall the operator Φ_λ defined on X for each $\lambda \in \mathbb{C}$ and $z \in X$ by

$$\Phi_\lambda(z)(s) := \Phi(e^{\lambda(.)}z)(s) = v(s)e^{-\lambda}z(s) + \int_0^\infty \int_{-1}^0 \beta(\sigma,s,b)e^{\lambda\sigma}z(b)d\sigma db.$$

We have the following result.

Lemma 4.2.7 ([53, Theorem 6.15]). *For each $\lambda \in \mathbb{R}$, if $s(A+\Phi_\lambda) < \lambda$, then $s(\mathcal{A}) < \lambda$.*

Lemma 4.2.8 ([227, Theorem 1.1, p. 334]). *Let B be the generator of a positive C_0-semigroup $(S(t))_{t\geq 0}$ on the Banach lattice $L^1(\Omega,\mu)$, where (Ω,μ) is a σ-finite measure space. Then $\omega_0(S) = s(B)$ holds.*

We now investigate when equation (4.28) has the Bohr–Neugebauer property.

Proposition 4.2.9. *Assume that*

$$v_\infty < \mu_\infty \tag{4.35}$$

and β satisfies the following conditions:

$$\lim_{a\to\infty} \sup_{\substack{-1\leq\sigma\leq 0 \\ b\geq 0}} \int_a^\infty \beta(\sigma,s,b)ds = 0 \tag{4.36}$$

and

$$\lim_{h\to 0} \sup_{\substack{-1\le\sigma\le 0\\ b\ge 0}} \int_0^\infty \|\beta(\sigma, s+h, b) - \beta(\sigma, s, b)\| \mathrm{d}s = 0. \tag{4.37}$$

Then equation (4.28) *has the Bohr–Neugebauer property.*

Proof. It suffices to prove that $\omega_{\mathrm{ess}}(\mathcal{T}) < 0$. Consider the following decomposition:

$$\Phi = \Phi^1 + \Phi^2,$$

where Φ_1 is defined for each $\varphi \in W^{1,1}([-1,0], X)$ and $s \ge 0$ by

$$(\Phi^1\varphi)(s) := \nu(s)\varphi(-1)(s)$$

and Φ_2 is defined for each $\varphi \in L^1([-1,0], X)$ and $s \ge 0$ by

$$(\Phi^2\varphi)(s) := \int_0^\infty \int_{-1}^0 \beta(\sigma, s, b)\varphi(\sigma)(b)\mathrm{d}\sigma\mathrm{d}b.$$

Note that the condition (4.29) implies that Φ^2 is a bounded operator.

Consider the following decomposition of the operator $\mathcal{A}$:

$$\mathcal{A} = \begin{pmatrix} A & \Phi \\ 0 & \frac{\mathrm{d}}{\mathrm{d}\sigma} \end{pmatrix} = \begin{pmatrix} A & \Phi^1 + \Phi^2 \\ 0 & \frac{\mathrm{d}}{\mathrm{d}\sigma} \end{pmatrix} = \begin{pmatrix} A & \Phi^1 \\ 0 & \frac{\mathrm{d}}{\mathrm{d}\sigma} \end{pmatrix} + \begin{pmatrix} 0 & \Phi^2 \\ 0 & 0 \end{pmatrix} = \mathcal{A}_1 + \mathcal{K},$$

where $\mathcal{A}_1$ is the operator defined by

$$\begin{cases} D(\mathcal{A}_1) := D(\mathcal{A}), \\ \mathcal{A}_1 := \begin{pmatrix} A & \Phi^1 \\ 0 & \frac{\mathrm{d}}{\mathrm{d}\sigma} \end{pmatrix} \end{cases}$$

and $\mathcal{K} : \mathcal{X} \to \mathcal{X}$ is the bounded operator given by

$$\mathcal{K} = \begin{pmatrix} 0 & \Phi^2 \\ 0 & 0 \end{pmatrix}.$$

Using again [53, Theorem 1.37] and [53, Theorem 6.10], $\mathcal{A}_1$ generates a positive semigroup $(\mathcal{T}_1(t))_{t\ge 0}$ on the Banach lattice $\mathcal{X}$. Using the Fréchet–Kolmogorov theorem [277, p. 275], one can see that conditions (4.36) and (4.37) imply that the operator $\mathcal{K}$ is compact. Hence by [136, Proposition IV.2.12],

$$\omega_{\mathrm{ess}}(\mathcal{T}) = \omega_{\mathrm{ess}}(\mathcal{T}_1) \le \omega_0(\mathcal{T}_1). \tag{4.38}$$

The space $L^1([-1,0],X)$ is canonically isomorphic to $L^1([-1,0]\times\mathbb{R}^+)$ and the space $X\times L^1([-1,0]\times\mathbb{R}^+)$ with norm $\|(z,\varphi)\| = \|z\|_{L^1(\mathbb{R}^+)} + \|\varphi\|_{L^1([-1,0]\times\mathbb{R}^+)}$ is again an L^1-space. So by Lemma 4.2.8, it follows that

$$\omega_0(\mathcal{T}_1) = s(\mathcal{A}_1). \tag{4.39}$$

In order to investigate when $s(\mathcal{A}_1) < 0$, one needs to investigate $s(A+\Phi_0^1)$. The operator $A+\Phi_0^1$ is given by

$$((A+\Phi_0^1)z)(s) = -z'(s) - (\mu(s)-\nu(s))z(s) \quad \text{for } z \in D(A). \tag{4.40}$$

Let $(T_1(t))_{t\geq 0}$ be the semigroup generated by $(A+\Phi_0^1, D(A))$. The semigroup $(T_1(t))_{t\geq 0}$ is given explicitly for each $z \in X$ by

$$(T_1(t)z)(s) = \begin{cases} 0 & \text{for } s < t, \\ e^{-\int_{s-t}^{s}(\mu(b)-\nu(b))db} z(s-t) & \text{for } s > t. \end{cases} \tag{4.41}$$

Using the fact that

$$\lim_{s\to\infty} \nu(s) - \mu(s) = \nu_\infty - \mu_\infty$$

exists, one can prove as in [161, Proposition 2.1] that

$$\sigma(A+\Phi_0^1) = \{\lambda \in \mathbb{C} : \operatorname{Re}\lambda \leq \nu_\infty - \mu_\infty\} \quad \text{and} \quad \sigma(T_1(t)) = \{\lambda \in \mathbb{C} : \|\lambda\| \leq e^{(\nu_\infty-\mu_\infty)t}\}. \tag{4.42}$$

Thus for all $t \geq 0$,

$$e^{s(A+\Phi_0^1)t} = e^{\omega_0(T_1)t} = r(T_1(t)) = e^{(\nu_\infty-\mu_\infty)t}. \tag{4.43}$$

Therefore by assumption (4.35),

$$s(A+\Phi_0^1) = \nu_\infty - \mu_\infty < 0.$$

It follows by Lemma 4.2.7 that $s(\mathcal{A}_1) < 0$ and thus, by (4.38) and (4.39), we have $\omega_{ess}(\mathcal{T}) < 0$. The proof is complete by applying Theorem 4.2.2. □

In the sequel, we will assume that the birth rate has the following form:

$$\beta(\sigma,s,b) = \beta_1(s)\beta_2(\sigma,b),$$

where $\beta_1 \neq 0$. Then condition (4.29) becomes

$$\sup_{\substack{-1\leq\sigma\leq 0 \\ b\geq 0}} \beta_2(\sigma,b) < \infty \quad \text{and} \quad \int_0^\infty \beta_1(s)ds < \infty. \tag{4.44}$$

In reality, individuals with large sizes cannot give birth, thus without loss of generality we can assume that the birth function component $\beta_2(\sigma, s)$ vanishes for large sizes s. Let $m > 0$ be the maximal size of fertility.

Proposition 4.2.10. *Assume that*

$$s \mapsto \int_0^s \beta_1(b)\mathrm{d}b \in L^1(\mathbb{R}^+), \tag{4.45}$$

$$\int_0^\infty \left(\int_{-1}^0 \beta_2(\sigma, s)\mathrm{d}\sigma \right) \left(\int_0^s \beta_1(b) e^{-\int_b^s (\nu_\infty - \mu_\infty + \mu(c) - \nu(c))\mathrm{d}c} \mathrm{d}b \right) \mathrm{d}s > 1, \tag{4.46}$$

and

$$\int_0^\infty \left(\int_{-1}^0 \beta_2(\sigma, s)\mathrm{d}\sigma \right) \left(\int_0^s \beta_1(b) e^{-\int_b^s (\mu(c) - \nu(c))\mathrm{d}c} \mathrm{d}b \right) \mathrm{d}s < 1. \tag{4.47}$$

Then the size-structured population equation (4.28) *has a unique almost periodic solution which is globally attractive. As a consequence, all other solutions are asymptotically almost periodic.*

Remark. One can interpret Proposition 4.2.10 in this way: If the birth rate and the density-dependent migration are small enough with respect to the mortality rate, then the behavior of the population is fully dictated by the oscillating behavior of the local environment (non-density-dependent migration) as time passes.

The following lemma is needed in the proof of Proposition 4.2.10.

Lemma 4.2.11 ([161, Corollary 1.7]). *Let $S(t)_{t\geq 0}$ be a positive C_0-semigroup on a Banach lattice and let B be its infinitesimal generator. If there exist t_0 and a compact operator K such that $r(S(t_0) - K) < r(S(t_0))$, then $s(B)$ is an eigenvalue of B.*

Proof of Proposition 4.2.10. We will prove that $\omega_0(\mathcal{T}) < 0$. Since $\omega_0(\mathcal{T}) = s(\mathcal{A})$ by Lemma 4.2.8, it is sufficient to prove that $s(\mathcal{A}) < 0$. To do this, we will use Lemma 4.2.7 and investigate $s(A + \Phi_0)$. The operator $A + \Phi_0$ is given by

$$((A + \Phi_0)z)(s) = -z'(s) - (\mu(s) - \nu(s))z(s) + \beta_1(s) \int_0^m \int_{-1}^0 \beta_2(\sigma, b)z(b)\mathrm{d}\sigma\mathrm{d}b.$$

We claim that $s(A + \Phi_0)$ is an eigenvalue of $A + \Phi_0$. In fact, consider the following decomposition:

$$A + \Phi_0 = (A + \Phi_0^1) + \Phi_0^2, \tag{4.48}$$

where $A + \Phi_0^1$ is defined by (4.40) and Φ_0^2 is given by

$$(\Phi_0^2 z)(s) = \beta_1(s) \int_0^m \int_{-1}^0 \beta_2(\sigma, b) z(b) \mathrm{d}\sigma \mathrm{d}b \quad \text{for } z \in X.$$

The operator Φ_0^2 is of finite rank and thus compact. Note that the semigroup $(T_1(t))_{t\geq 0}$ generated by the operator $A + \Phi_0^1$ is not compact because $\sigma_{\mathrm{ess}}(A + \Phi_0^1) = \sigma(A + \Phi_0^1) \neq \emptyset$ (see (4.42)). Being a bounded perturbation of the operator $A + \Phi_0^1$, the operator $A + \Phi_0$ generates a positive semigroup $(T_0(t))_{t\geq 0}$. Using [136, Proposition IV.2.12], we deduce from the decomposition (4.48) and the compactness of the operator Φ_0^2 that the operator $T_0(t) - T_1(t)$ is compact for $t > 0$. Let $K := T_0(t_0) - T_1(t_0)$ for some $t_0 > 0$. Thus from (4.43) we have

$$r(T_0(t_0) - K) = r(T_1(t_0)) = e^{(\nu_\infty - \mu_\infty)t_0}.$$

Since $r(T_0(t)) = e^{\omega_0(T_0)t}$ for all $t \geq 0$, to show that $r(T_0(t_0) - K) < r(T_0(t_0))$ it suffices to establish that $(\nu_\infty - \mu_\infty) < \omega_0(T_0)$. Note that $\omega_0(T_0) = s(A + \Phi_0)$, again by Lemma 4.2.8. Consider the function ξ defined by

$$\xi(\lambda) = \int_0^m \left(\int_{-1}^0 \beta_2(\sigma, s) \mathrm{d}\sigma \right) \left(\int_0^s \beta_1(b) e^{-\int_b^s (\lambda + \mu(c) - \nu(c)) \mathrm{d}c} \mathrm{d}b \right) \mathrm{d}s - 1.$$

We have $\lim_{\lambda\to\infty} \xi(\lambda) = -1$ and $\lim_{\lambda\to-\infty} \xi(\lambda) = \infty$ and ξ is decreasing. This implies that there exists a unique $\lambda_0 \in \mathbb{R}$ such that

$$\xi(\lambda_0) = 0. \tag{4.49}$$

We claim that λ_0 is an eigenvalue of $A + \Phi_0$ with an eigenvector given by

$$z_0(s) = \int_0^s \beta_1(b) e^{-\int_b^s (\lambda_0 + \mu(c) - \nu(c)) \mathrm{d}c} \mathrm{d}b.$$

In fact, we have

$$z_0'(s) = -(\lambda_0 + \mu(s) - \nu(s)) z_0(s) + \beta_1(s). \tag{4.50}$$

Using (4.49) and (4.50), we have

$$\begin{aligned} ((A + \Phi_0) z_0)(s) &= -z_0'(s) - (\mu(s) - \nu(s)) z_0(s) + \beta_1(s) \int_0^m \left(\int_{-1}^0 \beta_2(\sigma, b) \mathrm{d}\sigma \right) z_0(b) \mathrm{d}b \\ &= -z_0'(s) - (\mu(s) - \nu(s)) z_0(s) + \beta_1(s)(\xi(\lambda_0) + 1) \end{aligned}$$

$$= -z_0'(s) - (\mu(s) - \nu(s))z_0(s) + \beta_1(s)$$
$$= \lambda_0 z_0(s).$$

Note that (4.46) is equivalent to $\xi(\nu_\infty - \mu_\infty) > 0$, which implies by the monotony of ξ that

$$\nu_\infty - \mu_\infty < \lambda_0; \tag{4.51}$$

see Figure 4.3.

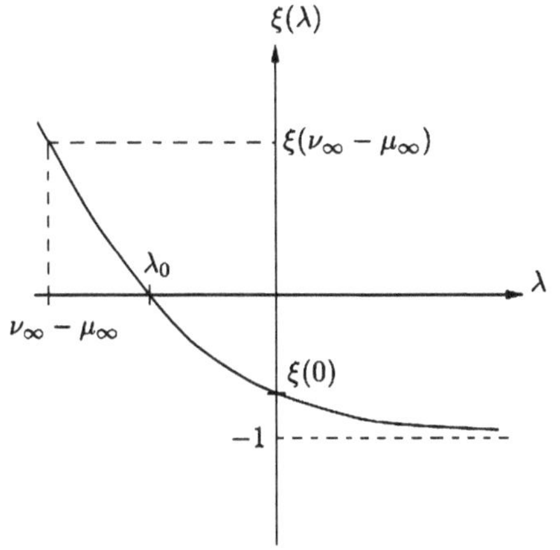

Figure 4.3: Graph of ξ.

One can see that (4.51), together with (4.45), ensures that $z_0 \in L^1(\mathbb{R}^+)$. Now by (4.44) and (4.50), we have $z_0' \in L^1(\mathbb{R}^+)$. We conclude that $z_0 \in W^{1,1}(\mathbb{R}^+)$ and thus $z_0 \in D(A+\Phi_0)$ because $z_0(0) = 0$. Since $z_0 \neq 0$, we deduce that λ_0 is an eigenvalue of $A + \Phi_0$. As a consequence, (4.51) implies that $\nu_\infty - \mu_\infty < s(A + \Phi_0)$ and thus $r(T_0(t_0) - K) < r(T_0(t_0))$. By applying Lemma 4.2.11, we deduce that $\lambda_1 := s(A+\Phi_0)$ is an eigenvalue of the operator $A + \Phi_0$. Thus there exists $z \in D(A)$ with $z \neq 0$ such that

$$((A + \Phi_0)z)(s) = \lambda_1 z(s),$$

that is,

$$z'(s) = -(\lambda_1 + \mu(s) - \nu(s))z(s) + \beta_1(s) \int_0^m \left(\int_{-1}^0 \beta_2(\sigma, b)\mathrm{d}\sigma \right) z(b)\mathrm{d}b. \tag{4.52}$$

By solving (4.52) and taking into account the fact that $z(0) = 0$, we get

$$z(s) = C_z \left(\int_0^s e^{-\int_b^s (\lambda_1 + \mu(c) - \nu(c))\mathrm{d}c} \beta_1(b)\mathrm{d}b \right), \tag{4.53}$$

where C_z is the constant given by $C_z := \int_0^m (\int_{-1}^0 \beta_2(\sigma, b)d\sigma)z(b)db$. We note that (4.53) implies $C_z \neq 0$ because $z \neq 0$. Multiplying (4.53) by $\int_{-1}^0 \beta_2(\sigma, s)d\sigma$, we have

$$\left(\int_{-1}^{0} \beta_2(\sigma, s)d\sigma\right)z(s) = C_z\left(\int_{-1}^{0} \beta_2(\sigma, s)d\sigma\right)\left(\int_{0}^{s} e^{-\int_b^s(\lambda_1+\mu(c)-\nu(c))dc}\beta_1(b)db\right).$$

Now, integrating the above equation and using the fact that $C_z \neq 0$, we get

$$1 = \int_0^m \left(\int_{-1}^{0} \beta_2(\sigma, s)d\sigma\right)\left(\int_{0}^{s} e^{-\int_b^s(\lambda_1+\mu(c)-\nu(c))dc}\beta_1(b)db\right)ds,$$

that is, $\xi(\lambda_1) = 0$. Thus $s(A + \Phi_0) = \lambda_1 = \lambda_0$ because λ_0 is the only real zero of ξ. Note that (4.47) is equivalent to $\xi(0) < 0$, which implies by monotony of ξ that $s(A+\Phi_0) = \lambda_0 < 0$ (see Figure 4.3). Therefore, using Lemma 4.2.7, we conclude that $\omega_0(\mathcal{T}) = s(\mathcal{A}) < 0$. The proof is complete by applying Corollary 4.2.3. □

4.3 Evolution equation using Favard condition

In [138], we investigated the existence of almost periodic solutions to the following differential equation:

$$\frac{d}{dt}x(t) = Ax(t) + f(t) \quad \text{for } t \in \mathbb{R}, \tag{4.54}$$

where $A : D(A) \to X$ is the generator of a C_0-group $(T(t))_{t\in\mathbb{R}}$. This time, we did not assume any kind of compactness or stability assumption on the semigroup $(T(t))_{t\in\mathbb{R}}$. On the other hand, we assumed that the phase space of solutions has a special geometric property, namely uniform convexity, and the C_0-group $(T(t))_{t\in\mathbb{R}}$ satisfies a separation property that appeared in earlier works of Favard for finite-dimensional ordinary differential equations. This class of C_0-groups turns out to contain a large class of C_0-groups. Thus, our results can be applied to a wide range of partial differential equations such as wave, Schrödinger, and transport equations. It is rewarding to see how a combination of the Favard property with geometric properties of the underlying Banach space allow us to overcome many of the typical difficulties encountered in the infinite-dimensional situation.

4.3.1 C_0-groups satisfying the Favard condition

The Favard condition plays an essential role in finding almost periodic solutions for differential equations. In this section, we study the class of C_0-groups which satisfy this condition.

Definition 4.3.1. We say that a C_0-group $(T(t))_{t\in\mathbb{R}}$ on a Banach space Z satisfies the Favard condition if each bounded trajectory $t \mapsto T(t)x$ satisfies

$$\inf_{t\in\mathbb{R}}|T(t)x| > 0 \quad \text{or} \quad x = 0.$$

Example 1. Consider the multiplication C_0-group defined on the Banach space $Z = L^1(\mathbb{R}, \mathbb{R})$ by

$$(T(t)z)(\xi) = e^{a(\xi)t}z(\xi) \quad \text{for } t, \xi \in \mathbb{R},$$

where

$$a(\xi) = \begin{cases} 1 & \text{if } \xi \geq 0, \\ 0 & \text{if } \xi < 0. \end{cases}$$

For each $z \in Z$, we have

$$|T(t)z| = \int_{-\infty}^{0} |z(\xi)|d\xi + e^t \int_{0}^{\infty} |z(\xi)|d\xi \quad \text{for } t \in \mathbb{R}.$$

If $t \mapsto T(t)z$ is a bounded trajectory, then there exists $M \geq 0$ such that

$$\int_{0}^{\infty} |z(\xi)|d\xi \leq Me^{-t} \quad \text{for all } t \in \mathbb{R}.$$

It follows by taking $t \to \infty$ that $\int_0^\infty |z(\xi)|d\xi = 0$, and we get $|T(t)z| = \int_{-\infty}^0 |z(\xi)|d\xi$. Consequently, if $\inf_{t\in\mathbb{R}} |T(t)z| = 0$, then $z = 0$. Thus, the C_0-group $(T(t))_{t\in\mathbb{R}}$ satisfies the Favard condition.

Now we give examples of C_0-groups which do not satisfy the Favard's condition.

Example 2. Consider the following C_0-group defined on the Banach space $Z = L^1(\mathbb{R}, \mathbb{R})$:

$$(T(t)z)(\xi) = e^{\int_\xi^{\xi+t} a(s)ds}z(\xi + t) \quad \text{for } t, \xi \in \mathbb{R},$$

where

$$a(\xi) = \begin{cases} 1 & \text{if } \xi \geq 0, \\ 0 & \text{if } \xi < 0. \end{cases}$$

For each $z \in Z$, we have

$$|T(t)z| = \begin{cases} \int_{-\infty}^t |z(\xi)|d\xi + e^t \int_t^0 e^{-\xi}|z(\xi)|d\xi + e^t \int_0^\infty |z(\xi)|d\xi & \text{if } t \leq 0, \\ \int_{-\infty}^0 |z(\xi)|d\xi + \int_0^t e^{\xi}|z(\xi)|d\xi + e^t \int_t^\infty |z(\xi)|d\xi & \text{if } t \geq 0. \end{cases}$$

Let z_0 be the function defined by

$$z_0(\xi) = \begin{cases} e^{2\xi} & \text{if } \xi \le 0, \\ 0 & \text{if } \xi \ge 0. \end{cases}$$

Then $z_0 \in Z$ and

$$|T(t)z_0| = \begin{cases} e^t - \frac{e^{2t}}{2} & \text{if } t \le 0, \\ \frac{1}{2} & \text{if } t \ge 0. \end{cases}$$

The trajectory $t \mapsto T(t)z_0$ is nontrivial and bounded on $\mathbb{R}$. However, we have $\lim_{t\to-\infty} |T(t)z_0| = 0$. This shows that $(T(t))_{t\in\mathbb{R}}$ does not satisfy the Favard condition.

Example 3. Let $\rho(s) = \frac{1}{1+s^2}$ for $s \in \mathbb{R}$. Consider the following Banach space:

$$Z = \{z : \mathbb{R} \to \mathbb{R} \text{ continuous such that } t \mapsto \rho(t)z(t) \in \mathrm{BUC}(\mathbb{R}, \mathbb{R})\}$$

equipped with the norm

$$|z|_\rho = \sup_{s\in\mathbb{R}} |\rho(s)z(s)| \quad \text{for each } z \in Z,$$

where $\mathrm{BUC}(\mathbb{R}, \mathbb{R})$ is the space of bounded uniformly continuous functions from $\mathbb{R}$ to $\mathbb{R}$. Now, we consider the left-shift group defined on Z by

$$(T(t)z)(s) = z(t + s) \quad \text{for all } t, s \in \mathbb{R}.$$

Then $(T(t))_{t\in\mathbb{R}}$ is a C_0-group on Z. In fact, if $z \in Z$, then from the uniform continuity of $s \mapsto \rho(s)z(s)$ and $s \mapsto \frac{\rho(s)}{\rho(t+s)}$ for each $t \in \mathbb{R}$, we get, for each $\varepsilon > 0$,

$$\begin{aligned} &|\rho(s)(T(t)z)(s) - \rho(s')(T(t)z)(s')| \\ &\quad = |\rho(s)z(t+s) - \rho(s')z(t+s')| \\ &\quad \le \left|\frac{\rho(s)}{\rho(t+s)}\rho(t+s)z(t+s) - \frac{\rho(s)}{\rho(t+s)}\rho(t+s')z(t+s')\right| \\ &\qquad + \left|\frac{\rho(s)}{\rho(t+s)}\rho(t+s')z(t+s') - \frac{\rho(s')}{\rho(t+s')}\rho(t+s')z(t+s')\right| \\ &\quad \le \left|\frac{\rho(s)}{\rho(t+s)}\right| |\rho(t+s)z(t+s) - \rho(t+s')z(t+s')| \\ &\qquad + \left|\frac{\rho(s)}{\rho(t+s)} - \frac{\rho(s')}{\rho(t+s')}\right| |\rho(t+s')z(t+s')| \\ &\quad \le n(t)\varepsilon + N\varepsilon, \end{aligned}$$

whenever $|s - s'|$ is small enough, where $n(t) = \sup_{s\in\mathbb{R}} |\frac{\rho(s)}{\rho(t+s)}| < \infty$ and $N = \sup_{s\in\mathbb{R}} |\rho(s)z(s)|$. Thus $s \mapsto \rho(s)(T(t)z)(s)$ is uniformly continuous, and then $T(t)z \in Z$

for each $t \in \mathbb{R}$. This shows that $T(t)$ maps Z into itself. The boundedness of the operator $T(t)$ follows from the following estimate:

$$\begin{aligned}
|T(t)z|_\rho &= \sup_{s\in\mathbb{R}} |\rho(s)z(t+s)| \\
&= \sup_{s\in\mathbb{R}} |\rho(s-t)z(s)| \le r(t) \sup_{s\in\mathbb{R}} |\rho(s)z(s)| = r(t)|z|_\rho,
\end{aligned}$$

where $r(t) = \sup_{s\in\mathbb{R}} \frac{\rho(s-t)}{\rho(s)} < \infty$. In the other hand, we have

$$\begin{aligned}
|T(t)z - z|_\rho &= \sup_{s\in\mathbb{R}} |\rho(s)z(t+s) - \rho(s)z(s)| \\
&\le \sup_{s\in\mathbb{R}} \left| \frac{\rho(s)}{\rho(s+t)} \rho(s+t)z(t+s) - \rho(s+t)z(t+s) \right| \\
&\quad + \sup_{s\in\mathbb{R}} |\rho(s+t)z(t+s) - \rho(s)z(s)| \\
&= \sup_{s\in\mathbb{R}} \left| \frac{\rho(s)}{\rho(s+t)} - 1 \right| N + \sup_{s\in\mathbb{R}} |\rho(s+t)z(t+s) - \rho(s)z(s)| \\
&\le N a(t) + \sup_{s\in\mathbb{R}} |\rho(s+t)z(t+s) - \rho(s)z(s)|,
\end{aligned}$$

where $a(t) = \sup_{s\in\mathbb{R}} |\frac{\rho(s)}{\rho(s+t)} - 1| \le \frac{t^2}{2} + \frac{1}{2}\sqrt{4t^2 + t^4} \to 0$ as $t \to 0$. From the uniform continuity of $s \mapsto \rho(s)z(s)$, we have $\sup_{s\in\mathbb{R}} |\rho(s+t)z(t+s) - \rho(s)z(s)| \to 0$ as $t \to 0$. Therefore, $(T(t))_{t\in\mathbb{R}}$ is a strongly continuous group on Z.

Let $z_0 : \mathbb{R} \to \mathbb{R}$ be a continuous function with compact support $[a, b]$. Then, the function $x(t) = T(t)z_0$ is a nontrivial bounded trajectory of $(T(t))_{t\in\mathbb{R}}$ and $\inf_{t\in\mathbb{R}} |x(t)|_\rho = 0$. In fact,

$$\begin{aligned}
|x(t)|_\rho &= \sup_{s\in\mathbb{R}} |\rho(s)z_0(s+t)| \\
&= \sup_{\theta\in\mathbb{R}} |\rho(\theta - t)z_0(\theta)| \\
&= \sup_{a\le\theta\le b} |\rho(\theta - t)z_0(\theta)| \\
&\le |z_0|_\infty \sup_{a\le\theta\le b} |\rho(\theta - t)| = |z_0|_\infty \sup_{a-t\le s\le b-t} |\rho(s)| \to 0,
\end{aligned}$$

as $t \to \pm\infty$. We conclude that $(T(t))_{t\in\mathbb{R}}$ does not satisfy the Favard condition.

Now we give an important class of C_0-groups which satisfy the Favard condition.

Proposition 4.3.2. *Every bounded C_0-group satisfies the Favard condition.*

Proof. Let $M = \sup_{t\in\mathbb{R}} |T(t)|$. If $t \mapsto T(t)x$ is a trajectory for some $x \in X$ such that $\inf_{t\in\mathbb{R}} |T(t)x| = 0$, then we have

$$|x| = \inf_{t\in\mathbb{R}} |T(-t)T(t)x| \le M \inf_{t\in\mathbb{R}} |T(t)x|.$$

It follows that $x = 0$. □

The following proposition gives a relationship between the Favard condition and the almost automorphic aspect of the trajectories.

Proposition 4.3.3. *Let $(T(t))_{t\in\mathbb{R}}$ be a C_0-group such that for each bounded trajectory $t \mapsto T(t)x$ on $\mathbb{R}$, the scalar function $t \mapsto \langle\varphi, T(t)x\rangle$ is almost automorphic for each $\varphi \in X^*$. Then $(T(t))_{t\in\mathbb{R}}$ satisfies the Favard condition.*

Proof. Let $t \mapsto T(t)x$ be a bounded trajectory for some $x \in X$ such that $\inf_{t\in\mathbb{R}} |T(t)x| = 0$. We claim that $x = 0$. In fact, consider a sequence $(t_n)_{n\ge0}$ such that $\lim_{n\to\infty} |T(t_n)x| = 0$. Then, for all $t \in \mathbb{R}$, we have that

$$T(t + t_n)x = T(t)T(t_n)x \to 0, \tag{4.55}$$

as $n \to \infty$. Let $\varphi \in X^*$. Then, by the almost automorphy of the function $t \mapsto \langle\varphi, T(t)x\rangle$, there exists a subsequence $(t'_n)_n \subset (t_n)_n$ and a scalar function $t \mapsto h(t)$ such that, for each $t \in \mathbb{R}$,

$$\langle\varphi, T(t + t'_n)x\rangle \to h(t) \tag{4.56}$$

and

$$h(t - t'_n) \to \langle\varphi, T(t)x\rangle, \tag{4.57}$$

as $n \to \infty$. It follows from (4.55) and (4.56) that $h(t) = 0$ for all $t \in \mathbb{R}$. Now from (4.57) we can see that $\langle\varphi, T(t)x\rangle = 0$ for all $t \in \mathbb{R}$. Since this holds for every $\varphi \in X^*$, we deduce that $x = 0$. □

An example of C_0-groups that satisfy the sufficient condition in Proposition 4.3.3 will be given in Corollary 4.3.5. First, let us recall the following result.

Theorem 4.3.4 ([199, Theorem 5, p. 94]). *Let $(U(t))_{t\in\mathbb{R}}$ be a C_0-semigroup, A its infinitesimal generator, and $f : \mathbb{R} \to X$ an almost periodic function. Suppose that $\sigma(A) \cap i\mathbb{R}$ is countable. If the function $u : \mathbb{R} \to X$ defined by*

$$u(t) = U(t - t_0)u(t_0) + \int_{t_0}^{t} U(t - s)f(s)ds \quad \textit{for all } t \ge t_0$$

is bounded on $\mathbb{R}$, then the scalar function $t \mapsto \langle\varphi, u(t)\rangle$ is almost periodic for each $\varphi \in X^*$.

By taking $f = 0$ in Theorem 4.3.4 and using Proposition 4.3.3, we have the following result.

Corollary 4.3.5. *If $(T(t))_{t\in\mathbb{R}}$ is a C_0-group such that $\sigma(A)\cap i\mathbb{R}$ is countable, where A is the infinitesimal generator, then $(T(t))_{t\in\mathbb{R}}$ satisfies the Favard condition.*

Remark 4.3.6. From Corollary 4.3.5, one can see that the Favard condition is natural for C_0-groups in finite-dimensional spaces.

4.3.1.1 Almost periodic and almost automorphic solutions of equation (4.54)

In this section, we give the main results of this chapter. Consider the following evolution equation:

$$\frac{\mathrm{d}}{\mathrm{d}t}x(t) = Ax(t) + f(t) \quad \text{for } t \in \mathbb{R}, \tag{4.58}$$

where $A : D(A) \to X$ is the generator of a C_0-group $(T(t))_{t\in\mathbb{R}}$. Our aim is to find sufficient conditions to guarantee the existence of almost periodic (resp. almost automorphic) solutions of equation (4.58), when the forcing f is S^1-almost periodic (resp. S^1-almost automorphic). We consider the following notion of solution:

Definition 4.3.7. We say that a function $x : \mathbb{R} \to X$ is a mild solution on $\mathbb{R}$ of equation (4.58) if, for all $t \in \mathbb{R}$,

$$x(t) = T(t)x(0) + \int_0^t T(t-s)f(s)\mathrm{d}s.$$

In the sequel of this chapter, a mild solution will be simply called a solution. We have the following Massera-type theorem.

Theorem 4.3.8. *Assume that X is uniformly convex, the C_0-group $(T(t))_{t\in\mathbb{R}}$ satisfies the Favard condition, and equation* (4.58) *has a bounded solution in $\mathbb{R}^+$. If f is S^1-almost automorphic, then equation* (4.58) *has at least one weakly almost automorphic solution.*

Proof. We divide the proof into three steps.

Step 1. We show that equation (4.58) has a bounded solution on $\mathbb{R}$.

Let u be a bounded solution of equation (4.58) on $\mathbb{R}^+$ and $(s_n)_n$ be a sequence of positive real numbers such that $\lim_{n\to\infty} s_n = \infty$. Using the reflexivity of X (Milman–Pettis theorem) and the S^1-almost automorphy of f, there exist a subsequence $(s'_n)_n \subset (s_n)_n$ and a function $\tilde{f} \in L^1_{\mathrm{loc}}(\mathbb{R}, X)$ such that, for all $t \in \mathbb{R}$,

$$\begin{cases} \int_t^{t+1} |f(s+s'_n) - \tilde{f}(s)|\mathrm{d}s \to 0, \\ \int_t^{t+1} |\tilde{f}(s-s'_n) - f(s)|\mathrm{d}s \to 0, \end{cases} \tag{4.59}$$

as $n \to \infty$, and

$$u(s'_n) \to c \tag{4.60}$$

in the weak sense as $n \to \infty$. Consider the function

$$y(t) := T(t)c + \int_0^t T(t-s)\tilde{f}(s)\mathrm{d}s \quad \text{for all } t \in \mathbb{R}. \tag{4.61}$$

For each $t \in \mathbb{R}$, there exists $N(t) \in \mathbb{N}$ such that for all $n \geq N(t)$ we have $t + s_n' \geq 0$ and

$$u(t+s_n') = T(t)u(s_n') + \int_0^t T(t-s)f(s+s_n')\mathrm{d}s.$$

We have, for all $t \in \mathbb{R}$,

$$\int_0^t T(t-s)f(s+s_n')\mathrm{d}s \to \int_0^t T(t-s)\tilde{f}(s)\mathrm{d}s, \tag{4.62}$$

as $n \to \infty$. In fact, for each $t \in \mathbb{R}$,

$$\begin{aligned}
\left|\int_0^t T(t-s)f(s+s_n')\mathrm{d}s - \int_0^t T(t-s)\tilde{f}(s)\mathrm{d}s\right| &\leq \int_{\min(0,t)}^{\max(0,t)} |T(t-s)||f(s+s_n') - \tilde{f}(s)|\mathrm{d}s \\
&\leq \sum_{k=\lfloor \min(0,t)\rfloor}^{\lfloor \max(0,t)\rfloor} \int_k^{k+1} |T(t-s)||f(s+s_n') - \tilde{f}(s)|\mathrm{d}s \\
&\leq \sum_{k=\lfloor \min(0,t)\rfloor}^{\lfloor \max(0,t)\rfloor} \left[M(k,t)\int_k^{k+1} |f(s+s_n') - \tilde{f}(s)|\mathrm{d}s\right],
\end{aligned}$$

where $M(k,t) = \sup_{k \leq s \leq k+1} |T(t-s)|$. Thus (4.62) follows from (4.59). We deduce that, for each $t \in \mathbb{R}$, $u(t+s_n') \to y(t)$ in the weak sense as $n \to \infty$. Since u is bounded on $\mathbb{R}^+$, y is bounded on $\mathbb{R}$. Now from the definition of the function y in (4.61), we have that, for all $n \in \mathbb{N}$ and $t \in \mathbb{R}$,

$$y(t-s_n') = T(t)y(-s_n') + \int_0^t T(t-s)\tilde{f}(s-s_n')\mathrm{d}s. \tag{4.63}$$

We can extract a subsequence $(s_n'')_n \subset (s_n')_n$ such that

$$y(-s_n'') \to d$$

in the weak sense as $n \to \infty$. Define the function z by

$$z(t) := T(t)d + \int_0^t T(t-s)f(s)ds \quad \text{for all } t \in \mathbb{R}.$$

It is clear that the function z is a solution of equation (4.58). From (4.59) and (4.63), we can show that, for all $t \in \mathbb{R}$,

$$y(t - s''_n) \to z(t)$$

in the weak sense as $n \to \infty$. Therefore z is a bounded solution of equation (4.58) on $\mathbb{R}$.

Step 2. Let Ω_f be the set of all bounded solutions of equation (4.58) on $\mathbb{R}$. From Step 1, we have $\Omega_f \neq \emptyset$. Let

$$\omega_f := \inf_{x \in \Omega_f} |x|_\infty,$$

where $|\cdot|_\infty$ is the supremum norm on $BC(\mathbb{R}, X)$. We will show that ω_f is reached in a unique point $x \in \Omega_f$.

Let $(x_n)_n$ be a sequence in Ω_f such that

$$\lim_{n\to\infty} |x_n|_\infty = \omega_f. \tag{4.64}$$

Then, for all $t \in \mathbb{R}$,

$$x_n(t) = T(t)x_n(0) + \int_0^t T(t-s)f(s)ds. \tag{4.65}$$

By (4.64), the sequence of functions $(x_n)_n$ is bounded in $BC(\mathbb{R}, X)$. It follows that $(x_n(0))_n$ is a bounded sequence in the reflexive space X. Hence, there exists a subsequence $(x'_n)_n \subset (x_n)_n$ such that

$$x'_n(0) \to x_0$$

in the weak sense as $n \to \infty$. Consider the function x defined for all $t \in \mathbb{R}$ by

$$x(t) := T(t)x_0 + \int_0^t T(t-s)f(s)ds. \tag{4.66}$$

From (4.65) and (4.66), we can see that, for all $t \in \mathbb{R}$,

$$x'_n(t) \to x(t) \tag{4.67}$$

in the weak sense as $n \to \infty$. For $\varphi \in X^*$ with $|\varphi| \le 1$, we have that

$$|\langle \varphi, x'_n(t)\rangle| \le |\varphi||x'_n(t)| \le |x'_n|_\infty. \tag{4.68}$$

Letting $n \to \infty$ in (4.68), using (4.64) and (4.67), we obtain that, for every $\varphi \in X^*$ with $|\varphi| \leq 1$ and $t \in \mathbb{R}$,

$$|\langle \varphi, x(t) \rangle| \leq \omega_f.$$

It follows that, for $t \in \mathbb{R}$,

$$|x(t)| \leq \omega_f.$$

Therefore, we have that $x \in \Omega_f$ and $|x|_\infty \leq \omega_f$. By definition of ω_f, we conclude that

$$|x|_\infty = \omega_f = \inf_{y \in \Omega_f} |y|_\infty. \tag{4.69}$$

We claim that x is the unique element of Ω_f which satisfies (4.69). In fact, if there is another solution $\tilde{x} \in \Omega_f$ such that

$$|\tilde{x}|_\infty = |x|_\infty = \omega_f = \inf_{y \in \Omega_f} |y|_\infty,$$

then $\frac{x(t)-\tilde{x}(t)}{2} = T(t)(\frac{x(0)-\tilde{x}(0)}{2})$ is a nonzero bounded trajectory of the C_0-group $(T(t))_{t \in \mathbb{R}}$. It follows from the Favard condition that

$$\inf_{t \in \mathbb{R}} \left| \frac{x(t) - \tilde{x}(t)}{2} \right| = c > 0. \tag{4.70}$$

One can observe that $\omega_f > 0$. In fact, if $\omega_f = 0$, then $x = \tilde{x} = 0$, which is not true. Since X is uniformly convex, there exists $\delta_{c,\omega_f} > 0$ such that, for any $u, v \in X$ with $|u| \leq 1$ and $|v| \leq 1$, if

$$|u - v| \geq \frac{2c}{\omega_f},$$

then

$$\left| \frac{u+v}{2} \right| \leq 1 - \delta_{c,\omega_f}.$$

From (4.70), we see that, for all $t \in \mathbb{R}$,

$$\left| \frac{x(t)}{\omega_f} - \frac{\tilde{x}(t)}{\omega_f} \right| \geq \frac{2c}{\omega_f}.$$

From the uniform convexity of X, we obtain that

$$\left| \frac{x(t) + \tilde{x}(t)}{2} \right| \leq \omega_f - \omega_f \delta_{c,\omega_f}.$$

Hence

$$\left|\frac{x+\tilde{x}}{2}\right|_\infty < \omega_f = \inf_{y\in\Omega_f} |y|_\infty,$$

which contradicts the definition of ω_f, since $\frac{x+\tilde{x}}{2} \in \Omega_f$. Thus x is the unique solution of equation (4.58) which satisfies (4.69).

Step 3. We prove that the solution x constructed in Step 2 is weakly almost automorphic. Let $(t_n)_n$ be an arbitrary real sequence. Since x is bounded, X is reflexive, and $f \in SAA^1(\mathbb{R}, X)$, we can find a subsequence $(t'_n)_n \subset (t_n)_n$ and a function $\tilde{f} \in L^1_{loc}(\mathbb{R}, X)$ such that

$$x(t'_n) \to y_0 \tag{4.71}$$

in the weak sense as $n \to \infty$ and, for all $t \in \mathbb{R}$,

$$\begin{cases} \int_t^{t+1} |f(s+t'_n) - \tilde{f}(s)| ds \to 0, \\ \int_t^{t+1} |\tilde{f}(s-t'_n) - f(s)| ds \to 0, \end{cases} \tag{4.72}$$

as $n \to \infty$. Define the function y for all $t \in \mathbb{R}$ by

$$y(t) := T(t)y_0 + \int_0^t T(t-s)\tilde{f}(s) ds. \tag{4.73}$$

For all $t \in \mathbb{R}$, we have

$$x(t+t'_n) = T(t)x(t'_n) + \int_0^t T(t-s)f(s+t'_n) ds.$$

Using the same arguments as in Step 1, we have that

$$\int_0^t T(t-s)f(s+t'_n) ds \to \int_0^t T(t-s)\tilde{f}(s) ds,$$

as $n \to \infty$. This, together with (4.71), implies that, for all $t \in \mathbb{R}$,

$$x(t+t'_n) \to y(t) \tag{4.74}$$

in the weak sense as $n \to \infty$. For each $\varphi \in X^*$ with $|\varphi| \le 1$, we have that

$$|\langle \varphi, x(t+t'_n)\rangle| \le |\varphi||x(t+t'_n)| \le |x|_\infty.$$

It follows by taking $n \to \infty$ that

$$|y(t)| \leq |x|_\infty \quad \text{for all } t \in \mathbb{R}.$$

Therefore $y \in \Omega_{\tilde{f}}$ and

$$|y|_\infty \leq |x|_\infty.$$

Now using the boundedness of the function y, we can extract another subsequence $(t''_n)_n \subset (t'_n)_n$ such that

$$y(-t''_n) \to z_0$$

in the weak sense. Consider the function

$$z(t) := T(t)z_0 + \int_0^t T(t-s)f(s)ds \quad \text{for } t \in \mathbb{R}.$$

Using the same arguments as in Step 1, we can show that, for all $t \in \mathbb{R}$,

$$y(t - t''_n) \to z(t) \tag{4.75}$$

in the weak sense. We obtain that $z \in \Omega_f$ and

$$|z|_\infty \leq |y|_\infty \leq |x|_\infty.$$

But since x is the only element in Ω_f such that

$$|x|_\infty = \inf_{u \in \Omega_f} |u|_\infty,$$

we get $z = x$ and the weak almost automorphy of x follows from (4.74) and (4.75). □

A similar result holds for almost periodic solutions.

Theorem 4.3.9. *Assume that X is uniformly convex, the C_0-group $(T(t))_{t\in\mathbb{R}}$ satisfies the Favard condition and equation* (4.58) *has a bounded solution in $\mathbb{R}^+$. If f is S^1-almost periodic, then equation* (4.58) *has at least one weakly almost periodic solution.*

Proof. Let x be the unique solution on $\mathbb{R}$ such that $|x|_\infty = \inf_{y\in\Omega_f} |y|_\infty$. This solution was constructed in the proof of Theorem 4.3.8. Let $(s_n)_n$ be a sequence of real numbers. Then, using the Bochner-type characterization in Proposition 2.3.18, there exist a subsequence $(s_n)_n \subset (s'_n)_n$ and a function $\tilde{f} \in L^1_{loc}(\mathbb{R}, X)$ such that

$$\sup_{t\in\mathbb{R}} \int_t^{t+1} |f(s + s'_n) - \tilde{f}(s)| ds \to 0 \tag{4.76}$$

and $x(s'_n) \to y_0$, where the second convergence holds in the weak sense. Consider the function

$$\widetilde{x}(t) := T(t)y_0 + \int_0^t T(t-s)\widetilde{f}(s)ds.$$

Using the same argument as in the proof of Theorem 4.3.8, we can prove that, for all $t \in \mathbb{R}$,

$$x(t+s'_n) \to \widetilde{x}(t)$$

in the weak sense as $n \to \infty$. We claim that this convergence is uniform for $t \in \mathbb{R}$. In fact, if the convergence is not uniform, then there exist $\varphi \in X^*$, $\rho > 0$, and three sequences $(p_n)_n$, $(q_n)_n$, $(t_n)_n$ with $p_n, q_n \geq n$ such that

$$|\langle \varphi, x(t_n + s'_{p_n}) - x(t_n + s'_{q_n})\rangle| > \rho. \tag{4.77}$$

Let $a_n := t_n + s'_{p_n}$ and $b_n := t_n + s'_{q_n}$. Since f is S^1-almost periodic and the solution x is bounded, there exist two subsequences, which for simplicity will be also denoted by $(a_n)_n$ and $(b_n)_n$, and two functions $\widetilde{f}_1, \widetilde{f}_2 \in L^1_{\text{loc}}(\mathbb{R}, X)$ such that

$$\begin{cases} \sup_{t\in\mathbb{R}} \int_t^{t+1} |f(s+a_n) - \widetilde{f}_1(s)|ds \to 0, \\ \sup_{t\in\mathbb{R}} \int_t^{t+1} |f(s+b_n) - \widetilde{f}_2(s)|ds \to 0 \end{cases}$$

as $n \to \infty$, and

$$\begin{cases} x(a_n) \to c_1, \\ x(b_n) \to c_2 \end{cases} \tag{4.78}$$

in the weak sense as $n \to \infty$. We have

$$\begin{aligned} \int_t^{t+1} |f(s+a_n) - f(s+b_n)|ds &= \int_t^{t+1} |f(s+t_n+s'_{p_n}) - f(s+t_n+s'_{q_n})|ds \\ &= \int_{t+t_n}^{t+t_n+1} |f(s+s'_{p_n}) - f(s+s'_{q_n})|ds \\ &\leq \sup_{t\in\mathbb{R}} \int_t^{t+1} |f(s+s'_{p_n}) - f(s+s'_{q_n})|ds \\ &\leq \sup_{t\in\mathbb{R}} \int_t^{t+1} |f(s+s'_{p_n}) - \widetilde{f}(s)|ds + \sup_{t\in\mathbb{R}} \int_t^{t+1} |\widetilde{f}(s) - f(s+s'_{q_n})|ds. \end{aligned}$$

It follows from (4.76) that

$$\int_t^{t+1} |f(s+a_n) - f(s+b_n)| \mathrm{d}s \to 0,$$

as $n \to \infty$. But we have

$$\int_t^{t+1} |\tilde{f}_1(s) - \tilde{f}_2(s)| \mathrm{d}s \le \int_t^{t+1} |\tilde{f}_1(s) - f(s+a_n)| \mathrm{d}s + \int_t^{t+1} |f(s+a_n) - f(s+b_n)| \mathrm{d}s + \int_t^{t+1} |f(s+b_n) - \tilde{f}_2(s)| \mathrm{d}s.$$

Letting $n \to \infty$, we conclude that, for all $t \in \mathbb{R}$,

$$\int_t^{t+1} |\tilde{f}_1(s) - \tilde{f}_2(s)| \mathrm{d}s = 0.$$

That is, $\tilde{f}_1 = \tilde{f}_2$ a. e. Now from (4.78), we have, for all $t \in \mathbb{R}$,

$$x(t+a_n) \to T(t)c_1 + \int_0^t T(t-s)\tilde{f}_1(s) \mathrm{d}s := \tilde{x}_1(t), \tag{4.79}$$

$$x(t+b_n) \to T(t)c_2 + \int_0^t T(t-s)\tilde{f}_1(s) \mathrm{d}s := \tilde{x}_2(t) \tag{4.80}$$

in the weak sense. This implies that

$$|\tilde{x}_1|_\infty \le |x|_\infty \quad \text{and} \quad |\tilde{x}_2|_\infty \le |x|_\infty. \tag{4.81}$$

Using the same method as in Step 3 in the proof of Theorem 4.3.8, we can prove that there exist two subsequences of $(a_n)_n$ and $(b_n)_n$ (denoted by the same symbol) and two solutions $y_1, y_2 \in \Omega_f$ such that

$$\begin{cases} \tilde{x}_1(t-a_n) \to y_1(t), \\ \tilde{x}_2(t-b_n) \to y_2(t) \end{cases}$$

in the weak sense. This implies that

$$|y_1|_\infty \le |\tilde{x}_1|_\infty \quad \text{and} \quad |y_2|_\infty \le |\tilde{x}_2|_\infty.$$

But since $|x|_\infty = \inf_{z \in \Omega_f} |z|_\infty$, we deduce that

$$|x|_\infty \leq |y_1|_\infty \leq |\tilde{x}_1|_\infty \quad \text{and} \quad |x|_\infty \leq |y_2|_\infty \leq |\tilde{x}_2|_\infty. \tag{4.82}$$

It follows from (4.81) and (4.82) that

$$|\tilde{x}_1|_\infty = |\tilde{x}_2|_\infty = |x|_\infty.$$

We claim that

$$|\tilde{x}_1|_\infty = |\tilde{x}_2|_\infty = \inf_{\tilde{z}\in\Omega_{\tilde{f}_1}} |\tilde{z}|_\infty.$$

In fact, on the one hand, we have $|\tilde{x}_1|_\infty = |\tilde{x}_2|_\infty \geq \inf_{\tilde{z}\in\Omega_{\tilde{f}_1}} |\tilde{z}|_\infty$ since $\tilde{x}_1, \tilde{x}_2 \in \Omega_{\tilde{f}_1}$. In the other hand, we know that there exists $\tilde{z}_0 \in \Omega_{\tilde{f}_1}$ such that $|\tilde{z}_0|_\infty = \inf_{\tilde{z}\in\Omega_{\tilde{f}_1}} |\tilde{z}|_\infty$. Let $(a_n')_n$ be a subsequence of $(a_n)_n$ such that $\tilde{z}_0(-a_n') \to c_0$ in the weak sense. Hence

$$\tilde{z}_0(t - a_n') \to T(t)c_0 + \int_0^t T(t-s)f(s)ds := z_0(t).$$

This implies that $|z_0|_\infty \leq |\tilde{z}_0|_\infty$. But since $z_0 \in \Omega_f$ and $|x|_\infty = \inf_{z\in\Omega_f} |z|_\infty$, then

$$|\tilde{x}_1|_\infty = |x|_\infty \leq |z_0|_\infty \leq |\tilde{z}_0|_\infty.$$

Therefore

$$|\tilde{x}_1|_\infty = |\tilde{x}_2|_\infty = \inf_{\tilde{z}\in\Omega_{\tilde{f}_1}} |\tilde{z}|_\infty.$$

Now using the uniform convexity of X and the fact that $(T(t))_{t\in\mathbb{R}}$ satisfies the Favard condition, we can prove using the same method as in Step 2 in the proof of Theorem 4.3.8 that $\tilde{x}_1 = \tilde{x}_2$. It follows from (4.79) and (4.80) that, for all $\varphi \in X^*$ and $t \in \mathbb{R}$,

$$\langle \varphi, x(t + a_n) - x(t + b_n)\rangle = \langle \varphi, x(t + t_n + s'_{p_n}) - x(t + t_n + s'_{q_n})\rangle \to 0,$$

which contradicts (4.77) for $t = 0$. □

Remark 4.3.10. Theorems 4.3.8 and 4.3.9 are generalizations of [152, Theorem 5.8] to infinite-dimensional spaces, since by Corollary 4.3.5, the Favard condition always holds for C_0-groups in finite-dimensional spaces.

In finite-dimensional spaces, the notion of almost automorphy is equivalent to the notion of weak almost automorphy which is not the case in infinite-dimensional spaces. The next result shows that both notions are equivalent for solutions of equation (4.58) if we assume that the C_0-group $(T(t))_{t\in\mathbb{R}}$ is bounded. We note that this condition on the C_0-group $(T(t))_{t\in\mathbb{R}}$ is stronger than the Favard condition (see Proposition 4.3.2).

Theorem 4.3.11. *Assume that X is uniformly convex, the C_0-group $(T(t))_{t\in\mathbb{R}}$ is bounded, and f is S^1-almost automorphic. Then, a solution of equation* (4.58) *is almost automorphic if and only if it is weakly almost automorphic.*

The following lemma is needed in the proof of Theorem 4.3.11.

Lemma 4.3.12. *Let Y be a Banach space. A subset $B \subset Y$ is not relatively compact if and only if there exist a constant $c > 0$ and a sequence $(x_n)_n \in B$ such that*

$$|x_p - x_q| > c \quad \text{for } p \neq q. \tag{4.83}$$

Proof of Lemma 4.3.12. It is clear that a sequence $(x_n)_n$ with the property (4.83) does not have a Cauchy subsequence. This implies automatically that B is not relatively compact. Now, assume that B is not relatively compact. Then by the Heine–Borel theorem, the closure $\overline{B}$ is not totally bounded. That is, there exists a radius $r > 0$ such that $\overline{B}$ cannot be covered by a finite number of balls of radius r. Let $x_1 \in B$. Then $\overline{B}$ cannot be covered by the ball $\overline{B}(x_1, r)$. That is, there exists $y_1 \in \overline{B}$ such that $|y_1 - x_1| > r$. Since $y_1 \in \overline{B}$, there exists $x_2 \in B$ such that $|y_1 - x_2| < \frac{r}{2}$. It follows by the triangle inequality that $|x_1 - x_2| > \frac{r}{2}$. Now $\overline{B}$ cannot be covered by $\overline{B}(x_1, r) \cup \overline{B}(x_2, r)$. That is, there exists $y_2 \in \overline{B}$ such that $|y_2 - x_i| > r$ for $i = 1, 2$. Since $y_2 \in \overline{B}$, there exists $x_3 \in B$ such that $|y_2 - x_3| < \frac{r}{2}$. It follows again by the triangle inequality that $|x_3 - x_i| > \frac{r}{2}$ for $i = 1, 2$. Thus, we can construct inductively a sequence $(x_n)_n$ such that, for all n,

$$|x_n - x_i| > \frac{r}{2} \quad \text{for } i = 1, \ldots, n - 1.$$

That is,

$$|x_p - x_q| > \frac{r}{2} \quad \text{for } p \neq q. \tag{4.84}$$

Therefore we get (4.83) for $c = \frac{r}{2}$. □

Proof of Theorem 4.3.11. Let x be a solution of equation (4.58) which is weakly almost automorphic. If $x = 0$, then x is almost automorphic. Suppose that $x \neq 0$. We claim that x has a relatively compact range. In fact, if the range $\{x(t) : t \in \mathbb{R}\}$ is not relatively compact, then by Lemma 4.3.12, there exist a constant $c > 0$ and a sequence $(t_n)_n$ such that

$$\left|x(t_p) - x(t_q)\right| > c \quad \text{for all } p \neq q. \tag{4.85}$$

We then have, for all $t \in \mathbb{R}$ and $n \in \mathbb{N}$,

$$x(t + t_n) = T(t)x(t_n) + \int_0^t T(t - s)f(s + t_n)ds.$$

It follows that, for all $p, q \in \mathbb{N}$,

$$x(t+t_p) - x(t+t_q) = T(t)(x(t_p) - x(t_q)) + \int_0^t T(t-s)(f(s+t_p) - f(s+t_q))\mathrm{d}s.$$

Hence

$$x(t_p) - x(t_q) = T(-t)(x(t+t_p) - x(t+t_q)) - \int_0^t T(-s)(f(s+t_p) - f(s+t_q))\mathrm{d}s.$$

Let $M_T = \sup_{t\in\mathbb{R}} |T(t)|$. Then we have

$$\left|x(t_p) - x(t_q)\right| \leq M_T\left|x(t+t_p) - x(t+t_q)\right| + M_T \int_0^t \left|f(s+t_p) - f(s+t_q)\right|\mathrm{d}s. \tag{4.86}$$

One can see that

$$\int_0^t \left|f(s+t_p) - f(s+t_q)\right|\mathrm{d}s \leq \sum_{k=0}^{\lfloor t\rfloor} \int_k^{k+1} \left|f(s+t_p) - f(s+t_q)\right|\mathrm{d}s.$$

From the S^1-almost automorphy of f, there exists a subsequence $(t'_n)_n \subset (t_n)_n$ such that

$$\int_k^{k+1} \left|f(s+t'_p) - f(s+t'_q)\right|\mathrm{d}s \to 0,$$

as $p, q \to \infty$. Hence for p, q large enough, we get

$$\int_0^t \left|f(s+t'_p) - f(s+t'_q)\right|\mathrm{d}s \leq \frac{c}{2M_T}. \tag{4.87}$$

Thus using (4.87) and (4.85) with the inequality (4.86), we obtain

$$\left|\frac{x(t+t'_p)}{|x|_\infty} - \frac{x(t+t'_q)}{|x|_\infty}\right| > \frac{c}{2M_T|x|_\infty}. \tag{4.88}$$

Since X is uniformly convex, there exists $\delta > 0$ such that for any $u, v \in X$ with $|u| \leq 1$ and $|v| \leq 1$, if $|u - v| \geq \frac{c}{2M_T|x|_\infty}$, then $|\frac{u+v}{2}| \leq 1 - \delta$. Thus (4.88) implies that

$$\left|\frac{x(t+t'_p) + x(t+t'_q)}{2}\right| \leq |x|_\infty - \delta|x|_\infty.$$

For all $\varphi \in X^*$ with $|\varphi| \leq 1$, we have

$$\left|\left\langle \varphi, \frac{x(t+t_p') + x(t+t_q')}{2} \right\rangle\right| \leq \left|\frac{x(t+t_p') + x(t+t_q')}{2}\right| \leq |x|_\infty - \delta |x|_\infty. \tag{4.89}$$

Since x is weakly almost automorphic, there exist a subsequence $(t_n'')_n \subset (t_n')_n$ and a function $\tilde{x}$ such that, for all $t \in \mathbb{R}$,

$$x(t + t_n'') \to \tilde{x}(t) \tag{4.90}$$

and

$$\tilde{x}(t - t_n'') \to x(t) \tag{4.91}$$

in the weak sense. It is clear that (4.90) and (4.91) imply

$$|x|_\infty = |\tilde{x}|_\infty.$$

Letting $p, q \to \infty$ in (4.89), we get, for all $\varphi \in X^*$ with $|\varphi| \leq 1$,

$$|\langle \varphi, \tilde{x}(t)\rangle| \leq |x|_\infty - \delta |x|_\infty = |\tilde{x}|_\infty - \delta |\tilde{x}|_\infty,$$

yielding

$$|\tilde{x}|_\infty \leq |\tilde{x}|_\infty - \delta |\tilde{x}|_\infty,$$

which is a contradiction since $|\tilde{x}|_\infty = |x|_\infty \neq 0$. We conclude that $\{x(t) : t \in \mathbb{R}\}$ is relatively compact. We deduce from Proposition 2.3.12 that x is almost automorphic. □

Theorem 4.3.9 provides at least one weakly almost periodic solution when the C_0-group $(T(t))_{t\in\mathbb{R}}$ satisfies the Favard condition. If we replace this condition with the stronger condition that $(T(t))_{t\in\mathbb{R}}$ is bounded, we obtain at least one almost periodic solution (in the strong sense).

Corollary 4.3.13. *Assume that X is uniformly convex, the C_0-group $(T(t))_{t\in\mathbb{R}}$ is bounded, and equation* (4.58) *has a bounded solution in $\mathbb{R}^+$. If f is S^1-almost periodic, then equation* (4.58) *has at least one almost periodic solution. In addition, every other solution with relatively compact range is almost periodic.*

Proof. Theorem 4.3.9 ensures the existence of a weakly almost periodic solution x which is also weakly almost automorphic. This solution has a relatively compact range by Theorem 4.3.11, implying by Proposition 2.3.11 that x is almost periodic. Now if y is another solution with relatively compact range, then $z := x - y$ has also relatively compact range. In addition, we have

$$z(t) = T(t)z_0 \quad \text{for all } t \in \mathbb{R},$$

where $z_0 = x(0) - y(0)$. Let $(t_n)_n$ be a real sequence. Then there exists a subsequence $(t'_n)_n \subset (t_n)_n$ such that $z(t'_n) \to z_1 \in X$. It follows that

$$\sup_{t\in\mathbb{R}} |z(t + t'_n) - T(t)z_1| = \sup_{t\in\mathbb{R}} |T(t)z(t'_n) - T(t)z_1| \le \sup_{t\in\mathbb{R}} |T(t)||z(t'_n) - z_1| \to 0.$$

Thus z is almost periodic, so that $y = x - z$ is also almost periodic. □

Lemma 4.3.14. *Assume that f is S^1-almost automorphic. Then, every solution of equation (4.58) having a relatively compact range is uniformly continuous.*

Proof. Let x be a solution of equation (4.58) such that $K = \overline{\{x(t) : t \in \mathbb{R}\}}$ is compact. If x is not uniformly continuous, then there exist $\varepsilon > 0$ and two real sequences $(s_n)_n$ and $(h_n)_n$ such that $\lim_{n\to\infty} h_n = 0$ and

$$|x(s_n + h_n) - x(s_n)| > \varepsilon \quad \text{for all } n \in \mathbb{N}. \tag{4.92}$$

For all $n \in \mathbb{N}$, we have

$$\begin{aligned} x(s_n + h_n) &= T(h_n)x(s_n) + \int_{s_n}^{s_n+h_n} T(s_n + h_n - s)f(s)ds \\ &= T(h_n)x(s_n) + \int_0^{h_n} T(h_n - s)f(s + s_n)ds. \end{aligned}$$

Let $M_0 \ge 1$ and $\omega_0 \in \mathbb{R}$ be such that $|T(t)| \le M_0 e^{\omega_0|t|}$ for all $t \in \mathbb{R}$. It follows that, for all $n \in \mathbb{N}$ such that $h_n \ge 0$,

$$\begin{aligned} |x(s_n + h_n) - x(s_n)| &\le |T(h_n)x(s_n) - x(s_n)| + \int_0^{h_n} |T(h_n - s)||f(s + s_n)|ds \\ &\le \sup_{y\in K} |T(h_n)y - y| + M_0 \int_0^{h_n} e^{\omega_0(h_n - s)}|f(s + s_n)|ds \\ &\le \sup_{y\in K} |T(h_n)y - y| + M_0 e^{|\omega_0|h_n} \int_0^{h_n} |f(s + s_n)|ds. \end{aligned}$$

In the other hand, for all $n \in \mathbb{N}$ such that $h_n < 0$, we have

$$
\begin{aligned}
|x(s_n + h_n) - x(s_n)| &\le |T(h_n)x(s_n) - x(s_n)| + \int_{h_n}^{0} |T(h_n - s)||f(s + s_n)|ds \\
&\le \sup_{y \in K} |T(h_n)y - y| + M_0 \int_{h_n}^{0} e^{\omega_0 (s - h_n)} |f(s + s_n)|ds \\
&\le \sup_{y \in K} |T(h_n)y - y| + M_0 e^{|\omega_0| h_n} \int_{h_n}^{0} |f(s + s_n)|ds.
\end{aligned}
$$

We deduce that, for all $n \in \mathbb{N}$,

$$
|x(s_n + h_n) - x(s_n)| \le \sup_{y \in K} |T(h_n)y - y| + M_0 e^{|\omega_0 h_n|} \int_{-|h_n|}^{|h_n|} |f(s + s_n)|ds. \tag{4.93}
$$

By the Banach–Steinhaus theorem, we get

$$
\sup_{y \in K} |T(h_n)y - y| \to 0, \tag{4.94}
$$

as $n \to \infty$. Now, from the S^1-almost automorphy of f, there exist a subsequence $(s'_n)_n \subset (s_n)_n$ and a function $\tilde{f} \in L^1_{\mathrm{loc}}(\mathbb{R}, X)$ such that, for each $t \in \mathbb{R}$,

$$
\int_{t}^{t+1} |f(s + s'_n) - \tilde{f}(s)|ds \to 0, \tag{4.95}
$$

as $n \to \infty$. Let $(h'_n)_n$ be the corresponding subsequence of $(h_n)_n$. We can assume that $|h_n| \le 1$ for all $n \in \mathbb{N}$. Then, from (4.95) and Lebesgue's dominated convergence theorem, we have

$$
\begin{aligned}
\int_{-|h'_n|}^{|h'_n|} |f(s + s'_n)|ds &\le \int_{-|h'_n|}^{|h'_n|} |f(s + s'_n) - \tilde{f}(s)|ds + \int_{-|h'_n|}^{|h'_n|} |\tilde{f}(s)|ds \\
&\le \int_{-1}^{1} |f(s + s'_n) - \tilde{f}(s)|ds + \int_{-|h'_n|}^{|h'_n|} |\tilde{f}(s)|ds \to 0,
\end{aligned} \tag{4.96}
$$

as $n \to \infty$. We deduce from (4.93), (4.94), and (4.96) that

$$
|x(s'_n + h'_n) - x(s'_n)| \to 0,
$$

as $n \to \infty$, which contradicts (4.92). Hence x is uniformly continuous. □

Using Proposition 2.3.6, Theorems 4.3.8, 4.3.11, Lemma 4.3.14, and the same idea as in the proof of Corollary 4.3.13, we have the following result:

Corollary 4.3.15. *Assume that X is uniformly convex, the C_0-group $(T(t))_{t\in\mathbb{R}}$ is bounded, and equation* (4.58) *has a bounded solution in $\mathbb{R}^+$. If f is S^1-almost automorphic, then equation* (4.58) *has at least one compact almost automorphic solution. In addition, every other solution with relatively compact range can be written as the sum of an almost periodic and a compact almost automorphic function.*

4.3.2 Applications

In this section, we provide two examples of partial differential equations to illustrate our theoretical results.

Example 1. Let Ω be an open subset of $\mathbb{R}^n$. We consider the following wave equation:

$$\begin{cases} \frac{\partial^2}{\partial t^2}u(t,x) = \Delta u(t,x) - u(t,x) + f(t)\psi(x) & \text{for } (t,x) \in \mathbb{R}\times\Omega, \\ u(t,x) = 0, & \text{for } (t,x) \in \mathbb{R}\times\partial\Omega, \end{cases} \tag{4.97}$$

where $\psi \in L^2(\Omega)$ and $f : \mathbb{R} \to \mathbb{R}$ is an S^1-almost periodic (resp. S^1-almost automorphic) function. We consider the Hilbert space $X = H_0^1(\Omega) \times L^2(\Omega)$ equipped with the inner product

$$\left\langle \begin{pmatrix} u_1 \\ u_2 \end{pmatrix}, \begin{pmatrix} v_1 \\ v_2 \end{pmatrix} \right\rangle = \int_\Omega (\nabla u_1 \cdot \nabla v_1 + u_1 v_1 + u_2 v_2)\,d\xi.$$

We define the abstract functions $x(t)$ and $F(t)$ by

$$x(t) = \begin{pmatrix} x_1(t) \\ x_2(t) \end{pmatrix} = \begin{pmatrix} u(t,\cdot) \\ \frac{\partial}{\partial t}u(t,\cdot) \end{pmatrix}$$

and

$$F(t) = \begin{pmatrix} 0 \\ f(t)\psi(\cdot) \end{pmatrix}.$$

Let A be the operator defined on X by

$$A = \begin{pmatrix} 0 & I \\ \Delta - I & 0 \end{pmatrix},$$

with a domain

$$D(A) = \left\{ x = \begin{pmatrix} x_1 \\ x_2 \end{pmatrix} \in X, \quad \Delta x_1 \in L^2(\Omega) \text{ and } x_2 \in H_0^1(\Omega) \right\}.$$

Then, equation (4.97) can be written in the following abstract form:

$$\frac{\mathrm{d}}{\mathrm{d}t} x(t) = Ax(t) + F(t) \quad \text{for } t \in \mathbb{R}. \tag{4.98}$$

The operator A is skew-adjoint in X. In fact, for $\left(\begin{smallmatrix} u_1 \\ u_2 \end{smallmatrix}\right), \left(\begin{smallmatrix} v_1 \\ v_2 \end{smallmatrix}\right) \in D(A)$, using integration by parts,

$$\begin{aligned}
\left\langle A\begin{pmatrix} u_1 \\ u_2 \end{pmatrix}, \begin{pmatrix} v_1 \\ v_2 \end{pmatrix} \right\rangle &= \left\langle \begin{pmatrix} u_2 \\ \Delta u_1 - u_1 \end{pmatrix}, \begin{pmatrix} v_1 \\ v_2 \end{pmatrix} \right\rangle \\
&= \int_\Omega (\nabla u_2.\nabla v_1 + u_2 v_1 + (\Delta u_1 - u_1) v_2) \mathrm{d}\xi \\
&= \int_\Omega (-u_2 \Delta v_1 + u_2 v_1 + (\Delta u_1 - u_1) v_2) \mathrm{d}\xi \\
&= \int_\Omega (-v_2 \Delta u_1 - v_2 u_1 - (\Delta v_1 - v_1) u_2) \mathrm{d}\xi \\
&= -\left\langle \begin{pmatrix} u_1 \\ u_2 \end{pmatrix}, A\begin{pmatrix} v_1 \\ v_2 \end{pmatrix} \right\rangle.
\end{aligned}$$

It follows by Stone's theorem [247, Theorem 10.8] that A generates a unitary group which is then a bounded C_0-group. The space X is uniformly convex. In addition, it is clear that $F \in SAP^1(\mathbb{R}, X)$ (resp. $F \in SAA^1(\mathbb{R}, X)$). It follows from Corollaries 4.3.13 and 4.3.15 that if equation (4.98) has a bounded solution on $\mathbb{R}^+$, then it has an almost periodic (resp. compact almost automorphic) solution.

Example 2. Consider the following one dimensional transport equation:

$$\frac{\partial}{\partial t} u(t,\xi) = \frac{\partial}{\partial \xi} u(t,\xi) + a(\xi) u(t,\xi) + f(t)\psi(\xi) \quad \text{for } (t,\xi) \in \mathbb{R}^2, \tag{4.99}$$

where $\psi \in L^p(\mathbb{R})$ for some $p > 1$ and $a : \mathbb{R} \to \mathbb{R}$ is the function defined by

$$a(\xi) = \begin{cases} 1 & \text{if } \xi \in [0,1], \\ 0 & \text{if } \xi \notin [0,1]. \end{cases}$$

Consider the space $X = L^p(\mathbb{R})$ and the operator A defined on X by

$$\begin{cases} D(A) = \{z \in L^p(\mathbb{R}) : z \text{ absolutely continuous and } z' \in L^p(\mathbb{R})\}, \\ (Az)(\xi) = z'(\xi) + a(\xi) z(\xi). \end{cases}$$

The operator A generates a C_0-group on X which is given by the following explicit form:

$$(T(t)z)(\xi) = e^{\int_\xi^{\xi+t} a(s)ds} z(\xi + t) \quad \text{for } t, \xi \in \mathbb{R}.$$

Let $z \in X$. Since $a(\xi) \leq 1$ for all $\xi \in \mathbb{R}$, then for all $t \leq 0$,

$$\begin{aligned}|T(t)z|^p &= \int_{-\infty}^{\infty} e^{-p\int_{\xi+t}^{\xi} a(s)ds} |z(\xi + t)|^p d\xi \\ &\leq e^{pt} \int_{-\infty}^{\infty} |z(\xi + t)|^p d\xi \leq |z|^p.\end{aligned}$$

Therefore $t \mapsto T(t)z$ is bounded on $(-\infty, 0]$. Now for $t \geq 1$, we have

$$\begin{aligned}|T(t)z|^p &= \int_{-\infty}^{\infty} e^{p\int_\xi^{\xi+t} a(s)ds} |z(\xi + t)|^p d\xi \\ &= \int_{-\infty}^{-t} e^{p\int_\xi^{\xi+t} a(s)ds} |z(\xi + t)|^p d\xi + \int_{-t}^{-t+1} e^{p\int_\xi^{\xi+t} a(s)ds} |z(\xi + t)|^p d\xi \\ &\quad + \int_{-t+1}^{0} e^{p\int_\xi^{\xi+t} a(s)ds} |z(\xi + t)|^p d\xi + \int_{0}^{1} e^{p\int_\xi^{\xi+t} a(s)ds} |z(\xi + t)|^p d\xi \\ &\quad + \int_{1}^{\infty} e^{p\int_\xi^{\xi+t} a(s)ds} |z(\xi + t)|^p d\xi \\ &\leq \int_{-\infty}^{0} |z(\xi)|^p d\xi + e^p \int_0^1 |z(\xi)|^p d\xi + e^p \int_1^t |z(\xi)|^p d\xi + e^p \int_t^{t+1} |z(\xi)|^p d\xi + \int_{t+1}^{\infty} |z(\xi)|^p d\xi \\ &\leq (2 + 3e^p)|z|^p.\end{aligned}$$

It follows by the Banach–Steinhaus theorem that $(T(t))_{t\in\mathbb{R}}$ is a bounded C_0-group on X. The space $X = L^p(\mathbb{R})$ is uniformly convex for $p > 1$, thus if $f \in SAP^1(\mathbb{R}, \mathbb{R})$ (resp. $f \in SAA^1(\mathbb{R}, \mathbb{R})$) and equation (4.99) has a bounded solution on $\mathbb{R}^+$, then it follows by Corollaries 4.3.13 and 4.3.15 that equation (4.99) has an almost periodic (resp. compact almost automorphic) solution.

4.4 Evolution equations through the minimizing for some subvariant functional

Our aim is to investigate the existence of bounded and compact almost automorphic solutions to the following evolution equation in a Banach space $(X, \|\cdot\|)$:

$$x'(t) = Ax(t) + f(t, x(t)) \quad \text{for } t \in \mathbb{R}, \tag{4.100}$$

where $A : D(A) \subset X \to X$ is the infinitesimal generator of a C_0-semigroup of bounded linear operators on X and $f : \mathbb{R} \times X \to X$ is an almost automorphic function in t uniformly with respect to the second argument. We give sufficient conditions for the existence of compact almost automorphic solutions when equation (4.100) admits at least a bounded solution on $\mathbb{R}^+$. Our main sufficient condition is the unicity of some solutions with values in some compact set which minimize a functional. First, we study the existence of bounded solutions and, second, we will show the existence of a compact almost automorphic solution to equation (4.100). Then we apply our results to nonlinear partial differential equations. We give sufficient conditions ensuring the existence of compact almost automorphic solutions to some heat and wave equations. Even in the almost periodic framework, our results on the wave equation are new (Corollary 4.4.34).

Early results concerning almost periodic solutions of partial differential equations were obtained starting in the 1950s by the Italian school, namely Amerio, Biroli, Prousem and others [28, 29, 29–66, 249, 250]. Amerio introduced the concept of *minimax principle* in the literature in his paper [28] as a generalization to the nonlinear case Favard theory [147, 148]. The minimax principle asserts that under suitable assumptions, the unique minimizer of the supremum norm among solutions with values in some compact set is almost periodic. Several applications to partial differential equations were given in [29, 30, 35]. Then in the spirit of the Italian school, Dafermos, Haraux, and Ishii gave important contributions to the question of almost periodic solutions [114, 172, 174, 184]. Haraux [172] and Ishii [184] established the existence of almost periodic solutions of contractive almost periodic processes on a Banach space. The main result of Ishii [184] ensures the existence of some almost periodic solution, if there exists a solution defined on $\mathbb{R}^+$ with a compact range. In [174], Haraux studied the existence of almost periodic solutions for the following evolution equation:

$$x'(t) + \widetilde{A}x(t) \ni f(t) \quad \text{for } t \in \mathbb{R}, \tag{4.101}$$

where $\widetilde{A}$ is a maximal monotone operator on $\mathbb{R}^2$ and f is almost periodic. He proved that every bounded solution on $\mathbb{R}^+$ of equation (4.101) is asymptotically almost periodic. The dimension 2 of the Euclidean space is essential for the latter result. Moreover, the author proved that every bounded solution on $\mathbb{R}$ of equation (4.101) is almost periodic (see [174, Theorem 2.1]) by using the minimax principle which is also valid for an arbitrary Banach space.

The existence of an almost automorphic solution to equation (4.100) has been extensively studied when the semigroup generated by A has an exponential dichotomy, since in that case equation (4.100) has a unique bounded solution on $\mathbb{R}$ which is almost automorphic when f is almost automorphic in t and Lipschitz with respect to the second argument, and the Lipschitz constant is sufficiently small (depending on the constant governing the exponential dichotomy). In [167] and [168], the authors investigated the existence and uniqueness of an almost periodic solution to equation (4.100) when $A = 0$

and f is dissipative with respect to the second argument, and they proposed, as an application, the following ordinary differential equation in a Banach space E:

$$x'(t) = -|x(t)|^{\alpha} x(t) + h(t) \quad \text{for } t \in \mathbb{R}, \tag{4.102}$$

where $\alpha \geq 0$ and $h : \mathbb{R} \to E$ is a continuous function. The authors showed if the input function h is almost periodic, then equation (4.102) has a unique bounded solution on $\mathbb{R}$ which is also almost periodic. Recently, in [142], the authors extended the works [167] and [168] to the almost automorphic case; in fact, they proved the existence and uniqueness of a bounded solution on $\mathbb{R}$ which is compact almost automorphic. In [96] and [143], the authors studied the existence, uniqueness, and attractiveness of a pseudo almost automorphic solution to some general dissipative systems in Banach spaces. Let us indicate the contribution of Zaidman on almost periodic and automorphic functions (see, for instance, [283–286]).

Fink introduced the concept of a *subvariant functional* in [150] to prove the existence of compact almost automorphic solutions to the following ordinary differential equation:

$$x'(t) = \varphi(t, x(t)) \quad \text{for } t \in \mathbb{R}, \tag{4.103}$$

where the function $\varphi : \mathbb{R} \times \mathbb{R}^n \to \mathbb{R}^n$ is compact almost automorphic in t uniformly with respect to the second argument. In the particular case of ordinary differential equations in finite-dimensional spaces, *the subvariant functional method* of Fink is a generalization of the minimax principle of Amerio, in the sense where the almost automorphic solution which is reached by minimizing a function, the so-called subvariant functional, is not necessarily the supremum norm. In [98], an extension in Banach spaces of the main results in [150] is given. In this work, we extend the results of [98, 150] to equation (4.100) and use the subvariant functional method to prove the existence of a compact almost automorphic solution. We prove that the existence of a K-bounded mild solution of equation (4.100) that minimizes some subvariant functional is compact almost automorphic.

Let us fix our notations and recall some definitions. Let $x_0 \in X$ and $f \in C(\mathbb{R} \times X, X)$ (continuous maps). We say that x is a *mild solution on* $[t_0, +\infty)$ (where $t_0 \in \mathbb{R}$) of equation (4.100) with the initial condition $x(t_0) = x_0$, if $x \in C([t_0, +\infty), X)$ satisfies

$$x(t) = T(t - t_0)x_0 + \int_{t_0}^{t} T(t - \sigma)f(\sigma, x(\sigma))d\sigma \quad \text{for } t \geq t_0.$$

Observe that for a mild solution on $[t_0, +\infty)$ of equation (4.100), one has

$$x(t) = T(t - s)x(s) + \int_{s}^{t} T(t - \sigma)f(\sigma, x(\sigma))d\sigma \quad \text{for } t \geq s \geq t_0. \tag{4.104}$$

We say that x is a *mild solution on* $\mathbb{R}$ of equation (4.100) for a given $f \in C(\mathbb{R} \times X, X)$, if $x \in C(\mathbb{R}, X)$ satisfies equation (4.104) for each $t \geq s$. For some preliminary results related to C_0-semigroups, we refer the reader to [247].

Let $\mathrm{BC}(\mathbb{R}, X)$ be the space of all bounded and continuous functions from $\mathbb{R}$ to a Banach space X, equipped with the uniform topology. Let $x \in \mathrm{BC}(\mathbb{R}, X)$ and $\tau \in \mathbb{R}$. We define the translation function x_τ by

$$x_\tau(s) = x(\tau + s) \quad \text{for } s \in \mathbb{R}.$$

We say that a mapping $f : \mathbb{R} \times X \to X$, $(t, x) \mapsto f(t, x)$ is *almost periodic in t uniformly with respect to x* when it satisfies the two following conditions:

(i) $f \in C(\mathbb{R} \times X, X)$;

(ii) For all compact subsets K of X and any $\varepsilon > 0$, there exists $\ell > 0$ such that, for all $a \in \mathbb{R}$, there exists $\tau \in [a, a + \ell]$ such that

$$\sup_{t \in \mathbb{R}} \|f(t + \tau, x) - f(t, x)\| \leq \varepsilon.$$

Denote by $\mathrm{AP}_u(\mathbb{R} \times X, X)$ the set of all such mappings.

We say that a mapping $f : \mathbb{R} \times X \to X$, $(t, x) \mapsto f(t, x)$ is *almost automorphic in t uniformly with respect to x* when it satisfies the two following conditions:

(i) $f \in C(\mathbb{R} \times X, X)$;

(ii) For all compact subsets K of X and all sequences of real numbers $(t'_n)_n$, there exist a map $g : \mathbb{R} \times K \to X$ and a subsequence of $(t'_n)_n$, denoted by $(t_n)_n$, such that

$$\forall t \in \mathbb{R}, \quad \lim_{n \to +\infty} \sup_{x \in K} \|f(t + t_n, x) - g(t, x)\| = 0, \tag{4.105}$$

$$\forall t \in \mathbb{R}, \quad \lim_{n \to +\infty} \sup_{x \in K} \|g(t - t_n, x) - f(t, x)\| = 0. \tag{4.106}$$

We denote by $\mathrm{AA}_u(\mathbb{R} \times X, X)$ the set of all such mappings. An example of an almost automorphic function in t uniformly with respect to x is $f(t, x) = F(x) + b(t)$, with $F \in C(X, X)$ and $b \in \mathrm{AA}(\mathbb{R}, X)$. We write $\mathcal{L}(X, X)$ for the space of linear and bounded maps from X to X. Another example is $f(t, x) = A(t)x + b(t)$, with $A \in \mathrm{AA}(\mathbb{R}, \mathcal{L}(X, X))$ and $b \in \mathrm{AA}(\mathbb{R}, X)$.

Remark 4.4.1. Our definition is different from that used by Fink in [150]. The author assumed that f is *compact almost automorphic in t uniformly with respect to x*, that is, for all compact subsets K of X and for all sequences of real numbers $(t'_n)_n$, there exist a map $g : \mathbb{R} \times K \to X$ and a subsequence of $(t'_n)_n$, denoted by $(t_n)_n$, such that, for all compact subsets I of $\mathbb{R}$, one has

$$\lim_{n \to +\infty} \sup_{t \in I} \sup_{x \in K} \|f(t + t_n, x) - g(t, x)\| = 0,$$

$$\lim_{n \to +\infty} \sup_{t \in I} \sup_{x \in K} \|g(t - t_n, x) - f(t, x)\| = 0.$$

Remark 4.4.2. Here we recall a characterization of the almost automorphic functions in t uniformly with respect to x (see [99, Theorem 3.14]): $f \in AA_u(\mathbb{R} \times X, X)$ if and only if $\forall x \in X$, the partial function $t \mapsto f(t,x) \in AA(\mathbb{R}, X)$ and f is uniformly continuous on each compact K of X with respect to t, that is, for all compact subsets K of X, $\forall \varepsilon > 0$, $\exists \delta > 0$, and $\forall x_1, x_2 \in X$, one has

$$\|x_1 - x_2\| \leq \delta \quad \Longrightarrow \quad \sup_{t \in \mathbb{R}} \|f(t,x_1) - f(t,x_2)\| \leq \varepsilon.$$

We have a similar characterization for almost periodic functions: $f \in AP_u(\mathbb{R} \times X, X)$ if and only if $\forall x \in X$, $t \mapsto f(t,x) \in AP(\mathbb{R}, X)$ and f is uniformly continuous on each compact K of X with respect to t (see [99, Lemma 2.6]).

Theorem 4.4.3 ([120]). *Let $f \in AA_c(\mathbb{R} \times X, Y)$ be Lipschitz with respect to the second argument. If $x \in AA_c(\mathbb{R}, X)$, then the composition function $t \mapsto f(t, x(t))$ belongs to $AA_c(\mathbb{R}, Y)$.*

4.4.1 Almost automorphic solution minimizing a subvariant functional

In this section, we use the subvariant functional method which has been introduced for the first time by Fink in [150] to prove the existence of almost periodic and compact almost automorphic solutions for some ordinary differential equations. Let K be a compact subset of X. Let $C_K(\mathbb{R}, X)$ denote the set

$$C_K(\mathbb{R}, X) = \{x \in C(\mathbb{R}, X) : \text{for all } t \in \mathbb{R},\ x(t) \in K\}.$$

A mapping $\lambda_K : C_K(\mathbb{R}, X) \to \mathbb{R}$ is called a *subvariant functional* associated to the compact set K if λ_K satisfies the two following conditions:

(i) λ_K is invariant with respect to translation: $\lambda_K(x_\tau) = \lambda_K(x)$ for each $\tau \in \mathbb{R}$, where $x_\tau(\cdot) = x(\tau + \cdot)$;

(ii) λ_K is lower semicontinuous for the topology of compact convergence: if $\lim_{n \to +\infty} x_n = y$ uniformly on each compact subset of $\mathbb{R}$, then $\lambda_K(y) \leq \liminf_{n \to +\infty} \lambda_K(x_n)$.

Remark 4.4.4. In the above definition of a subvariant functional, note that conditions (i) and (ii) imply the following:

(iii) if $\lim_{n \to +\infty} x_{\tau_n} = y$ uniformly on each compact subset of $\mathbb{R}$, then $\lambda_K(y) \leq \lambda_K(x)$.

Let us denote by $\mathcal{F}_K$ the set of mild solutions x on $\mathbb{R}$ of equation (4.100) such that $x(t) \in K$ for each $t \in \mathbb{R}$. A solution x_* is called a *minimal K-valued solution* of equation (4.100) if

$$x_* \in \mathcal{F}_K \quad \text{and} \quad \lambda_K(x_*) = \inf_{x \in \mathcal{F}_K} \lambda_K(x).$$

An example of a subvariant functional is

$$\lambda_K(x) = \sup_{t \in \mathbb{R}} \|x(t)\|$$

or, more generally,

$$\lambda_K(x) = \sup_{t \in \mathbb{R}} \Phi(x(t)), \quad \text{where } \Phi \in C(K, \mathbb{R}).$$

Another example which is given in [150] is the following:

$$\lambda_K(x) = \sup_{t \in \mathbb{R}} x(t) - \inf_{t \in \mathbb{R}} x(t),$$

where x is $\mathbb{R}$-valued.

Remark 4.4.5. Our definition of a subvariant functional is a modification of the definition due to Fink. Our is slightly stronger than that used in [150]. Fink defines the subvariant functional by only (iii) instead of (i) and (ii). For this reason, contrary to our results, in [150], the existence of a minimal K-valued solution is assumed, except in the particular case of the subvariant functional $\lambda_K(x) = \sup_{t \in \mathbb{R}} \|x(t)\|$.

The following hypotheses will be used in the main results:

(H1) Operator A *is the infinitesimal generator of a* C_0*-semigroup* $(T(t))_{t \geq 0}$.

(H2) *The* C_0*-semigroup* $(T(t))_{t \geq 0}$ *is compact.*

(H3) There exists $t_0 \in \mathbb{R}$ *such that* for all $R > 0$, *one has* $\sup_{t \geq t_0} \sup_{\|x\| \leq R} \|f(t,x)\| < +\infty$.

(H4) $f \in AA_u(\mathbb{R} \times X, X)$.

(H5) $f \in AP_u(\mathbb{R} \times X, X)$.

Theorem 4.4.6. *Assume that* **(H1)–(H4)** *hold. In addition, suppose that equation* (4.100) *admits at least one mild solution* x_0 *defined and bounded on* $[t_0, +\infty)$. *Then:*

(i) *The set* $\{x_0(t) : t \geq t_0\}$ *is relatively compact.*

(ii) *Let* K *be a compact subset of* X *such that* $\{x_0(t) : t \geq t_0\} \subset K$. *If* λ_K *is a subvariant functional associated to the compact set* K, *then equation* (4.100) *admits at least one minimal* K*-valued solution.*

(iii) *If equation* (4.100) *has a unique minimal* K*-valued solution, then the minimizing solution is compact almost automorphic.*

The proof of Theorem 4.4.6 will be given in Section 4.4.3.

Corollary 4.4.7. *Assume that* **(H1)–(H4)** *hold. In addition, suppose that equation* (4.100) *admits at least one mild solution* x_0 *defined and bounded on* $[t_0, +\infty)$. *Let* $K = \overline{\{x_0(t)\,;\, t \geq t_0\}}$. *If equation* (4.100) *has at most one* K*-mild solution on* $\mathbb{R}$, *then this solution is compact almost automorphic.*

Proof. Take $\lambda_K(x) = 1$. □

From Corollary 4.4.7, we easily deduce the following result.

Corollary 4.4.8. *Assume that* **(H1)–(H4)** *hold. In addition, suppose that equation* (4.100) *admits a unique mild solution which is bounded on* $\mathbb{R}$*. Then this solution is compact almost automorphic.*

When the C_0-semigroup $(T(t))_{t\geq 0}$ is not compact, Theorem 4.4.6 becomes

Theorem 4.4.9. *Assume that* **(H1)** *and* **(H4)** *hold. In addition, suppose that equation* (4.100) *admits at least one mild solution* x_0 *defined on* $[t_0, +\infty)$ *such that* $\{x_0(t) : t \geq t_0\}$ *is relatively compact. Then the following statements hold:*
(i) *Let* K *be a compact subset of* X *such that* $\{x_0(t) : t \geq t_0\} \subset K$*. If* λ_K *is a subvariant functional associated to the compact set* K*, then equation* (4.100) *admits at least a minimal* K*-valued solution.*
(ii) *If equation* (4.100) *has a unique minimal* K*-valued solution, then the minimizing solution is compact almost automorphic.*

The proof of Theorem 4.4.9 will be given in Section 4.4.3.

Now we give some corollaries of the main results in the almost periodic case.

For $f \in \mathrm{AP}_u(\mathbb{R} \times X, X)$, we denote by $\mathcal{H}(f)$ the set of functions $g : \mathbb{R} \times X \to X$ such that, for each compact subset K in X, there exists a real sequence $(t_n)_n$ satisfying

$$\lim_{n\to+\infty} \sup_{t\in} \sup_{x\in K} \|f(t + t_n, x) - g(t, x)\| = 0.$$

For $g \in \mathcal{H}(f)$, we consider the following equation:

$$x'(t) = Ax(t) + g(t, x(t)) \quad \text{for } t \in \mathbb{R}. \tag{4.107}$$

Corollary 4.4.10. *Assume that* **(H1)–(H3)** *and* **(H5)** *hold. In addition, suppose that equation* (4.100) *admits at least one mild solution* x_0 *defined and bounded on* $[t_0, +\infty)$*. Then:*
(i) *The set* $\{x_0(t) : t \geq t_0\}$ *is relatively compact.*
(ii) *Let* K *be a compact subset of* X *such that* $\{x_0(t) : t \geq t_0\} \subset K$*. If* λ_K *is a subvariant functional associated to the compact set* K*, then equation* (4.100) *admits at least one minimal* K*-valued solution.*
(iii) *If, for all* $g \in \mathcal{H}(f)$*, equation* (4.107) *has at most one minimal* K*-valued solution, then the unique minimal* K*-valued solution of equation* (4.100) *is almost periodic.*

Corollary 4.4.11. *Assume that* **(H1)** *and* **(H5)** *hold. In addition, suppose that equation* (4.100) *admits at least one mild solution* x_0 *defined on* $[t_0, +\infty)$ *such that* $\{x_0(t) : t \geq t_0\}$ *is relatively compact. Then the following statements are true:*
(i) *Let* K *be a compact subset of* X *such that* $\{x_0(t) : t \geq t_0\} \subset K$*. If* λ_K *is a subvariant functional associated to the compact set* K*, then equation* (4.100) *admits at least one minimal* K*-valued solution.*

(ii) *If, for all $g \in \mathcal{H}(f)$, equation* (4.107) *has at most one minimal K-valued solution, then the unique minimal K-valued solution of equation* (4.100) *is almost periodic.*

The proof of Corollaries 4.4.10 and 4.4.11 will be given in Section 4.4.3.

4.4.2 Boundedness and compactness of solutions

The objective of this section is to state some results on the bounded solutions and those which have a relatively compact range. The following conditions will be used in this section:

(C1) *K is a compact subset of X.*

(C2) *$F : \mathbb{R} \times K \to X$ is a measurable function.*

Let I be the interval $[t_0, +\infty)$ or the whole real line $\mathbb{R}$. Assume that the function F satisfies, for each a and $b \in I$ such that $a \leq b$,

$$\sup_{a \leq t \leq b} \sup_{x \in K} \|F(t,x)\| < +\infty. \tag{4.108}$$

We say that u is a K-*mild solution on I* of

$$u'(t) = Au(t) + F(t, u(t)), \tag{4.109}$$

if $u \in C(I, X)$ satisfies

$$u(t) \in K \quad \text{for all } t \in I$$

and

$$u(t) = T(t-s)u(s) + \int_s^t T(t-\sigma)F(\sigma, u(\sigma))\, d\sigma \quad \text{for } s, t \in I \text{ such that } t \geq s. \tag{4.110}$$

Note that, due to (4.108), the map $\sigma \mapsto F(\sigma, u(\sigma)) \in L^1((s,t), X)$.

Lemma 4.4.12. *Let I be the interval $[t_0, +\infty)$ or the whole real line $\mathbb{R}$. Assume that* **(H1)**, **(C1)**, *and* **(C2)** *hold. In addition, we assume that*

$$k = \sup_{t \in I} \sup_{x \in K} \|F(t,x)\| < +\infty. \tag{4.111}$$

Then there exists a function $\phi : [0, +\infty) \to [0, +\infty)$ satisfying

$$\lim_{\tau \to 0} \phi(\tau) = 0 \tag{4.112}$$

and such that every K-mild solution u on I of equation (4.109) *verifies, for all $t, s \in I$,*

$$\|u(t) - u(s)\| \le \phi(|t - s|). \tag{4.113}$$

Remark 4.4.13. Lemma 4.4.12 means that every K-mild solution of equation (4.109) is uniformly continuous with the uniform continuity modulus ϕ. A consequence of the latter lemma is the following result: if u is a K-mild solution on $\mathbb{R}$ and $(t_n)_n$ is a sequence of real numbers, then the family $\{t \mapsto u(t + t_n) : n \in \mathbb{N}\}$ is uniformly equicontinuous on $\mathbb{R}$.

Proof. Since $(T(t))_{t\ge0}$ is a C_0-semigroup, there exist $\omega \ge 0$ and $M \ge 1$ such that

$$\|T(t)\| \le Me^{\omega t} \quad \text{for } t \ge 0 \tag{4.114}$$

(cf. [247, Theorem 2.2, p. 4]). Denote by $\phi : [0, +\infty) \to [0, +\infty)$ the function defined by

$$\phi(\tau) = \sup_{x\in K}\|T(\tau)x - x\| + kM \int_0^\tau e^{\omega\sigma}\, d\sigma. \tag{4.115}$$

Recall that a C_0-semigroup is strongly continuous, i. e., $T(\cdot)x : [0, +\infty) \to X$ is continuous for each $x \in X$, and then

$$\lim_{t\to0} T(t)x = x \quad \text{for } x \in K.$$

In view of Banach–Steinhaus theorem (see [258, p. 327]), we obtain

$$\lim_{t\to0} \sup_{x\in K}\|T(t)x - x\| = 0. \tag{4.116}$$

Using (4.115) and (4.116), we deduce (4.112). Since u is a K-mild solution on I of equation (4.109), equation (4.110) holds, therefore

$$\|u(t) - u(s)\| \le \|T(t - s)u(s) - u(s)\| + \int_s^t \|T(t - \sigma)\|\|F(\sigma, u(\sigma))\|\, d\sigma,$$

for each s and $t \in I$ such that $t \ge s$. By using (4.111) and (4.114), we obtain

$$\|u(t) - u(s)\| \le \sup_{x\in K}\|T(t - s)x - x\| + kM \int_s^t e^{\omega(t-\sigma)}\, d\sigma.$$

Now, by interchanging s and t, we deduce that

$$\|u(t) - u(s)\| \le \sup_{x\in K}\|T(| t - s |)x - x\| + kM \int_0^{|t-s|} e^{\omega\sigma}\, d\sigma \quad \text{for } t, s \in I,$$

therefore (4.113) holds with the function ϕ defined by (4.115). □

Lemma 4.4.14. *Assume that* **(H1)**, **(C1)**, *and* **(C2)** *hold. In addition, we assume that, for all* $t \in \mathbb{R}$,

$$F(t, \cdot) \in C(K, X), \tag{4.117}$$

$$k = \sup_{t \in \mathbb{R}} \sup_{x \in K} \|F(t, x)\| < +\infty. \tag{4.118}$$

We assume that u *is a* K*-mild solution on* $\mathbb{R}$ *of equation* (4.109). *If there exist a sequence* $(t'_n)_n$ *of real numbers and a map* $G : \mathbb{R} \times K \to X$ *such that*

$$\lim_{n \to +\infty} \sup_{x \in K} \|F(t + t'_n, x) - G(t, x)\| = 0 \quad \text{for all } t \in \mathbb{R}, \tag{4.119}$$

then there exists a subsequence of $(t'_n)_n$, *denoted by* $(t_n)_n$, *such that*

$$u(t + t_n) \to v(t) \quad \text{as } n \to +\infty \tag{4.120}$$

uniformly on each compact subset of $\mathbb{R}$, *where* v *is a* K*-mild solution on* $\mathbb{R}$ *of*

$$v'(t) = Av(t) + G(t, v(t)). \tag{4.121}$$

Remark 4.4.15.
(i) The mild solution v satisfies $v(t) \in K$ for each $t \in \mathbb{R}$.
(ii) Function G is measurable on $\mathbb{R} \times K$ and satisfies (4.117) and (4.118).

Proof. If we denote by $u_n(t) = u(t + t'_n)$, then for each $n \in \mathbb{N}$, $u_n \in C(\mathbb{R}, X)$ and $u_n(t) \in K$ for $t \in \mathbb{R}$, therefore, for each $t \in \mathbb{R}$, the set $\{u_n(t) : n \in \mathbb{N}\}$ is a relatively compact subset of X. By Lemma 4.4.12, the solution u satisfies (4.113), thus the sequence of functions $(u_n)_n$ satisfies

$$\|u_n(t) - u_n(s)\| \le \phi(|t - s|) \quad \text{for } t, s \in \mathbb{R} \text{ and } n \in \mathbb{N},$$

where ϕ satisfies (4.112). Therefore the sequence $(u_n)_n$ is uniformly equicontinuous on $\mathbb{R}$. In view of Arzelà–Ascoli theorem (see [258, p. 312]), we can assert that $\{u_n : n \in \mathbb{N}\}$ is a relatively compact subset of $C(\mathbb{R}, X)$ endowed with the topology of compact convergence. From the sequence $(t'_n)_n$, we can extract a subsequence $(t_n)_n$ such that there exists $v \in C(\mathbb{R}, X)$ and (4.120) holds.

It remains to prove that v is a K-mild solution on $\mathbb{R}$ of equation (4.121). Since u is a K-mild solution of equation (4.109), for each $n \in \mathbb{N}$, one has

$$u(t + t_n) \in K \quad \text{for all } t \in \mathbb{R}, \tag{4.122}$$

$$u(t + t_n) = T(t - s)u(s + t_n) + \int_s^t T(t - \sigma)F(\sigma + t_n, u(\sigma + t_n))\, d\sigma \quad \text{for } t \ge s. \tag{4.123}$$

From (4.120) and (4.122), we obtain

$$v(t) \in K \quad \text{for all } t \in \mathbb{R}, \tag{4.124}$$

and, from (4.117) and (4.119), it follows that $G(t,\cdot) \in C(K,X)$ for each $t \in \mathbb{R}$. By the triangle inequality,

$$\begin{aligned}
&\|F(t+t_n,u(t+t_n)) - G(t,v(t))\| \\
&\quad \le \|F(t+t_n,u(t+t_n)) - G(t,u(t+t_n))\| + \|G(t,u(t+t_n)) - G(t,v(t))\| \\
&\quad \le \sup_{x\in K}\|F(t+t_n,x) - G(t,x)\| + \|G(t,u(t+t_n)) - G(t,v(t))\|,
\end{aligned}$$

and, from (4.119) and (4.120), we deduce that

$$\lim_{n\to+\infty} F(\sigma+t_n,u(\sigma+t_n)) = G(\sigma,v(\sigma)) \quad \text{for } s \le \sigma \le t,$$

since $G(t,\cdot) \in C(K,X)$ for each $t \in \mathbb{R}$. Therefore,

$$\lim_{n\to+\infty} T(t-\sigma)F(\sigma+t_n,u(\sigma+t_n)) = T(t-\sigma)G(\sigma,v(\sigma)) \quad \text{for } s \le \sigma \le t.$$

Moreover, using (4.114), we have

$$\|T(t-\sigma)F(\sigma+t_n,u(\sigma+t_n))\| \le kMe^{\omega(t-\sigma)} \quad \text{for } s \le \sigma \le t,$$

where k is the constant defined by (4.118) and $\sigma \mapsto kMe^{\omega(t-\sigma)} \in L^1(s,t)$. In view of the Lebesgue's dominated convergence theorem, we obtain

$$\lim_{n\to+\infty} \int_s^t T(t-\sigma)F(\sigma+t_n,u(\sigma+t_n))\,d\sigma = \int_s^t T(t-\sigma)G(\sigma,v(\sigma))\,d\sigma. \tag{4.125}$$

Using (4.120), (4.123), and (4.125), we deduce

$$v(t) = T(t-s)v(s) + \int_s^t T(t-\sigma)G(\sigma,v(\sigma))\,d\sigma \quad \text{for } t \ge s. \tag{4.126}$$

The continuous function v satisfies (4.124) and (4.126), therefore v is a K-mild solution of equation (4.121). □

Lemma 4.4.16. *Let K be a compact subset of X. If $f \in AA_u(\mathbb{R}\times X, X)$, then the function g defined by* (4.105) *satisfies*

(i) *$g : \mathbb{R}\times K \to X$ is measurable;*
(ii) *for all $t \in \mathbb{R}$, $g(t,\cdot) \in C(K,X)$.*

Proof. Claim (i) is obvious and (ii) is a consequence of $f(t+t_n,\cdot) \in C(K,X)$ and $f(t+t_n,\cdot) \to g(t,\cdot)$ uniformly on K. □

Lemma 4.4.17 ([98]). *Let K be a compact subset of X. If $f \in AA_u(\mathbb{R}\times X, X)$, then*

$$\sup_{t\in\mathbb{R}}\sup_{x\in K}\|f(t,x)\| < +\infty \quad \text{and} \quad \sup_{t\in\mathbb{R}}\sup_{x\in K}\|g(t,x)\| < +\infty,$$

where g is the function defined by (4.105).

Lemma 4.4.18. *Assume that* **(H1)** *and* **(H4)** *hold. If equation* (4.100) *admits at least one mild solution x_0 on $[t_0,+\infty)$ such that $\{x_0(t) : t \in \mathbb{R}\}$ is relatively compact, then there exists a mild solution x on $\mathbb{R}$ of equation* (4.100) *such that*

$$\{x(t) : t \in \mathbb{R}\} \subset \overline{\{x_0(t) : t \geq t_0\}}. \tag{4.127}$$

Proof. Denote by $K = \overline{\{x_0(t) : t \geq t_0\}}$ the compact subset of X. The mild solution x_0 satisfies

$$x_0(t) \in K \quad \text{for all } t \geq t_0, \tag{4.128}$$

$$x_0(t) = T(t-s)x_0(s) + \int_s^t T(t-\sigma)f(\sigma, x_0(\sigma))\,d\sigma \quad \text{for } t \geq s \geq t_0. \tag{4.129}$$

By hypothesis **(H4)**, we have $k = \sup_{t\in\mathbb{R}}\sup_{x\in K}\|f(t,x)\| < +\infty$ (cf. Lemma 4.4.17). Using Lemma 4.4.12 with $F : [t_0,+\infty)\times K \to X$ defined by $F(t,x) = f(t,x)$ for each $(t,x) \in [t_0,+\infty)\times K$, we obtain the existence of a function $\phi : [0,+\infty) \to [0,+\infty)$ satisfying (4.112) and

$$\|x_0(t) - x_0(s)\| \leq \phi(|t-s|) \quad \text{for } t,s \geq t_0. \tag{4.130}$$

Let $(t'_n)_n$ be a sequence of real numbers such that

$$\lim_{n\to+\infty} t'_n = +\infty.$$

Since f is almost automorphic in t uniformly with respect to x, there exist a map $g : \mathbb{R}\times K \to X$ and a subsequence of $(t'_n)_n$, denoted by $(t_n)_n$, such that, for all $t \in \mathbb{R}$, we have

$$\lim_{n\to+\infty}\sup_{x\in K}\|f(t+t_n,x) - g(t,x)\| = 0, \tag{4.131}$$

$$\lim_{n\to+\infty}\sup_{x\in K}\|g(t-t_n,x) - f(t,x)\| = 0. \tag{4.132}$$

Given any interval $(\tau,+\infty)$, for $n \in \mathbb{N}$ sufficiently large ($\tau + t_n \geq t_0$), the function $t \mapsto x_0(\cdot + t_n)$ is defined on $(\tau,+\infty)$. Moreover, (4.128) and (4.130) imply

$$\begin{aligned} x_0(t+t_n) &\in K \quad \text{for } t \geq \tau, \\ \|x_0(t+t_n) - x_0(s+t_n)\| &\leq \phi(|t-s|) \quad \text{for } t, s \geq \tau. \end{aligned} \tag{4.133}$$

Taking τ as a sequence tending to $-\infty$, applying Arzelà–Ascoli theorem, and using a diagonal argument, we can assert that there exist $x_* \in C(\mathbb{R}, X)$ and a subsequence of $(t_n)_n$ such that

$$x_0(t+t_n) \to x_*(t) \quad \text{as } n \to +\infty, \tag{4.134}$$

uniformly on each compact subset of $\mathbb{R}$. Since x_0 satisfies (4.129), for each $t \geq s$ and for $n \in \mathbb{N}$ sufficiently large, we have

$$x_0(t+t_n) = T(t-s)x_0(s+t_n) + \int_s^t T(t-\sigma)f(\sigma + t_n, x_0(\sigma + t_n))\, d\sigma. \tag{4.135}$$

By using (4.131), (4.133)–(4.135), we deduce that x_* is a K-mild solution on $\mathbb{R}$ of

$$x_*'(t) = Ax_*(t) + g(t, x_*(t)).$$

The proof is similar to that of Lemma 4.4.14. From Lemmas 4.4.16 and 4.4.17, we obtain that the function $g : \mathbb{R} \times K \to X$ satisfies hypotheses (4.117) and (4.118) of Lemma 4.4.14. Applying Lemma 4.4.14 with $F = g$, $u = x_*$, and the sequence $(-t_n)_n$ (cf. (4.132)), we obtain the existence of a K-mild solution x of equation (4.100). Consequently, x is a mild solution satisfying (4.127). □

Proposition 4.4.19. *Assume that* (H1)–(H4) *hold. If equation* (4.100) *admits at least one mild solution x_0 defined and bounded on* $[t_0, +\infty)$, *i. e.,* $\sup_{t \geq t_0} \|x_0(t)\| < +\infty$, *then*
(i) *its range $\{x_0(t) : t \geq t_0\}$ is relatively compact in X,*
(ii) *there exists a mild solution x on $\mathbb{R}$ of equation* (4.100) *satisfying* (4.127).

Proof. (i) Let $\theta \in (0,1)$. Then the mild solution x_0 verifies

$$x_0(t) = T(\theta)x_0(t-\theta) + \int_{t-\theta}^t T(t-\sigma)f(\sigma, x_0(\sigma))\, d\sigma \quad \text{for } t \geq t_0 + 1. \tag{4.136}$$

Here x_0 is the sum of two terms. The first satisfies

$$T(\theta)x_0(t-\theta) \in T(\theta)(\overline{B}(0, \|x_0\|_\infty)), \tag{4.137}$$

where $\|x_0\|_\infty = \sup_{t \geq t_0} \|x_0(t)\| < +\infty$ and $\overline{B}(0, \|x_0\|_\infty)$ is the closed ball with radius $\|x_0\|_\infty$ centered at the origin. The set $T(\theta)(\overline{B}(0, \|x_0\|_\infty)$ is relatively compact in X since the operator $T(\theta)$ is compact for each $\theta > 0$. For the second term, by (4.114), we obtain

$$\left\| \int_{t-\theta}^{t} T(t-\sigma)f(\sigma, x_0(\sigma))\, d\sigma \right\| \le kM \int_{t-\theta}^{t} e^{\omega(t-\sigma)}\, d\sigma,$$

with

$$k = \sup_{t \ge t_0} \sup_{\|\xi\| \le \|x_0\|_\infty} \|f(t,\xi)\| < +\infty.$$

Using the fact that $\int_{t-\theta}^{t} e^{\omega(t-\sigma)}\, d\sigma = \int_0^{\theta} e^{\omega\sigma}\, d\sigma$, we obtain

$$\int_{t-\theta}^{t} T(t-\sigma)f(\sigma, x_0(\sigma))\, d\sigma \in \overline{B}(0, \delta(\theta)), \tag{4.138}$$

where

$$\delta(\theta) = kM \int_0^{\theta} e^{\omega\sigma}\, d\sigma.$$

By (4.136)–(4.138), for each $\theta \in (0,1)$, we deduce the existence a compact subset K_θ of X such that

$$\{x_0(t) : t \ge t_0 + 1\} \subset K_\theta + \overline{B}(0, \delta(\theta)). \tag{4.139}$$

For the sequel, we introduce the Kuratowski's measure of noncompactness $\alpha(\cdot)$ of a bounded subset B in the Banach space X defined by

$$\alpha(B) = \inf\{\varepsilon > 0 : B \text{ has a finite cover of balls of diameter } < \varepsilon\}.$$

The Kuratowski's measure of noncompactness verifies the following two properties:

$$\begin{aligned} &\alpha(B_1 + B_2) \le \alpha(B_1) + \alpha(B_2),\\ &\alpha(B) = 0 \quad \Longleftrightarrow \quad B \text{ is relatively compact in } X. \end{aligned}$$

For further properties of $\alpha(\cdot)$, see [194, Section 1.4]. From the inclusion (4.139), we obtain the inequality

$$\alpha(\{x_0(t) : t \ge t_0 + 1\}) \le \alpha(K_\theta) + \alpha(\overline{B}(0, \delta(\theta))).$$

Since $\alpha(\overline{B}(0, \delta(\theta))) \le 2\delta(\theta)$ and K_θ is a compact subset of X, it follows that

$$\alpha(\{x_0(t) : t \ge t_0 + 1\}) \le 2\delta(\theta) \quad \text{for } 0 < \theta < 1,$$

which implies $\alpha(\{x_0(t) : t \ge t_0 + 1\}) = 0$ since $\lim_{\theta \to +\infty} \delta(\theta) = 0$. From properties of the Kuratowski's measure of noncompactness, it follows that $\{x_0(t) : t \ge t_0 + 1\}$ is relatively compact in X, and then the range of x_0, namely $\{x_0(t) : t \geq t_0\}$, is also relatively compact in X because x_0 is continuous.

(ii) The proof is straightforward using Lemma 4.4.18. □

4.4.3 Proofs of main results

The object of this section is to prove Theorems 4.4.6, 4.4.9, and Corollaries 4.4.10, 4.4.11.

Proof of Theorem 4.4.6. (i) It is a consequence of Proposition 4.4.19.

(ii) Let $\delta = \inf_{x \in \mathcal{F}_K} \lambda_K(x)$ the greatest lower bound (infimum) of $\{\lambda_K(x) : x \in \mathcal{F}_K\}$ in $\overline{\mathbb{R}} = \mathbb{R} \cup \{-\infty, +\infty\}$. By Lemma 4.4.18, we obtain the existence of a mild solution x such that $\{x(t) : t \in \mathbb{R}\} \subset K$, therefore $\mathcal{F}_K$ is nonempty, so δ exists in $\mathbb{R} \cup \{-\infty\}$. Then there exists a sequence $(x_n)_n$ with values in $\mathcal{F}_K$ such that

$$\lim_{n \to +\infty} \lambda_K(x_n) = \delta. \tag{4.140}$$

By definition of $\mathcal{F}_K$, we have that $\{x_n(t) : n \in \mathbb{N}\}$ is a subset of the compact K for each $t \in \mathbb{R}$. By Lemma 4.4.17, we can assert that $\sup_{t \in \mathbb{R}} \sup_{x \in K} \|f(t,x)\| < +\infty$. Now using Lemma 4.4.12 with $F : \mathbb{R} \times K \to X$ defined by $F(t,x) = f(t,x)$ for each $(t,x) \in \mathbb{R} \times K$, we obtain that there exists a function $\phi : [0,+\infty) \to [0,+\infty)$ satisfying (4.112) such that

$$\|x_n(t) - x_n(s)\| \leq \phi(|t-s|) \quad \text{for } n \in N \text{ and } t,s \in I.$$

Therefore the sequence $(x_n)_n$ is equicontinuous. In view of Arzelà–Ascoli theorem, we can assert that $\{x_n : n \in \mathbb{N}\}$ is a relatively compact subset of $C(\mathbb{R}, X)$ endowed with the topology of compact convergence, hence there exists a subsequence of $(x_n)_n$ such that

$$x_n(t) \to x_*(t) \quad \text{as } n \to +\infty, \tag{4.141}$$

uniformly on each compact subset of $\mathbb{R}$. Obviously, $x_* \in C_K(\mathbb{R}, X)$. From (4.141), we deduce that $f(t, x_n(t)) \to f(t, x_*(t))$ as $n \to +\infty$ uniformly on each compact subset of $\mathbb{R}$. Since x_n is a mild solution on $\mathbb{R}$ of equation (4.100), we have

$$x_n(t) = T(t-s)x_n(s) + \int_s^t T(t-\sigma)f(\sigma, x_n(\sigma))\,d\sigma \quad \text{for } t \geq s$$

and, letting $n \to +\infty$,

$$x_*(t) = T(t-s)x_*(s) + \int_s^t T(t-\sigma)f(\sigma, x_*(\sigma))\,d\sigma \quad \text{for } t \geq s,$$

which shows that x_* is a mild solution on $\mathbb{R}$ of equation (4.100). Then $x_* \in \mathcal{F}_K$, and so

$$\delta \le \lambda_K(x_*). \tag{4.142}$$

From (4.141) and the definition of a subvariant functional, we have

$$\lambda_K(x_*) \le \liminf_{n\to+\infty} \lambda_K(x_n). \tag{4.143}$$

From (4.140), (4.142), and (4.143), we deduce that

$$\lambda_K(x_*) = \delta,$$

therefore x_* is a minimal K-valued solution,

$$\lambda_K(x_*) = \inf_{x\in\mathcal{F}_K} \lambda_K(x), \tag{4.144}$$

which implies the existence of the minimal K-valued solution.

(iii) Let us denote by x_* the unique minimal K-valued solution of equation (4.100). To check that x_* is compact almost automorphic, we have to prove that if $(t_n)_n$ is any sequence of real numbers, then one can pick a subsequence of $(t_n)_n$ such that

$$x_*(t+t_n) \to y_*(t) \quad \text{as } n \to +\infty, \tag{4.145}$$

$$y_*(t-t_n) \to x_*(t) \quad \text{as } n \to +\infty, \tag{4.146}$$

uniformly on each compact subset of $\mathbb{R}$. In fact, by assumption, we can choose a subsequence of $(t_n)_n$ such that, for all $t \in \mathbb{R}$,

$$\lim_{n\to+\infty} \sup_{x\in K} \|f(t+t_n,x) - g(t,x)\| = 0, \tag{4.147}$$

$$\lim_{n\to+\infty} \sup_{x\in K} \|g(t-t_n,x) - f(t,x)\| = 0, \tag{4.148}$$

where g is a map from $\mathbb{R}\times K$ to X. Since $\sup_{t\in\mathbb{R}} \sup_{x\in K} \|f(t,x)\| < +\infty$ (cf. Lemma 4.4.17) and (4.145) holds, applying Lemma 4.4.14 with $u = x_*$, $F = f$, and the sequence $(t_n)_n$, we obtain (4.145) where y_* is a mild solution on $\mathbb{R}$ of equation

$$x'(t) = Ax(t) + g(t, x(t)),$$

which satisfies all the hypotheses of Lemma 4.4.14 (cf. Lemmas 4.4.16 and 4.4.17). From (4.145), by the definition of the subvariant λ_K, we obtain that $\lambda_K(y_*) \le \lambda_K(x_*)$ and, with (4.144), we deduce that

$$\lambda_K(y_*) \le \inf_{x\in\mathcal{F}_K} \lambda_K(x). \tag{4.149}$$

Since x_* is K-valued, by (4.145), we obtain that y_* is K-valued. Applying again Lemma 4.4.14 to $u = y_*$, $F = g$, and the sequence $(-t_n)_n$, we obtain that

$$y_*(t-t_n) \to z_*(t) \quad \text{as } n \to +\infty, \tag{4.150}$$

(for a subsequence) where z_* is a mild solution on $\mathbb{R}$ of equation (4.100), because (4.148) holds. Moreover, from (4.150), we deduce that $\lambda_K(z_*) \le \lambda_K(y_*)$ and, with (4.149), obtain

$$\lambda_K(z_*) \le \inf_{x\in\mathcal{F}_K} \lambda_K(x). \tag{4.151}$$

Since y_* is K-valued, from (4.150) we obtain that z_* is K-valued, and then $z_* \in \mathcal{F}_K$. From (4.151), we obtain that $\lambda_K(z_*) = \inf_{x\in\mathcal{F}_K} \lambda_K(x)$, therefore z_* is a minimal K-valued solution of equation (4.100). By the uniqueness of the minimal K-valued solution of equation (4.100), we deduce that $x_* = z_*$, therefore (4.146) is fulfilled, thus x_* is compact almost automorphic. □

Proof of Theorem 4.4.9. In the proof of Theorem 4.4.6, assumptions **(H2)** and **(H3)** are only used to get that the set $\{x_0(t) : t \ge t_0\}$ is relatively compact (cf. Proposition 4.4.19). □

Proof of Corollary 4.4.10. Since $\mathrm{AP}_u(\mathbb{R}\times X, X) \subset \mathrm{AA}_u(\mathbb{R}\times X, X)$, claims (i) and (ii) result due to Theorem 4.4.6. For the same reason, equation (4.100) admits a unique minimal K-valued solution x_*. To check that x_* is almost periodic, we have to prove that if $(t_n)_n$ and $(s_n)_n$ are two arbitrary sequences of real numbers, then one can pick up common subsequences of $(t_n)_n$ and $(s_n)_n$ such that

$$\forall t \in \mathbb{R}, \quad \lim_{p\to\infty}\lim_{n\to\infty} x_*(t+t_n+s_p) = \lim_{m\to\infty} x_*(t+t_m+s_m), \tag{4.152}$$

(see [154, Theorem 1.17, p. 12]), instead of (4.145) and (4.146). In fact, by hypothesis **(H5)**, for each compact subset K of X, we can choose common subsequences of $(t_n)_n$ and $(s_n)_n$ such that

$$\lim_{p\to\infty}\lim_{n\to\infty} \sup_{t\in}\sup_{x\in K}\|f(t+t_n+s_p,x) - g(t,x)\| = 0, \tag{4.153}$$

$$\lim_{m\to\infty} \sup_{t\in}\sup_{x\in K}\|f(t+t_m+s_m,x) - g(t,x)\| = 0, \tag{4.154}$$

where g is a function from $\mathbb{R}\times X$ to X. Observe that $g \in \mathcal{H}(f)$ and $f \in \mathrm{AP}_u(\mathbb{R}\times X, X)$. Denoting $y_*(t) = \lim_{p\to\infty}\lim_{n\to\infty} x_*(t+t_n+s_p)$ and $z_*(t) = \lim_{m\to\infty} x_*(t+t_m+s_m)$, we deduce that y_* and z_* are two minimal K-valued solution of equation (4.107), where g is the function defined by (4.153); the proof of this assertion is similar to that given in the proof of Theorem 4.4.6. By the uniqueness of the minimal K-valued solution of equation (4.107), for each $g \in \mathcal{H}(f)$, we deduce that $y_* = z_*$, therefore (4.152) holds, thus x_* is almost periodic. This ends the proof. □

Proof of Corollary 4.4.11. By using a similar reasoning as that in the proof of Corollary 4.4.10, we deduce Corollary 4.4.11 from Theorem 4.4.9. □

4.4.4 Heat and wave equations with nonlinearities

In this section, we provide two examples of partial differential equations to illustrate our theoretical results. We give several existence theorems of compact almost automorphic solutions for a heat equation and a wave equation defined on a domain of $\mathbb{R}^n$. We also give some existence results in the almost periodic context.

4.4.4.1 Heat equation

To apply Theorem 4.4.6, we consider the following heat equation in a bounded open subset Ω of $\mathbb{R}^n$ with a smooth boundary $\partial\Omega$:

$$\begin{cases} \frac{\partial}{\partial t}v(t,x) = \sum_{i=1}^{n} \frac{\partial^2}{\partial x^2}v(t,x) + g(v(t,x)) + h(t,x) & \text{in } \mathbb{R}\times\Omega, \\ v(t,x) = 0 \quad \text{on } \mathbb{R}\times\partial\Omega, \end{cases} \tag{4.155}$$

where $g : \mathbb{R} \to \mathbb{R}$ and $h : \mathbb{R}\times\overline{\Omega} \to \mathbb{R}$ are continuous functions.

In order to rewrite equation (4.155) in the abstract form, we introduce the space $X = C_0(\Omega)$ of all continuous functions from $\overline{\Omega}$ (the closure of Ω) to $\mathbb{R}$ vanishing on $\partial\Omega$, endowed with the uniform norm topology. Define the operator $A : D(A) \subset X \to X$ by

$$\begin{cases} D(A) = \{x \in C_0(\Omega) \cap H_0^1(\Omega) : \Delta x \in C_0(\Omega)\}, \\ Ax = \Delta x, \end{cases}$$

where Δ is the Laplacian operator. Let us denote by λ_1 the smallest eigenvalue of $-\Delta$ in $H_0^1(\Omega)$ ($\lambda_1 > 0$ since Ω is bounded).

Lemma 4.4.20. *The linear operator A generates a compact C_0-semigroup $(T(t))_{t\geq 0}$ on X such that*

$$\|T(t)\| \leq M\exp(-\lambda_1 t) \quad \textit{for } t \geq 0, \tag{4.156}$$

with $M = \exp(\lambda_1|\Omega|^{\frac{2}{n}}(4\pi)^{-1})$. Consequently, hypotheses **(H1)** *and* **(H2)** *are satisfied.*

Proof. The bound (4.156) is a consequence of (see [90, Proposition 3.5.10, p. 47]). Using a result in [84], we deduce that the C_0-semigroup $(T(t))_{t\geq 0}$ is compact. □

In order to study the existence of an almost automorphic solution of equation (4.155), we suppose that:

(E1) The function $h : \mathbb{R}\times\overline{\Omega} \to \mathbb{R}$ is continuous, h satisfies $h(t,\xi) = 0$ on $\mathbb{R}\times\partial\Omega$, and h is $C_0(\Omega)$-almost automorphic, which means that for any sequence of real numbers $(s_n')_n$, there exist a subsequence $(s_n)_n$ and a measurable function $k : \mathbb{R}\times\Omega \to \mathbb{R}$ such that

$$\forall t \in \mathbb{R}, \quad \lim_{n\to\infty} \sup_{\xi\in\Omega} |h(t+s_n,\xi) - k(t,\xi)| = 0$$

and

$$\forall t \in \mathbb{R}, \quad \lim_{n\to\infty} \sup_{\xi\in\Omega} |k(t - s_n, \xi) - h(t, \xi)| = 0.$$

Remark 4.4.21. As the function $t \mapsto h(t, \cdot)$ belongs to $AA(\mathbb{R}, X)$, it is bounded on $\mathbb{R}$, therefore the function h is bounded on $\mathbb{R} \times \overline{\Omega}$. An example of function h satisfying **(E1)** is the following:

$$h(t, \xi) = \sin\left(\frac{1}{2 + \cos t + \cos(\sqrt{2}t)}\right) h_0(\xi) \quad \text{for } t \in \mathbb{R} \text{ and } \xi \in \Omega,$$

where $h_0 \in X = C_0(\Omega)$. It is well known that each almost periodic function is uniformly continuous. Since the function $t \mapsto \sin(\frac{1}{2+\cos t+\cos \sqrt{(2t)}})$ is not uniformly continuous (see [234]), it is not almost periodic.

Moreover, we suppose that:

(E2) g is locally Lipschitz, g satisfies $g(0) = 0$, and $\limsup_{|r|\to+\infty} \frac{g(r)}{r} < \lambda_1$.

Let $G : X \to X$ be the superposition operator of g defined by

$$G(x)(\xi) = g(x(\xi)) \quad \text{for } x \in X \text{ and } \xi \in \Omega. \tag{4.157}$$

Let $H : \mathbb{R} \to X$ be defined by

$$H(t)(\xi) = h(t, \xi) \quad \text{for } t \in \mathbb{R} \text{ and } \xi \in \Omega. \tag{4.158}$$

Let us denote by $f : \mathbb{R}\times X \to X$ the map defined by

$$f(t, x) = G(x) + H(t) \quad \text{for } t \in \mathbb{R} \text{ and } x \in X. \tag{4.159}$$

If **(E1)** and **(E2)** hold, then $G \in C(X, X)$ and $H \in AA(\mathbb{R}, X)$, therefore f satisfies hypothesis **(H4)**. With these notations, equation (4.155) takes the following abstract form:

$$x'(t) = Ax(t) + f(t, x(t)) \quad \text{for } t \in \mathbb{R}. \tag{4.160}$$

For the initial value problem

$$\begin{cases} x'(t) = Ax(t) + f(t, x(t)) & \text{for } t \geq 0, \\ x(0) = x_0, \end{cases} \tag{4.161}$$

we have the following result:

Proposition 4.4.22. *Assume that* **(E1)** *and* **(E2)** *hold. Then for every* $x_0 \in X$*, equation* (4.161) *has a unique mild solution* x *on* $[0, +\infty)$*. Moreover, this solution* x *is bounded on* $[0, +\infty)$*.*

Proof. By **(E1)** and **(E2)**, we have $H \in L^\infty(\mathbb{R}, X)$, g is locally Lipschitz, $g(0) = 0$, and there exists $M > 0$ such that $rg(r) \le Cr^2$ for $|r| \ge M$, with $C < \lambda_1$. The existence and uniqueness of a global mild solution of equation (4.161) which is bounded on $[0, +\infty)$ results from [90, Proposition 8.3.7, p. 117]. □

In order to prove the existence of a compact almost automorphic solution, we need to assume the following:

(E3) The function $r \mapsto g(r) - \lambda_1 r$ is nonincreasing on $\mathbb{R}$.

Remark 4.4.23. The functions g defined by $g(r) = \lambda_1 \sin(r)$ and $g(r) = \lambda_1 r - r^3$ satisfy **(E2)** and **(E3)**. If g is locally Lipschitz, $g(0) = 0$, and the function $r \mapsto g(r) - \lambda r$ $(\lambda < \lambda_1)$ is nonincreasing on $\mathbb{R}$, then hypotheses **(E2)** and **(E3)** hold, and the function $r \mapsto g(r) - \lambda_1 r$ is strictly decreasing. An example of a function satisfying **(E2)** and **(E3)** such that $r \mapsto g(r) - \lambda_1 r$ is not strictly decreasing is the following: $g(r) = \lambda_1 |r|$ on $[-\pi, \pi]$ and g is 2π-periodic on $\mathbb{R}$.

To prove our result of existence of an almost automorphic solution, we use the two following lemmas. Let us denote by $E : X \to \mathbb{R}$ the map defined by

$$E(x) = \frac{1}{2} \int_\Omega |x(\xi)|^2 \, d\xi. \tag{4.162}$$

It is obvious that E is continuous on X.

Lemma 4.4.24. *Let $p \in C(\mathbb{R}, X)$.*

(i) *If $w \in C(\mathbb{R}, X)$ satisfies*

$$w(t) = T(t-s)w(s) + \int_s^t T(t-\sigma)p(\sigma)\, d\sigma \quad \text{for } t \ge s, \tag{4.163}$$

then

$$E(w(t)) \le E(w(s)) + \int_s^t \int_\Omega \{p(t,\xi) - \lambda_1 w(t,\xi)\} w(t,\xi)\, d\xi \, d\sigma \quad \text{for } t \ge s. \tag{4.164}$$

(ii) *If $w \in C(\mathbb{R}, X)$ satisfies*

$$w(t) = T(t-s)w(s) + \lambda_1 \int_s^t T(t-\sigma)w(\sigma)\, d\sigma \quad \text{for } t \ge s \tag{4.165}$$

and $t \mapsto E(w(t))$ is constant on $\mathbb{R}$, then there exists $w_0 \in X$ such that

$$\forall t \in \mathbb{R}, \quad w(t) = w_0. \tag{4.166}$$

Proof. To prove Lemma 4.4.24, we use the L^2-theory. Consider $Y = L^2(\Omega)$, the Lebesgue space of order 2 from Ω to $\mathbb{R}$, endowed with its usual scalar product $(\cdot|\cdot)_Y$. The associated norm is denoted by $\|\cdot\|_Y$. Define the operator $B : D(B) \subset Y \to Y$ by $D(B) = H^2(\Omega) \cap H_0^1(\Omega)$ and $By = \Delta y$. Denote by $(S(t))_{t\geq 0}$ the C_0-semigroup generated by B in Y. Recall that

$$\forall y \in H_0^1(\Omega), \quad |\nabla y|_Y^2 \geq \lambda_1 |y|_Y^2, \tag{4.167}$$

and

$$y \in D(B) \text{ and } |\nabla y|_Y^2 = \lambda_1 |y|_Y^2 \quad \Longrightarrow \quad By + \lambda_1 y = 0. \tag{4.168}$$

(i) Let $w \in C(\mathbb{R}, X)$ satisfy (4.163). By using the fact that $X \subset Y$ with a continuous injection and $T(t)\phi = S(t)\phi$ for all $t \geq 0$ and $\phi \in X$, we deduce that $w \in C(\mathbb{R}, Y)$ and w satisfies

$$w(t) = S(t-s)w(s) + \int_s^t S(t-\sigma)p(\sigma)\,\mathrm{d}\sigma \quad \text{for } t \geq s. \tag{4.169}$$

Let $s \in \mathbb{R}$ and $T > 0$. Assume that $w(s) \in D(B)$ and $p \in C^1([s, s+T], Y)$. In this case, we have $w \in C([s, s+T], D(B)) \cap C^1([s, s+T], Y)$ and

$$w'(t) = Bw(t) + p(t) \quad \text{for } s \leq t \leq s+T.$$

It follows that

$$\frac{1}{2}\frac{\mathrm{d}}{\mathrm{d}t}[|w(t)|_Y^2] = (w'(t)|w(t))_Y = (Bw(t) + \lambda_1 w(t)|w(t))_Y + (p(t) - \lambda_1 w(t)|w(t))_Y.$$

Hence by Green's formula and (4.167), we get

$$(By + \lambda_1 y|y)_Y = -|\nabla y|_Y^2 + \lambda_1 |y|_Y^2 \leq 0 \quad \text{for } y \in D(B),$$

therefore

$$\frac{1}{2}\frac{\mathrm{d}}{\mathrm{d}t}[|w(t)|_Y^2] \leq (p(t) - \lambda_1 w(t)|w(t))_Y \tag{4.170}$$

and, by integrating (4.170) on $[s, s+T]$, we obtain

$$\frac{1}{2}|w(t)|_Y^2 \leq \frac{1}{2}|w(s)|_Y^2 + \int_s^t (p(\sigma) - \lambda_1 w(\sigma)|w(\sigma))_Y \mathrm{d}\sigma \quad \text{for } s \leq t \leq s+T. \tag{4.171}$$

In the general case, we consider two sequences $(w_n^s)_n \subset D(B)$ and $(p_n)_n \subset C^1([s, s+T], Y)$ such that $(w_n^s)_n$ converges to $w(s)$ in Y and $(p_n)_n$ converges to p in $L^1((s, s+T), Y)$. Denote by w_n the corresponding solution of (4.169). It follows that the sequence $(w_n)_n$ converges uniformly on $[s, s+T]$ to w and w_n satisfies

$$\frac{1}{2}|w_n(t)|_Y^2 \le \frac{1}{2}|w_n^s|_Y^2 + \int_s^t (p_n(\sigma) - \lambda_1 w_n(\sigma)|w_n(\sigma))_Y \mathrm{d}\sigma \quad \text{for } s \le t \le s+T,$$

then w satisfies (4.171) on each interval $[s, s+T]$. Consequently, w satisfies (4.164) on $\mathbb{R}$.

(ii) Let $w \in C(\mathbb{R}, X)$ satisfy (4.165). We fix $s \in \mathbb{R}$ and $T > 0$. Using [90, Proposition 5.1.1, p. 61], we obtain $w \in C((s, s+T], D(B)) \cap C^1((s, s+T], Y)$ and

$$w'(t) = Bw(t) + \lambda_1 w(t) \quad \text{for } s < t \le s+T. \tag{4.172}$$

Moreover, $t \mapsto E(w(t))$ is constant, thus

$$\frac{\mathrm{d}}{\mathrm{d}t} E(w(t)) = (w'(t)|w(t))_Y = 0,$$

therefore

$$(Bw(t) + \lambda_1 w(t)|w(t))_Y = 0 \quad \text{for } s < t \le s+T.$$

By using Green's formula and (4.168), we have $Bw(t) + \lambda_1 w(t) = 0$ and, from (4.172), deduce that $w(t)$ does not depend of t. Consequently, there exists $w_0 \in X$ such that (4.166) holds on each interval $[s, s+T]$. This ends the proof. □

Lemma 4.4.25. *Assume that* **(E1)–(E3)** *hold. Let u and v be two mild solutions on $\mathbb{R}$ of equation* (4.160)*. Then the following assertions hold:*
(i) *The function $t \mapsto E(u(t) - v(t))$ is nonincreasing on $\mathbb{R}$.*
(ii) *If the function $t \mapsto E(u(t) - v(t))$ is constant on $\mathbb{R}$, then there exists $w_0 \in X$ such that*

$$u(t) - v(t) = w_0 \quad \textit{for } t \in \mathbb{R}, \tag{4.173}$$

$$G(u(t)) - G(v(t)) = \lambda_1 w_0 \quad \textit{for } t \in \mathbb{R}. \tag{4.174}$$

Moreover, for all $\theta \in [0,1]$, $\theta u + (1-\theta)v$ is also a mild solution on $\mathbb{R}$ of equation (4.160).

Proof. (i) Let $w = u - v$. Then $w \in C(\mathbb{R}, X)$ satisfies (4.163) with

$$p(t) = f(t, u(t)) - f(t, v(t)) = G(u(t)) - G(v(t)),$$

where f and G are functions defined by (4.157)–(4.159). Using Lemma 4.4.24, we obtain

$$q(t) \le q(s) + \int_s^t \int_\Omega \{\phi(u(t,\xi)) - \phi(v(t,\xi))\}\{u(t,\xi) - v(t,\xi)\}\, \mathrm{d}\xi\, \mathrm{d}\sigma \quad \text{for } t \ge s, \tag{4.175}$$

where

$$q(t) = E(u(t) - v(t)),$$
$$\phi(r) = g(r) - \lambda_1 r \quad \text{for } r \in \mathbb{R}. \tag{4.176}$$

By **(E3)**, ϕ is nonincreasing, thus

$$\{\phi(r_1) - \phi(r_2)\}\{r_1 - r_2\} \le 0 \quad \text{for } r_1 \text{ and } r_2 \in \mathbb{R}, \tag{4.177}$$

and, from (4.175) and (4.177), we obtain that the function q is nonincreasing on $\mathbb{R}$.

(ii) Assume that the function q is constant on $\mathbb{R}$. By (4.175) and (4.177), we obtain that

$$\{\phi(u(t,\xi)) - \phi(v(t,\xi))\}\{u(t,\xi) - v(t,\xi)\} = 0 \quad \text{for } t \in \mathbb{R} \text{ and } \xi \in \Omega.$$

Since ϕ is nonincreasing, it follows that

$$\phi(u(t,\xi)) = \phi(v(t,\xi)) \quad \text{for } t \in \mathbb{R} \text{ and } \xi \in \Omega, \tag{4.178}$$

and then, from (4.157) and (4.176), we deduce that

$$G(u(t)) - \lambda_1 u(t) = G(v(t)) - \lambda_1 v(t) \quad \text{for } t \in \mathbb{R}.$$

Since

$$f(t,u(t)) - f(t,v(t)) = G(u(t)) - G(v(t)) = \lambda_1(u(t) - v(t)) \quad \text{for } t \in \mathbb{R}, \tag{4.179}$$

the function w satisfies (4.165). Since the function $t \mapsto E(w(t))$ is constant on $\mathbb{R}$, by Lemma 4.4.24, we deduce the existence of $w_0 \in X$ such that (4.173) holds on $\mathbb{R}$ and, by using (4.179), obtain (4.174). Since ϕ is nonincreasing, we deduce from (4.178) that

$$\phi(\theta u(t,\xi)) + (1-\theta)v(t,\xi)) = \phi(u(t,\xi)) \quad \text{for } t \in \mathbb{R}, \xi \in \Omega, \text{and } \theta \in [0,1].$$

It follows that

$$\phi(\theta u(t,\xi)) + (1-\theta)v(t,\xi)) = \theta\phi(u(t,\xi)) + (1-\theta)\phi(v(t,\xi)) \quad \text{for } t \in \mathbb{R},\ \xi \in \Omega, \text{and } \theta \in [0,1],$$

hence

$$g(\theta u(t,\xi)) + (1-\theta)v(t,\xi)) = \theta g(u(t,\xi)) + (1-\theta)g(v(t,\xi)) \quad \text{for } t \in \mathbb{R},\ \xi \in \Omega, \text{and } \theta \in [0,1],$$

and so

$$f(t,\theta u(t)) + (1-\theta)v(t)) = \theta f(t,u(t)) + (1-\theta)f(t,v(t)) \quad \text{for } t \in \mathbb{R} \text{ and } \theta \in [0,1].$$

Consequently, $\theta u + (1-\theta)v$ is a mild solution of equation (4.160). This ends the proof. □

Proposition 4.4.26. *Assume that* **(E1)–(E3)** *hold. Then equation* (4.160) *has a at least one mild solution x_1 which is compact almost automorphic. If x_2 is a mild solution which is only almost automorphic, then there exists $w_0 \in X$ such that*

$$x_2(t) = x_1(t) + w_0 \quad \textit{for } t \in \mathbb{R}. \tag{4.180}$$

Consequently, x_2 is compact almost automorphic. Moreover, if the function $r \mapsto g(r) - \lambda_1 r$ is strictly decreasing on $\mathbb{R}$, then the compact almost automorphic mild solution is unique.

Proof. To prove the existence of a compact almost automorphic solution of (4.160), we use Theorem 4.4.6. By Lemma 4.4.20, assumptions **(H1)** and **(H2)** hold. Hypothesis **(H4)** results from **(E1)** and **(E2)**. For hypothesis **(H3)**, since g is locally Lipschitz, G is Lipschitz continuous on bounded sets of X, therefore G is bounded on the bounded sets of X. It follows that, for all $t \in \mathbb{R}$ and $x \in X$ such that $\|x\| \le R$,

$$\|f(t,x)\| \le \sup_{\|x\| \le R} \|G(x)\| + \sup_{t \in \mathbb{R}} \|H(t)\| < +\infty,$$

therefore **(H3)** holds. By Proposition 4.4.22, equation (4.160) admits at least one mild solution x_0 defined and bounded on $[0, +\infty)$. By Theorem 4.4.6, the set $\overline{\{x_0(t) : t \ge 0\}}$ is compact. Let us denote by K the closed convex hull of the compact set $\overline{\{x_0(t) : t \ge 0\}}$. Then K is a convex and compact subset of X. To prove the existence of a compact almost automorphic solution, we use Theorem 4.4.6 with the subvariant functional $\lambda_K(x) = \sup_{t \in \mathbb{R}} E(x(t))$ associated to the compact set K. It remains to prove the uniqueness of the minimal K-valued solution. Let u and v be two minimal K-valued solutions of equation (4.160). Denote

$$\delta = \sup_{t \in \mathbb{R}} E(u(t)) = \sup_{t \in \mathbb{R}} E(v(t)). \tag{4.181}$$

Case 1: $E(u(t) - v(t)) = c$ for all $t \in \mathbb{R}$.

In this case, by Lemma 4.4.25, $\frac{1}{2}u + \frac{1}{2}v$ is a mild solution of equation (4.160). Moreover, $\frac{1}{2}u + \frac{1}{2}v \in \mathcal{F}_K$, because K is a convex set. Then, by the definition of δ, we obtain

$$\delta \le \sup_{t \in \mathbb{R}} E\left(\frac{1}{2}u(t) + \frac{1}{2}v(t)\right). \tag{4.182}$$

Using the parallelogram law,

$$E\left(\frac{1}{2}u(t) + \frac{1}{2}v(t)\right) + E\left(\frac{1}{2}u(t) - \frac{1}{2}v(t)\right) = \frac{1}{2}E(u(t)) + \frac{1}{2}E(v(t)),$$

we obtain

$$\sup_{t \in \mathbb{R}} E\left(\frac{1}{2}u(t) + \frac{1}{2}v(t)\right) + \frac{1}{4}c \le \frac{1}{2}\sup_{t \in \mathbb{R}} E(u(t)) + \frac{1}{2}\sup_{t \in \mathbb{R}} E(v(t)). \tag{4.183}$$

By (4.181)–(4.183), we obtain $\delta + \frac{1}{4}c \leq \frac{1}{2}\delta + \frac{1}{2}\delta$, and then $E(u(t) - v(t)) = c \leq 0$ for all $t \in \mathbb{R}$. By the definition of E, we obtain $u = v$, therefore equation (4.160) admits at most one minimal K-valued solution.

Case 2: General case. Let $(t_n)_n$ be a sequence of real numbers such that $\lim_{n\to+\infty} t_n = -\infty$. In fact, due to hypothesis **(H4)**, we can choose a subsequence of $(t_n)_n$ such that, for all $t \in \mathbb{R}$, we have

$$\lim_{n\to+\infty} \sup_{x\in K} \|f(t + t_n, x) - f_*(t, x)\| = 0,$$
$$\lim_{n\to+\infty} \sup_{x\in K} \|f_*(t - t_n, x) - f(t, x)\| = 0,$$

where f_* is a map from $\mathbb{R} \times K$ to X. There exists a subsequence of $(t_n)_n$ such that

$$u(t + t_n) \to u_1(t) \quad \text{as } n \to +\infty, \tag{4.184}$$
$$u_1(t - t_n) \to u_2(t) \quad \text{as } n \to +\infty, \tag{4.185}$$
$$v(t + t_n) \to v_1(t) \quad \text{as } n \to +\infty, \tag{4.186}$$
$$v_1(t - t_n) \to v_2(t) \quad \text{as } n \to +\infty, \tag{4.187}$$

uniformly on each compact subset of $\mathbb{R}$, where u_2 and v_2 are two minimal K-valued solutions of equation (4.160). The proof of this assertion is similar to that given in the proof of Theorem 4.4.6, by using Lemma 4.4.14 twice. From (4.184)–(4.187), we deduce that, for $t \in \mathbb{R}$,

$$\lim_{m\to+\infty} \lim_{n\to+\infty} E(u(t + t_n - t_m) - v(t + t_n - t_m)) = E(u_2(t) - v_2(t))$$

and

$$\lim_{n\to+\infty} E(u(t + t_n) - v(t + t_n)) = \sup_{\tau\in\mathbb{R}} E(u(\tau) - v(\tau)),$$

since the function $t \mapsto E(u(t) - v(t))$ is nonincreasing on $\mathbb{R}$ (cf. Lemma 4.4.25) and $\lim_{n\to+\infty} t_n = -\infty$. Therefore, for all $t \in \mathbb{R}$,

$$E(u_2(t) - v_2(t)) = \sup_{\tau\in\mathbb{R}} E(u(\tau) - v(\tau)). \tag{4.188}$$

By (4.188) and Case 1, we obtain that $u_2 = v_2$. From the definition of E, we obtain

$$\sup_{\tau\in\mathbb{R}} E(u(\tau) - v(\tau)) = 0,$$

therefore $u = v$. In conclusion, we have proved that equation (4.160) admits at most one minimal K-valued solution. In view of Theorem 4.4.6, we assert that equation (4.160) has at least one compact almost automorphic solution.

Let x_1 be a mild solution which is compact almost automorphic. If x_2 is a mild solution which is almost automorphic, then, by Lemma 4.4.25, the function $t \mapsto E(x_1(t) - x_2(t))$ is almost automorphic and nonincreasing, therefore it is constant on $\mathbb{R}$. Again with help of Lemma 4.4.25, we obtain (4.180). If the function $r \mapsto g(r) - \lambda_1 r$ is strictly decreasing, the uniqueness of the compact almost automorphic solution results from (4.173) and (4.174). This ends the proof. □

Now we give a corollary of Proposition 4.4.26 in the almost periodic case. For that, we need the following hypothesis:

(E4) The function $h : \mathbb{R} \times \overline{\Omega} \to \mathbb{R}$ is continuous, h satisfies $h(t, \xi) = 0$ on $\mathbb{R} \times \partial\Omega$, and h is $C_0(\Omega)$-almost periodic, which means that for all $\varepsilon > 0$ there exists $\ell > 0$ such that for all $a \in \mathbb{R}$ there exists $\tau \in [a, a + \ell]$ such that

$$\sup_{t \in \mathbb{R}} \sup_{\xi \in \Omega} \left| h(t + \tau, \xi) - h(t, \xi) \right| \leq \varepsilon.$$

Corollary 4.4.27. *Assume that* (E2)–(E4) *hold. Then equation* (4.160) *has at least one mild solution x_1 which is almost periodic. If x_2 is a mild solution which is almost periodic, then there exists $w_0 \in X$ such that*

$$x_2(t) = x_1(t) + w_0 \quad \textit{for } t \in \mathbb{R}.$$

Moreover, if the function $r \mapsto g(r) - \lambda_1 r$ is strictly decreasing on $\mathbb{R}$, then the almost periodic mild solution is unique.

Proof. The proof of Corollary 4.4.27 is similar to that given for Proposition 4.4.26 – we use Corollary 4.4.10 instead of Theorem 4.4.6. In fact, equation (4.160) can be written as

$$x'(t) = Ax(t) + G(x(t)) + H(t) \quad \text{for } t \in \mathbb{R}, \tag{4.189}$$

where G and H are defined by (4.157) and (4.158). Let us denote by $\mathcal{H}(H)$ the set of functions $H_* \in AP(\mathbb{R}, X)$ such that there exists a real sequence $(t_n)_n$ satisfying

$$\lim_{n \to +\infty} \sup_{t \in} \left\| H(t + t_n) - H_*(t) \right\| = 0.$$

Then, for $H_* \in \mathcal{H}(H)$, equation (4.107) can be written as

$$x'(t) = Ax(t) + G(x(t)) + H_*(t) \quad \text{for } t \in \mathbb{R}. \tag{4.190}$$

Since equation (4.190) is just like (4.189), equation (4.190) has a unique minimal K-valued solution for the compact K and the subvariant functional defined in the proof of Proposition 4.4.26. Using Corollary 4.4.10, we obtain the existence of a mild solution x_1 which is almost periodic. The proof of other assertions of Corollary 4.4.27 results from Proposition 4.4.26 since an almost periodic function is almost automorphic. □

4.4.4.2 Wave equation

To apply Theorem 4.4.9, we considers the following wave equation in an arbitrary open subset Ω of $\mathbb{R}^n$:

$$\begin{cases} \frac{\partial^2}{\partial t^2} v(t,x) = \sum_{i=1}^n \frac{\partial^2}{\partial x_i^2} v(t,x) + g(\frac{\partial}{\partial t} v(t,x)) + \lambda v(t,x) + h(t,x) & \text{in } \mathbb{R} \times \Omega, \\ v(t,x) = 0 & \text{on } \mathbb{R} \times \partial\Omega, \end{cases} \tag{4.191}$$

where $g : \mathbb{R} \to \mathbb{R}$, $h : \mathbb{R} \times \overline{\Omega} \to \mathbb{R}$ are continuous functions, and λ is an arbitrary real number.

In order to rewrite equation (4.191) in the abstract form, we introduce the Hilbert space $X = H_0^1(\Omega) \times L^2(\Omega)$, endowed with the inner product denoted by $\langle \cdot | \cdot \rangle$, associated to the norm

$$\|x\| = \left(\int_\Omega |\nabla x_1(\xi)|^2 + |x_1(\xi)|^2 + |x_2(\xi)|^2 \, d\xi \right)^{\frac{1}{2}} \quad \text{for } x = [x_1, x_2] \in X.$$

Define the operator $A : D(A) \subset X \to X$ by

$$\begin{cases} D(A) = \{[x_1, x_2] \in X : \Delta x_1 \in L^2(\Omega) \text{ and } x_2 \in H_0^1(\Omega)\}, \\ A[x_1, x_2] = [x_2, \Delta x_1 + \lambda x_1], \end{cases}$$

where Δ is the Laplacian operator.

Lemma 4.4.28. *The linear operator A generates a C_0-semigroup $(T(t))_{t \geq 0}$ on X. Moreover, the operator A satisfies*

$$\langle Ax | x \rangle = (1 + \lambda) \int_\Omega x_1(\xi) x_2(\xi) \, d\xi \quad \text{for } x = [x_1, x_2] \in D(A). \tag{4.192}$$

Proof. Define the operator $B : D(B) \subset X \to X$ by

$$\begin{cases} D(B) = D(A), \\ B[x_1, x_2] = [x_2, \Delta x_1 - x_1]. \end{cases}$$

The linear operator B is skew-adjoint on X, therefore B generates a group of isometries [90, Proposition 2.6.9, p. 33]. Define the operator $L : X \to X$ by

$$Lx = [0, x_1] \quad \text{for } x = [x_1, x_2] \in X.$$

Then $L \in \mathcal{L}(X, X)$ and $A = B + (1 + \lambda)L$, therefore A generates a C_0-semigroup $(T(t))_{t \geq 0}$ on X [247, Theorem 1.1, p. 76]. Since the linear operator B is skew-adjoint on X, for all $x \in D(A)$, $\langle Ax | x \rangle = (1 + \lambda)\langle Lx | x \rangle$, therefore (4.192) holds. □

In order to study the existence of an almost automorphic solution of equation (4.191), we suppose that:

(E5) The function $h : \mathbb{R} \times \overline{\Omega} \to \mathbb{R}$ is continuous and $L^2(\Omega)$-almost automorphic, which means that for any sequence of real numbers $(s_n')_n$, there exist a subsequence $(s_n)_n$ and a measurable function $k : \mathbb{R} \times \Omega \to \mathbb{R}$ such that

$$\forall t \in \mathbb{R}, \quad \lim_{n\to\infty} \int_\Omega |h(t+s_n,\xi) - k(t,\xi)|^2 \, d\xi = 0$$

and

$$\forall t \in \mathbb{R}, \quad \lim_{n\to\infty} \int_\Omega |k(t-s_n,\xi) - h(t,\xi)|^2 \, d\xi = 0.$$

Remark 4.4.29. The function $t \mapsto h(t,\cdot)$ belongs to $AA(\mathbb{R}, L^2(\Omega))$.

Moreover, we suppose that:

(E6) g is Lipschitz continuous.

(E7) $\lambda \in \mathbb{R}$ is such that $u \in H_0^1(\Omega)$ and $\Delta u + \lambda u = 0$ imply $u = 0$.

(E8) The function g is nonincreasing on $\mathbb{R}$.

Remark 4.4.30. In the particular case where the domain Ω is bounded and its boundary $\partial\Omega$ is such that the Dirichlet problem $\Delta u + \lambda u = 0$ in Ω, $u = 0$ on $\partial\Omega$, has a sequence of eigenvalues $\{\lambda_n : n \geq n\}$ such that $0 < \lambda_1 \leq \lambda_2 \leq \cdots \leq \lambda_n \leq \lambda_{n+1} \leq \cdots$ and $\lim_{n\to\infty} \lambda_n = \infty$, if λ is not an eigenvalue of $-\Delta$ in $H_0^1(\Omega)$, then λ satisfies hypothesis **(E7)**.

Let $G : X \to X$ be the map defined by

$$G(x)(\xi) = [0, g(x_2(\xi))] \quad \text{for } \xi \in \Omega \text{ and } x = [x_1, x_2] \in X. \tag{4.193}$$

Let $H : \mathbb{R} \to X$ be defined by

$$H(t)(\xi) = [0, h(t,\xi)] \quad \text{for } t \in \mathbb{R} \text{ and } \xi \in \Omega. \tag{4.194}$$

Let $f : \mathbb{R}\times X \to X$ denote the mapping defined by

$$f(t,x) = G(x) + H(t) \quad \text{for } t \in \mathbb{R} \text{ and } x \in X. \tag{4.195}$$

Since **(E5)** and **(E6)** hold, $G \in C(X,X)$ and $H \in AA(\mathbb{R}, X)$, therefore f satisfies hypothesis **(H4)**. With those notations, equation (4.191) takes the following abstract form:

$$x'(t) = Ax(t) + f(t, x(t)) \quad \text{for } t \in \mathbb{R}. \tag{4.196}$$

Let $q : X \to \mathbb{R}$ and $\Phi : X \to \mathbb{R}$ denote the quadratic forms defined by

$$q(x) = \frac{1}{2}\int_{\Omega} |x_1(\xi)|^2 \, d\xi, \tag{4.197}$$

$$\Phi(x) = \frac{1}{2}\|x\|^2 - (1+\lambda)q(x). \tag{4.198}$$

Then,

$$\Phi(x) = \frac{1}{2}\int_{\Omega} \left(|\nabla x_1(\xi)|^2 + |x_2(\xi)|^2 - \lambda |x_1(\xi)|^2\right) d\xi \quad \text{for } x = [x_1, x_2] \in X.$$

It is obvious that q and Φ are continuous on X. With assumption **(E7)**, the quadratic form Φ is not in general positive or negative definite. To prove our result of existence of an almost automorphic solution, we use Theorem 4.4.9, with the following subvariant functional:

$$\lambda_K(x) = \sup_{t\in\mathbb{R}} \Phi(x(t)) \quad \text{for } x \in C_K(\mathbb{R}, X).$$

To state our result, we use two lemmas below. We consider the following linear equation:

$$x'(t) = Ax(t) + p(t) \quad \text{for } t \in \mathbb{R}, \tag{4.199}$$

where $p \in C(\mathbb{R}, X)$.

Lemma 4.4.31. *Let $p = [0, p_2] \in C(\mathbb{R}, X)$.*

(i) *If w is a mild solution on $\mathbb{R}$ of the linear equation (4.199), then the function $t \mapsto \Phi(w(t))$ is of class C^1 on $\mathbb{R}$ and*

$$\frac{d}{dt}[\Phi(w(t))] = \langle p(t)|w(t)\rangle = \int_{\Omega} p_2(t,\xi)w_2(t,\xi)\, d\xi \quad \text{for } t \in \mathbb{R}. \tag{4.200}$$

(ii) *Moreover, if $p(t) = 0$ for all $t \in \mathbb{R}$, then the function $t \mapsto \Phi(w(t))$ is constant on $\mathbb{R}$ and the function $t \mapsto q(w(t))$ is of class C^2 on $\mathbb{R}$ with*

$$\frac{d^2}{dt^2}[q(w(t))] = 2\int_{\Omega} |w_2(t,\xi)|^2 \, d\xi - 2\Phi(w(0)) \quad \text{for } t \in \mathbb{R}. \tag{4.201}$$

Proof. Let us denote by $b_i : X \times X \to \mathbb{R}$ $(i = 1, 2)$ the two bilinear forms defined by

$$b_1(x,y) = \int_{\Omega} x_1(\xi)y_1(\xi)\, d\xi \quad \text{and} \quad b_2(x,y) = \int_{\Omega} x_1(\xi)y_2(\xi)\, d\xi$$

for $x = [x_1, x_2]$ and $y = [y_1, y_2] \in X$. Then b_1 and b_1 are continuous, b_1 is symmetric and satisfies $b_1(x,x) = 2q(x)$, where q is the function defined by (4.197).

(i) Let w be a mild solution on $\mathbb{R}$ of equation (4.199). We fix $s \in \mathbb{R}$ and $T > 0$. Assume that $w(s) \in D(A)$ and $p_2 \in C^1([s, s+T], L^2(\Omega))$. Then w is a classical solution on $[s, s+T]$ of (4.199), namely $w \in C([s, s+T], D(A)) \cap C^1([s, s+T], X)$ and

$$w'(t) = Aw(t) + p(t) \quad \text{for } s \le t \le s+T.$$

It follows that $t \mapsto q(w(t))$ is of class C^1 and

$$\frac{\mathrm{d}}{\mathrm{d}t}[q(w(t))] = \frac{\mathrm{d}}{\mathrm{d}t}\left[\frac{1}{2} b_1(w(t), w(t))\right] = b_1(w(t), w'(t)) = b_1(w(t), Aw(t) + p(t)),$$

thus

$$\frac{\mathrm{d}}{\mathrm{d}t}[q(w(t))] = \int_\Omega w_1(t,\xi) w_2(t,\xi)\,\mathrm{d}\xi = b_2(w(t), w(t)). \tag{4.202}$$

Consequently, $t \mapsto \Phi(w(t))$ is also of class C^1,

$$\begin{aligned}\frac{\mathrm{d}}{\mathrm{d}t}[\Phi(w(t))] &= \langle w'(t) | w(t)\rangle - (1+\lambda)\frac{\mathrm{d}}{\mathrm{d}t} q(w(t)) \\ &= \langle Aw(t) | w(t)\rangle + \langle p(t) | w(t)\rangle - (1+\lambda)\int_\Omega w_1(t,\xi) w_2(t,\xi)\,\mathrm{d}\xi\end{aligned}$$

and, by using (4.192), we obtain (4.200) on $[s, s+T]$. Then by integrating (4.202) and (4.200) on $[s, s+T]$, we obtain

$$q(w(t)) = q(w(s)) + \int_s^t b_2(w(\sigma), w(\sigma))\,\mathrm{d}\sigma, \tag{4.203}$$

$$\Phi(w(t)) = \Phi(w(s)) + \int_s^t \langle p(\sigma) | w(\sigma)\rangle\,\mathrm{d}\sigma \tag{4.204}$$

for $s \le t \le s+T$. For the proof of (4.203) and (4.204) in the general case, we choose two sequences $(w_n^s)_n \subset D(A)$ and $(p_n)_n \subset C^1([s, s+T], X)$ (with $p_n = [0, p_n^2]$) such that $(w_n^s)_n$ converges to $w(s)$ in X and $(p_n)_n$ converges to p in $L^1((s, s+T), X)$. Consequently, w satisfies (4.203) and (4.204) on each interval $[s, s+T]$. Since w, p, and b_2 are continuous, we deduce that $t \mapsto q(w(t))$, $t \mapsto \Phi(w(t))$ are of class C^1 and (4.200), (4.202) hold.

(ii) Moreover, if $p(t) = 0$ for all $t \in \mathbb{R}$, using (4.200), we deduce that the function $t \mapsto \Phi(w(t))$ is constant on $\mathbb{R}$, therefore

$$\frac{1}{2}\|w(t)\|^2 - (1+\lambda) q(w(t)) = \Phi(w(0)) \quad \text{for all } t \in \mathbb{R}. \tag{4.205}$$

We fix $s \in \mathbb{R}$ and $T > 0$. Assume that $w(s) \in D(A)$. In this case, we have

$$w \in C([s, s+T], D(A)) \cap C^1([s, s+T], X)$$

and

$$w'(t) = Aw(t) \quad \text{for } s \le t \le s+T. \tag{4.206}$$

From (4.202), it follows that $t \mapsto q(w(t))$ is of class C^2 and

$$\frac{d^2}{dt^2}[q(w(t))] = b_2(w'(t), w(t)) + b_2(w(t), w'(t)) = b_2(Aw(t), w(t)) + b_2(w(t), Aw(t)),$$

thus

$$\frac{d^2}{dt^2}[q(w(t))] = \int_\Omega (|w_2(t,\xi)|^2 + w_1(t,\xi)(\Delta w_1(t,\xi) + \lambda w_1(t,\xi)))d\xi,$$

and, by using Green's formula, we obtain

$$\frac{d^2}{dt^2}[q(w(t))] = \int_\Omega (|w_2(t,\xi)|^2 - |\nabla w_1(t,\xi)|^2 + \lambda |w_1(t,\xi)|^2)d\xi.$$

From (4.205), we deduce that (4.201) holds. By integrating (4.201) on $[s, s+T]$, we get

$$\frac{d}{dt}[q(w(t))] = \frac{d}{dt}[q(w(s))] + 2\int_s^t \int_\Omega |w_2(\sigma,\xi)|^2 \, d\xi \, d\sigma - 2(t-s)\Phi(w(0)) \tag{4.207}$$

for $s \le t \le s+T$. For the proof of (4.207) in the general case, we choose a sequence $(w_n^s)_n \subset D(A)$ such that $(w_n^s)_n$ converges to $w(s)$ in X. Since the function $t \mapsto \int_\Omega |w_2(t,\xi)|^2 \, d\xi$ is continuous, we deduce that the function $t \mapsto q(w(t))$ is of class C^2 on each interval $[s, s+T]$ and satisfies (4.201). □

Lemma 4.4.32. *Assume that* (E5)–(E8) *hold. Let u and v be two mild solutions on $\mathbb{R}$ of equation* (4.196). *Then the following assertions hold:*
(i) *The function $t \mapsto \Phi(u(t) - v(t))$ is nonincreasing on $\mathbb{R}$.*
(ii) *If the function $t \mapsto \Phi(u(t) - v(t))$ is constant on $\mathbb{R}$, then*

$$G(u(t)) = G(v(t)) \quad \text{for } t \in \mathbb{R} \tag{4.208}$$

and, for all $\theta \in [0,1]$, $\theta u + (1-\theta)v$ is also a mild solution on $\mathbb{R}$ of equation (4.196).
(iii) *If u and v are bounded on $\mathbb{R}$ and if $\Phi(u(t) - v(t)) = c$ for all $t \in \mathbb{R}$ with $c \le 0$, then $c = 0$ and $u = v$.*

Proof. (i) Denote $w = u - v$. Then $w \in C(\mathbb{R}, X)$ satisfies (4.199) with

$$p(t) = f(t, u(t)) - f(t, v(t)) = G(u(t)) - G(v(t)), \tag{4.209}$$

where f and G are functions defined by (4.193)–(4.195). By Lemma 4.4.31, we obtain

$$\frac{\mathrm{d}}{\mathrm{d}t}\Phi(w(t)) = \langle G(u(t)) - G(v(t))|w(t)\rangle,$$

therefore

$$\frac{\mathrm{d}}{\mathrm{d}t}\Phi(w(t)) = \int_\Omega \{g(u_2(t,\xi)) - g(v_2(t,\xi))\}\{u_2(t,\xi) - v_2(t,\xi)\}\,\mathrm{d}\xi. \tag{4.210}$$

Since g is nonincreasing, it follows that

$$\{g(r_1) - g(r_2)\}\{r_1 - r_2\} \le 0 \quad \text{for } r_1, r_2 \in \mathbb{R} \tag{4.211}$$

and, from (4.210) and (4.211), we obtain that the function $t \mapsto \Phi(u(t) - v(t))$ is nonincreasing on $\mathbb{R}$.

(ii) Assume that the function $t \mapsto \Phi(u(t) - v(t))$ is constant on $\mathbb{R}$. Using (4.210) and (4.211), we obtain

$$\{g(u_2(t,\xi)) - g(v_2(t,\xi))\}\{u_2(t,\xi) - v_2(t,\xi)\} = 0 \quad \text{for } t \in \mathbb{R} \text{ and } \xi \in \Omega.$$

Employing **(E8)**, we obtain

$$g(u_2(t,\xi)) = g(v_2(t,\xi)) = 0 \quad \text{for } t \in \mathbb{R} \text{ and } \xi \in \Omega,$$

therefore (4.208) holds, and then we deduce

$$g(\theta u_2(t,\xi) + (1-\theta)v_2(t,\xi)) = g(u_2(t,\xi)) \quad \text{for } t \in \mathbb{R},\ \xi \in \Omega, \text{ and } \theta \in [0,1].$$

It follows that

$$G(\theta u(t) + (1-\theta)v(t)) = \theta G(u(t)) + (1-\theta)G(v(t)) \quad \text{for } t \in \mathbb{R} \text{ and } \theta \in [0,1],$$

therefore

$$f(t, \theta u(t) + (1-\theta)v(t)) = \theta f(t, u(t)) + (1-\theta)f(t, v(t)) \quad \text{for } t \in \mathbb{R} \text{ and } \theta \in [0,1],$$

and so $\theta u + (1-\theta)v$ is also a mild solution on $\mathbb{R}$ of equation (4.196).

(iii) From (4.208) and (4.209), we deduce that $w \in C(\mathbb{R}, X)$ satisfies the homogeneous equation associated to (4.199), i. e., $p(t) = 0$ for all $t \in \mathbb{R}$. By Lemma 4.4.31, the function $t \mapsto q(w(t))$ is of class C^2 on $\mathbb{R}$ and satisfies (4.201). By assumption, $c = \Phi(w(0)) \le 0$ and, due to (4.201), we obtain $\frac{\mathrm{d}^2}{\mathrm{d}t^2}q(w(t)) \ge 0$. Thus $t \mapsto q(w(t))$ is convex and bounded on $\mathbb{R}$, and therefore $t \mapsto q(w(t))$ is constant, i. e., $\frac{\mathrm{d}^2}{\mathrm{d}t^2}q(w(t)) = 0$. From (4.201), we obtain $c = 0$ and $w_2(t) = 0$ for all $t \in \mathbb{R}$. We fix $s \in \mathbb{R}$ and $T > 0$. Assume that $w_1(s) \in H_0^1(\Omega)$ and $\Delta w_1(s) \in L^2(\Omega)$. In this case, w is a classical solution of the homogeneous equation associated to (4.199), we deduce that $w_1(t) = w_1(s)$ for all $t \in [s, s+T]$ and $\Delta w_1(s) + \lambda w_1(s) = 0$. From hypothesis **(E7)**, we deduce that $w_1(t) = 0$, thus $w(t) = 0$ for all $t \in [s, s+T]$. For the general case, we choose a sequence $(w_n^s)_n \subset D(A)$ such that $(w_n^s)_n$ converges to $w(s)$ in X, and then obtain that $w(t) = 0$ for each interval $[s, s+T]$, therefore $u(t) = v(t)$ for all $t \in \mathbb{R}$. □

Proposition 4.4.33. *Assume that* **(E5)–(E8)** *hold. In addition, suppose that equation* (4.196) *admits at least one mild solution* x_0 *on* $[t_0, +\infty)$ *such that* $\{x_0(t) : t \geq t_0\}$ *is relatively compact. Then equation* (4.196) *has at least one mild solution which is compact almost automorphic. Moreover, if the function g is strictly decreasing on* $\mathbb{R}$, *then the mild solution, which is compact almost automorphic, is unique.*

Proof. Hypothesis **(H1)** is a consequence of Lemma 4.4.28 and hypothesis **(H4)** results from **(E5)** and **(E6)**. Therefore all the hypotheses of Theorem 4.4.9 are fulfilled. Let K be the closed convex hull of the compact set $\overline{x([t_0, +\infty))}$. Then K is convex and compact. To prove the existence of a compact almost automorphic solution of equation (4.196), we use Theorem 4.4.9 with the subvariant functional $\lambda_K(x) = \sup_{t\in\mathbb{R}} \Phi(x(t))$ associated to the compact set K. It remains to prove the uniqueness of the minimal K-valued solution of equation (4.196). Let u and v be two minimal K-valued solutions of equation (4.196). Denote

$$\delta = \sup_{t\in\mathbb{R}} \Phi(u(t)) = \sup_{t\in\mathbb{R}} \Phi(v(t)). \tag{4.212}$$

Case 1: $\Phi(u(t) - v(t)) = c$ for all $t \in \mathbb{R}$.

In this case, using Lemma 4.4.32, $\frac{1}{2}u + \frac{1}{2}v$ is a solution of equation (4.196). Since K is convex, we obtain $\frac{1}{2}u + \frac{1}{2}v \in \mathcal{F}_K$, and then

$$\delta \leq \sup_{t\in\mathbb{R}} \Phi\left(\frac{1}{2}u(t) + \frac{1}{2}v(t)\right). \tag{4.213}$$

From the parallelogram law applied to the quadratic form Φ, we deduce

$$\sup_{t\in\mathbb{R}} \Phi\left(\frac{1}{2}u(t) + \frac{1}{2}v(t)\right) + \frac{1}{4}c \leq \frac{1}{2}\sup_{t\in\mathbb{R}} \Phi(u(t)) + \frac{1}{2}\sup_{t\in\mathbb{R}} \Phi(v(t)). \tag{4.214}$$

By (4.212)–(4.214), we obtain $\delta + \frac{1}{4}c \leq \frac{1}{2}\delta + \frac{1}{2}\delta$, and then $\Phi(u(t) - v(t)) = c \leq 0$ for all $t \in \mathbb{R}$. By Lemma 4.4.32, we obtain $u = v$, therefore equation (4.196) admits at most one minimal K-valued solution.

Case 2: *General case.*

Let $(t_n)_n$ be a sequence of real numbers such that $\lim_{n\to+\infty} t_n = -\infty$. Then there exists a subsequence of $(t_n)_n$ such that

$$u(t + t_n) \to u_1(t) \quad \text{as } n \to +\infty, \tag{4.215}$$

$$u_1(t - t_n) \to u_2(t) \quad \text{as } n \to +\infty, \tag{4.216}$$

$$v(t+t_n) \to v_1(t) \quad \text{as } n \to +\infty, \tag{4.217}$$

$$v_1(t-t_n) \to v_2(t) \quad \text{as } n \to +\infty, \tag{4.218}$$

uniformly on each compact subset of $\mathbb{R}$, where u_2 and v_2 are two minimal K-valued solutions of equation (4.196). The proof of this assertion is similar to that given for Proposition 4.4.26. By Lemma 4.4.32, the function $t \mapsto \Phi(u(t)-v(t))$ is nonincreasing on $\mathbb{R}$ and, from (4.215)–(4.218) and $\lim_{n\to+\infty} t_n = -\infty$, we deduce, for all $t \in \mathbb{R}$,

$$\Phi(u_2(t)-v_2(t)) = \sup_{\tau\in\mathbb{R}} \Phi(u(\tau)-v(\tau)). \tag{4.219}$$

By (4.219) and Case 1, we obtain that $u_2 = v_2$, thus $\Phi(u_2(t)-v_2(t)) = 0$ for all $t \in \mathbb{R}$. Therefore

$$\sup_{\tau\in\mathbb{R}} \Phi(u(\tau)-v(\tau)) = 0. \tag{4.220}$$

By taking a sequence of real numbers $(t_n)_n$ such that $\lim_{n\to+\infty} t_n = +\infty$, we similarly conclude that

$$\inf_{\tau\in\mathbb{R}} \Phi(u(\tau)-v(\tau)) = 0. \tag{4.221}$$

From (4.220) and (4.221), we have, for all $t \in \mathbb{R}$,

$$\Phi(u(t)-v(t)) = 0,$$

therefore, using Case 1, we deduce that $u = v$. In conclusion, we have proved that equation (4.196) admits at most one minimal K-valued solution. In view of Theorem 4.4.9, we assert that equation (4.196) has at least one compact almost automorphic solution.

Now assume that g is strictly decreasing. If u and v are two mild solutions which are compact almost automorphic, then the function $t \mapsto \Phi(u(t)-v(t))$ is almost automorphic and, by Lemma 4.4.32, it is nonincreasing, therefore it is constant on $\mathbb{R}$. Again with the help of Lemma 4.4.32, we obtain (4.208) and, since g is strictly decreasing, $u_2 = v_2$. Denote $w = u - v$. It follows that w is a mild solution on $\mathbb{R}$ of the linear equation (4.199) and, by the same argument as at the end of the proof of Lemma 4.4.32, we obtain $u = v$. □

Now we give a corollary of Proposition 4.4.33 in the almost periodic case. For that, we need the following hypothesis:

(E9) The function $h : \mathbb{R}\times\overline{\Omega} \to \mathbb{R}$ is continuous and $L^2(\Omega)$-almost periodic, which means that for all $\varepsilon > 0$, there exists $\ell > 0$ such that for all $a \in \mathbb{R}$ there exists $\tau \in [a, a+\ell]$ such that

$$\sup_{t\in\mathbb{R}} \int_{\Omega} |h(t+\tau,\xi) - h(t,\xi)|^2 \, d\xi \le \varepsilon.$$

Corollary 4.4.34. *Assume that* (E6)–(E9) *hold. In addition, suppose that equation (*4.196*) admits at least one mild solution x_0 on $[t_0, +\infty)$ such that $\{x_0(t) : t \geq t_0\}$ is relatively compact. Then equation (*4.196*) has at least one mild solution which is almost periodic. Moreover, if the function g is strictly decreasing on $\mathbb{R}$, then the mild solution, which is almost periodic, is unique.*

Proof. The proof of Corollary 4.4.34 is similar to that given for Corollary 4.4.27, by using Corollary 4.4.11 instead of Corollary 4.4.10. □

4.5 Nonautonomous evolution equations: A spectral countability condition

Let us consider equations of the form

$$\frac{\mathrm{d}x}{\mathrm{d}t} = A(t)x + f(t), \tag{4.222}$$

where $A(t)$ is a (generally unbounded) linear operator on a Banach space $\mathbb{X}$ which is periodic, and f is an $\mathbb{X}$-valued almost automorphic function on $\mathbb{R}$. We are interested in conditions under which every bounded mild solution of this equation is almost automorphic.

It is well known (see, for example, [39, 94, 199, 255] and the references therein) that if, for the t-independent operator A that is the generator of a C_0-semigroup, its imaginary part of spectrum is countable and $\mathbb{X}$ does not contain any subspace isomorphic to c_0, then every bounded and uniformly continuous mild solution of (4.222) is almost periodic. It is natural to raise a question if this is true for almost automorphic solutions, that is, if f is almost automorphic and $\mathbb{X}$ does not contain any subspace isomorphic to c_0, then every bounded mild solution of (4.222) almost automorphic.

In this work we will prove that this is true. In fact, we will prove a little more general assertion in which $A(t)$ generates a periodic evolutionary process. As an almost automorphic function is not necessarily uniformly continuous, the methods of proving this standard result for almost periodic solutions, which are based on the uniform continuity of the bounded mild solutions and of the forcing term f, are no longer available. We refer the reader to [39, 199] for more details on the methods of proving this result. To overcome this difficulty, we will discretize the equations. We will first solve the discrete analog of the above problem, and then apply it to solve our main one. In order to handle the discrete analog, we will introduce a new concept of a uniform spectrum of a bounded sequence with respect to a subspace that allows us to give a short proof of the claim for discrete equations. The main results are Theorems 4.5.8 and 4.5.10.

4.5.1 Kadets theorem

In this section we will use the standard notation c_0 for the Banach space of all numerical sequences $\{a_n\}_{n=1}^{\infty}$ such that $\lim_{n\to\infty} a_n = 0$, equipped with the supremum norm. In the simplest case, the problem we are considering becomes the following: when is am integral of an almost automorphic function also almost automorphic? We can take the same counterexample as in [199] to show that additional conditions should be imposed on the space $\mathbb{X}$.

Example 4.5.1. Consider the function $f(t)$ with values in c_0 defined by

$$f(t) = \{(1/n)\cos(t/n)\}_{n=1}^{\infty} \quad \forall t \in \mathbb{R}.$$

The integral $F(t) = \int_0^t f(\xi)d\xi$ of $f(t)$ is $F(t) = \{\sin(t/n)\}_{n=1}^{\infty}$. Obviously, f is almost periodic (so it is almost automorphic), and F is bounded. However, the range of F, as shown in [199, pp. 81–82], is not precompact, so F cannot be almost automorphic.

The Kadets theorem (see, e. g., [199, Theorem 2, p. 86]) says that if f is almost periodic and F is bounded, then F is almost periodic if and only if $\mathbb{X}$ does not contain any subspace isomorphic to c_0. An extension of the Kadets theorem to almost automorphic functions was given in [48].

The following extension of the Kadets theorem to sequences will be used in this section.

Lemma 4.5.2. *Assume that $x = \{x_n\}_{n\in\mathbb{Z}}$ is a sequence in a Banach space $\mathbb{X}$ that does not contain any subspace isomorphic to c_0, and the difference*

$$x - Sx = y \tag{4.223}$$

is almost automorphic. Then, the sequence x itself is almost automorphic.

Proof. This lemma is a special case of [48, Theorem 1]. □

As is well known (see, e. g., [199]), a convex Banach space does not contain any subspace isomorphic to c_0. In particular, every finite-dimensional space does not contain any subspace isomorphic to c_0.

4.5.2 Spectral theory of bounded sequences

For each bounded sequence $g := \{g_n\}_{n\in\mathbb{Z}}$ in $\mathbb{X}$, we will denote by $S(k)g$ the k-translation of g in $l^\infty(\mathbb{X})$, i. e., $(S(k)g)_n = g_{n+k}$, $\forall n \in \mathbb{Z}$. And S will stand for $S(1)$.

Definition 4.5.3. Let $\mathcal{A}$ be a closed subspace of $l^\infty(\mathbb{X})$. We say that $\mathcal{A}$ satisfies *Condition H* if the following conditions are satisfied:

1. Every sequence of the form (constant sequence) $\{a\}_{n\in\mathbb{Z}}$ is in $\mathcal{A}$;
2. If $\{x_n\}_{n\in\mathbb{Z}} \in \mathcal{A}$ and $q \in \Gamma$, then the sequence $\{q^n x_n\}_{n\in\mathbb{Z}}$ is in $\mathcal{A}$;
3. If B is a bounded linear operator in $\mathbb{X}$, then $\{Bx_n\}_{n\in\mathbb{Z}} \in \mathcal{A}$ whenever $\{x_n\}_{n\in\mathbb{Z}}$ is in $\mathcal{A}$;
4. $S\mathcal{A} = \mathcal{A}$.

As an example of a subspace of $l^\infty(\mathbb{X})$ that satisfies Condition H we can take $aa(\mathbb{X})$.

In the rest of this section, we will always assume that $\mathcal{A}$ is a closed subspace of $l^\infty(\mathbb{X})$ that satisfies Condition H. Then, in the quotient space $l^\infty(\mathbb{X})/\mathcal{A}$ acts the operator $\tilde{S}$ defined by $\tilde{S}(x+\mathcal{A}) = Sx+\mathcal{A}, \forall x \in l^\infty(\mathbb{X})$. The operator $\tilde{S}$ is an isometric operator because of Condition H (4). Let π be the canonical projection from $l^\infty(\mathbb{X})$ onto $l^\infty(\mathbb{X})/\mathcal{A}$, and denote $\pi x = \tilde{x}$. Below, the notation $\mathcal{M}_{\hat{x}}$ means the closure of the subspace of $l^\infty(\mathbb{X})/\mathcal{A}$ that spans the set $\{\hat{S}(n)\hat{x} : n \in \mathbb{Z}\}$.

Definition 4.5.4. The uniform spectrum of $x \in l^\infty(\mathbb{X})$ with respect to $\mathcal{A}$, which is denoted by $\mathrm{sp}_{\mathcal{A}}(x)$, is defined to be

$$\sigma(\tilde{S}|_{\mathcal{M}_{\tilde{x}}}).$$

It may be seen that $\mathrm{sp}_{\mathcal{A}}(x)$ is part of the unit circle Γ.

Lemma 4.5.5. *Let $x = \{x_n\}_{n\in\mathbb{Z}} \in l^\infty(\mathbb{X})$ and $\mathcal{A}$ be a subspace of $l^\infty(\mathbb{X})$ that satisfies Condition H. Then $\mathrm{sp}_{\mathcal{A}}(x)$ consists of all points z_0 of the unit circle Γ such that the Carleman transform*

$$\hat{\tilde{x}}(\lambda) := \begin{cases} \sum_{n=0}^{\infty} \lambda^{-n-1}\tilde{S}(n)\tilde{x}, & \forall|\lambda| > 1, \\ -\sum_{n=1}^{\infty} \lambda^{n-1}\tilde{S}(-n)\tilde{x}, & \forall|\lambda| < 1 \end{cases}$$

has no holomorphic extension to any neighborhood of z_0.

Proof. The proof follows the lines of that in [228, Lemma 2.4], so we omit the details. □

Consider the linear difference equation

$$x_{n+1} = Bx_n + f_n, \quad n \in \mathbb{Z}, \tag{4.224}$$

where B is a bounded linear operator. Below we will denote by $\sigma_\Gamma(B)$ the part of spectrum of B on the unit circle Γ of the complex plane.

Lemma 4.5.6. *Let $x \in l^\infty(\mathbb{X})$ be a solution of* (4.224)*, and let $f \in \mathcal{A}$. Then*

$$\mathrm{sp}_{\mathcal{A}}(x) \subset \sigma_\Gamma(B). \tag{4.225}$$

Proof. First, we note that the operator of multiplication by B in $l^\infty(\mathbb{X})$ has the same spectrum as B. And for the reader's convenience, we will use the same notation B to indicate this multiplication operator if it does not cause any danger of confusion. So, by Condition

H, since the operator of multiplication by B leaves $\mathcal{A}$ invariant, B induces a bounded linear operator $\tilde{B}$ in the quotient space $l^\infty(\mathbb{X})/\mathcal{A}$. Taking the Carleman transform of (4.224), we have

$$\begin{aligned}\widehat{\tilde{S}\tilde{x}}(\lambda) &= \tilde{B}\hat{\tilde{x}}(\lambda),\\ \lambda\hat{\tilde{x}}(\lambda) - \tilde{x} &= \tilde{B}\hat{\tilde{x}}(\lambda).\end{aligned}$$

Therefore,

$$\begin{aligned}\lambda\hat{\tilde{x}}(\lambda) - \tilde{B}\hat{\tilde{x}}(\lambda) &= \tilde{x},\\ (\lambda - \tilde{B})\hat{\tilde{x}}(\lambda) &= \tilde{x}.\end{aligned}$$

Let $\eta_0 \in \rho(\tilde{B})$. For η close to η_0 such that $\eta \notin i\mathbb{R}$,

$$\hat{\tilde{x}}(\eta) = (\eta - \tilde{B})^{-1}\tilde{x}.$$

Using the analyticity of the resolvent $(\eta-\tilde{B})^{-1}$ with respect to η in a neighborhood of $\eta_0 \in \rho(B)$, we see that $\hat{\tilde{x}}(\eta)$ can be extended analytically to a neighborhood of η_0. Therefore, by Lemma 4.5.5, $\mathrm{sp}_{\mathcal{A}}(x) \subset \sigma_\Gamma(\tilde{B}) := \Gamma \cap \sigma(\tilde{B})$. Now it suffices to show that $\sigma(\tilde{B}) \subset \sigma(B)$ to complete the proof. To this end, supposing that $z_0 \in \rho(B)$, we will show that z_0 is in $\rho(\tilde{B})$. In fact, by definition, for every $y \in l^\infty(\mathbb{X})$, there is a unique $w \in l^\infty(\mathbb{X})$ such that $z_0 w_n - Bw_n = y_n$ for all $n \in \mathbb{Z}$. This shows that, given $\tilde{y} \in l^\infty(\mathbb{X})/\mathcal{A}$, there exists a solution $\tilde{w} \in l^\infty(\mathbb{X})/\mathcal{A}$ such that $z_0\tilde{w} - \tilde{B}\tilde{w} = \tilde{y}$. Next, we show that such $\tilde{w}$ is unique, that is,

$$z_0(w + \mathcal{A}) - B(w + \mathcal{A}) = y + \mathcal{A}.$$

And this is obvious from Condition H. The lemma is proved. □

Lemma 4.5.7. *Let $\mathcal{A}$ be the space of all almost automorphic sequences in $\mathbb{X}$, and let x be in $l^\infty(\mathbb{X})$ such that $\mathrm{sp}_{\mathcal{A}}(x)$ is countable. Moreover, assume that the space $\mathbb{X}$ does not contain any subspace isomorphic to c_0. Then, $x \in \mathcal{A}$.*

Proof. If $\mathrm{sp}_{\mathcal{A}}(x)$ is empty, due to the boundedness of the linear operator $\tilde{S}|_{\mathcal{M}_{\tilde{x}}}$, the space $\mathcal{M}_{\tilde{x}}$ must be trivial, that is, $\tilde{x} = 0$, or in other words, x is almost automorphic.

Suppose that $\mathrm{sp}_{\mathcal{A}}(x)$ is not empty. Since $\tilde{S}|_{\mathcal{M}_{\tilde{x}}}$ is an isometry in $\mathcal{M}_{\tilde{x}}$ and $\mathrm{sp}_{\mathcal{A}}(x) = \sigma(\tilde{S}|_{\mathcal{M}_{\tilde{x}}})$ is countable, by the Gelfand theorem, there is a point z_0 in its spectrum that is an eigenvalue of $\tilde{S}|_{\mathcal{M}_{\tilde{x}}}$ (for the Gelfand theorem, see, e. g., [39, 50]). So, we can find an element $\tilde{y} \in \mathcal{M}_{\tilde{x}}$ such that $z_0\tilde{y} = \tilde{S}\tilde{y}$. To complete the proof, we will show that, for each bounded sequence $x \in l^\infty(\mathbb{X})$, if $z_0 x - Sx$ is in $\mathcal{A}$, then $x \in \mathcal{A}$.

Let us consider the isomorphism $V_\lambda : l^\infty(\mathbb{X}) \to l^\infty(\mathbb{X})$ defined as

$$V_\lambda x(n) = \lambda^n x_n, \quad \forall n \in \mathbb{Z},$$

where $\lambda \in \Gamma$. Note that V_λ leaves $\mathcal{A}$ invariant and

$$V_\lambda S V_\lambda^{-1} = \lambda^{-1} S.$$

Thus, we have

$$z_0 x - Sx = z_0(x - z_0^{-1} Sx) = z_0(x - V_{z_0} S V_{z_0}^{-1} x).$$

Therefore, $z_0 x - Sx$ is in $\mathcal{A}$ means that $x - V_{z_0} S V_{z_0}^{-1} x$ is in $\mathcal{A}$. In turn, this is equivalent to saying that $V_{z_0}^{-1} x - S V_{z_0}^{-1} x$ is in $\mathcal{A}$ because $x - V_{z_0} S V_{z_0}^{-1} x = V_{z_0}(I - S) V_{z_0}^{-1} x$ and both V_{z_0}, $V_{z_0}^{-1}$ are isomorphisms on $\mathcal{A}$. Now we are ready to apply Lemma 4.5.2 to show that $V_{z_0}^{-1} x$ must be almost automorphic, so x must be almost automorphic. (Obviously, if x is almost automorphic, then $\tilde{x} = 0$. So, $\mathrm{sp}_{\mathcal{A}}(x)$ must be empty.) The lemma is proved. □

The main result of this section will be proved based on the following.

Theorem 4.5.8. *Let B be a bounded linear operator in $\mathbb{X}$ with $\sigma_\Gamma(B)$ being countable, and let $\mathbb{X}$ not contain any subspace isomorphic to c_0. Assume further that $\{x_n\}_{n\in\mathbb{Z}}$ is a bounded sequence that satisfies the equation*

$$x_{n+1} = Bx_n + y_n, \quad n \in \mathbb{Z}, \tag{4.226}$$

where $\{y_n\}_{n\in\mathbb{Z}}$ is in $aa(\mathbb{X})$. Then $\{x_n\}$ is almost automorphic.

Proof. This theorem is an immediate consequence of Lemmas 4.5.6 and 4.5.7. □

4.5.3 Almost automorphy of bounded solutions of evolution equations

Let us consider equations of the form

$$\frac{du(t)}{dt} = A(t)u(t) + f(t), \quad t \in \mathbb{R}, \tag{4.227}$$

where f is an almost automorphic function with values in $\mathbb{X}$, and $A(t)$ generates a 1-periodic evolutionary process $(U(t,s))_{t\geq s}$ in a Banach space $\mathbb{X}$, that is, a two-parameter family of bounded linear operators that satisfies the following conditions:

1. $U(t,t) = I$ for all $t \in \mathbb{R}$;
2. $U(t,s)U(s,r) = U(t,r)$ for all $t \geq s \geq r$;
3. The map $(t,s) \mapsto U(t,s)x$ is continuous for every fixed $x \in \mathbb{X}$;
4. $U(t+1, s+1) = U(t,s)$ for all $t \geq s$ (1-*periodicity*);
5. $\|U(t,s)\| \leq Ne^{\omega(t-s)}$ for some positive N, ω independent of $t \geq s$.

We emphasize that the above choice of the period of the equations is merely for the simplification of the notation, but does not mean a restriction. We refer the reader to [39, 176] for more information on the applications of this concept of evolutionary processes to partial differential equations.

An $\mathbb{X}$-valued continuous function u on $\mathbb{R}$ is said to be a mild solution of (4.227) if

$$u(t) = U(t,s)u(s) + \int_s^t U(t,\xi)f(\xi)\mathrm{d}\xi, \quad \forall t \geq s; t,s \in \mathbb{R}. \tag{4.228}$$

Lemma 4.5.9. *Let u be a bounded mild solution of* (4.227) *on $\mathbb{R}$ and f be almost automorphic. Then, u is almost automorphic if and only if the sequence $\{u(n)\}_{n\in\mathbb{Z}}$ is almost automorphic.*

Proof. (Necessity) Obviously, if u is almost automorphic, the sequence $\{u(n)\}_{n\in\mathbb{Z}}$ is almost automorphic.

(Sufficiency) Let the sequence $\{u(n)\}_{n\in\mathbb{Z}}$ be almost automorphic. We now prove that u is almost automorphic. The proof is divided into several steps:

Step 1: We first suppose that $\{n_k'\}$ is a given sequence of integers. Then there exist a subsequence $\{n_k\}$ and a sequence $\{v(n)\}$ such that

$$\lim_{k\to\infty} u(n+n_k) = v(n), \qquad \lim_{k\to\infty} v(n-n_k) = u(n), \quad \forall n \in \mathbb{Z}, \tag{4.229}$$

$$\lim_{k\to\infty} f(t+n_k) = g(t), \qquad \lim_{k\to\infty} g(t-n_k) = f(t), \quad \forall t \in \mathbb{R}. \tag{4.230}$$

For every fixed $t \in \mathbb{R}$, let us denote by $[t]$ the integer part of t. Then, define

$$v(\eta) := U(\eta,[t])v([t]) + \int_{[t]}^{\eta} U(\eta,\xi)g(\xi)\mathrm{d}\xi, \quad \eta \in [[t],[t]+1).$$

In this way, we can define v on the whole line $\mathbb{R}$. Now we show that $\lim_{k\to\infty} u(t+n_k) = v(t)$. In fact,

$$\begin{aligned}
\lim_{k\to\infty}\|u(t+n_k)-v(t)\| &\leq \lim_{k\to\infty}\|U(t+n_k,[t]+n_k)u([t]+n_k) - U(t,[t])v([t])\| \\
&\quad + \lim_{k\to\infty}\int_{[t]}^{t}\|U(t,\eta)\|\|f(\eta+n_k)-g(\eta)\|\mathrm{d}\eta \\
&= \lim_{k\to\infty}\|U(t,[t])u([t]+n_k) - U(t,[t])v([t])\| \\
&\quad + \lim_{k\to\infty}\int_{[t]}^{t}\|U(t,\eta)\|\|f(\eta+n_k)-g(\eta)\|\mathrm{d}\eta = 0.
\end{aligned}$$

Similarly, we can show that

$$\lim_{k\to\infty}\|v(t-n_k)-u(t)\| = 0.$$

Step 2: Now we consider the general case where $\{s'_k\}_{k\in\mathbb{Z}}$ may not be an integer sequence. The main lines are similar to those in Step 1 combined with the strong continuity of the process and the precompactness of the range of the function f.

Set $n'_k = [s'_k]$ for every k. Since $\{t_k\}_{k\in\mathbb{Z}}$, where $t_k := s'_k - [s'_k]$, is a sequence in $[0,1)$, we can choose a subsequence $\{n_k\}$ from $\{n'_k\}$ such that $\lim_{k\to\infty} t_k = t_0 \in [0,1]$ and (4.229) holds for a function v, as shown in Step 1.

Let us first consider the case $0 < t_0 + t - [t_0 + t]$. We show that

$$\lim_{k\to\infty} u(t_k + t + n_k) = \lim_{k\to\infty} u(t_0 + t + n_k) = v(t_0 + t). \tag{4.231}$$

In fact, for sufficiently large k, from the above assumption we have $[t_0 + t] = [t_k + t]$. Using the 1-periodicity of the process $(U(t,s))_{t\geq s}$ we have

$$\|u(t_k + t + n_k) - u(t_0 + t + n_k)\| \leq A(k) + B(k), \tag{4.232}$$

where $A(k)$ and $B(k)$ are defined and estimated below. From the 1-periodicity of the process $(U(t,s))_{t\geq s}$, we have

$$\begin{aligned} A(k) :=& \|U(t_k + t + n_k, [t_k + t] + n_k)u([t_k + t] + n_k) \\ &- U(t_0 + t + n_k, [t_0 + t] + n_k)u([t_0 + t] + n_k)\| \\ =& \|U(t_k + t, [t_0 + t])u([t_0 + t] + n_k) - U(t_0 + t, [t_0 + t])u([t_0 + t] + n_k)\|. \end{aligned}$$

Using the strong continuity of the process $(U(t,s))_{t\geq s}$ and the precompactness of the range of the sequence $\{u(n)\}_{n\in\mathbb{Z}}$, we have $\lim_{k\to\infty} A(k) = 0$. Next, we define

$$B(k) := \left\| \int_{[t_k+t]+n_k}^{t_k+t+n_k} U(t_k + t + n_k, \eta)f(\eta)d\eta - \int_{[t_0+t]+n_k}^{t_0+t+n_k} U(t_0 + t + n_k, \eta)f(\eta)d\eta \right\|.$$

From the 1-periodicity of the process $(U(t,s))_{t\geq s}$ and since $[t_0 + t] = [t_k + t]$, we have

$$\begin{aligned} B(k) =& \left\| \int_0^{t_k+t-[t_k+t]} U(t_k + t + n_k, [t_0 + t] + n_k + \theta)f([t_0 + t] + n_k + \theta)d\theta \right. \\ &\left. - \int_0^{t_0+t-[t_0+t]} U(t_0 + t + n_k, [t_0 + t] + n_k + \theta)f([t_0 + t] + n_k + \theta)d\theta \right\| \\ =& \left\| \int_0^{t_k+t-[t_0+t]} U(t_k + t - [t_0 + t], \theta)f([t_0 + t] + n_k + \theta)d\theta \right. \\ &\left. - \int_0^{t_0+t-[t_0+t]} U(t_0 + t - [t_0 + t], \theta)f([t_0 + t] + n_k + \theta)d\theta \right\|. \end{aligned}$$

From the strong continuity of the process $(U(t,s))_{t\geq s}$ and the precompactness of the range of f, it follows that $\lim_{k\to\infty} B(k) = 0$. So, in view of Step 1, we see that (4.231) holds.

Next, we consider the case when $t_0 + t - [t_0 + t] = 0$, that is, $t_0 + t$ is an integer. If $t_k + t \geq t_0 + t$, we can repeat the above argument. So, we omit the details. Now suppose that $t_k + t < t_0 + t$. Then

$$\|u(t_k + t + n_k) - u(t_0 + t + n_k)\| \leq C(k) + D(k), \tag{4.233}$$

where $C(k)$ and $D(k)$ are defined and estimated as follows:

$$\begin{aligned} C(k) &:= \|U(t_k + t + n_k, [t_k + t] + n_k)u([t_k + t] + n_k) \\ &\quad - U(t_0 + t + n_k, t_0 + t - 1 + n_k)u(t_0 + t - 1 + n_k)\| \\ &= \|U(t_k + t, t_0 + t - 1)u(t_0 + t - 1 + n_k) \\ &\quad - U(t_0 + t, t_0 + t - 1)u(t_0 + t - 1 + n_k)\|. \end{aligned}$$

Now using the strong continuity of the process $(U(t,s))_{t\geq s}$ and the precompactness of the range of the sequence $\{u(n)\}_{n\in\mathbb{Z}}$, we obtain $\lim_{k\to\infty} C(k) = 0$.

As for $D(k)$, we have

$$\begin{aligned} D(k) &:= \Bigg\| \int_{[t_k+t]+n_k}^{t_k+t+n_k} U(t_k + t + n_k, \eta)f(\eta)\mathrm{d}\eta \\ &\qquad - \int_{[t_0+t]+n_k-1}^{t_0+t+n_k} U(t_0 + t + n_k, \eta)f(\eta)\mathrm{d}\eta \Bigg\| \\ &= \Bigg\| \int_{[t_0+t]+n_k-1}^{t_k+t+n_k} U(t_k + t + n_k, \eta)f(\eta)\mathrm{d}\eta \\ &\qquad - \int_{[t_0+t]+n_k-1}^{t_0+t+n_k} U(t_0 + t + n_k, \eta)f(\eta)\mathrm{d}\eta \Bigg\| \\ &= \Bigg\| \int_0^{t_k+1-t_0} U(t_k + t, t_0 + t - 1 + \theta)f(t_0 + t + n_k - 1 + \theta)\mathrm{d}\theta \\ &\qquad - \int_0^1 U(t_0 + t, t_0 + t - 1 + \theta)f(t_0 + t + n_k - 1 + \theta)\mathrm{d}\theta \Bigg\|. \end{aligned}$$

From the strong continuity of the process $(U(t,s))_{t\geq s}$ and the precompactness of the range of f, it follows that $\lim_{k\to\infty} D(k) = 0$. This finishes the proof of the lemma. □

As a main result of this section we have the following:

Theorem 4.5.10. *Let $A(t)$ in equation (4.227) generate a 1-periodic strongly continuous evolutionary process, and let f be almost automorphic. Assume further that the space $\mathbb{X}$ does not contain any subspace isomorphic to c_0 and the part of spectrum of the monodromy operator $U(1,0)$ on the unit circle is countable. Then, every bounded mild solution of equation (4.227) on the real line is almost automorphic.*

Proof. The theorem is an immediate consequence of Lemmas 4.5.6, 4.5.7, and 4.5.9. In fact, we need only to prove the sufficiency. Let us consider the discrete equation

$$u(n+1) = U(n+1,n)u(n) + \int_n^{n+1} U(n+1,\xi)f(\xi)\mathrm{d}\xi, \quad n \in \mathbb{Z}.$$

From the 1-periodicity of the process $(U(t,s))_{t\geq s}$, this equation can be rewritten in the form

$$u(n+1) = Bu(n) + y_n, \quad n \in \mathbb{Z}, \tag{4.234}$$

where

$$B := U(1,0); \quad y_n := \int_n^{n+1} U(n+1,\xi)f(\xi)\mathrm{d}\xi, \quad n \in \mathbb{Z}.$$

We are going to show that the sequence $\{y_n\}_{n\in\mathbb{Z}}$ defined above is almost automorphic. In fact, since f is automorphic, for any given sequence $\{n'_k\}$ there are a subsequence $\{n_k\}$ and a measurable function g such that $\lim_{k\to\infty} f(t+n_k) = g(t)$ and $\lim_{m\to\infty} g(t-n_m) = f(t)$ for every $t \in \mathbb{R}$. Therefore, if we set

$$w_n = \lim_{k\to\infty} \int_{n+n_k}^{n+n_k+1} U(n+n_k,\xi)f(\xi)\mathrm{d}\xi, \quad n \in \mathbb{Z},$$

then, by the 1-periodicity of $(U(t,s))_{t\geq s}$ and the Lebesgue dominated convergence theorem, we have

$$w_n = \lim_{k\to\infty} \int_n^{n+1} U(n,\eta)f(n_k+\eta)\mathrm{d}\eta = \int_n^{n+1} U(n,\eta)g(\eta)\mathrm{d}\eta.$$

Therefore, $\lim_{k\to\infty} y_{n+n_k} = w_n$ for every $n \in \mathbb{Z}$. Similarly, we can show that $\lim_{k\to\infty} w_{n-n_k} = y_n$.

By Lemma 4.5.7, since $\{u(n)\}$ is a bounded solution of (4.234), $\mathbb{X}$ does not contain any subspace isomorphic to c_0, and the part of spectrum of $U(1,0)$ on the unit circle is countable, $\{u(n)\}$ is almost automorphic. By Lemma 4.5.9, this yields that the solution u itself is almost automorphic. □

4.6 Semilinear fractional differential equations

As a natural extension of almost automorphy, the concept of asymptotic almost automorphy, which is the central issue to be discussed in this section, was introduced in the literature [232] by N'Guérékata in the early 1980s. Since then, this notion has found several developments and has been generalized into different directions. Until now, the asymptotically almost automorphic functions, as well as the asymptotically almost automorphic solutions for differential systems, have been investigated by many mathematicians, see [85] by Bugajewski and N'Guérékata, [121] by Diagana, Hernández, and dos Santos, [128] by Ding, Xiao, and Liang for the asymptotically almost automorphic solutions to integrodifferential equations, [293] by Zhao, Chang, and N'Guérékata for the asymptotically almost automorphic solutions to the nonlinear delay integral equations, and [92] by Chang and Tang, [294] by Zhao, Chang, and Nieto for the asymptotically almost automorphic solutions to stochastic differential equations. The existence of asymptotically almost automorphic solutions has become one of the most attractive topics in the qualitative theory of differential equations due to its significance and applications in physics, mathematical biology, control theory, and so on. We refer the reader to the monograph of N'Guérékata [239] for the recent theory and applications of asymptotically almost automorphic functions.

With motivation coming from a wide range of engineering and physical applications, fractional differential equations have recently attracted great attention of mathematicians and other scientists. Such equations generalize ordinary differential equations to arbitrary noninteger orders. Fractional differential equations find numerous applications in the field of viscoelasticity, feedback amplifiers, electrical circuits, electroanalytical chemistry, fractional multipoles, neuron modeling, encompassing different branches of physics, chemistry, and biological sciences [126, 179, 188, 219, 248, 256, 296]. Many physical processes appear to exhibit fractional order behavior that may vary with time or space. In recent years, there has been a significant development in ordinary and partial differential equations involving fractional derivatives, we only enumerate here the monographs of Kilbas et al. [188, 256], Diethelm [126], Hilfer [179], Podlubny [248], Miller [219], Zhou [296], and the papers of Agarwal et al. [10, 14], Benchohra et al. [11, 55], El-Borai [135], Lakshmikantham et al. [192, 193, 195, 196], Mophou et al. [220, 222–224], N'Guérékata [238], Zhou et al. [297–300]; see also the references therein.

The study of almost periodic and almost automorphic type solutions to fractional differential equations was initiated by Araya and Lizama [38]. In their work, the authors investigated the existence and uniqueness of an almost automorphic mild solution of the semilinear fractional differential equation

$$\mathrm{D}_t^{\alpha} x(t) = Ax(t) + F(t, x(t)), \quad t \in \mathbb{R}, 1 < \alpha < 2,$$

when A is a generator of an α-resolvent family and D_t^{α} is the Riemann–Liouville fractional derivative. In [110], Cuevas and Lizama considered the fractional differential equation

$$D_t^{\alpha}x(t) = Ax(t) + D_t^{\alpha-1}F(t, x(t)), \quad t \in \mathbb{R}, 1 < \alpha < 2, \tag{4.235}$$

where A is a linear operator of sectorial negative type on a complex Banach space X, while the fractional derivative is understood in the Riemann–Liouville sense. Under suitable conditions on $F(t, x)$, the authors proved the existence and uniqueness of an almost automorphic mild solution to equation (4.235). Cuevas et al. [12, 13] studied respectively the pseudo almost periodic and pseudo almost periodic of class infinity mild solutions to equation (4.235) assuming that $F : \mathbb{R} \times X \to X$, $(t, x) \to F(t, x)$ is a pseudo almost periodic or pseudo almost periodic of class infinity function satisfying suitable conditions in $x \in X$. Agarwal et al. [15] studied the existence and uniqueness of a weighted pseudo almost periodic mild solution to equation (4.235). Ding et al. [127] investigated the existence and uniqueness of an almost automorphic solution to equation (4.235) assuming that $F : \mathbb{R} \times X \to X$, $(t, x) \to F(t, x)$ is Stepanov-like almost automorphic in $t \in \mathbb{R}$ satisfying some kind of Lipschitz conditions. Cuevas et al. [112] studied the existence of almost periodic (resp., pseudo almost periodic) mild solutions to equation (4.235) assuming that $F : \mathbb{R} \times X \to X$, $(t, x) \to F(t, x)$ is Stepanov almost (resp., Stepanov-like pseudo almost) periodic in $t \in \mathbb{R}$ uniformly for $x \in X$. Chang et al. [93] studied the existence and uniqueness of weighted pseudo almost automorphic solution to equation (4.235) with Stepanov-like weighted pseudo almost automorphic coefficient. He et al. [175] studied also the existence and uniqueness of weighted Stepanov-like pseudo almost automorphic mild solution to equation (4.235). Cao et al. [88] studied the existence and uniqueness of antiperiodic mild solution to equation (4.235). In [207], Cuevas et al. showed sufficient conditions to ensure the existence and uniqueness of a mild solution to the equation (4.235) in the following classes of vector-valued function spaces: periodic functions, asymptotically periodic functions, pseudo periodic functions, almost periodic functions, asymptotically almost periodic functions, pseudo almost periodic functions, almost automorphic functions, asymptotically almost automorphic functions, pseudo almost automorphic functions, compact almost automorphic functions, asymptotically compact almost automorphic functions, pseudo compact almost automorphic functions, S-asymptotically ω-periodic functions, decay functions, and mean decay functions.

Recently, Xia et al. [274] established several sufficient criteria for the existence and uniqueness of (μ, ν)-pseudo almost automorphic solution to the semilinear fractional differential equation

$$D_t^{\alpha}x(t) = Ax(t) + D_t^{\alpha-1}F(t, Bx(t)), \quad t \in \mathbb{R},$$

where $1 < \alpha < 2$, A is a sectorial operator of type $\omega < 0$ on a complex Banach space X, while B is a bounded linear operator. The fractional derivative is understood in the Riemann–Liouville sense. Their discussion is divided into two cases, i. e., $F : \mathbb{R} \times X \to X$, $(t, x) \to F(t, x)$ is (μ, ν)-pseudo almost automorphic and $F : \mathbb{R} \times X \to X$, $(t, x) \to F(t, x)$ is Stepanov-like (μ, ν)-pseudo almost automorphic. Kavitha et al. [186] studied weighted pseudo almost automorphic solutions of the fractional integro-differential equation

$$D_t^\alpha x(t) = Ax(t) + D_t^{\alpha-1}F(t, x(t), Kx(t)), \quad t \in \mathbb{R},$$

where $1 < \alpha < 2$ and

$$Kx(t) = \int_{-\infty}^{t} k(t-s)h(s, x(s))ds,$$

A is a linear densely defined sectorial operator on a complex Banach space X, $F : \mathbb{R}\times X\times X \to X$, $(t, x, y) \to F(t, x, y)$ is a weighted pseudo almost automorphic function in $t \in \mathbb{R}$ for each $x, y \in X$ satisfying suitable conditions. The fractional derivative is understood in the Riemann–Liouville sense. Mophou [221] investigated the existence and uniqueness of a weighted pseudo almost automorphic mild solution to the fractional differential equation

$$D_t^\alpha x(t) = Ax(t) + D_t^{\alpha-1}F(t, x(t), Bx(t)), \quad t \in \mathbb{R}, 1 < \alpha < 2, \tag{4.236}$$

where $A : D(A) \subset X \to X$ is a linear densely operator of sectorial type on a complex Banach space X, $B : X \to X$ is a bounded linear operator, and $F : \mathbb{R} \times X \times X \to X$, $(t, x, y) \to F(t, x, y)$ is a weighted pseudo almost automorphic function in $t \in \mathbb{R}$ for each $x, y \in X$ satisfying suitable conditions. The fractional derivative D_t^α is understood in the Riemann–Liouville sense. Chang et al. [91] investigated some existence results of μ-pseudo almost automorphic mild solutions to equation (4.236) assuming that $F : \mathbb{R} \times X \times X \to X$, $(t, x, y) \to F(t, x, y)$ is a μ-pseudo almost automorphic function in $t \in \mathbb{R}$ for each $x, y \in X$ satisfying suitable conditions.

Equation (4.236) is motivated by physical problems. Indeed, due to their applications in fields of science where characteristics of anomalous diffusion are present, equations as in (4.236) are attracting increasing interest (cf. [54, 134, 160] and the references therein). For example, anomalous diffusion in fractals [134] or in macroeconomics [16] has been recently well studied in the setting of fractional Cauchy problems for equation (4.236). For this reason, equation (4.236) has gotten considerable attention in recent years (cf. [12, 13, 15, 16, 54, 88, 91, 93, 110, 112, 127, 134, 160, 175, 186, 207, 221, 274] and the references therein).

To the best of our knowledge, much less is known about the existence of asymptotically almost automorphic mild solutions to equation (4.236) when the nonlinearity $F(t, x, y)$ as a whole loses the Lipschitz continuity with respect to x and y. Motivated by the above-mentioned works, the purpose of this section is to establish some new existence results of asymptotically almost automorphic mild solutions to equation (4.236). In our results, the nonlinearity $F : \mathbb{R} \times X \times X \to X$, $(t, x, y) \to F(t, x, y)$ does not have to satisfy a (locally) Lipschitz condition (see Remark 4.6.19). However, in many papers (for instance, [12, 13, 15, 38, 88, 91, 93, 110, 112, 127, 175, 186, 207, 221, 274]) on almost periodic and almost automorphic solutions to fractional differential equations, to be able to apply the well-known Banach contraction principle, a (locally) Lipschitz condition for the nonlinearity of corresponding fractional differential equations is needed. As can be seen, our results generalize those, as well as related research, and have more broad applications. In particular, as an application and to illustrate our main results, we will examine some sufficient conditions for the existence of asymptotically almost automorphic mild solutions to the fractional relaxation–oscillation equation given by

$$\begin{aligned}\partial_t^{\alpha} u(t,x) &= \partial_x^2 u(t,x) - pu(t,x)\\ &\quad + \partial_t^{\alpha-1}\Big[\mu a(t)\sin\Big(\frac{1}{2+\cos t+\cos(\sqrt{2}t)}\Big)[\sin u(t,x)+u(t,x)]\\ &\quad + \nu e^{-|t|}[u(t,x)+\sin u(t,x)]\Big], \quad t\in\mathbb{R},\ x\in[0,\pi],\end{aligned}$$

with boundary conditions $u(t,0) = u(t,\pi) = 0$, $t \in \mathbb{R}$, where $a(t) \in \mathrm{BC}(\mathbb{R},\mathbb{R}^+)$ is a function, while p, μ, and ν are positive constants.

4.6.1 Preliminaries

This section is concerned with some notations, definitions, lemmas, and preliminary facts which are used in what follows.

From now on, let $(X,\|\cdot\|)$, $(Y,\|\cdot\|_Y)$ be two Banach spaces, $\mathrm{BC}(\mathbb{R},X)$ (resp., $\mathrm{BC}(\mathbb{R}\times Y\times Y,X)$) is the space of all X-valued bounded continuous functions (resp., jointly bounded continuous functions $F:\mathbb{R}\times Y\times Y\to X$). Furthermore, $C_0(\mathbb{R},X)$ (resp., $C_0(\mathbb{R}\times Y\times Y,X)$) is the closed subspace of $\mathrm{BC}(\mathbb{R},X)$ (resp., $\mathrm{BC}(\mathbb{R}\times Y\times Y,X)$) consisting of functions vanishing at infinity (vanishing at infinity uniformly in any compact subset of $Y\times Y$, in other words,

$$\lim_{|t|\to+\infty}\|g(t,x,y)\| = 0 \quad \text{uniformly for } (x,y)\in\mathbb{K},$$

where $\mathbb{K}$ is an any compact subset of $Y\times Y$). Let also $\mathbb{L}(X)$ be the Banach space of all bounded linear operators from X into itself endowed with the norm

$$\|T\|_{\mathbb{L}(X)} = \sup\{\|Tx\| : x\in X,\ \|x\|=1\}.$$

For a bounded linear operator $A\in\mathbb{L}(X)$, let $\rho(A)$ and $D(A)$ stand for the resolvent and domain of A, respectively.

First, let us recall some basic definitions and results on almost automorphic and asymptotically almost automorphic functions.

Definition 4.6.1 ([237]). A continuous function $F:\mathbb{R}\times Y\times Y\to X$ is said to be almost automorphic in $t\in\mathbb{R}$ uniformly for all $(x,y)\in K$, where K is any bounded subset of $Y\times Y$, if for every sequence of real numbers $\{s_n'\}$, there exists a subsequence $\{s_n\}$ such that

$$\lim_{n\to\infty} F(t+s_n, x, y) = \Theta(t,x,y) \quad \text{exists for each } t \in \mathbb{R} \text{ and each } (x,y) \in K$$

and

$$\lim_{n\to\infty} \Theta(t-s_n, x, y) = F(t,x,y) \quad \text{exists for each } t \in \mathbb{R} \text{ and each } (x,y) \in K.$$

The collection of such functions is denoted by $AA(\mathbb{R}\times Y\times Y, X)$.

Remark 4.6.2. The function $F : \mathbb{R}\times X\times X \to X$ given by

$$F(t,x,y) = \sin\left(\frac{1}{2+\cos t+\cos(\sqrt{2}t)}\right)[\sin(x)+y]$$

is almost automorphic in $t \in \mathbb{R}$ uniformly for all $(x,y) \in K$, where K is any bounded subset of $X\times X$, $X = L^2[0,\pi]$.

Similar to the Lemma 2.2 of [203] and Proposition 3.2 of [221], we have the following result on almost automorphic functions.

Lemma 4.6.3. *Let $F : \mathbb{R}\times X\times X \to X$ be almost automorphic in $t \in \mathbb{R}$ uniformly for all $(x,y) \in K$, where K is any bounded subset of $X\times X$, and assume that $F(t,x,y)$ is uniformly continuous on K uniformly for $t \in \mathbb{R}$, that is, for any $\varepsilon > 0$, there exists $\delta > 0$ such that $x_1, x_2, y_1, y_2 \in K$ and $\|x_1-y_1\| + \|x_2-y_2\| < \delta$ imply that*

$$\|F(t,x_1,x_2) - F(t,y_1,y_2)\| < \varepsilon \quad \text{for all } t \in \mathbb{R}.$$

Let $x, y : \mathbb{R}\to X$ be almost automorphic. Then the function $\Upsilon : \mathbb{R}\to X$ defined by $\Upsilon(t) = F(t,x(t),y(t))$ is almost automorphic.

Proof. Suppose that $\{s_n\}$ is a sequence of real numbers. Then by the definition of almost automorphic functions, we can extract a subsequence $\{\tau_n\}$ of $\{s_n\}$ such that

$$(P_1)\ \lim_{n\to\infty} x(t+\tau_n) = \tilde{x}(t) \quad \text{for each } t\in\mathbb{R}, \quad (P_2)\ \lim_{n\to\infty} \tilde{x}(t-\tau_n) = x(t) \quad \text{for each } t\in\mathbb{R},$$

$$(P_3)\ \lim_{n\to\infty} y(t+\tau_n) = \tilde{y}(t) \quad \text{for each } t\in\mathbb{R}, \quad (P_4)\ \lim_{n\to\infty} \tilde{y}(t-\tau_n) = y(t) \quad \text{for each } t\in\mathbb{R},$$

$$(P_5)\ \lim_{n\to\infty} F(t+\tau_n, x, y) = \tilde{F}(t,x,y) \quad \text{for each } t\in\mathbb{R}, x,y\in X,$$

$$(P_6)\ \lim_{n\to\infty} \tilde{F}(t-\tau_n, x, y) = F(t,x,y) \quad \text{for each } t\in\mathbb{R}, x,y\in X.$$

Write

$$\tilde{\Upsilon}(t) := \tilde{F}(t,\tilde{x}(t),\tilde{y}(t)), \quad t\in\mathbb{R}.$$

Then

$$
\begin{aligned}
\|\Upsilon(t+\tau_n)-\widetilde{\Upsilon}(t)\| &= \|F(t+\tau_n,x(t+\tau_n),y(t+\tau_n))-\widetilde{F}(t,\widetilde{x}(t),\widetilde{y}(t))\| \\
&\le \|F(t+\tau_n,x(t+\tau_n),y(t+\tau_n))-F(t+\tau_n,\widetilde{x}(t),\widetilde{y}(t))\| \\
&\quad + \|F(t+\tau_n,\widetilde{x}(t),\widetilde{y}(t))-\widetilde{F}(t,\widetilde{x}(t),\widetilde{y}(t))\|.
\end{aligned}
$$

Since $x(t)$ and $y(t)$ are almost automorphic, $x(t), y(t)$ and $\widetilde{x}(t), \widetilde{y}(t)$ are bounded. Therefore we can choose a bounded subset $K \subset X \times X$ such that

$$(x(t),y(t)) \in K,\ (\widetilde{x}(t),\widetilde{y}(t)) \in K \quad \text{for all } t \in \mathbb{R}.$$

By (P_1), (P_3), and the uniform continuity of $F(t,x,y)$ in $(x(t),y(t)) \in K$, we have

$$\lim_{n\to\infty}\|F(t+\tau_n,x(t+\tau_n),y(t+\tau_n))-F(t+\tau_n,\widetilde{x}(t),\widetilde{y}(t))\| = 0.$$

Moreover, by (P_5),

$$\lim_{n\to\infty}\|F(t+\tau_n,\widetilde{x}(t),\widetilde{y}(t))-\widetilde{F}(t,\widetilde{x}(t),\widetilde{y}(t))\| = 0,$$

so remembering the triangle inequality, we deduce that

$$\lim_{n\to\infty}\|\Upsilon(t+\tau_n)-\widetilde{\Upsilon}(t)\| = 0 \quad \text{for each } t \in \mathbb{R}.$$

Using the same arguments, we can prove that

$$\lim_{n\to\infty}\|\widetilde{\Upsilon}(t-\tau_n)-\Upsilon(t)\| = 0 \quad \text{for each } t \in \mathbb{R}.$$

This proves that $\Upsilon(t)$ is almost automorphic by definition. □

Remark 4.6.4. If $F(t,x,y)$ satisfies Lipschitz condition with respect to x and y uniformly in $t \in \mathbb{R}$, i. e., for each pair $x_1, x_2, y_1, y_2 \in X$,

$$\|F(t,x_1,x_2)-F(t,y_1,y_2)\| \le L(\|x_1-y_1\|+\|x_2-y_2\|)$$

uniformly in $t \in \mathbb{R}$, where $L > 0$ is called the Lipschitz constant for the function $F(t,x,y)$, then $F(t,x,y)$ is uniformly continuous on K uniformly in $t \in \mathbb{R}$, where K is any bounded subset of $X \times X$.

Remark 4.6.5. If $F(t,x,y)$ satisfies a local Lipschitz condition with respect to x and y uniformly in $t \in \mathbb{R}$, i. e., for each pair $x_1, x_2, y_1, y_2 \in X$, $t \in \mathbb{R}$,

$$\|F(t,x_1,x_2)-F(t,y_1,y_2)\| \le L(t)(\|x_1-y_1\|+\|x_2-y_2\|),$$

where $L(t) \in BC(\mathbb{R},\mathbb{R}^+)$, then $F(t,x,y)$ is uniformly continuous on K uniformly in $t \in \mathbb{R}$, where K is any bounded subset of $X \times X$.

Definition 4.6.6 ([237]). A continuous function $F : \mathbb{R} \to X$ is said to be asymptotically almost automorphic if it can be decomposed as $F(t) = G(t) + \Phi(t)$, where

$$G(t) \in \mathrm{AA}(\mathbb{R}, X), \quad \Phi(t) \in C_0(\mathbb{R}, X).$$

Denote by $\mathrm{AAA}(\mathbb{R}, X)$ the set of all such functions.

Remark 4.6.7. The function $F : \mathbb{R} \to \mathbb{R}$ defined by

$$F(t) = G(t) + \Phi(t) = \sin\left(\frac{1}{2 + \cos t + \cos(\sqrt{2}t)}\right) + e^{-|t|}$$

is an asymptotically almost automorphic function with

$$G(t) = \sin\left(\frac{1}{2 + \cos t + \cos(\sqrt{2}t)}\right) \in \mathrm{AA}(\mathbb{R}, \mathbb{R}), \quad \Phi(t) = e^{-|t|} \in C_0(\mathbb{R}, \mathbb{R}).$$

Lemma 4.6.8 ([237]). *The set* $\mathrm{AAA}(\mathbb{R}, X)$ *is also a Banach space with the supremum norm* $\|\cdot\|_\infty$.

Definition 4.6.9 ([237]). A continuous function $F : \mathbb{R} \times Y \times Y \to X$ is said to be asymptotically almost automorphic if it can be decomposed as $F(t, x, y) = G(t, x, y) + \Phi(t, x, y)$, where

$$G(t, x, y) \in \mathrm{AA}(\mathbb{R} \times Y \times Y, X), \ \Phi(t, x, y) \in C_0(\mathbb{R} \times Y \times Y, X).$$

Denote by $\mathrm{AAA}(\mathbb{R} \times Y \times Y, X)$ the set of all such functions.

Remark 4.6.10. The function $F : \mathbb{R} \times X \times X \to X$ given by

$$F(t, x, y) = G(t, x, y) + \Phi(t, x, y) = \sin\left(\frac{1}{2 + \cos t + \cos(\sqrt{2}t)}\right)[\sin(x) + y] + e^{-|t|}[x + \sin(y)]$$

is asymptotically almost automorphic uniformly in $t \in \mathbb{R}$ for any $(x, y) \in K$, where K is any bounded subset of $X \times X$, $X = L^2[0, \pi]$, and

$$G(t, x, y) = \sin\left(\frac{1}{2 + \cos t + \cos(\sqrt{2}t)}\right)[\sin(x) + y] \in \mathrm{AA}(\mathbb{R} \times X \times X, X),$$
$$\Phi(t, x, y) = e^{-|t|}[x + \sin(y)] \in C_0(\mathbb{R} \times X \times X, X).$$

Next we give some basic definitions and properties of the fractional calculus theory which are used further in this chapter.

Definition 4.6.11 ([188]). The fractional integral of order $\alpha > 0$ with the lower limit t_0 for a function f is defined as

$$I^{\alpha}f(t) = \frac{1}{\Gamma(\alpha)}\int_{t_0}^{t}(t-s)^{\alpha-1}f(s)\mathrm{d}s, \quad t > t_0,\ \alpha > 0,$$

provided the right-hand side is pointwise defined on $[t_0, \infty)$, where Γ is the Gamma function.

Definition 4.6.12 ([188]). The Riemann–Liouville derivative of order $\alpha > 0$ with the lower limit t_0 for a function $f : [t_0, \infty) \to \mathbb{R}$ can be written as

$$D_t^{\alpha}f(t) = \frac{1}{\Gamma(n-\alpha)}\frac{\mathrm{d}^n}{\mathrm{d}t^n}\int_{t_0}^{t}(t-s)^{-\alpha}f(s)\mathrm{d}s, \quad t > t_0,\ n-1 < \alpha < n.$$

The first and maybe the most important property of the Riemann–Liouville fractional derivative is that, for $t > t_0$ and $\alpha > 0$, one has $D_t^{\alpha}(I^{\alpha}f(t)) = f(t)$, which means that the Riemann–Liouville fractional differentiation operator is a left inverse to the Riemann–Liouville fractional integration operator of the same order α.

It is important to define a sectorial operator for the definition of a mild solution of any fractional abstract equation. So let us now give the definitions of sectorial linear operators and their associated solution operators.

Definition 4.6.13 (Sectorial operator, [162]). A closed and linear operator A is said to be sectorial of type ω and angle θ if there exist $0 < \theta < \frac{\pi}{2}$, $M > 0$, and $\omega \in \mathbb{R}$ such that its resolvent $\rho(A)$ exists outside the sector $\omega + S_\theta := \{\omega + \lambda : \lambda \in \mathbb{C}, |\arg(-\lambda)| < \theta\}$ and

$$\|(\lambda - A)^{-1}\| \le \frac{M}{|\lambda - \omega|}, \quad \lambda \notin \omega + S_\theta.$$

Sectorial operators are well studied in the literature, usually for the case $\omega = 0$. For a recent reference including several examples and properties, we refer the reader to [162]. Note that an operator A is sectorial of type ω if and only if $\omega I - A$ is sectorial of type 0.

Definition 4.6.14 ([109]). Let A be a closed and linear operator with domain $D(A)$ defined on a Banach space X. We call A is the generator of a solution operator if there are $\omega \in \mathbb{R}$ and a strongly continuous function $S_\alpha : \mathbb{R}^+ \to \mathbb{L}(X)$ such that $\{\lambda^{\alpha} : \operatorname{Re}\lambda > \omega\} \subseteq \rho(A)$ and

$$\lambda^{\alpha-1}(\lambda^{\alpha} - A)^{-1}x = \int_{0}^{\infty} e^{-\lambda t}S_\alpha(t)x\mathrm{d}t, \quad \operatorname{Re}\lambda > \omega, x \in X.$$

In this case, $S_\alpha(t)$ is called the solution operator generated by A.

Note that if A is sectorial of type ω with $0 \le \theta \le \pi(1 - \frac{\alpha}{2})$, then A is the generator of a solution operator given by

$$S_\alpha(t) := \frac{1}{2\pi i}\int_\gamma e^{-\lambda t}\lambda^{\alpha-1}(\lambda^\alpha - A)^{-1}\mathrm{d}\lambda,$$

where γ is a suitable path lying outside the sector $\omega + \Sigma_\theta$ (cf. [162]).

Very recently, Cuesta [162, Theorem 1] has proved that if A is a sectorial operator of type $\omega < 0$ for some $M > 0$ and $0 \leq \theta < \pi(1 - \frac{\alpha}{2})$, then there exists $C > 0$ such that

$$\|S_\alpha(t)\|_{\mathbb{L}(X)} \leq \frac{CM}{1 + |\omega|t^\alpha} \quad \text{for } t \geq 0.$$

In the border case $\alpha = 1$, this is analogous to saying that A is the generator of an exponentially stable C_0-semigroup. The main difference is that in the case $\alpha > 1$ the solution family $S_\alpha(t)$ decays like $t^{-\alpha}$. Cuesta's result proves that $S_\alpha(t)$ is, in fact, integrable.

In the following, we present the following compactness criterion, which is a special case of the general compactness result of Theorem 2.1 in [254].

Lemma 4.6.15 ([254]). *A set $D \subset C_0(\mathbb{R}, X)$ is relatively compact if*
(1) *D is equicontinuous;*
(2) $\lim_{|t|\to+\infty} x(t) = 0$ *uniformly for $x \in D$;*
(3) *the set $D(t) := \{x(t) : x \in D\}$ is relatively compact in X for every $t \in \mathbb{R}$.*

The following Krasnoselskii's fixed point theorem plays a key role in the proofs of our main results, which can be found in many books.

Lemma 4.6.16 ([261]). *Let U be a bounded closed and convex subset of X, and J_1, J_2 be maps of U into X such that $J_1x + J_2y \in U$ for every pair $x, y \in U$. If J_1 is a contraction and J_2 is completely continuous, then the equation $J_1x + J_2x = x$ has a solution on U.*

4.6.2 Asymptotically almost automorphic mild solutions

In this section, we study the existence of asymptotically almost automorphic mild solutions for the semilinear fractional differential equations of the form

$$\mathrm{D}_t^\alpha x(t) = Ax(t) + \mathrm{D}_t^{\alpha-1}F(t, x(t), Bx(t)), \quad t \in \mathbb{R},\ 1 < \alpha < 2, \tag{4.237}$$

where $A : D(A) \subset X \to X$ is a linear densely defined operator of sectorial type $\omega < 0$ on a complex Banach space X, $B : X \to X$ is a bounded linear operator, and $F : \mathbb{R}\times X\times X \to X$, $(t, x, y) \to F(t, x, y)$ is a given function to be specified later. The fractional derivative D_t^α is to be understood in the Riemann–Liouville sense.

We recall the following definition that will be essential for us.

Definition 4.6.17 ([221]). Assume that A generates an integrable solution operator $S_\alpha(t)$. A continuous function $x : \mathbb{R} \to X$ satisfying the integral equation

$$x(t) = \int_{-\infty}^{t} S_\alpha(t-\sigma)F(\sigma, x(\sigma), Bx(\sigma))d\sigma, \quad t \in \mathbb{R}$$

is called a mild solution on $\mathbb{R}$ to equation (4.237).

In the proofs of our results, we need the following auxiliary result.

Lemma 4.6.18. *Consider $Y(t) \in AA(\mathbb{R}, X)$ and $Z(t) \in C_0(\mathbb{R}, X)$. Let*

$$\Phi_1(t) := \int_{-\infty}^{t} S_\alpha(t-s)Y(s)ds, \quad \Phi_2(t) := \int_{-\infty}^{t} S_\alpha(t-s)Z(s)ds, \quad t \in \mathbb{R}.$$

Then $\Phi_1(t) \in AA(\mathbb{R}, X)$, $\Phi_2(t) \in C_0(\mathbb{R}, X)$.

Proof. First, note that

$$\int_0^\infty \frac{1}{1+|\omega|s^\alpha} ds = \frac{\omega^{-\frac{1}{\alpha}}\pi}{\alpha \sin\frac{\pi}{\alpha}} \quad \text{for } 1 < \alpha < 2.$$

Then

$$\begin{aligned} \|\Phi_1(t)\| &= \left\| \int_{-\infty}^{t} S_\alpha(t-s)Y(s)ds \right\| = \left\| \int_0^{+\infty} S_\alpha(\tau)Y(t-\tau)d\tau \right\| \\ &\le CM\|Y\|_\infty \int_0^\infty \frac{1}{1+|\omega|\tau^\alpha} d\tau = \frac{CM\omega^{-\frac{1}{\alpha}}\pi}{\alpha \sin\frac{\pi}{\alpha}} \|Y\|_\infty, \end{aligned}$$

which implies $\Phi_1(t)$ is well defined and continuous on $\mathbb{R}$. Since $Y(t) \in AA(\mathbb{R}, X)$, for any $\varepsilon > 0$ and every sequence of real numbers $\{s_n'\}$, there exist a subsequence $\{s_n\}$, a function $\widetilde{Y}(t)$, and $N \in \mathbb{N}$ such that

$$\|Y(s+s_n) - \widetilde{Y}(s)\| < \varepsilon \quad \text{for each } n > N \text{ and every } s \in \mathbb{R}.$$

Define

$$\widetilde{\Phi_1}(t) := \int_{-\infty}^{t} T(t-s)\widetilde{Y}(s)ds.$$

Then

$$\begin{aligned}\|\Phi_1(t+s_n)-\widetilde{\Phi_1}(t)\| &= \left\|\int_{-\infty}^{t+s_n} S_\alpha(t+s_n-s)Y(s)\mathrm{d}s - \int_{-\infty}^{t} S_\alpha(t-s)Y(s)\mathrm{d}s\right\| \\ &= \left\|\int_{0}^{+\infty} S_\alpha(s)Y(t+s_n-s)\mathrm{d}s - \int_{0}^{+\infty} S_\alpha(s)Y(t-s)\mathrm{d}s\right\| \\ &\le CM\int_0^\infty \frac{1}{1+|\omega|s^\alpha}\|Y(s+s_n)-\widetilde{Y}(s)\|\mathrm{d}s \le \frac{CM\omega^{-\frac{1}{\alpha}}\pi\varepsilon}{\alpha\sin\frac{\pi}{\alpha}}\end{aligned}$$

for each $n > N$ and every $t \in \mathbb{R}$. This implies

$$\widetilde{\Phi_1}(t) = \lim_{n\to\infty}\Phi_1(t+s_n)$$

is well defined for each $t \in \mathbb{R}$.

By a similar argument, one can obtain

$$\lim_{n\to\infty}\widetilde{\Phi_1}(t-s_n) = \Phi_1(t) \quad \text{for each } t \in \mathbb{R}.$$

Thus $\Phi_1(t) \in AA(\mathbb{R}, X)$.

Since $Z(t) \in C_0(\mathbb{R}, X)$, one can choose an $N_1 > 0$ such that $\|Z(t)\| < \varepsilon$ for all $t > N_1$. This enables us to conclude that, for all $t > N_1$,

$$\begin{aligned}\|\Phi_2(t)\| &\le \left\|\int_{-\infty}^{N_1} S_\alpha(t-s)Z(s)\mathrm{d}s\right\| + \left\|\int_{N_1}^{t} S_\alpha(t-s)Z(s)\mathrm{d}s\right\| \\ &\le CM\|Z\|_\infty\int_{-\infty}^{N_1}\frac{1}{1+|\omega|(t-s)^\alpha}\mathrm{d}s + \varepsilon CM\int_{N_1}^{t}\frac{1}{1+|\omega|(t-s)^\alpha}\mathrm{d}s \\ &\le \frac{CM\|Z\|_\infty}{|\omega|}\int_{-\infty}^{N_1}\frac{1}{(t-s)^\alpha}\mathrm{d}s + \frac{CM\omega^{-\frac{1}{\alpha}}\pi\varepsilon}{\alpha\sin\frac{\pi}{\alpha}} \\ &\le \frac{CM\|Z\|_\infty}{|\omega|}\frac{1}{(\alpha-1)(t-N_1)^{\alpha-1}} + \frac{CM\omega^{-\frac{1}{\alpha}}\pi\varepsilon}{\alpha\sin\frac{\pi}{\alpha}},\end{aligned}$$

which implies

$$\lim_{t\to+\infty}\|\Phi_2(t)\| = 0.$$

On the other hand, from $Z(t) \in C_0(\mathbb{R}, X)$ it follows that there exists an $N_2 > 0$ such that $\|Z(t)\| < \varepsilon$ for all $t < -N_2$. This enables us to conclude that, for all $t < -N_2$,

$$\begin{aligned}\|\Phi_2(t)\| &= \left\| \int_{-\infty}^{t} S_\alpha(t-s)Z(s)ds \right\| \le \int_{-\infty}^{t} \|S_\alpha(t-s)\| \|Z(s)\| ds \\ &\le CM\varepsilon \int_{-\infty}^{t} \frac{1}{1+|\omega|(t-s)^\alpha} ds = \frac{CM\omega^{-\frac{1}{\alpha}}\pi\varepsilon}{\alpha \sin\frac{\pi}{\alpha}},\end{aligned}$$

which implies

$$\lim_{t\to-\infty} \|\Phi_2(t)\| = 0.$$

□

Now we are in position to state and prove our first main result. But first, let us introduce the following assumptions:

(H_1) $F(t,x,y) = F_1(t,x,y) + F_2(t,x,y) \in AAA(\mathbb{R}\times X\times X, X)$ with

$$F_1(t,x,y) \in AA(\mathbb{R}\times X\times X, X), \quad F_2(t,x,y) \in C_0(\mathbb{R}\times X\times X, X)$$

and there exists a constant $L > 0$ such that, for all $t \in \mathbb{R}$ and $x_1, x_2, y_1, y_2 \in X$,

$$\|F_1(t,x_1,x_2) - F_1(t,y_1,y_2)\| \le L(\|x_1 - y_1\| + \|x_2 - y_2\|); \tag{4.238}$$

(H_2) There exist a function $\beta(t) \in C_0(\mathbb{R},\mathbb{R}^+)$ and a nondecreasing function $\Phi : \mathbb{R}^+ \to \mathbb{R}^+$ such that, for all $t \in \mathbb{R}$ and $x, y \in X$ with $\|x\| + \|y\| \le r$,

$$\|F_2(t,x,y)\| \le \beta(t)\Phi(r) \quad \text{and} \quad \liminf_{r\to+\infty} \frac{\Phi(r)}{r} = \rho_1. \tag{4.239}$$

Remark 4.6.19. Assuming that $F(t,x,y)$ satisfies assumption (H_1), it is noted that $F(t,x,y)$ does not have to possess Lipschitz continuity with respect to x and y. Such asymptotically almost automorphic functions $F(t,x,y)$ are more complicated than those with Lipschitz continuity with respect to x and y and little is known about them.

Let $\beta(t)$ be the function involved in assumption (H_2). Define

$$\sigma(t) := \int_{-\infty}^{t} \frac{\beta(s)}{1+|\omega|(t-s)^\alpha} ds, \quad t \in \mathbb{R}.$$

Lemma 4.6.20. $\sigma(t) \in C_0(\mathbb{R},\mathbb{R}^+)$.

Proof. Since $\beta(t) \in C_0(\mathbb{R},\mathbb{R}^+)$, one can choose a $T_1 > 0$ such that $\|\beta(t)\| < \varepsilon$ for all $t > T_1$. This enables us to conclude that, for all $t > T_1$,

$$\|\sigma(t)\| \le \left\| \int_{-\infty}^{T_1} \frac{\beta(s)}{1+|\omega|(t-s)^\alpha} \mathrm{d}s \right\| + \left\| \int_{T_1}^{t} \frac{\beta(s)}{1+|\omega|(t-s)^\alpha} \mathrm{d}s \right\|$$
$$\le \|\beta\|_\infty \int_{-\infty}^{T_1} \frac{1}{1+|\omega|(t-s)^\alpha} \mathrm{d}s + \varepsilon \int_{T_1}^{t} \frac{1}{1+|\omega|(t-s)^\alpha} \mathrm{d}s$$
$$\le \frac{\|\beta\|_\infty}{|\omega|} \int_{-\infty}^{T_1} \frac{1}{(t-s)^\alpha} \mathrm{d}s + \frac{\omega^{-\frac{1}{\alpha}} \pi \varepsilon}{\alpha \sin \frac{\pi}{\alpha}}$$
$$\le \frac{\|\beta\|_\infty}{|\omega|} \frac{1}{(\alpha-1)(t-T_1)^{\alpha-1}} + \frac{\omega^{-\frac{1}{\alpha}} \pi \varepsilon}{\alpha \sin \frac{\pi}{\alpha}},$$

which implies

$$\lim_{t \to +\infty} \|\sigma(t)\| = 0.$$

On the other hand, from $\beta(t) \in C_0(\mathbb{R}, \mathbb{R}^+)$ it follows that there exists a $T_2 > 0$ such that $\|\beta(t)\| < \varepsilon$ for all $t < -T_2$. This enables us to conclude that, for all $t < -T_2$,

$$\|\sigma(t)\| = \left\| \int_{-\infty}^{t} \frac{\beta(s)}{1+|\omega|(t-s)^\alpha} \mathrm{d}s \right\| \le \varepsilon \int_{-\infty}^{t} \frac{1}{1+|\omega|(t-s)^\alpha} \mathrm{d}s = \frac{\omega^{-\frac{1}{\alpha}} \pi \varepsilon}{\alpha \sin \frac{\pi}{\alpha}},$$

which implies

$$\lim_{t \to -\infty} \|\sigma(t)\| = 0.$$ □

Theorem 4.6.21. *Assume that A is sectorial of type $\omega < 0$. Let $F : \mathbb{R} \times X \times X \to X$ satisfy the hypotheses (H_1) and (H_2). Put $\rho_2 := \sup_{t \in \mathbb{R}} \sigma(t)$. Then equation (4.237) has at least one asymptotically almost automorphic mild solution provided that*

$$\frac{CML(1+\|B\|_{\mathbb{L}(X)})\omega^{-\frac{1}{\alpha}}\pi}{\alpha \sin \frac{\pi}{\alpha}} + CM(1+\|B\|_{\mathbb{L}(X)})\rho_1\rho_2 < 1. \tag{4.240}$$

Proof. The proof is divided into the following five steps:

Step 1. Define a mapping Λ on $AA(\mathbb{R}, X)$ by

$$(\Lambda v)(t) = \int_{-\infty}^{t} S_\alpha(t-s) F_1(s, v(s), Bv(s)) \mathrm{d}s, \quad t \in \mathbb{R}, \tag{4.241}$$

and prove Λ has a unique fixed point $v(t) \in AA(\mathbb{R}, X)$.

First, since the function $s \to F_1(s, v(s), Bv(s))$ is bounded in $\mathbb{R}$ and

$$
\begin{aligned}
\|[\Lambda v](t)\| &\le \int_{-\infty}^{t} \|S_\alpha(t-s)\| \|F_1(s, v(s), Bv(s))\| ds \\
&\le CM \int_{-\infty}^{t} \frac{1}{1 + |\omega|(t-s)^\alpha} \|F_1(s, v(s), Bv(s))\| ds \\
&\le CM \|F_1\|_\infty \int_{-\infty}^{t} \frac{1}{1 + |\omega|(t-s)^\alpha} ds = \frac{CML\omega^{-\frac{1}{\alpha}} \pi \|F_1\|_\infty}{\alpha \sin \frac{\pi}{\alpha}},
\end{aligned}
$$

we obtain that $(\Lambda v)(t)$ exists. Moreover, from $F_1(t, x, y) \in AA(\mathbb{R} \times X \times X, X)$ satisfying (4.238), together with Lemma 4.6.3 and Remark 4.6.4, it follows that

$$
F_1(\cdot, v(\cdot), Bv(\cdot)) \in AA(\mathbb{R}, X) \quad \text{for every } v(\cdot) \in AA(\mathbb{R}, X).
$$

This, together with Lemma 4.6.18, implies that Λ is well defined and maps $AA(\mathbb{R}, X)$ into itself.

In the sequel, we verify Λ is continuous.

Let $v_n(t)$, $v(t)$ be in $AA(\mathbb{R}, X)$ with $v_n(t) \to v(t)$ as $n \to \infty$, then one has

$$
\begin{aligned}
\|[\Lambda v_n](t) - [\Lambda v](t)\| &= \left\| \int_{-\infty}^{t} S_\alpha(t-s)[F_1(s, v_n(s), Bv_n(s)) - F_1(s, v(s), Bv(s))] ds \right\| \\
&\le L \int_{-\infty}^{t} \|S_\alpha(t-s)\| [\|v_n(s) - v(s)\| + \|Bv_n(s) - Bv(s)\|] ds \\
&\le CML \int_{-\infty}^{t} \frac{1}{1 + |\omega|(t-s)^\alpha} (1 + \|B\|_{\mathbb{L}(X)}) \|v_n(s) - v(s)\| ds \\
&\le CML(1 + \|B\|_{\mathbb{L}(X)}) \|v_n - v\|_\infty \int_{-\infty}^{t} \frac{1}{1 + |\omega|(t-s)^\alpha} ds \\
&= \frac{CML(1 + \|B\|_{\mathbb{L}(X)}) \omega^{-\frac{1}{\alpha}} \pi}{\alpha \sin \frac{\pi}{\alpha}} \|v_n - v\|_\infty.
\end{aligned}
$$

Therefore, as $n \to \infty$, $\Lambda v_n \to \Lambda v$, hence Λ is continuous.

Next, we prove that Λ is a contraction on $AA(\mathbb{R}, X)$ and has a unique fixed point $v(t) \in AA(\mathbb{R}, X)$.

In fact, letting $v_1(t)$, $v_2(t)$ be in $AA(\mathbb{R}, X)$, similarly as in the above proof of the continuity of Λ one has

$$\|[\Lambda v_1](t) - [\Lambda v_2](t)\| \le \frac{CML(1 + \|B\|_{\mathbb{L}(X)})\omega^{-\frac{1}{\alpha}}\pi}{\alpha \sin\frac{\pi}{\alpha}}\|v_1 - v_2\|_\infty,$$

which implies

$$\|[\Lambda v_1](t) - [\Lambda v_2](t)\|_\infty \le \frac{CML(1 + \|B\|_{\mathbb{L}(X)})\omega^{-\frac{1}{\alpha}}\pi}{\alpha \sin\frac{\pi}{\alpha}}\|v_1 - v_2\|_\infty.$$

Together with (4.240), this proves that Λ is a contraction on $AA(\mathbb{R}, X)$. Thus, the Banach's fixed point theorem implies that Λ has a unique fixed point $v(t) \in AA(\mathbb{R}, X)$.

Step 2. Set

$$\Omega_r := \{\omega(t) \in C_0(\mathbb{R}, X) : \|\omega\|_\infty \le r\}.$$

For the above $v(t)$, define $\Gamma := \Gamma^1 + \Gamma^2$ on $C_0(\mathbb{R}, X)$ as

$$\begin{aligned}(\Gamma^1\omega)(t) &= \int_{-\infty}^{t} S_\alpha(t-s)[F_1(s, v(s) + \omega(s), B(v(s) + \omega(s))) - F_1(s, v(s), Bv(s))]ds,\\ (\Gamma^2\omega)(t) &= \int_{-\infty}^{t} S_\alpha(t-s)F_2(s, v(s) + \omega(s), B(v(s) + \omega(s)))ds,\end{aligned} \tag{4.242}$$

and prove that Γ maps Ω_{k_0} into itself, where k_0 is a given constant.

First, from (4.238) it follows that, for all $s \in \mathbb{R}$, $\omega(s) \in X$,

$$\begin{aligned}\|F_1(s, v(s) + \omega(s), B(v(s) + \omega(s))) - F_1(s, v(s), Bv(s))\| &\le L[\|\omega(s)\| + \|B\omega(s)\|]\\ &\le L(1 + \|B\|_{\mathbb{L}(X)})\|\omega(s)\|,\end{aligned}$$

which implies that

$$F_1(\cdot, v(\cdot) + \omega(\cdot), B(v(\cdot) + \omega(\cdot))) - F_1(\cdot, v(\cdot), Bv(\cdot)) \in C_0(\mathbb{R}, X) \quad \text{for every } \omega(\cdot) \in C_0(\mathbb{R}, X).$$

According to (4.239), one has

$$\begin{aligned}\|F_2(s, v(s) + \omega(s), B(v(s) + \omega(s)))\| &\le \beta(s)\Phi\Big(\|\omega(s) + B\omega(s)\| + \sup_{s\in\mathbb{R}}\|v(s) + Bv(s)\|\Big)\\ &\le \beta(s)\Phi\Big((1 + \|B\|_{\mathbb{L}(X)})\|\omega(s)\| + (1 + \|B\|_{\mathbb{L}(X)})\sup_{s\in\mathbb{R}}\|v(s)\|\Big)\\ &= \beta(s)\Phi\Big((1 + \|B\|_{\mathbb{L}(X)})\Big[\|\omega(s)\| + \sup_{s\in\mathbb{R}}\|v(s)\|\Big]\Big)\end{aligned}$$

for all $s \in \mathbb{R}$ and $\omega(s) \in X$ with $\|\omega(s)\| \le r$, then

$$F_2(\cdot, v(\cdot) + \omega(\cdot), B(v(\cdot) + \omega(\cdot))) \in C_0(\mathbb{R}, X) \quad \text{as } \beta(\cdot) \in C_0(\mathbb{R}, \mathbb{R}^+).$$

This, together with Lemma 4.6.18, yields that Γ is well defined and maps $C_0(\mathbb{R}, X)$ into itself.

On the other hand, in view of (4.239) and (4.240), it is not difficult to see that there exists a constant $k_0 > 0$ such that

$$\frac{CML(1+\|B\|_{\mathbb{L}(X)})\omega^{-\frac{1}{\alpha}}\pi}{\alpha \sin\frac{\pi}{\alpha}} k_0 + CM\rho_2\Phi\Big((1+\|B\|_{\mathbb{L}(X)})\Big(k_0 + \sup_{s\in\mathbb{R}}\|v(s)\|\Big)\Big) \leq k_0.$$

This enables us to conclude that, for any $t \in \mathbb{R}$ and $\omega_1(t), \omega_2(t) \in \Omega_{k_0}$,

$$\begin{aligned}
&\|(\Gamma^1\omega_1)(t) + (\Gamma^2\omega_2)(t)\| \\
&\leq \left\| \int_{-\infty}^{t} S_\alpha(t-s)[F_1(s, v(s)+\omega_1(s), B(v(s)+\omega_1(s))) - F_1(s, v(s), Bv(s))]ds \right\| \\
&\quad + \left\| \int_{-\infty}^{t} S_\alpha(t-s)F_2(s, v(s)+\omega_2(s), B(v(s)+\omega_2(s)))ds \right\| \\
&\leq \int_{-\infty}^{t} \|S_\alpha(t-s)\| \|F_1(s, v(s)+\omega_1(s), B(v(s)+\omega_1(s))) - F_1(s, v(s), Bv(s))\| ds \\
&\quad + \int_{-\infty}^{t} \|S_\alpha(t-s)\| \|F_2(s, v(s)+\omega_2(s), B(v(s)+\omega_2(s)))\| ds \\
&\leq CM \int_{-\infty}^{t} \frac{1}{1+|\omega|(t-s)^\alpha}[\|\omega_1(s)\| + \|B\omega_1(s)\|]ds \\
&\quad + CM \int_{-\infty}^{t} \frac{\beta(s)}{1+|\omega|(t-s)^\alpha}\Phi(\|\omega_2(s)\| + \|B\omega_2(s)\| + \|v(s)\| + \|Bv(s)\|)ds \\
&\leq CML(1+\|B\|_{\mathbb{L}(X)})\|\omega_1\|_\infty \int_{-\infty}^{t} \frac{1}{1+|\omega|(t-s)^\alpha} ds \\
&\quad + CM\sigma(t)\Phi((1+\|B\|_{\mathbb{L}(X)})(\|\omega_2\|_\infty + \|v(s)\|_\infty)) \\
&= \frac{CML\omega^{-\frac{1}{\alpha}}\pi(1+\|B\|_{\mathbb{L}(X)})}{\alpha\sin\frac{\pi}{\alpha}}\|\omega\|_\infty + CM\rho_2\Phi((1+\|B\|_{\mathbb{L}(X)})(\|\omega_2\|_\infty + \|v(s)\|_\infty)) \\
&\leq \frac{CML(1+\|B\|_{\mathbb{L}(X)})\omega^{-\frac{1}{\alpha}}\pi}{\alpha\sin\frac{\pi}{\alpha}} k_0 + CM\rho_2\Phi((1+\|B\|_{\mathbb{L}(X)})(k_0 + \|v(s)\|_\infty)) \leq k_0,
\end{aligned}$$

which implies that $(\Gamma^1\omega_1)(t) + (\Gamma^2\omega_2)(t) \in \Omega_{k_0}$. Thus Γ maps Ω_{k_0} into itself.

Step 3. Show that Γ^1 is a contraction on Ω_{k_0}.

In fact, for any $\omega_1(t), \omega_2(t) \in \Omega_{k_0}$ and $t \in \mathbb{R}$, from (4.238) it follows that

$$\begin{aligned}
&\|[F_1(s, v(s) + \omega_1(s), B(v(s) + \omega_1(s))) - F_1(s, v(s), Bv(s))] \\
&\quad - [F_1(s, v(s) + \omega_2(s), B(v(s) + \omega_2(s))) - F_1(s, v(s), Bv(s))]\| \\
&\le L[\|\omega_1(s) - \omega_2(s)\| + \|B\omega_1(s) - B\omega_2(s)\|] \\
&\le L(1 + \|B\|_{\mathbb{L}(X)})\|\omega_1(s) - \omega_2(s)\|.
\end{aligned}$$

Thus

$$\begin{aligned}
&\|(\Gamma^1\omega_1)(t) - (\Gamma^1\omega_2)(t)\| \\
&= \left\| \int_{-\infty}^{t} S_\alpha(t-s)[(F_1(s, v(s) + \omega_1(s), B(v(s) + \omega_1(s))) - F_1(s, v(s), Bv(s))) \right. \\
&\quad \left. - (F_1(s, v(s) + \omega_2(s), B(v(s) + \omega_2(s))) - F_1(s, v(s), Bv(s)))]\mathrm{d}s \right\| \\
&\le L \int_{-\infty}^{t} \|S_\alpha(t-s)\|(1 + \|B\|_{\mathbb{L}(X)})\|\omega_1(s) - \omega_2(s)\|\mathrm{d}s \\
&\le CML(1 + \|B\|_{\mathbb{L}(X)})\|\omega_1 - \omega_2\|_\infty \int_{-\infty}^{t} \frac{1}{1 + |\omega|(t-s)^\alpha}\mathrm{d}s \\
&= \frac{CML(1 + \|B\|_{\mathbb{L}(X)})\omega^{-\frac{1}{\alpha}}\pi}{\alpha \sin\frac{\pi}{\alpha}}\|\omega_1 - \omega_2\|_\infty,
\end{aligned}$$

which implies that

$$\|(\Gamma^1\omega_1)(t) - (\Gamma^1\omega_2)(t)\|_\infty \le \frac{CML(1 + \|B\|_{\mathbb{L}(X)})\omega^{-\frac{1}{\alpha}}\pi}{\alpha \sin\frac{\pi}{\alpha}}\|\omega_1 - \omega_2\|_\infty.$$

Thus, in view of (4.240), one obtains the conclusion.

Step 4. Show that Γ^2 is completely continuous on Ω_{k_0}.

Given $\varepsilon > 0$. Let $\{\omega_k\}_{k=1}^{+\infty} \subset \Omega_{k_0}$ with $\omega_k \to \omega_0$ in $C_0(\mathbb{R}, X)$ as $k \to +\infty$. Since $\sigma(t) \in C_0(\mathbb{R}, \mathbb{R}^+)$, one may choose a $t_1 > 0$ big enough such, that for all $t \ge t_1$,

$$\Phi((1 + \|B\|_{\mathbb{L}(X)})(k_0 + \|v\|_\infty))\sigma(t) < \frac{\varepsilon}{3CM}.$$

Also, in view of (H_1), we have

$$F_2(s, v(s) + \omega_k(s), B(v(s) + \omega_k(s))) \to F_2(s, v(s) + \omega_0(s), B(v(s) + \omega_0(s)))$$

for all $s \in (-\infty, t_1]$ as $k \to +\infty$, and

$$\|F_2(\cdot, v(\cdot) + \omega_k(\cdot), B(v(\cdot) + \omega_k(\cdot))) - F_2(\cdot, v(\cdot) + \omega_0(\cdot), B(v(\cdot) + \omega_0(\cdot)))\|$$
$$\le 2\Phi((1 + \|B\|_{\mathbb{L}(X)})(k_0 + \|v\|_\infty))\beta(\cdot) \in L^1(-\infty, t_1].$$

Hence, using the Lebesgue dominated convergence theorem, we deduce that there exists an $N > 0$ such that

$$CM \int_{-\infty}^{t_1} \frac{1}{1 + |\omega|(t-s)^\alpha} \|F_2(s, v(s) + \omega_k(s), B(v(s) + \omega_k(s)))$$
$$- F_2(s, v(s) + \omega_0(s), B(v(s) + \omega_0(s)))\| ds \le \frac{\varepsilon}{3}$$

whenever $k \ge N$. Thus

$$\begin{aligned}
\|(\Gamma^2\omega_k)(t) - (\Gamma^2\omega_0)(t)\| &= \Bigg\| \int_{-\infty}^{t} S_\alpha(t-s) F_2(s, v(s) + \omega_k(s), B(v(s) + \omega_k(s))) ds \\
&\quad - \int_{-\infty}^{t} S_\alpha(t-s) F_2(s, v(s) + \omega_0(s), B(v(s) + \omega_0(s))) ds \Bigg\| \\
&\le CM \int_{-\infty}^{t_1} \frac{1}{1 + |\omega|(t-s)^\alpha} \|F_2(s, v(s) + \omega_k(s), B(v(s) + \omega_k(s))) \\
&\quad - F_2(s, v(s) + \omega_0(s), B(v(s) + \omega_0(s)))\| ds \\
&\quad + 2CM\Phi((1 + \|B\|_{\mathbb{L}(X)})(k_0 + \|v\|_\infty)) \int_{t_1}^{\max\{t, t_1\}} \frac{\beta(s)}{1 + |\omega|(t-s)^\alpha} ds \\
&\le CM \int_{-\infty}^{t_1} \frac{1}{1 + |\omega|(t-s)^\alpha} \|F_2(s, v(s) + \omega_k(s), B(v(s) + \omega_k(s))) \\
&\quad - F_2(s, v(s) + \omega_0(s), B(v(s) + \omega_0(s)))\| ds \\
&\quad + 2CM\Phi((1 + \|B\|_{\mathbb{L}(X)})(k_0 + \|v\|_\infty))\sigma(t) \\
&\le \frac{\varepsilon}{3} + \frac{2\varepsilon}{3} = \varepsilon
\end{aligned}$$

whenever $k \ge N$. Accordingly, Γ^2 is continuous on Ω_{k_0}.

In the sequel, we consider the compactness of Γ^2.

Set $B_r(X)$ for the closed ball with center at 0 and radius r in X, $V = \Gamma^2(\Omega_{k_0})$, and $z(t) = \Gamma^2(u(t))$ for $u(t) \in \Omega_{k_0}$. First, for all $\omega(t) \in \Omega_{k_0}$ and $t \in \mathbb{R}$,

$$\begin{aligned}
\|(\Gamma^2\omega)(t)\| &= \Bigg\| \int_{-\infty}^{t} S_\alpha(t-s) F_2(s, v(s) + \omega(s), B(v(s) + \omega(s))) ds \Bigg\| \\
&\le CM\sigma(t)\Phi((1 + \|B\|_{\mathbb{L}(X)})(k_0 + \|v\|_\infty)),
\end{aligned}$$

in view of $\sigma(t) \in C_0(\mathbb{R}, \mathbb{R}^+)$ which follows from Lemma 4.6.20, one concludes that

$$\lim_{|t|\to+\infty} (\Gamma^2\omega)(t) = 0 \quad \text{uniformly for } \omega(t) \in \Omega_{k_0}.$$

As

$$\begin{aligned}\|(\Gamma^2\omega)(t)\| &= \left\| \int_{-\infty}^{t} S_\alpha(t-s)F_2(s, v(s)+\omega(s), B(v(s)+\omega(s)))\mathrm{d}s \right\| \\ &= \left\| \int_{0}^{+\infty} S_\alpha(\tau)F_2\big(t-\tau, v(t-\tau)+\omega(t-\tau), B\big(v(t-\tau)+\omega(t-\tau)\big)\big)\mathrm{d}\tau \right\|,\end{aligned}$$

for a given $\varepsilon_0 > 0$, one can choose a $\xi > 0$ such that

$$\left\| \int_{\xi}^{+\infty} S_\alpha(\tau)F_2\big(t-\tau, v(t-\tau)+\omega(t-\tau), B\big(v(t-\tau)+\omega(t-\tau)\big)\big)\mathrm{d}\tau \right\| < \varepsilon_0.$$

Thus we get

$$\begin{aligned}z(t) \in\; & \overline{\xi c(\{S_\alpha(\tau)F_2(\lambda, v(\lambda)+\omega(\lambda), B(v(\lambda)+\omega(\lambda))) : 0 \le \tau \le \xi, t-\xi \le \lambda \le \xi, \|\omega\|_\infty \le r\})} \\ & + B_{\varepsilon_0}(X),\end{aligned}$$

where $c(K)$ denotes the convex hull of K. Using that $S_\alpha(\cdot)$ is strongly continuous, we infer that

$$K = \{S_\alpha(\tau)F_2(\lambda, v(\lambda)+\omega(\lambda), B(v(\lambda)+\omega(\lambda))) : 0 \le \tau \le \xi, t-\xi \le \lambda \le \xi, \|\omega\|_\infty \le r\}$$

is a relatively compact set, and $V \subset \overline{\xi c(K)} + B_{\varepsilon_0}(X)$, which implies that V is a relatively compact subset of X.

Next, we verify the equicontinuity of the set $\{(\Gamma^2\omega)(t) : \omega(t) \in \Omega_{k_0}\}$.

Let $k > 0$ be small enough and $t_1, t_2 \in \mathbb{R}$, $\omega(t) \in \Omega_{k_0}$. Then by (4.239), we have

$$\begin{aligned}&\|(\Gamma^2\omega)(t_2) - (\Gamma^2\omega)(t_1)\| \\ &\quad \le \int_{t_1}^{t_2} \|S_\alpha(t_2-s)F_2(s, v(s)+\omega(s), B(v(s)+\omega(s)))\|\mathrm{d}s \\ &\quad + \int_{-\infty}^{t_1-k} \|[S_\alpha(t_2-s)-S_\alpha(t_1-s)]F_2(s, v(s)+\omega(s), B(v(s)+\omega(s)))\|\mathrm{d}s \\ &\quad + \int_{t_1-k}^{t_1} \|[S_\alpha(t_2-s)-S_\alpha(t_1-s)]F_2(s, v(s)+\omega(s), B(v(s)+\omega(s)))\|\mathrm{d}s\end{aligned}$$

$$\leq CM\Phi((1+\|B\|_{\mathbb{L}(X)})(k_0+\|v\|_\infty))\int_{t_1}^{t_2}\frac{\beta(s)}{1+|\omega|(t_2-s)^\alpha}ds$$

$$+\Phi((1+\|B\|_{\mathbb{L}(X)})(k_0+\|v\|_\infty))\sup_{s\in[-\infty,t_1-k]}\|S_\alpha(t_2-s)-S_\alpha(t_1-s)\|\int_{-\infty}^{t_1-k}\beta(s)ds$$

$$+CM\Phi((1+\|B\|_{\mathbb{L}(X)})(k_0+\|v\|_\infty))\int_{t_1-k}^{t_1}\left(\frac{\beta(s)}{1+|\omega|(t_2-s)^\alpha}+\frac{\beta(s)}{1+|\omega|(t_1-s)^\alpha}\right)ds$$

$$\to 0 \quad \text{as } t_2-t_1\to 0, k\to 0,$$

which implies the equicontinuity of the set $\{(\Gamma^2\omega)(t) : \omega(t)\in\Omega_{k_0}\}$.

Now an application of Lemma 4.6.15 justifies the compactness of Γ^2.

Step 5. Show that equation (4.237) has at least one asymptotically almost automorphic mild solution.

First, the complete continuity of Γ^2, together with the results of Steps 2 and 3, as well as Lemma 4.6.16, yields that Γ has at least one fixed point $\omega(t)\in\Omega_{k_0}$. Furthermore, $\omega(t)\in C_0(\mathbb{R},X)$.

Then, consider the following coupled system of integral equations

$$\begin{cases} v(t)=\int_{-\infty}^t S_\alpha(t-s)F_1(s,v(s),Bv(s))ds, \quad t\in\mathbb{R},\\ \omega(t)=\int_{-\infty}^t S_\alpha(t-s)[F_1(s,v(s)+\omega(s),B(v(s)+\omega(s)))-F_1(s,v(s),Bv(s))]ds\\ \qquad +\int_{-\infty}^t S_\alpha(t-s)F_2(s,v(s)+\omega(s),B(v(s)+\omega(s)))ds, \quad t\in\mathbb{R}.\end{cases} \tag{4.243}$$

From the result of Step 1, together with the above fixed point $\omega(t)\in C_0(\mathbb{R},X)$, it follows that

$$(v(t),\omega(t))\in AA(\mathbb{R},X)\times C_0(\mathbb{R},X)$$

is a solution to system (4.243). Thus

$$x(t):=v(t)+\omega(t)\in AAA(\mathbb{R},X)$$

and it is a solution to the integral equation

$$x(t)=\int_{-\infty}^t S_\alpha(t-s)F(s,x(s),Bx(s))ds, \quad t\in\mathbb{R},$$

that is, $x(t)$ is an asymptotically almost automorphic mild solution to equation (4.237). □

Taking $A=-\rho^\alpha I$ with $\rho>0$ in equation (4.237), the above theorem gives the following corollary.

Corollary 4.6.22. *Let $F : \mathbb{R} \times X \times X \to X$ satisfy (H_1) and (H_2). Put $\rho_2 := \sup_{t\in\mathbb{R}} \sigma(t)$. Then equation* (4.237) *admits at least one asymptotically almost automorphic mild solution whenever*

$$\frac{CL(1 + \|B\|_{\mathbb{L}(X)})\rho\pi}{\alpha \sin\frac{\pi}{\alpha}} + C(1 + \|B\|_{\mathbb{L}(X)})\rho_1\rho_2 < 1.$$

Remark 4.6.23. It is interesting to note that the function $\alpha \to \frac{\alpha\sin(\frac{\pi}{\alpha})}{\rho\pi}$ is increasing from 0 to $\frac{2}{\rho\pi}$ in the interval $1 < \alpha < 2$. Therefore, with respect to the condition (4.240), the class of admissible terms $F_1(t, x(t), Bx(t))$ is the best in the case $\alpha = 2$ and the worst in the case $\alpha = 1$.

Theorem 4.6.21 can be extended to the case of $F_1(t, x, y)$ being locally Lipschitz continuous with respect to x and y, and we assume:
(H_1') $F(t, x, y) = F_1(t, x, y) + F_2(t, x, y) \in \mathrm{AAA}(\mathbb{R} \times X \times X, X)$ with

$$F_1(t, x, y) \in \mathrm{AA}(\mathbb{R} \times X \times X, X), \quad F_2(t, x, y) \in C_0(\mathbb{R} \times X \times X, X)$$

and for all $x_1, x_2, y_1, y_2 \in X$, $t \in \mathbb{R}$,

$$\|F_1(t, x_1, x_2) - F_1(t, y_1, y_2)\| \le L(t)(\|x_1 - y_1\| + \|x_2 - y_2\|), \tag{4.244}$$

where $L(t)$ is a function on $\mathbb{R}$.

Then we have the following result:

Theorem 4.6.24. *Assume that A is sectorial of type $\omega < 0$. Let $F : \mathbb{R} \times X \times X \to X$ satisfy the hypotheses (H_1') and $H_2)$ with $L(t) \in \mathrm{BC}(\mathbb{R}, \mathbb{R}^+)$. Put $\rho_2 := \sup_{t\in\mathbb{R}} \sigma(t)$. Let $\|L\| = \sup_{t\in\mathbb{R}} \int_t^{t+1} L(s)ds$. Then equation* (4.237) *has at least one asymptotically almost automorphic mild solution provided that*

$$\frac{CM\|L\|\omega^{-\frac{1}{\alpha}}\pi(1 + \|B\|_{\mathbb{L}(X)})}{\alpha \sin\frac{\pi}{\alpha}} + CM\rho_1\rho_2(1 + \|B\|_{\mathbb{L}(X)}) < 1. \tag{4.245}$$

Proof. The proof is divided into the following five steps:

Step 1. Define a mapping Λ on $\mathrm{AA}(\mathbb{R}, X)$ by (4.241), and prove Λ has a unique fixed point $v(t) \in \mathrm{AA}(\mathbb{R}, X)$.

First, similar to the proof in Step 1 of Theorem 4.6.21, we can prove that $(\Lambda v)(t)$ exists. Moreover, from $F_1(t, x, y) \in \mathrm{AA}(\mathbb{R}\times X\times X, X)$ satisfying (4.244), together with Lemma 4.6.3 and Remark 4.6.5, it follows that

$$F_1(\cdot, v(\cdot), Bv(\cdot)) \in \mathrm{AA}(\mathbb{R}, X) \quad \text{for every } v(\cdot) \in \mathrm{AA}(\mathbb{R}, X).$$

This, together with Lemma 4.6.18, implies that Λ is well defined and maps $AP(\mathbb{R},X)$ into itself.

In the sequel, we verify Λ is continuous.

Let $v_n(t)$, $v(t)$ be in $AA(\mathbb{R},X)$ with $v_n(t) \to v(t)$ as $n \to \infty$, then one has

$$\begin{aligned}
\|[\Lambda v_n](t) - [\Lambda v](t)\| &= \left\| \int_{-\infty}^{t} S_\alpha(t-s)[F_1(s,v_n(s),Bv_n(s)) - F_1(s,v(s),Bv(s))]ds \right\| \\
&\le \int_{-\infty}^{t} L(s)\|S_\alpha(t-s)\|[\|v_n(s)-v(s)\| + \|Bv_n(s) - Bv(s)\|]ds \\
&\le CM \int_{-\infty}^{t} \frac{L(s)}{1+|\omega|(t-s)^\alpha}(1+\|B\|_{\mathbb{L}(X)})\|v_n(s)-v(s)\|ds \\
&\le CM(1+\|B\|_{\mathbb{L}(X)})\left(\sum_{m=0}^{+\infty} \int_{t-(m+1)}^{t-m} \frac{L(s)}{1+|\omega|(t-s)^\alpha}ds\right)\|v_n - v\|_\infty \\
&\le CM(1+\|B\|_{\mathbb{L}(X)})\left(\sum_{m=0}^{+\infty} \frac{1}{1+|\omega|m^\alpha} \int_{t-(m+1)}^{t-m} L(s)ds\right)\|v_n - v\|_\infty \\
&\le \frac{CM\|L\|\omega^{-\frac{1}{\alpha}}\pi(1+\|B\|_{\mathbb{L}(X)})}{\alpha \sin\frac{\pi}{\alpha}}\|v_n - v\|_\infty.
\end{aligned}$$

Therefore, as $n \to \infty$, $\Lambda v_n \to \Lambda v$, hence Λ is continuous.

Next, we prove that Λ is a contraction on $AA(\mathbb{R},X)$ and has a unique fixed point $v(t) \in AA(\mathbb{R},X)$.

In fact, for $v_1(t)$, $v_2(t)$ in $AA(\mathbb{R},X)$, similar to the above proof of the continuity of Λ, one has

$$\|(\Lambda v_1)(t) - (\Lambda v_2)(t)\| \le \frac{CM\|L\|\omega^{-\frac{1}{\alpha}}\pi(1+\|B\|_{\mathbb{L}(X)})}{\alpha \sin\frac{\pi}{\alpha}}\|v_1 - v_2\|_\infty,$$

which implies that

$$\|(\Lambda v_1)(t) - (\Lambda v_2)(t)\|_\infty \le \frac{CM\|L\|\omega^{-\frac{1}{\alpha}}\pi(1+\|B\|_{\mathbb{L}(X)})}{\alpha \sin\frac{\pi}{\alpha}}\|v_1 - v_2\|_\infty.$$

Hence, by (4.245), together with the contraction principle, Λ has a unique fixed point $v(t) \in AA(\mathbb{R},X)$.

Step 2. Set

$$\Omega_r := \{\omega(t) \in C_0(\mathbb{R},X) : \|\omega\|_\infty \le r\}.$$

For the above $v(t)$, define $\Gamma := \Gamma^1 + \Gamma^2$ on $C_0(\mathbb{R}, X)$ as in (4.242), and prove that Γ maps Ω_{k_0} into itself, where k_0 is a given constant.

First, from (4.244) it follows that, for all $s \in \mathbb{R}$, $\omega(s) \in X$,

$$\begin{aligned}\|F_1(s, v(s) + \omega(s), B(v(s) + \omega(s))) - F_1(s, v(s), Bv(s))\| &\le L(s)[\|\omega(s)\| + \|B\omega(s)\|] \\ &\le L(s)(1 + \|B\|_{\mathbb{L}(X)})\|\omega(s)\|,\end{aligned}$$

which, together with $L(s) \in BC(\mathbb{R}, \mathbb{R}^+)$, implies that

$$F_1(\cdot, v(\cdot) + \omega(\cdot), B(v(\cdot) + \omega(\cdot))) - F_1(\cdot, v(\cdot), Bv(\cdot)) \in C_0(\mathbb{R}, X) \quad \text{for every } \omega(\cdot) \in C_0(\mathbb{R}, X).$$

According to (4.239), one has

$$\begin{aligned}\|F_2(s, v(s) + \omega(s), B(v(s) + \omega(s)))\| &\le \beta(s)\Phi\Big(\|\omega(s) + B\omega(s)\| + \sup_{s\in\mathbb{R}}\|v(s) + Bv(s)\|\Big) \\ &\le \beta(s)\Phi\Big((1 + \|B\|_{\mathbb{L}(X)})\|\omega(s)\| + (1 + \|B\|_{\mathbb{L}(X)})\sup_{s\in\mathbb{R}}\|v(s)\|\Big) \\ &\le \beta(s)\Phi\Big((1 + \|B\|_{\mathbb{L}(X)})\Big[\|\omega(s)\| + \sup_{s\in\mathbb{R}}\|v(s)\|\Big]\Big)\end{aligned}$$

for all $s \in \mathbb{R}$ and $\omega(s) \in X$ with $\|\omega(s)\| \le r$, and thus

$$F_2(\cdot, v(\cdot) + \omega(\cdot), B(v(\cdot) + \omega(\cdot))) \in C_0(\mathbb{R}, X) \quad \text{as } \beta(\cdot) \in C_0(\mathbb{R}, \mathbb{R}^+).$$

This, together with Lemma 4.6.18, yields that Γ is well defined and maps $C_0(\mathbb{R}, X)$ into itself.

On the other hand, in view of (4.239) and (4.245), it is not difficult to see that there exists a constant $k_0 > 0$ such that

$$\frac{CM\|L\|\omega^{-\frac{1}{\alpha}}\pi(1 + \|B\|_{\mathbb{L}(X)})}{\alpha \sin\frac{\pi}{\alpha}}k_0 + CM\rho_2\Phi\Big((1 + \|B\|_{\mathbb{L}(X)})\Big(k_0 + \sup_{s\in\mathbb{R}}\|v(s)\|\Big)\Big) \le k_0.$$

This enables us to conclude that, for any $t \in \mathbb{R}$ and $\omega_1(t), \omega_2(t) \in \Omega_{k_0}$,

$$\begin{aligned}&\|(\Gamma^1\omega_1)(t) + (\Gamma^2\omega_2)(t)\| \\ &\quad\le \Big\|\int_{-\infty}^{t} S_\alpha(t-s)[F_1(s, v(s) + \omega_1(s), B(v(s) + \omega_1(s))) - F_1(s, v(s), Bv(s))]ds\Big\| \\ &\qquad + \Big\|\int_{-\infty}^{t} S_\alpha(t-s)F_2(s, v(s) + \omega_2(s), B(v(s) + \omega_2(s)))ds\Big\| \\ &\quad\le \int_{-\infty}^{t} L(s)\|S_\alpha(t-s)\|[\|\omega_1(s)\| + \|B\omega_1(s)\|]ds\end{aligned}$$

$$
\begin{aligned}
&+ CM \int_{-\infty}^{t} \frac{\beta(s)}{1 + |\omega|(t-s)^{\alpha}} \Phi(\|\omega_2(s)\| + \|B\omega_2(s)\| + \|v(s)\| + \|Bv(s)\|) \mathrm{d}s \\
&\le CM \int_{-\infty}^{t} \frac{L(s)}{1 + |\omega|(t-s)^{\alpha}} (1 + \|B\|_{\mathbb{L}(X)}) \|\omega_1(s)\| \mathrm{d}s \\
&+ CM \int_{-\infty}^{t} \frac{\beta(s)}{1 + |\omega|(t-s)^{\alpha}} \Phi((1 + \|B\|_{\mathbb{L}(X)})(\|\omega_2(s)\| + \|v(s)\|)) \mathrm{d}s \\
&\le CM(1 + \|B\|_{\mathbb{L}(X)}) \|\omega_1\|_{\infty} \int_{-\infty}^{t} \frac{L(s)}{1 + |\omega|(t-s)^{\alpha}} \mathrm{d}s \\
&+ CM\sigma(t)\Phi\Big((1 + \|B\|_{\mathbb{L}(X)})\Big(\|\omega_2\|_{\infty} + \sup_{s\in\mathbb{R}} \|v(s)\|\Big)\Big) \\
&\le CM\left(\sum_{m=0}^{+\infty} \int_{t-(m+1)}^{t-m} \frac{L(s)}{1 + |\omega|(t-s)^{\alpha}} \mathrm{d}s\right)(1 + \|B\|_{\mathbb{L}(X)}) \|\omega_1\|_{\infty} \\
&+ CM\rho_2\Phi\Big((1 + \|B\|_{\mathbb{L}(X)})\Big(\|\omega_2\|_{\infty} + \sup_{s\in\mathbb{R}} \|v(s)\|\Big)\Big) \\
&\le CM\left(\sum_{m=0}^{+\infty} \frac{1}{1 + |\omega| m^{\alpha}} \int_{t-(m+1)}^{t-m} L(s) \mathrm{d}s\right)(1 + \|B\|_{\mathbb{L}(X)}) \|\omega_1\|_{\infty} \\
&+ CM\rho_2\Phi\Big((1 + \|B\|_{\mathbb{L}(X)})\Big(k_0 + \sup_{s\in\mathbb{R}} \|v(s)\|\Big)\Big) \\
&\le CM\left(\sum_{m=0}^{+\infty} \frac{1}{1 + |\omega| m^{\alpha}}\right)\|L\|(1 + \|B\|_{\mathbb{L}(X)}) k_0 + CM\rho_2\Phi\Big((1 + \|B\|_{\mathbb{L}(X)})\Big(k_0 + \sup_{s\in\mathbb{R}} \|v(s)\|\Big)\Big) \\
&= \frac{CM\|L\| \omega^{-\frac{1}{\alpha}} \pi (1 + \|B\|_{\mathbb{L}(X)})}{\alpha \sin \frac{\pi}{\alpha}} k_0 + CM\rho_2\Phi\Big((1 + \|B\|_{\mathbb{L}(X)})\Big(k_0 + \sup_{s\in\mathbb{R}} \|v(s)\|\Big)\Big) \le k_0,
\end{aligned}
$$

which implies that $(\Gamma^1\omega_1)(t) + (\Gamma^2\omega_2)(t) \in \Omega_{k_0}$. Thus Γ maps Ω_{k_0} into itself.

Step 3. Show Γ^1 is a contraction on Ω_{k_0}.

In fact, for any $\omega_1(t), \omega_2(t) \in \Omega_{k_0}$ and $t \in \mathbb{R}$, from (4.244) it follows that

$$
\begin{aligned}
&\|[F_1(s, v(s) + \omega_1(s), B(v(s) + \omega_1(s))) - F_1(s, v(s), Bv(s))] \\
&\quad - [F_1(s, v(s) + \omega_2(s), B(v(s) + \omega_2(s))) - F_1(s, v(s), Bv(s))]\| \\
&\quad \le L(s)[\|\omega_1(s) - \omega_2(s)\| + \|B\omega_1(s) - B\omega_2(s)\|] \\
&\quad \le L(s)(1 + \|B\|_{\mathbb{L}(X)}) \|\omega_1(s) - \omega_2(s)\|.
\end{aligned}
$$

Thus

$$
\begin{aligned}
&\|(\Gamma^1\omega_1)(t)-(\Gamma^1\omega_2)(t)\| \\
&= \Bigg\| \int_{-\infty}^{t} S_\alpha(t-s)[(F_1(s,v(s)+\omega_1(s),B(v(s)+\omega_1(s)))-F_1(s,v(s),Bv(s))) \\
&\quad -(F_1(s,v(s)+\omega_2(s),B(v(s)+\omega_2(s)))-F_1(s,v(s),Bv(s)))]\mathrm{d}s\Bigg\| \\
&\le \int_{-\infty}^{t} L(s)\|S_\alpha(t-s)\|(1+\|B\|_{\mathbb{L}(X)})\|\omega_1(s)-\omega_2(s)\|\mathrm{d}s \\
&\le CM\int_{-\infty}^{t} \frac{L(s)}{1+|\omega|(t-s)^\alpha}(1+\|B\|_{\mathbb{L}(X)})\|\omega_1(s)-\omega_2(s)\|\mathrm{d}s \\
&\le CM\left(\sum_{m=0}^{+\infty}\int_{t-(m+1)}^{t-m}\frac{L(s)}{1+|\omega|(t-s)^\alpha}\mathrm{d}s\right)(1+\|B\|_{\mathbb{L}(X)})\|\omega_1-\omega_2\|_\infty \\
&\le CM\left(\sum_{m=0}^{+\infty}\frac{1}{1+|\omega|m^\alpha}\int_{t-(m+1)}^{t-m}L(s)\mathrm{d}s\right)(1+\|B\|_{\mathbb{L}(X)})\|\omega_1-\omega_2\|_\infty \\
&\le CM\left(\sum_{m=0}^{+\infty}\frac{1}{1+|\omega|m^\alpha}\right)\|L\|(1+\|B\|_{\mathbb{L}(X)})\|\omega_1-\omega_2\|_\infty \\
&= \frac{CM\|L\|\omega^{-\frac{1}{\alpha}}\pi(1+\|B\|_{\mathbb{L}(X)})}{\alpha\sin\frac{\pi}{\alpha}}\|\omega_1-\omega_2\|_\infty,
\end{aligned}
$$

which implies that

$$
\|(\Gamma^1\omega_1)(t)-(\Gamma^1\omega_2)(t)\|_\infty \le \frac{CM\|L\|\omega^{-\frac{1}{\alpha}}\pi(1+\|B\|_{\mathbb{L}(X)})}{\alpha\sin\frac{\pi}{\alpha}}\|\omega_1-\omega_2\|_\infty.
$$

Thus, in view of (4.245), one obtains the conclusion.

Step 4. Show that Γ^2 is completely continuous on Ω_{k_0}.

The proof is similar to that in Step 4 of Theorem 4.6.21.

Step 5. Show that equation (4.237) has at least one asymptotically almost automorphic mild solution.

The proof is similar to that in Step 5 of Theorem 4.6.21. □

Taking $A = -\rho^\alpha I$ with $\rho > 0$ in equation (4.237), Theorem 4.6.24 gives the following corollary.

Corollary 4.6.25. *Let $F : \mathbb{R}\times X\times X \to X$ satisfy (H_1') and (H_2) with $L(t) \in \mathrm{BC}(\mathbb{R},\mathbb{R}^+)$. Put $\rho_2 := \sup_{t\in\mathbb{R}}\sigma(t)$. Let $\|L\| = \sup_{t\in\mathbb{R}}\int_t^{t+1}L(s)\mathrm{d}s$. Then equation* (4.237) *admits at least one asymptotically almost automorphic mild solution whenever*

$$\frac{C\|L\|\rho\pi(1+\|B\|_{\mathbb{L}(X)})}{\alpha\sin\frac{\pi}{\alpha}}+C\rho_1\rho_2(1+\|B\|_{\mathbb{L}(X)})<1.$$

Now we consider a more general case of equations introducing a new class of functions $L(t)$. We assume the following condition:

(H_2') There exists a function $\beta(t)\in C_0(\mathbb{R},\mathbb{R}^+)$ such that, for all $t\in\mathbb{R}$ and $x,y\in X$,

$$\|F_2(t,x,y)\|\le\beta(t)(\|x\|+\|y\|). \tag{4.246}$$

Theorem 4.6.26. *Assume that A is sectorial of type $\omega<0$. Let $F:\mathbb{R}\times X\times X\to X$ satisfy the hypotheses (H_1') and (H_2') with $L(t)\in \mathrm{BC}(\mathbb{R},\mathbb{R}^+)$. Moreover, suppose the integral $\int_{-\infty}^{t}\max\{L(s),\beta(s)\}\mathrm{d}s$ exists for all $t\in\mathbb{R}$. Then equation (4.237) has at least one asymptotically almost automorphic mild solution.*

Proof. The proof is divided into the following five steps:

Step 1. Define a mapping Λ on $\mathrm{AA}(\mathbb{R},X)$ by (4.241), and prove Λ has a unique fixed point $v(t)\in\mathrm{AA}(\mathbb{R},X)$.

First, similar to the proof in Step 1 of Theorem 4.6.24, we can prove that Λ is well defined and maps $\mathrm{AP}(\mathbb{R},X)$ into itself; moreover, Λ is continuous.

Next, we prove that Λ is a contraction on $\mathrm{AA}(\mathbb{R},X)$ and has a unique fixed point $v(t)\in\mathrm{AA}(\mathbb{R},X)$.

In fact, consider $v_1(t)$, $v_2(t)$ in $\mathrm{AA}(\mathbb{R},X)$ and define a new norm

$$|||x|||:=\sup_{t\in\mathbb{R}}\{\mu(t)\|x(t)\|\},$$

where $\mu(t):=e^{-k\int_{-\infty}^{t}\max\{L(s),\beta(s)\}\mathrm{d}s}$ and k is a fixed positive constant. Let $C_\alpha:=\sup_{t\in\mathbb{R}}\|S_\alpha(t)\|$. Then we have

$$\begin{aligned}
\mu(t)\|\Lambda v_1(t)-\Lambda v_2(t)\| &= \mu(t)\left\|\int_{-\infty}^{t}S_\alpha(t-\sigma)[F_1(\sigma,v_1(\sigma),Bv_1(\sigma))-F_1(\sigma,v_2(\sigma),Bv_2(\sigma))]\mathrm{d}\sigma\right\|\\
&\le C_\alpha\int_{-\infty}^{t}\mu(t)L(\sigma)[\|v_1(\sigma)-v_2(\sigma)\|+\|Bv_1(\sigma)-Bv_2(\sigma)\|]\mathrm{d}\sigma\\
&= C_\alpha\int_{-\infty}^{t}\mu(t)\mu(\sigma)L(\sigma)\mu(\sigma)^{-1}(1+\|B\|_{\mathbb{L}(X)})\|v_1(\sigma)-v_2(\sigma)\|\mathrm{d}\sigma\\
&\le C_\alpha(1+\|B\|_{\mathbb{L}(X)})|||v_1-v_2|||\int_{-\infty}^{t}\mu(t)\mu(\sigma)^{-1}L(\sigma)\mathrm{d}\sigma\\
&= \frac{C_\alpha(1+\|B\|_{\mathbb{L}(X)})}{k}|||v_1-v_2|||\int_{-\infty}^{t}ke^{-k\int_\sigma^t\max\{L(\tau),\beta(\tau)\}\mathrm{d}\tau}L(\sigma)\mathrm{d}\sigma
\end{aligned}$$

$$\leq \frac{C_\alpha(1+\|B\|_{\mathbb{L}(X)})}{k}|||v_1 - v_2||| \int_{-\infty}^{t} ke^{-k\int_\sigma^t L(\tau)d\tau} L(\sigma)d\sigma$$
$$= \frac{C_\alpha(1+\|B\|_{\mathbb{L}(X)})}{k}|||v_1 - v_2||| \int_{-\infty}^{t} \frac{d}{d\sigma}(e^{k\int_t^\sigma L(\tau)d\tau})d\sigma$$
$$= \frac{C_\alpha(1+\|B\|_{\mathbb{L}(X)})}{k}(1-e^{-k\int_{-\infty}^t L(\tau)d\tau})|||v_1 - v_2|||$$
$$\leq \frac{C_\alpha(1+\|B\|_{\mathbb{L}(X)})}{k}|||v_1 - v_2|||,$$

which implies that

$$|||\Lambda x(t) - \Lambda y(t)||| \leq \frac{C_\alpha(1+\|B\|_{\mathbb{L}(X)})}{k}|||x-y|||.$$

Hence Λ has a unique fixed point $x \in AA(\mathbb{R}, X)$ when k is greater than $C_\alpha(1+\|B\|_{\mathbb{L}(X)})$.

Step 2. Set $\Theta_r := \{\omega(t) \in C_0(\mathbb{R}, X) : |||\omega||| \leq r\}$. For the above $v(t)$, define $\Gamma := \Gamma^1 + \Gamma^2$ on $C_0(\mathbb{R}, X)$ as in (4.242), and prove that Γ maps Θ_{k_0} into itself, where k_0 is a given constant.

First, from (4.244) it follows that, for all $s \in \mathbb{R}$, $\omega(s) \in X$,

$$\begin{aligned}&\|F_1(s, v(s)+\omega(s), B(v(s)+\omega(s))) - F_1(s, v(s), Bv(s))\| \\ &\quad \leq L(s)[\|\omega(s)\| + \|B\omega(s)\|] \leq L(s)(1+\|B\|_{\mathbb{L}(X)})\|\omega(s)\| + \|B\omega(s)\|,\end{aligned}$$

which, together with $L(s) \in BC(\mathbb{R}, \mathbb{R}^+)$, implies that

$$F_1(\cdot, v(\cdot)+\omega(\cdot), B(v(\cdot)+\omega(\cdot))) - F_1(\cdot, v(\cdot), Bv(\cdot)) \in C_0(\mathbb{R}, X) \quad \text{for every } \omega(\cdot) \in C_0(\mathbb{R}, X).$$

According to (4.246), one has

$$\begin{aligned}\|F_2(s, v(s)+\omega(s), B(v(s)+\omega(s)))\| &\leq \beta(s)(\|\omega(s)+B\omega(s)\| + \|v(s)+Bv(s)\|) \\ &\leq \beta(s)((1+\|B\|_{\mathbb{L}(X)})\|\omega(s)\| + (1+\|B\|_{\mathbb{L}(X)})\|v(s)\|) \\ &\leq \beta(s)((1+\|B\|_{\mathbb{L}(X)})[\|\omega(s)\| + \|v(s)\|])\end{aligned}$$

for all $s \in \mathbb{R}$ and $\omega(s) \in X$ with $\|\omega(s)\| \leq r$, and thus

$$F_2(\cdot, v(\cdot)+\omega(\cdot), B(v(\cdot)+\omega(\cdot))) \in C_0(\mathbb{R}, X) \quad \text{as } \beta(\cdot) \in C_0(\mathbb{R}, \mathbb{R}^+).$$

This, together with Lemma 4.6.18, yields that Γ is well defined and maps $C_0(\mathbb{R}, X)$ into itself.

On the other hand, it is not difficult to see that there exists a constant $k_0 > 0$ such that

$$\frac{2C_\alpha(1+\|B\|_{\mathbb{L}(X)})}{k}k_0 + \frac{C_\alpha(1+\|B\|_{\mathbb{L}(X)})}{k}|||v(s)||| \leq k_0,$$

when k is large enough. This enables us to conclude that, for any $t \in \mathbb{R}$ and $\omega_1(t), \omega_2(t) \in \Theta_{k_0}$,

$$\begin{aligned}
&\mu(t)\|(\Gamma^1\omega_1)(t) + (\Gamma^2\omega_2)(t)\| \\
&\le \mu(t)\left\|\int_{-\infty}^{t} S_\alpha(t-s)[F_1(s, v(s)+\omega_1(s), B(v(s)+\omega_1(s))) - F_1(s, v(s), Bv(s))]\mathrm{d}s\right\| \\
&\quad + \mu(t)\left\|\int_{-\infty}^{t} S_\alpha(t-s)F_2(s, v(s)+\omega_2(s), B(v(s)+\omega_2(s)))\mathrm{d}s\right\| \\
&\le C_\alpha \int_{-\infty}^{t} \mu(t)L(s)(\|\omega_1(s)\| + \|B\omega_1(s)\|)\mathrm{d}s \\
&\quad + C_\alpha \int_{-\infty}^{t} \mu(t)\beta(s)(\|\omega_2\| + \|v(s)\| + \|B\omega_2\| + \|Bv(s)\|)\mathrm{d}s \\
&= C_\alpha \int_{-\infty}^{t} \mu(t)\mu(s)L(s)\mu(s)^{-1}(1 + \|B\|_{\mathbb{L}(X)})\|\omega_1(s)\|\mathrm{d}s \\
&\quad + C_\alpha \int_{-\infty}^{t} \mu(t)\mu(s)\beta(s)\mu(s)^{-1}(1 + \|B\|_{\mathbb{L}(X)})(\|\omega_2\| + \|v(s)\|)\mathrm{d}s \\
&\le C_\alpha(1 + \|B\|_{\mathbb{L}(X)})|||\omega_1||| \int_{-\infty}^{t} \mu(t)\mu(s)^{-1}L(s)\mathrm{d}s \\
&\quad + C_\alpha(1 + \|B\|_{\mathbb{L}(X)})(|||\omega_2||| + |||v(s)|||) \int_{-\infty}^{t} \mu(t)\mu(s)^{-1}\beta(s)\mathrm{d}s \\
&= \frac{C_\alpha(1 + \|B\|_{\mathbb{L}(X)})}{k}|||\omega_1||| \int_{-\infty}^{t} ke^{-k\int_t^s \max\{L(\tau),\beta(\tau)\}\mathrm{d}\tau}L(s)\mathrm{d}s \\
&\quad + C_\alpha(1 + \|B\|_{\mathbb{L}(X)})(|||\omega_2||| + |||v(s)|||) \int_{-\infty}^{t} ke^{-k\int_t^s \max\{L(\tau),\beta(\tau)\}\mathrm{d}\tau}\beta(s)\mathrm{d}s \\
&\le \frac{C_\alpha(1 + \|B\|_{\mathbb{L}(X)})}{k}|||\omega_1||| \int_{-\infty}^{t} ke^{-k\int_t^s L(\tau)\mathrm{d}\tau}L(s)\mathrm{d}s \\
&\quad + C_\alpha(1 + \|B\|_{\mathbb{L}(X)})(|||\omega_2||| + |||v(s)|||) \int_{-\infty}^{t} ke^{-k\int_t^s \beta(\tau)\mathrm{d}\tau}\beta(s)\mathrm{d}s \\
&= \frac{C_\alpha(1 + \|B\|_{\mathbb{L}(X)})}{k}|||\omega_1||| \int_{-\infty}^{t} \frac{\mathrm{d}}{\mathrm{d}s}(e^{k\int_t^s L(\tau)\mathrm{d}\tau})\mathrm{d}s
\end{aligned}$$

$$
\begin{aligned}
&+ C_a(1 + \|B\|_{\mathbb{L}(X)})(|||\omega_2||| + |||v(s)|||) \int_{-\infty}^{t} \frac{\mathrm{d}}{\mathrm{d}s}(e^{k\int_t^s \beta(\tau)\mathrm{d}\tau})\mathrm{d}s \\
&= \frac{C_a(1 + \|B\|_{\mathbb{L}(X)})}{k}(1 - e^{-k\int_{-\infty}^t L(\tau)\mathrm{d}\tau})|||\omega_1||| \\
&\quad + \frac{C_a(1 + \|B\|_{\mathbb{L}(X)})}{k}(1 - e^{-k\int_{-\infty}^t \beta(\tau)\mathrm{d}\tau})(|||\omega_2||| + |||v(s)|||) \\
&\le \frac{C_a(1 + \|B\|_{\mathbb{L}(X)})}{k}|||\omega_1||| + \frac{C_a}{k}(1 + \|B\|_{\mathbb{L}(X)})(|||\omega_2||| + |||v(s)|||) \le k_0,
\end{aligned}
$$

which implies that $(\Gamma^1\omega_1)(t) + (\Gamma^2\omega_2)(t) \in \Theta_{k_0}$. Thus Γ maps Θ_{k_0} into itself.

Step 3. Show Γ^1 is a contraction on Θ_{k_0}.

In fact, for any $\omega_1(t), \omega_2(t) \in \Theta_{k_0}$ and $t \in \mathbb{R}$, from (4.244) it follows that

$$
\begin{aligned}
&\big\|[F_1(s, v(s) + \omega_1(s), B(v(s) + \omega_1(s))) - F_1(s, v(s), Bv(s))] \\
&\quad - [F_1(s, v(s) + \omega_2(s), B(v(s) + \omega_2(s))) - F_1(s, v(s), Bv(s))]\big\| \\
&\le L(s)[\|\omega_1(s) - \omega_2(s)\| + \|B\omega_1(s) - B\omega_2(s)\|] \\
&\le L(s)(1 + \|B\|_{\mathbb{L}(X)})\|\omega_1(s) - \omega_2(s)\|.
\end{aligned}
$$

Thus

$$
\begin{aligned}
&\mu(t)\|(\Gamma^1\omega_1)(t) - (\Gamma^1\omega_2)(t)\| \\
&= \mu(t)\Bigg\|\int_{-\infty}^{t} S_a(t-s)[(F_1(s, v(s) + \omega_1(s), B(v(s) + \omega_1(s))) - F_1(s, v(s), Bv(s))) \\
&\quad - (F_1(s, v(s) + \omega_2(s), B(v(s) + \omega_2(s))) - F_1(s, v(s), Bv(s)))]\mathrm{d}s\Bigg\| \\
&\le C_a \int_{-\infty}^{t} \mu(t)L(\sigma)(1 + \|B\|_{\mathbb{L}(X)})\|\omega_1(\sigma) - \omega_2(\sigma)\|\mathrm{d}\sigma \\
&= C_a \int_{-\infty}^{t} \mu(t)\mu(\sigma)L(\sigma)\mu(\sigma)^{-1}(1 + \|B\|_{\mathbb{L}(X)})\|\omega_1(\sigma) - \omega_2(\sigma)\|\mathrm{d}\sigma \\
&\le C_a(1 + \|B\|_{\mathbb{L}(X)})|||\omega_1 - \omega_2||| \int_{-\infty}^{t} \mu(t)\mu(\sigma)^{-1}L(\sigma)\mathrm{d}\sigma \\
&= \frac{C_a(1 + \|B\|_{\mathbb{L}(X)})}{k}|||\omega_1 - \omega_2||| \int_{-\infty}^{t} ke^{-k\int_\sigma^t \max\{L(\tau),\beta(\tau)\}\mathrm{d}\tau}L(\sigma)\mathrm{d}\sigma \\
&\le \frac{C_a(1 + \|B\|_{\mathbb{L}(X)})}{k}|||\omega_1 - \omega_2||| \int_{-\infty}^{t} ke^{-k\int_\sigma^t L(\tau)\mathrm{d}\tau}L(\sigma)\mathrm{d}\sigma
\end{aligned}
$$

$$
\begin{aligned}
&= \frac{C_\alpha(1+\|B\|_{\mathbb{L}(X)})}{k}|||\omega_1-\omega_2||| \int_{-\infty}^{t} \frac{\mathrm{d}}{\mathrm{d}\sigma}\big(e^{k\int_t^\sigma L(\tau)\mathrm{d}\tau}\big)\mathrm{d}\sigma \\
&= \frac{C_\alpha(1+\|B\|_{\mathbb{L}(X)})}{k}\big(1-e^{-k\int_{-\infty}^{t} L(\tau)\mathrm{d}\tau}\big)|||\omega_1-\omega_2||| \\
&\le \frac{C_\alpha(1+\|B\|_{\mathbb{L}(X)})}{k}|||\omega_1-\omega_2|||,
\end{aligned}
$$

which implies

$$
|||(\Gamma^1\omega_1)(t)-(\Gamma^1\omega_2)(t)||| \le \frac{C_\alpha(1+\|B\|_{\mathbb{L}(X)})}{k}|||\omega_1-\omega_2|||.
$$

Thus, when k is greater than $C_\alpha(1+\|B\|_{\mathbb{L}(X)})$, one obtains the conclusion.

Step 4. Show that Γ^2 is completely continuous on Θ_{k_0}.

Take any $\varepsilon > 0$. Let $\{\omega_n\}_{n=1}^{+\infty} \subset \Theta_{k_0}$ with $\omega_n \to \omega_0$ in Θ_{k_0} as $n \to +\infty$. Since $\sigma(t) \in C_0(\mathbb{R},\mathbb{R}^+)$, one may choose a $t_1 > 0$ big enough such that, for all $t \ge t_1$,

$$
(1+\|B\|_{\mathbb{L}(X)})(k_0+|||v|||)\sigma(t) < \frac{\varepsilon}{3CM}.
$$

Also, in view of (H_1'), we have

$$
F_2(s,v(s)+\omega_k(s),B(v(s)+\omega_k(s))) \to F_2(s,v(s)+\omega_0(s),B(v(s)+\omega_0(s)))
$$

for all $s \in (-\infty,t_1]$ as $k \to +\infty$, and

$$
\begin{aligned}
&\mu(\cdot)\|F_2(\cdot,v(\cdot)+\omega_n(\cdot),B(v(\cdot)+\omega_n(\cdot)))-F_2(\cdot,v(\cdot)+\omega_0(\cdot),Bv(\cdot)+\omega_0(\cdot))\| \\
&\quad \le \mu(\cdot)\beta(\cdot)(\|\omega_n(\cdot)\|+\|v(\cdot)\|+\|B\omega_n(\cdot)\|+\|Bv(\cdot)\|+\|\omega_0(\cdot)\|+\|v(\cdot)\|+\|B\omega_0(\cdot)\|+\|Bv(\cdot)\|) \\
&\quad \le \beta(\cdot)(|||\omega_n|||+|||v|||+|||B\omega_n|||+|||Bv|||+|||\omega_0|||+|||v|||+|||B\omega_0|||+|||Bv|||) \\
&\quad \le \beta(\cdot)(2(1+\|B\|_{\mathbb{L}(X)})(k_0+|||v|||)) \in L^1(-\infty,t_1].
\end{aligned}
$$

Hence, by the Lebesgue dominated convergence theorem, we deduce that there exists an $N > 0$ such that

$$
\begin{aligned}
CM\int_{-\infty}^{t_1} \frac{1}{1+|\omega|(t-s)^\alpha}\mu(t)\|F_2(s,v(s)+\omega_k(s),B(v(s)+\omega_k(s))) \\
- F_2(s,v(s)+\omega_0(s),B(v(s)+\omega_0(s)))\|\mathrm{d}s \le \frac{\varepsilon}{3}
\end{aligned}
$$

whenever $k \ge N$. Thus

$$\begin{aligned}
&\mu(t)\|(\Gamma^2\omega_k)(t)-(\Gamma^2\omega_0)(t)\|\\
&=\mu(t)\Bigg\|\int_{-\infty}^{t} S_\alpha(t-s)F_2(s,v(s)+\omega_k(s),B(v(s)+\omega_k(s)))ds\\
&\quad-\int_{-\infty}^{t} S_\alpha(t-s)F_2(s,v(s)+\omega_0(s),B(v(s)+\omega_0(s)))ds\Bigg\|\\
&\le CM\int_{-\infty}^{t_1}\frac{1}{1+|\omega|(t-s)^\alpha}\mu(t)\|F_2(s,v(s)+\omega_k(s),B(v(s)+\omega_k(s)))\\
&\quad-F_2(s,v(s)+\omega_0(s),B(v(s)+\omega_0(s)))\|ds\\
&\quad+CM(2(1+\|B\|_{\mathbb{L}(X)})(k_0+|||v|||))\int_{t_1}^{\max\{t,t_1\}}\frac{\beta(s)}{1+|\omega|(t-s)^\alpha}ds\\
&\le CM\int_{-\infty}^{t_1}\frac{1}{1+|\omega|(t-s)^\alpha}\mu(t)\|F_2(s,v(s)+\omega_k(s),B(v(s)+\omega_k(s)))\\
&\quad-F_2(s,v(s)+\omega_0(s),B(v(s)+\omega_0(s)))\|ds\\
&\quad+CM\sigma(t)(2(1+\|B\|_{\mathbb{L}(X)})(k_0+|||v|||))\\
&\le\frac{\varepsilon}{3}+\frac{2\varepsilon}{3}=\varepsilon
\end{aligned}$$

whenever $k\ge N$. Accordingly, Γ^2 is continuous on Θ_{k_0}.

In the sequel, we consider the compactness of Γ^2.

Set $B_r(X)$ for the closed ball with center at 0 and radius r in X, $V=\Gamma^2(\Theta_{k_0})$ and $z(t)=\Gamma^2(u(t))$ for $u(t)\in\Theta_{k_0}$. First, for all $\omega(t)\in\Theta_{k_0}$ and $t\in\mathbb{R}$,

$$\begin{aligned}
\mu(t)\|(\Gamma^2\omega)(t)\|&=\mu(t)\Bigg\|\int_{-\infty}^{t} S_\alpha(t-s)F_2(s,v(s)+\omega(s),B(v(s)+\omega(s)))ds\Bigg\|\\
&\le CM\int_{-\infty}^{t}\frac{1}{1+|\omega|(t-s)^\alpha}\mu(t)\|F_2(s,v(s)+\omega(s),B(v(s)+\omega(s)))\|ds\\
&\le CM\int_{-\infty}^{t}\frac{\beta(s)}{1+|\omega|(t-s)^\alpha}\mu(t)(\|v(s)\|+\|\omega(s)\|+\|Bv(s)\|+\|B\omega(s)\|)ds\\
&\le CM\int_{-\infty}^{t}\frac{\beta(s)}{1+|\omega|(t-s)^\alpha}\mu(t)(1+\|B\|_{\mathbb{L}(X)})(\|v(s)\|+\|\omega(s)\|)ds\\
&\le CM\sigma(t)(1+\|B\|_{\mathbb{L}(X)})(k_0+|||v(s)|||),
\end{aligned}$$

in view of $\sigma(t)\in C_0(\mathbb{R},\mathbb{R}^+)$ which follows from Lemma 4.6.20. One concludes that

$$\lim_{|t|\to+\infty}(\Gamma^2\omega)(t)=0 \quad \text{uniformly for } \omega(t)\in\Theta_{k_0}.$$

As

$$\begin{aligned}(\Gamma^2\omega)(t) &= \int_{-\infty}^{t} S_\alpha(t-s)F_2(s,v(s)+\omega(s),B(v(s)+\omega(s)))\mathrm{d}s\\ &= \int_{0}^{+\infty} S_\alpha(\tau)F_2(t-\tau,v(t-\tau)+\omega(t-\tau),B(v(t-\tau)+\omega(t-\tau)))\mathrm{d}\tau,\end{aligned}$$

for a given $\varepsilon_0>0$, one can choose a $\xi>0$ such that

$$\left|\left|\left|\int_{\xi}^{+\infty} S_\alpha(\tau)F_2(t-\tau,v(t-\tau)+\omega(t-\tau),B(v(t-\tau)+\omega(t-\tau)))\mathrm{d}\tau\right|\right|\right|<\varepsilon_0.$$

Thus we get

$$\begin{aligned}z(t)\in{}&\xi\overline{c(\{S_\alpha(\tau)F_2(\lambda,v(\lambda)+\omega(\lambda),B(v(\lambda)+\omega(\lambda))):0\le\tau\le\xi,t-\xi\le\lambda\le\xi,|||\omega|||\le k_0\})}\\ &+B_{\varepsilon_0}(\Theta_{k_0}),\end{aligned}$$

where $c(K)$ denotes the convex hull of K. Since $S_\alpha(\cdot)$ is strongly continuous, we infer that

$$K=\{S_\alpha(\tau)F_2(\lambda,v(\lambda)+\omega(\lambda),B(v(\lambda)+\omega(\lambda))):0\le\tau\le\xi,t-\xi\le\lambda\le\xi,|||\omega|||\le k_0\}$$

is a relatively compact set, and $V\subset\xi\overline{c(K)}+B_{\varepsilon_0}(\Theta_{k_0})$, which implies that V is a relatively compact subset of Θ_{k_0}.

Next, we verify the equicontinuity of the set $\{(\Gamma^2\omega)(t):\omega(t)\in\Theta_{k_0}\}$.

Let $k>0$ be small enough and $t_1,t_2\in\mathbb{R}$, $\omega(t)\in\Theta_{k_0}$. Then by (4.246) we have

$$\begin{aligned}&|||(\Gamma^2\omega)(t_2)-(\Gamma^2\omega)(t_1)|||\\ &\quad=\left|\left|\left|\int_{-\infty}^{t_2} S_\alpha(t_2-s)F_2(s,v(s)+\omega(s),B(v(s)+\omega(s)))\mathrm{d}s\right.\right.\right.\\ &\qquad\left.\left.\left.-\int_{-\infty}^{t_1} S_\alpha(t_1-s)F_2(s,v(s)+\omega(s),B(v(s)+\omega(s)))\mathrm{d}s\right|\right|\right|\\ &\quad\le\int_{t_1}^{t_2}|||S_\alpha(t_2-s)F_2(s,v(s)+\omega(s),B(v(s)+\omega(s)))|||\mathrm{d}s\\ &\qquad+\int_{-\infty}^{t_1-k}|||[S_\alpha(t_2-s)-S_\alpha(t_1-s)]F_2(s,v(s)+\omega(s),B(v(s)+\omega(s)))|||\mathrm{d}s\end{aligned}$$

$$+\int_{t_1-k}^{t_1} |||[S_\alpha(t_2-s)-S_\alpha(t_1-s)]F_2(s,v(s)+\omega(s),B(v(s)+\omega(s)))|||\mathrm{d}s$$

$$\leq CM(1+\|B\|_{\mathbb{L}(X)})(k_0+|||v|||)\int_{t_1}^{t_2}\frac{\beta(s)}{1+|\omega|(t_2-s)^\alpha}\mathrm{d}s$$

$$+(1+\|B\|_{\mathbb{L}(X)})(k_0+|||v|||)\sup_{s\in[-\infty,t_1-k]}\|S_\alpha(t_2-s)-S_\alpha(t_1-s)\|\int_{-\infty}^{t_1-k}\beta(s)\mathrm{d}s$$

$$+CM(1+\|B\|_{\mathbb{L}(X)})(k_0+|||v|||)\int_{t_1-k}^{t_1}\left(\frac{\beta(s)}{1+|\omega|(t_2-s)^\alpha}+\frac{\beta(s)}{1+|\omega|(t_1-s)^\alpha}\right)\mathrm{d}s$$

$$\to 0 \quad \text{as } t_2-t_1\to 0, k\to 0,$$

which implies the equicontinuity of the set $\{(\Gamma^2\omega)(t) : \omega(t)\in\Theta_{k_0}\}$.

Now an application of Lemma 4.6.15 justifies the compactness of Γ^2.

Step 5. Show that equation (4.237) has at least one asymptotically almost automorphic mild solution.

The proof is similar to that in Step 5 of Theorem 4.6.21. □

Taking $A=-\rho^\alpha I$ with $\rho>0$ in equation (4.237), Theorem 4.6.26 gives the following corollary.

Corollary 4.6.27. *Let $F:\mathbb{R}\times X\times X\to X$ satisfy (H_1') and (H_2') with $L(t)\in BC(\mathbb{R},\mathbb{R}^+)$. Moreover, assume the integral $\int_{-\infty}^t \max\{L(s),\beta(s)\}\mathrm{d}s$ exists for all $t\in\mathbb{R}$. Then equation (4.237) has at least one asymptotically almost automorphic mild solution.*

4.6.3 Applications

In this section we give an example to illustrate the above results.

Consider the following fractional relaxation-oscillation equation:

$$\begin{cases}\partial_t^\alpha u(t,x)=\partial_x^2 u(t,x)-pu(t,x)\\ \qquad +\partial_t^{\alpha-1}[\mu a(t)\sin(\frac{1}{2+\cos t+\cos(\sqrt{2}t)})[\sin u(t,x)+u(t,x)]\\ \qquad +\nu e^{-|t|}[u(t,x)+\sin u(t,x)]], \quad t\in\mathbb{R},\ x\in[0,\pi],\\ u(t,0)=u(t,\pi)=0, \quad t\in\mathbb{R},\end{cases} \tag{4.247}$$

where $a(t)\in BC(\mathbb{R},\mathbb{R}^+)$ is a function, p, μ and ν are positive constants.

Take $X=L^2([0,\pi])$ and define the operator A by

$$A\varphi := \varphi''-p\varphi, \quad \varphi\in D(A),$$

where

$$D(A) := \{\varphi \in X : \varphi'' \in X, \varphi(0) = \varphi(\pi)\} \subset X.$$

It is well known that $Au = u''$ is self-adjoint, with compact resolvent, and is the infinitesimal generator of an analytic semigroup on X. Hence, $pI - A$ is sectorial of type $\omega = -p < 0$. Let

$$F_1(t, x(\xi), y(\xi)) := \mu a(t) \sin\left(\frac{1}{2 + \cos t + \cos(\sqrt{2}t)}\right)[\sin x(\xi) + y(\xi)],$$
$$F_2(t, x(\xi), y(\xi)) := \nu e^{-|t|}[x(\xi) + \sin y(\xi)].$$

Then it is easy to verify that $F_1, F_2 : \mathbb{R} \times X \times X \to X$ are continuous, $F_1(t, x, y) \in AA(\mathbb{R} \times X \times X, X)$ satisfying

$$\begin{aligned}
&\|F_1(t, x_1, y_1) - F_1(t, x_2, y_2)\|_2^2 \\
&\leq \int_0^\pi \mu^2 \left| a(t) \sin\left(\frac{1}{2 + \cos t + \cos(\sqrt{2}t)}\right)\right|^2 |[\sin x_1(s) + y_1(s)] - [\sin x_2(s) + y_2(s)]| ds \\
&\leq \mu^2 a^2(t) \left|\sin\left(\frac{1}{2 + \cos t + \cos(\sqrt{2}t)}\right)\right|^2 (\|x_1 - x_2\|_2^2 + \|y_1 - y_2\|_2^2),
\end{aligned}$$

that is,

$$\|F_1(t, x_1, y_1) - F_1(t, x_2, y_2)\|_2 \leq \mu a(t)(\|x_1 - x_2\|_2 + \|y_1 - y_2\|_2) \\ \text{for all } t \in \mathbb{R}, x_1, y_1, x_2, y_2 \in X.$$

Furthermore,

$$\|F_1(t, x_1, y_1) - F_1(t, x_2, y_2)\|_2 \leq \mu \|a\|_\infty (\|x_1 - x_2\|_2 + \|y_1 - y_2\|_2) \\ \text{for all } t \in \mathbb{R}, x_1, y_1, x_2, y_2 \in X$$

and

$$\|F_2(t, x, y)\|_2^2 \leq \int_0^\pi \nu^2 e^{-2|t|} |x(s) + \sin y(s)| ds \leq \nu^2 e^{-2|t|} (\|x\|_2^2 + \|y\|_2^2),$$

that is,

$$\|F_2(t, x, y)\|_2 \leq \nu e^{-|t|}(\|x\|_2 + \|y\|_2) \quad \text{for all } t \in \mathbb{R},\ x, y \in X,$$

which implies $F_2(t, x, y) \in C_0(\mathbb{R} \times X \times X, X)$. Furthermore,

$$F(t, x, y) = F_1(t, x, y) + F_2(t, x, y) \in AAA(\mathbb{R} \times X \times X, X).$$

Thus, equation (4.247) can be reformulated as the abstract problem (4.237) and the assumptions (H_1) and (H_2) hold with

$$L = \mu\|a\|_\infty, \quad \Phi(r) = r, \quad \beta(t) = \nu e^{-|t|}, \quad \rho_1 = 1, \quad \rho_2 \leq \nu.$$

The assumption (H_1') holds with $L(t) = \mu a(t)$, and the assumption (H_2') holds, too.

As a consequence, the fractional relaxation–oscillation equation (4.247) has at least one asymptotically almost automorphic mild solutions if either

$$\frac{\mu CM\|a\|_\infty \pi |p|^{-\frac{1}{\alpha}}}{\alpha \sin(\frac{\pi}{\alpha})} + CM\nu < \frac{1}{2}$$

(by Theorem 4.6.21) or

$$\frac{\mu CM\|a\|\pi |p|^{-\frac{1}{\alpha}}}{\alpha \sin(\frac{\pi}{\alpha})} + CM\nu < \frac{1}{2}$$

(by Theorem 4.6.24), where $\|a\| = \sup_{t\in\mathbb{R}} \int_t^{t+1} a(s)ds$, or the integral

$$\int_{-\infty}^{t} \max\{\mu a(s), \nu e^{-|t|}\}ds$$

exists for all $t \in \mathbb{R}$ (by Theorem 4.6.26).

4.7 Some monotone differential inclusions: applications to parabolic and hyperbolic equations

Let $\mathcal{H}$ be a real Hilbert space and $\mathcal{A} : D(\mathcal{A}) \subset \mathcal{H} \to 2^{\mathcal{H}}$ be a multivalued operator with domain $D(\mathcal{A})$. We consider the following differential inclusions:

$$u'(t) + \mathcal{A}u(t) \ni f(t) \quad \text{for } t \in \mathbb{R}, \tag{4.248}$$

$$u'(t) + \mathcal{A}u(t) \ni g(t, u(t)) \quad \text{for } t \in \mathbb{R}, \tag{4.249}$$

where $f : \mathbb{R} \to \mathcal{H}$ and $g : \mathbb{R} \times \mathcal{H} \to \mathcal{H}$ are continuous functions. Many studies have been devoted to the existence of periodic and almost periodic solutions for the differential inclusion (4.248) when the operator $\mathcal{A}$ is maximal monotone on $\mathcal{H}$ and the forcing term f is periodic or almost periodic. Brézis [83, Theorem 3.4, p. 65] proved that for any $f \in L^1([a,b], \mathcal{H})$ and $u_0 \in \overline{D(\mathcal{A})}$ there exists a unique weak solution of the following differential inclusion:

$$\begin{cases} u'(t) + \mathcal{A}u(t) \ni f(t) & \text{for } t \in [a,b], \\ u(a) = u_0. \end{cases}$$

Brézis [83, Theorem 3.15, p. 95] showed that if $\mathcal{A}$ is maximal monotone, then for each $f \in L^1([0,T],\mathcal{H})$ the differential inclusion

$$\begin{cases} u'(t) + \mathcal{A}u(t) \ni f(t), \\ u(0) = u(T) \end{cases}$$

has at least a weak solution. Baillon and Haraux [44] studied the following differential inclusion:

$$u'(t) + \partial\phi(u(t)) \ni f(t) \quad \text{for } t \in [0,+\infty), \tag{4.250}$$

where ϕ is a proper, convex, and lower semicontinuous function, and $f \in L^2([0,+\infty),\mathcal{H})$ is T-periodic. They proved that if a T-periodic solution of (4.250) exists on $\mathbb{R}$, then for each solution u of (4.250) on $\mathbb{R}^+$ there exists a periodic strong solution w of (4.250) on $\mathbb{R}$ such that

$$u(t) \rightharpoonup w(t) \quad \text{as } t \to +\infty.$$

Haraux [169] proved that if the forcing term $f : \mathbb{R} \to \mathcal{H}$ is S^2-almost periodic, then each weak solution of (4.250) on $\mathbb{R}^+$ is asymptotic to an almost periodic weak solution of (4.250) on $\mathbb{R}$. Haraux [171] also proved that if (4.248) has a uniformly continuous weak solution on $\mathbb{R}^+$ and its range over $\mathbb{R}^+$ is relatively compact, then it has an almost periodic weak solution on $\mathbb{R}$ when f is almost periodic [171, Theorem 1, p. 295]. Furthermore, Haraux [174] proved that if the forcing term $f : \mathbb{R} \to \mathbb{R}^2$ is S^1-almost periodic, then all bounded solutions on $\mathbb{R}$ of (4.248) are almost periodic.

The aim of this section is to study the existence of compact almost automorphic weak solutions for (4.248) and (4.249). If $\mathcal{A}$ is maximal monotone and f is compact almost automorphic, we prove that if (4.248) has a uniformly continuous weak solution on $\mathbb{R}^+$ having a relatively compact range over $\mathbb{R}^+$, then it has at least one compact almost automorphic weak solution on $\mathbb{R}$. Our main result is proved by using the minimax principle due of Amerio [28]. As an application, we study the following partial differential inclusion:

$$\begin{cases} \frac{\partial^2}{\partial t^2}u(t,x) - \Delta u(t,x) + \beta(\frac{\partial}{\partial t}u(t,x)) \ni f(t,x) \quad \text{for } (t,x) \in \mathbb{R}\times\Omega, \\ u(t,x) = 0 \quad \text{for } (t,x) \in \mathbb{R}\times\partial\Omega, \end{cases}$$

where Ω is a bounded open set in $\mathbb{R}^N$ with smooth boundary $\partial\Omega$ such that $\dim(\Omega) \geq 2$, β is a strongly monotone graph in $\mathbb{R}\times\mathbb{R}$, and f is a compact almost automorphic function in $L^2(\Omega)$.

If $\mathcal{A}$ is strongly maximal monotone and f is compact almost automorphic, we prove that (4.248) has a unique bounded weak solution that is compact almost automorphic and globally attractive. Moreover, we use the contraction principle to prove the existence and uniqueness of compact almost automorphic weak solution for (4.249) where

g is compact almost automorphic in t and Lipschitz with respect to the second argument.

4.7.1 Maximal monotone operators and differential inclusions

Throughout this section, $\mathcal{H}$ is a real Hilbert space endowed with its norm $|\cdot|$ and inner product $(\cdot,\cdot)_{\mathcal{H}}$, and $2^{\mathcal{H}}$ is the powerset of $\mathcal{H}$. In this section we give some tools about maximal monotone operators. More details can be found in the monograph by Brézis [83].

Definition 4.7.1. Let $\mathcal{A} : D(\mathcal{A}) \subset \mathcal{H} \to 2^{\mathcal{H}}$ be a multivalued operator. Its domain is defined by

$$D(\mathcal{A}) = \{x \in \mathcal{H} : \mathcal{A}x \text{ is nonempty in } \mathcal{H}\}.$$

(i) The range of $\mathcal{A}$ is defined by

$$R(\mathcal{A}) = \bigcup_{x \in \mathcal{H}} \mathcal{A}x.$$

(ii) The graph of $\mathcal{A}$ is defined by

$$G(\mathcal{A}) = \{(x,y) \in \mathcal{H}^2 : y \in \mathcal{A}x\}.$$

(iii) Operator $\mathcal{A}$ is *monotone* if

$$(\mathcal{A}x - \mathcal{A}y, x - y)_{\mathcal{H}} \geq 0 \quad \text{for all } x,y \in D(\mathcal{A}),$$

which means that for each $x_1 \in D(A)$ and $x_2 \in D(A)$, one has

$$(y_1 - y_2, x_1 - x_2)_{\mathcal{H}} \geq 0 \quad \text{for all } y_1 \in \mathcal{A}x_1 \text{ and } y_2 \in \mathcal{A}x_2.$$

(iv) Operator $\mathcal{A}$ is *maximal monotone* if it is monotone and $G(\mathcal{A})$ is maximal with respect to inclusion among the graphs of all monotone operators.

(v) Operator $\mathcal{A}$ is *α-strongly maximal monotone* ($\alpha > 0$) if it is maximal monotone and

$$(\mathcal{A}x - \mathcal{A}y, x - y)_{\mathcal{H}} \geq \alpha|x - y|^2 \quad \text{for all } x,y \in D(\mathcal{A}).$$

Remark 4.7.2. Let $\mathcal{A} : D(\mathcal{A}) \subset \mathcal{H} \to 2^{\mathcal{H}}$ be a multivalued operator and $\alpha > 0$. Then,

$\mathcal{A}$ is maximal monotone if and only if $\mathcal{A} + \alpha I$ is α-strongly maximal monotone.

Theorem 4.7.3 ([83]). *If $\mathcal{A} : D(\mathcal{A}) \subset \mathcal{H} \to 2^{\mathcal{H}}$ is maximal monotone, then $\overline{D(\mathcal{A})}$ is convex.*

Theorem 4.7.4 ([171]). *Let $\mathcal{A}$ be a maximal monotone operator on $\mathcal{H}$. Then, for every $x \in D(\mathcal{A})$, the set $\mathcal{A}x$ is closed and convex.*

Remark 4.7.5. From the last theorem, the following operator is well defined:

$$\mathcal{A}^0 x = \mathrm{Proj}_{\mathcal{A}x}(0) \quad \text{for } x \in D(\mathcal{A}),$$

where $\mathrm{Proj}_C(x)$ is the projection of x on a closed convex $C \subset \mathcal{H}$. The operator $\mathcal{A}^0$ plays a key role in many questions related to solving nonlinear equations.

Consider the following differential inclusion:

$$\begin{cases} u'(t) + \mathcal{A}u(t) \ni f(t) & \text{for } t \in [a,b], \\ u(a) = u_0 \in \mathcal{H}. \end{cases} \tag{4.251}$$

Definition 4.7.6 ([83]). Let $f \in L^1([a,b],\mathcal{H})$. A continuous function $u : [a,b] \to \mathcal{H}$ is a *strong solution* of the differential inclusion (4.251) if u is absolutely continuous on each compact of $]a,b[$, $u(t) \in D(\mathcal{A})$ a. e. on $[a,b]$, and (4.251) is satisfied a. e. on $[a,b]$.

Definition 4.7.7 ([83]). A function u is a *weak solution* of (4.251) if there exist $f_n \in L^1([a,b],\mathcal{H})$ and $u_n \in C([a,b],\mathcal{H})$ such that u_n is a strong solution of

$$u_n'(t) + \mathcal{A}u_n(t) \ni f_n(t)$$

on $[a,b]$, $f_n \to f$ in $L^1([a,b],\mathcal{H})$ and $u_n \to u$ in $C([a,b],\mathcal{H})$.

Lemma 4.7.8 ([83]). *Let $f, f_n \in L^1([a,b],\mathcal{H})$. Assume that x_n is a weak solution of $x_n'(t) + \mathcal{A}x_n(t) \ni f_n(t)$ on $[a,b]$. If $x_n \to x$ uniformly on $[a,b]$ and $f_n \to f$ in $L^1([a,b],\mathcal{H})$, then x is a weak solution of* (4.251) *on $[a,b]$.*

Definition 4.7.9. Let $f \in L^1_{\mathrm{loc}}(\mathbb{R},\mathcal{H})$. A function u is a weak solution of (4.251) on $\mathbb{R}$ if it is a weak solution of (4.251) on every compact interval of $\mathbb{R}$.

We cite the following results regarding the existence and estimates of weak solutions.

Theorem 4.7.10 ([171, Theorem 36, p. 76]). *Assume that $\mathcal{A}$ is maximal monotone and $f \in L^1([a,b],\mathcal{H})$. If $u_0 \in \overline{D(\mathcal{A})}$, then there exists a unique weak solution of* (4.251). *Moreover, if u and v are two weak solutions of $u'(t) + \mathcal{A}u(t) \ni f(t)$ and $v'(t) + \mathcal{A}v(t) \ni g(t)$, respectively, then*

$$|u(t) - v(t)| \le |u(s) - v(s)| + \int_s^t |f(\sigma) - g(\sigma)|\,d\sigma \quad \textit{for } a \le s \le t \le b. \tag{4.252}$$

Theorem 4.7.11 ([271]). *Assume that $\mathcal{A}$ is α-strongly maximal monotone. Let I be an interval of $\mathbb{R}$ and $\tilde{f}, \hat{f} \in L^1_{loc}(I, \mathcal{H})$. If $\tilde{u}$ and $\hat{u}$ are weak solutions on I of $\tilde{u}'(t) + \mathcal{A}\tilde{u}(t) \ni \tilde{f}(t)$ and $\hat{u}'(t) + \mathcal{A}\hat{u}(t) \ni \hat{f}(t)$, respectively, then for any s and t in I, $s \le t$, we have*

$$|\tilde{u}(t) - \hat{u}(t)| \le e^{-\alpha(t-s)}|\tilde{u}(s) - \hat{u}(s)| + \int_s^t e^{-\alpha(t-\sigma)}|\tilde{f}(\sigma) - \hat{f}(\sigma)|d\sigma.$$

We close this section with the following lemma.

Lemma 4.7.12 ([171, Lemma 30, p. 220]). *Let S be a contraction defined on a closed convex subset of a real Hilbert space $(\mathcal{H}, |\cdot|)$. Then*

$$|Sx - Sy| = |x - y| \quad \Longrightarrow \quad S\left(\frac{x+y}{2}\right) = \frac{1}{2}(S(x) + S(y)).$$

4.7.2 Compact almost automorphic weak solutions of (4.248) where $\mathcal{A}$ is maximal monotone

In the sequel, we prove the existence of compact almost automorphic weak solutions of (4.248) where the operator $\mathcal{A}$ is maximal monotone.

Theorem 4.7.13. *Suppose that f is compact almost automorphic and $\mathcal{A}$ is maximal monotone. If* (4.248) *has a uniformly continuous weak solution on $\mathbb{R}^+$ having a relatively compact range over $\mathbb{R}^+$, then* (4.248) *has at least a compact almost automorphic weak solution.*

For the proof of Theorem 4.7.13, we need the following lemmas.

Lemma 4.7.14. *Let $F \in L^1_{loc}(\mathbb{R}, \mathcal{H})$ and x be a weak solution on $\mathbb{R}$ of the following differential inclusion:*

$$x'(t) + \mathcal{A}x(t) \ni F(t).$$

Assume that x is uniformly continuous on $\mathbb{R}$ and there exists a compact set K of $\mathcal{H}$ such that

$$x(t) \in K \quad \text{for all } t \in \mathbb{R}. \tag{4.253}$$

If there exist a sequence $(t_n)_n \subset \mathbb{R}$ and a function $G : \mathbb{R} \to \mathcal{H}$ such that

$$F(t + t_n) \to G(t) \quad \text{in } L^1_{loc}(\mathbb{R}, \mathcal{H}) \text{ as } n \to +\infty,$$

then there exists a subsequence of $(t_n)_n$ denoted by $(s_n)_n$ such that

$$x(t + s_n) \to y(t) \quad \text{as } n \to +\infty \tag{4.254}$$

uniformly on any compact subset of $\mathbb{R}$, *where* y *is a weak solution on* $\mathbb{R}$ *of the following differential inclusion:*

$$y'(t) + \mathcal{A}y(t) \ni G(t). \tag{4.255}$$

Furthermore, y *is uniformly continuous on* $\mathbb{R}$ *and* $y(t) \in K$ *for all* $t \in \mathbb{R}$.

Proof. For each $n \in \mathbb{N}$, we define x_n and F_n on $\mathbb{R}$ by $x_n(t) = x(t+t_n)$ and $F_n(t) = F(t+t_n)$. By (4.253), $(x_n)_n$ satifies $x_n(t) \in K$ for each $t \in \mathbb{R}$ and $n \in \mathbb{N}$. Consequently, $\{x_n(t) : n \in \mathbb{N}\}$ is a relatively compact set in $\mathcal{H}$ for each $t \in \mathbb{R}$. Since x is uniformly continuous on $\mathbb{R}$, the sequence $(x_n)_n$ is uniformly equicontinuous on $\mathbb{R}$. By Arzelà–Ascoli theorem, $\{x_n : n \in \mathbb{N}\}$ is a relatively compact subset of $BC(\mathbb{R}, \mathcal{H})$ endowed with the topology of compact convergence. From the sequence $(t_n)_n$, we can extract a subsequence $(s_n)_n$ such that there exists $y \in BC(\mathbb{R}, \mathcal{H})$ such that $(x_n)_n$ converges to y uniformly on each compact subset of $\mathbb{R}$ and hence (4.254) holds. Furthermore, since x_n is a weak solution of (4.248) with F_n and F_n converges to G in $L^1_{loc}(\mathbb{R}, \mathcal{H})$, Lemma 4.7.8 allows us to conclude that y is a weak solution of (4.255); moreover, $y(t) \in K$ for all $t \in \mathbb{R}$. Therefore, y is uniformly continuous on $\mathbb{R}$ because it is the limit of the sequence $(x_n)_n$, which is uniformly equicontinuous on $\mathbb{R}$. □

Lemma 4.7.15. *Suppose that* f *is compact almost automorphic. If* (4.248) *has a uniformly continuous weak solution* u_0 *on* $\mathbb{R}^+$ *having a relatively compact range over* $\mathbb{R}^+$, *then it has a uniformly continuous weak solution* u^* *on* $\mathbb{R}$; *moreover, its range over* $\mathbb{R}$ *is relatively compact.*

Proof. Let $(t_n)_n \subset \mathbb{R}$ be such that

$$\lim_{n\to+\infty} t_n = +\infty.$$

If $t \in [-1,1]$, then for sufficiently large n the sequence of functions $u_n : t \mapsto u_0(t+t_n)$ is well defined and uniformly equicontinuous. By Arzelà–Ascoli theorem, there exist a function v and a subsequence $(t_n^1)_n \subset (t_n)_n$ such that

$$u_0(t+t_n^1) \to v(t) \quad \text{as } n \to +\infty$$

uniformly on $[-1,1]$. Using the same argument, we deduce that for each $p \in \mathbb{N}^*$ there exists a subsequence $(t_n^p)_n \subset (t_n^{p-1})_n \subset \cdots \subset (t_n)_n$ such that

$$u_0(t+t_n^p) \to v(t) \quad \text{as } n \to +\infty$$

uniformly on $[-p,p]$. Let $(t_n')_n := (t_n^n)_n$ be Cantor's diagonal sequence. Then,

$$u_0(t+t_n') \to v(t) \quad \text{as } n \to +\infty \tag{4.256}$$

uniformly on any compact subset of $\mathbb{R}$. Since f is compact almost automorphic, there exist a continuous function g and a subsequence $(t''_n)_n \subset (t'_n)_n$ such that

$$f(t + t''_n) \to g(t) \quad \text{as } n \to +\infty, \tag{4.257}$$

$$g(t - t''_n) \to f(t) \quad \text{as } n \to +\infty \tag{4.258}$$

uniformly on any compact subset of $\mathbb{R}$. Moreover, using (4.256) and (4.257), we find that v is a weak solution on $\mathbb{R}$ of (4.248) with the found g. Furthermore, v is uniformly continuous on $\mathbb{R}$. By applying the above argument to the returning sequence $(-t''_n)_n$, we obtain a subsequence $(t'''_n)_n \subset (t''_n)_n$ and a function u^* such that

$$v(t - t'''_n) \to u^*(t) \quad \text{as } n \to +\infty \tag{4.259}$$

uniformly on any compact subset of $\mathbb{R}$. By (4.258), (4.259) and Lemma 4.7.8, we deduce that u^* is a weak solution on $\mathbb{R}$ of (4.248). Furthermore, the function u^* is uniformly continuous on $\mathbb{R}$ and its range is contained in the closure of the range of u_0; hence, it is relatively compact. □

Proof of Theorem 4.7.13. We use Amerio's principle. Let $K = \overline{\mathrm{Co}}(u^*(\mathbb{R}))$ be the closed convex hull of $u^*(\mathbb{R})$ in $\mathcal{H}$, where u^* is given in Lemma 4.7.15. Let Λ and Γ be the sets defined by

$$\Lambda = \Big\{u \in C(\mathbb{R}, \mathcal{H}) : u(\mathbb{R}) \subset K \text{ and } \sup_{t\in\mathbb{R}} \big|u(t+\sigma) - u(t)\big| \le \sup_{t\in\mathbb{R}} \big|u^*(t+\sigma) - u^*(t)\big| \\ \text{for all } \sigma \in \mathbb{R}\Big\},$$

$$\Gamma = \{u \in \Lambda : u \text{ is a weak solution of the differential inclusion (4.248) on } \mathbb{R}\}.$$

We define the operator $J : \Lambda \to \mathbb{R}^+$ by

$$J(u) = \sup_{t\in\mathbb{R}} \big|u(t)\big| \quad \text{for } u \in \Lambda.$$

We say that $\tilde{u}$ is a *minimal weak solution* of (4.248) if

$$\tilde{u} \in \Gamma \quad \text{and} \quad J(\tilde{u}) = \inf_{u\in\Gamma} J(u).$$

We divide the proof into three steps:

Step 1. We claim that (4.248) has a minimal weak solution $\hat{u}$ on $\mathbb{R}$. In fact, let

$$\delta = \inf_{u\in\Gamma} J(u).$$

Then, by Lemma 4.7.15, Γ is nonempty since $u^* \in \Gamma$. Hence, δ exists in $\mathbb{R}$. Consequently, there exists a sequence $(u_n)_n$ in Γ such that

$$\lim_{n\to+\infty} J(u_n) = \delta. \tag{4.260}$$

By the definition of Γ, for each $t \in \mathbb{R}$, $\{u_n(t) : n \in \mathbb{N}\}$ is a subset of the compact K and $(u_n)_n$ is uniformly equicontinuous on $\mathbb{R}$. Using Arzelà–Ascoli theorem, we assert that $\{u_n : n \in \mathbb{N}\}$ is a relatively compact subset of $\mathrm{BC}(\mathbb{R}, \mathcal{H})$ endowed with the topology of compact convergence. Thus, there exists a subsequence of $(u_n)_n$ denoted also by $(u_n)_n$ such that

$$u_n(t) \to \hat{u}(t) \quad \text{as } n \to +\infty \tag{4.261}$$

uniformly on any compact subset of $\mathbb{R}$. Since $u_n'(t) + \mathcal{A}u_n(t) \ni f(t)$ in the sense of weak solutions, using (4.261) together with Lemma 4.7.8 implies that $\hat{u}$ is a weak solution on $\mathbb{R}$ of (4.248) and $\hat{u} \in \Lambda$; consequently, $\hat{u} \in \Gamma$. Hence, we obtain that

$$\delta \le J(\hat{u}). \tag{4.262}$$

We note that J is lower semicontinuous with respect to the topology of compact convergence; namely, if $\lim_{n\to+\infty} x_n = x$ uniformly on compact subsets of $\mathbb{R}$, then $J(x) \le \liminf_{n\to+\infty} J(x_n)$. By (4.261), we get that

$$J(\hat{u}) \le \liminf_{n\to+\infty} J(u_n). \tag{4.263}$$

From (4.260), (4.262), and (4.263), we deduce that

$$J(\hat{u}) = \delta = \inf_{u\in\Gamma} J(u).$$

Step 2. We claim that the minimal weak solution $\hat{u}$ is unique. In fact, let $u, v \in \Gamma$ be such that

$$J(u) = J(v) = \delta. \tag{4.264}$$

Let $(t_n)_n \subset \mathbb{R}$ be such that

$$\lim_{n\to+\infty} t_n = -\infty. \tag{4.265}$$

From the compact almost automorphy of f, there exist a continuous function g and a subsequence of $(t_n)_n$ denoted also by $(t_n)_n$ such that

$$f(t + t_n) \to g(t) \quad \text{as } n \to +\infty,$$
$$g(t - t_n) \to f(t) \quad \text{as } n \to +\infty$$

uniformly on any compact subset of $\mathbb{R}$. Now, let us prove that

$$u(t+t_n) \to u_1(t) \quad \text{as } n \to +\infty, \tag{4.266}$$

$$u_1(t-t_n) \to u_2(t) \quad \text{as } n \to +\infty, \tag{4.267}$$

$$v(t+t_n) \to v_1(t) \quad \text{as } n \to +\infty, \tag{4.268}$$

$$v_1(t-t_n) \to v_2(t) \quad \text{as } n \to +\infty \tag{4.269}$$

uniformly on any compact subset of $\mathbb{R}$, where u_2 and v_2 are two minimal weak solutions on $\mathbb{R}$ of (4.248). Since $u \in \Gamma$, it is uniformly continuous on $\mathbb{R}$ and $u(\mathbb{R}) \subset K$. Applying Lemma 4.7.14 to $x = u$, $F = f$, and the sequence $(t_n)_n$, we obtain (4.266) where u_1 is a weak solution on $\mathbb{R}$ of the following differential inclusion:

$$u_1'(t) + \mathcal{A}u_1(t) \ni g(t).$$

Moreover, u_1 is uniformly continuous on $\mathbb{R}$ and $u_1(\mathbb{R}) \subset K$ which implies that $u_1 \in \Lambda$. Applying again Lemma 4.7.14 to $x = u_1$, $F = g$, and the returning sequence $(-t_n)_n$, we obtain (4.267) where u_2 is a weak solution on $\mathbb{R}$ of (4.248) with $u_2 \in \Gamma$. It follows from (4.266) and (4.267) that

$$J(u_2) \le J(u_1) \le J(u).$$

Using (4.264), we obtain $J(u_2) = \delta$ and, consequently, u_2 is a weak minimal solution on $\mathbb{R}$ of (4.248). Applying the same argument to v, we obtain (4.268) and (4.269) where v_2 is a weak minimal solution on $\mathbb{R}$ of (4.248). Since $u'(t)+\mathcal{A}u(t) \ni f(t)$ and $v'(t)+\mathcal{A}v(t) \ni f(t)$ in the sense of weak solutions and the operator $\mathcal{A}$ is monotone, we find by using inequality (4.252) that the function $t \mapsto |u(t) - v(t)|$ is nonincreasing. By (4.265), we obtain that

$$\lim_{n\to+\infty} \left|u(t+t_n) - v(t+t_n)\right| = \sup_{\sigma\in\mathbb{R}} \left|u(\sigma) - v(\sigma)\right|. \tag{4.270}$$

It follows from (4.266)–(4.269) that, for each $t \in \mathbb{R}$,

$$\begin{aligned} \lim_{m\to+\infty} \lim_{n\to+\infty} \left|u(t+t_n-t_m) - v(t+t_n-t_m)\right| &= \lim_{m\to+\infty} \left|u_1(t-t_m) - v_1(t-t_m)\right| \\ &= \left|u_2(t) - v_2(t)\right|. \end{aligned} \tag{4.271}$$

Combining (4.270) and (4.271), we obtain that, for each $t \in \mathbb{R}$,

$$\left|u_2(t) - v_2(t)\right| = \sup_{\sigma\in\mathbb{R}} \left|u(\sigma) - v(\sigma)\right| = c. \tag{4.272}$$

Consequently, we have

$$\left|u_2(t) - v_2(t)\right| = \left|u_2(0) - v_2(0)\right| \quad \text{for } t \in \mathbb{R}. \tag{4.273}$$

Let $S_t : \overline{D(\mathcal{A})} \to \overline{D(\mathcal{A})}$ be the operator defined for each $x_0 \in \overline{D(\mathcal{A})}$ by

$$S_t x_0 = x(t),$$

where x is the unique weak solution on $\mathbb{R}$ of (4.248) with initial data $x(0) = x_0$. Taking $f = g$ in (4.252), we deduce that the operator S_t is contractive on the closed convex set $\overline{D(\mathcal{A})}$. It follows from (4.273) that

$$|S_t u_2(0) - S_t v_2(0)| = |u_2(0) - v_2(0)|.$$

Using Lemma 4.7.12, we obtain that

$$S_t\left(\frac{u_2(0) + v_2(0)}{2}\right) = \frac{1}{2}(S_t u_2(0) + S_t v_2(0)) = \frac{u_2(t) + v_2(t)}{2}.$$

We conclude that $\frac{u_2+v_2}{2}$ is also a weak solution on $\mathbb{R}$ of (4.248). Since $u_2(\mathbb{R}) \subset K, v_2(\mathbb{R}) \subset K$, and K is convex, then $(\frac{u_2+v_2}{2})(\mathbb{R}) \subset K$ and $\frac{u_2+v_2}{2} \in \Gamma$. Hence,

$$\delta = \inf_{u\in\Gamma} J(u) \leq J\left(\frac{1}{2}u_2 + \frac{1}{2}v_2\right) = \sup_{t\in\mathbb{R}}\left|\frac{1}{2}u_2(t) + \frac{1}{2}v_2(t)\right|. \tag{4.274}$$

From the parallelogram law, we get that

$$\sup_{t\in\mathbb{R}}\left|\frac{1}{2}u_2(t) + \frac{1}{2}v_2(t)\right|^2 + \frac{1}{4}c^2 \leq \frac{1}{2}\sup_{t\in\mathbb{R}}|u_2(t)|^2 + \frac{1}{2}\sup_{t\in\mathbb{R}}|v_2(t)|^2. \tag{4.275}$$

By (4.274) and (4.275), we obtain that $\delta^2 + \frac{1}{4}c^2 \leq \frac{1}{2}\delta^2 + \frac{1}{2}\delta^2$. Hence, $|u_2(t) - v_2(t)| = c \leq 0$ for all $t \in \mathbb{R}$; consequently, $u_2 = v_2$, which implies by (4.272) that $u = v$.

Step 3. We claim that the unique minimal weak solution $\hat{u}$ is compact almost automorphic. Let $(t_n')_n \subset \mathbb{R}$. We have to prove that there exist a subsequence $(t_n)_n$ of $(t_n')_n$ and a continuous function v such that

$$\hat{u}(t + t_n) \to v(t) \quad \text{as } n \to +\infty, \tag{4.276}$$

$$v(t - t_n) \to \hat{u}(t) \quad \text{as } n \to +\infty \tag{4.277}$$

uniformly on any compact subset of $\mathbb{R}$. From the compact almost automorphy of f, there exists a subsequence $(t_n)_n \subset (t_n')_n$ such that

$$f(t + t_n) \to g(t) \quad \text{as } n \to +\infty,$$
$$g(t - t_n) \to f(t) \quad \text{as } n \to +\infty$$

uniformly on any compact subset of $\mathbb{R}$. Since $\hat{u} \in \Gamma$, it is uniformly continuous on $\mathbb{R}$ and $\hat{u}(t) \in K$ for all $t \in \mathbb{R}$. Applying Lemma 4.7.14 to $x = \hat{u}$, $F = f$, and the sequence $(t_n)_n$, we obtain (4.276), where v is a weak solution on $\mathbb{R}$ of the following differential inclusion:

$$v'(t) + \mathcal{A}v(t) \ni g(t).$$

Furthermore, v is uniformly continuous on $\mathbb{R}$ and $v(t) \in K$ for all $t \in \mathbb{R}$ since $v \in \Lambda$. Using (4.276), we obtain that

$$J(v) \leq J(\hat{u}). \tag{4.278}$$

Applying Lemma 4.7.14 to $x = v$, $F = g$, and the returning sequence $(-t_n)_n$, we find for a subsequence that

$$v(t - t_n) \to \omega(t) \quad \text{as } n \to +\infty \tag{4.279}$$

uniformly on any compact subset of $\mathbb{R}$ where $\omega \in \Gamma$. From (4.279) we obtain that

$$J(\omega) \leq J(v). \tag{4.280}$$

By (4.278) and (4.280), we get that

$$J(\omega) \leq J(\hat{u}) = \inf_{u \in \Gamma} J(u).$$

Consequently,

$$J(\omega) = J(\hat{u}) = \inf_{u \in \Gamma} J(u).$$

By uniqueness of the minimal weak solution of (4.248) (from Steps 1 and 2), we deduce that $\omega = \hat{u}$, (4.277) holds, and $\hat{u}$ is compact almost automorphic. $\quad\square$

4.7.3 Compact almost automorphic weak solutions of (4.248) and (4.249) where $\mathcal{A}$ is strongly maximal monotone

In the sequel, we prove the existence and uniqueness of compact almost automorphic weak solutions for the differential inclusions (4.248) and (4.249) where the operator $\mathcal{A}$ is strongly maximal monotone.

Theorem 4.7.16. *Assume that $\mathcal{A}$ is α-strongly maximal monotone ($\alpha > 0$) with $0 \in \mathcal{A}0$ and $f \in AA_c(\mathbb{R}, \mathcal{H})$. Then, (4.248) has a unique compact almost automorphic weak solution u_f that is globally attractive.*

Proof. The proof is divided in five steps:

Step 1. We claim that the differential inclusion (4.248) has a bounded weak solution u_f on $\mathbb{R}$. Let $n \in \mathbb{N}$ and consider the following problem:

$$\begin{cases} u'(t) + \mathcal{A}u(t) \ni f(t), \\ u(-n) = 0. \end{cases} \tag{4.281}$$

Then, (4.281) has a unique weak solution u_n on $[-n, +\infty)$. Since $\mathcal{A}0 \ni 0$, it follows by Theorem 4.7.11, with $\tilde{u} = u_n, \tilde{f} = f, \hat{u} = 0$, and $\hat{f} = 0$, that

$$|u_n(t)| \leq \int_{-n}^{t} e^{-a(t-\sigma)}|f(\sigma)|d\sigma \quad \text{for } t \in [-n, +\infty).$$

The compact almost automorphy of the function f implies its boundedness. Let $M_f = \sup_{t\in\mathbb{R}} |f(t)|$. Then, the latter inequality gives

$$|u_n(t)| \leq \frac{M_f}{a}(1 - e^{-a(t+n)}) \quad \text{for } t \in [-n, +\infty);$$

consequently,

$$|u_n(t)| \leq \frac{M_f}{a} \quad \text{for } t \in [-n, +\infty). \tag{4.282}$$

Let $I = [a, b]$ and n and m be such that $-n \leq -m \leq a$. Using Theorem 4.7.11 for $\tilde{u} = u_n$, $\tilde{f} = f, \hat{u} = u_m$, and $\hat{f} = f$, we get that

$$|u_n(t) - u_m(t)| \leq e^{-a(t+m)}|u_n(-m) - u_m(-m)| = e^{-a(t+m)}|u_n(-m)| \quad \text{for } t \in I. \tag{4.283}$$

Inequalities (4.282) and (4.283) imply

$$|u_n(t) - u_m(t)| \leq \frac{M_f}{a}e^{-a(a+m)} \quad \text{for } t \in I,$$

which yields that $(u_n)_n$ is a Cauchy sequence in $C(I, \mathcal{H})$ and hence converges to u_f in $C(I, \mathcal{H})$. By Lemma 4.7.8, the function u_f is a weak solution of (4.248) on I. Since I is arbitrary, by (4.282), u_f is a bounded weak solution on $\mathbb{R}$ of (4.248).

Step 2. We claim that u_f is unique. Suppose that v is another bounded weak solution on $\mathbb{R}$ of (4.248). By Theorem 4.7.11, we have for $t, \sigma \in \mathbb{R}$, $t \geq \sigma$,

$$|u_f(t) - v(t)| \leq e^{-a(t-\sigma)}|u_f(\sigma) - v(\sigma)|. \tag{4.284}$$

Since u_f and v are bounded, by letting $\sigma \to -\infty$ in (4.284), we obtain that $u_f = v$. The global attractivity also follows from (4.284) by taking $\sigma = 0$ and letting $t \to +\infty$.

Step 3. We claim that the range of u_f is relatively compact. Let $(t'_n)_n \subset \mathbb{R}$. From the almost automorphy of f, there exists a subsequence $(t_n)_n \subset (t'_n)_n$ such that, for each $t \in \mathbb{R}$,

$$|f(t + t_p) - f(t + t_q)| \to 0$$

as $p, q \to +\infty$. Let

$$\begin{cases} u_n(t) := u_f(t+t_n) & \text{for } t \in \mathbb{R}, \\ f_n(t) := f(t+t_n) & \text{for } t \in \mathbb{R}. \end{cases}$$

We claim that $(u_n(t))_n$ is a Cauchy sequence in $\mathcal{H}$ for each $t \in \mathbb{R}$. In fact, u_p and u_q are weak solutions of the following differential inclusions:

$$\begin{cases} u_p'(t) + \mathcal{A}u_p(t) \ni f_p(t) & \text{for } t \in \mathbb{R}, \\ u_q'(t) + \mathcal{A}u_q(t) \ni f_q(t) & \text{for } t \in \mathbb{R}. \end{cases} \tag{4.285}$$

Applying Theorem 4.7.11 to (4.285), we obtain that, for $t \geq \sigma$,

$$|u_p(t) - u_q(t)| \leq e^{-\alpha(t-\sigma)}|u_p(\sigma) - u_q(\sigma)| + \int_\sigma^t e^{-\alpha(t-s)}|f_p(s) - f_q(s)|\mathrm{d}s.$$

Using the boundedness of u_f and letting $\sigma \to -\infty$, we deduce that, for each $t \in \mathbb{R}$,

$$\begin{aligned} |u_p(t) - u_q(t)| &\leq \int_{-\infty}^t e^{-\alpha(t-s)}|f_p(s) - f_q(s)|\mathrm{d}s \\ &= \int_{-\infty}^t e^{-\alpha(t-s)}|f(s+t_p) - f(s+t_q)|\mathrm{d}s. \end{aligned}$$

Using Lebesgue's dominated convergence theorem, we conclude that $(u_n(t))_n$ is a Cauchy sequence in $\mathcal{H}$ for each $t \in \mathbb{R}$. Therefore, u_f has a relatively compact range.

Step 4. We claim that the solution u_f is uniformly continuous on $\mathbb{R}$. In fact, by Theorem 4.7.11 with $\tilde{u} = u_f(\cdot + h)$, $\hat{u} = u_f(\cdot)$, $\tilde{f} = f(\cdot + h)$, and $\hat{f} = f(\cdot)$, we obtain that, for $t \geq \sigma$,

$$|u_f(t+h) - u_f(t)| \leq e^{-\alpha(t-\sigma)}|u_f(\sigma+h) - u_f(\sigma)| + \int_\sigma^t e^{-\alpha(t-s)}|f(s+h) - f(s)|\mathrm{d}s.$$

Since u_f is bounded, we obtain by letting $\sigma \to -\infty$ that, for each $t \in \mathbb{R}$,

$$\begin{aligned} |u_f(t+h) - u_f(t)| &\leq \int_{-\infty}^t e^{-\alpha(t-s)}|f(s+h) - f(s)|\mathrm{d}s \\ &\leq \frac{1}{\alpha}\sup_{t\in\mathbb{R}}|f(t+h) - f(t)|, \end{aligned}$$

which implies that

$$\sup_{t\in\mathbb{R}}|u_f(t+h) - u_f(t)| \leq \frac{1}{\alpha}\sup_{t\in\mathbb{R}}|f(t+h) - f(t)|.$$

By Theorem 2.3.6, we get that u_f is uniformly continuous on $\mathbb{R}$.

Step 5. We claim that u_f is compact almost automorphic. Let $(t_n)_n \subset \mathbb{R}$. Since f is compact almost automorphic, then there exist a subsequence $(t'_n)_n \subset (t_n)_n$ and a continuous function $g : \mathbb{R} \to \mathcal{H}$ such that

$$\left| f(t + t'_n) - g(t) \right| \to 0 \quad \text{as } n \to +\infty,$$
$$\left| g(t - t'_n) - f(t) \right| \to 0 \quad \text{as } n \to +\infty$$

uniformly on any compact subset of $\mathbb{R}$. By Lemma 4.7.14, one can extract another subsequence $(t''_n)_n \subset (t'_n)_n \subset (t_n)_n$ such that

$$u_f(t + t''_n) \to y(t) \quad \text{as } n \to +\infty \tag{4.286}$$

uniformly on any compact subset of $\mathbb{R}$, where y is a weak solution on $\mathbb{R}$ of the following differential inclusion:

$$y'(t) + \mathcal{A}y(t) \ni g(t).$$

Since y is also uniformly continuous and its range is relatively compact, applying the same procedure to the function y using the returning sequence $(-t''_n)_n$, we have another subsequence $(t'''_n)_n \subset (t''_n)_n \subset (t'_n)_n \subset (t_n)_n$ such that

$$y(t - t'''_n) \to z(t) \quad \text{as } n \to +\infty \tag{4.287}$$

uniformly on any compact subset of $\mathbb{R}$, where z is a bounded weak solution of (4.248) on $\mathbb{R}$. From the uniqueness of the bounded weak solution (Step 2), we conclude that $z = u_f$. Thus, it follows from (4.286) and (4.287) that u_f is compact almost automorphic. $\square$

Theorem 4.7.17. *Assume that $\mathcal{A}$ is α-strongly maximal monotone ($\alpha > 0$) with $0 \in \mathcal{A}0$ and $g : \mathbb{R} \times \mathcal{H} \to \mathcal{H}$ is compact almost automorphic in t and Lipschitz with respect to the second argument. Then, (4.249) has a unique compact almost automorphic weak solution provided that $\mathrm{Lip}(g) < \alpha$ where $\mathrm{Lip}(g)$ is the Lipschitz constant of g.*

Proof. Let $v : \mathbb{R} \to \mathcal{H}$ be a compact almost automorphic function. Consider the following differential inclusion:

$$u'(t) + \mathcal{A}u(t) \ni g(t, v(t)) \quad \text{for } t \in \mathbb{R}. \tag{4.288}$$

By Theorem 4.4.3, the function $t \mapsto g(t, v(t))$ is compact almost automorphic. It follows from Theorem 4.7.16 that the differential inclusion (4.288) has a unique compact almost automorphic weak solution u_v. Let T be defined by

$$T : AA_c(\mathbb{R}, \mathcal{H}) \to AA_c(\mathbb{R}, \mathcal{H}),$$
$$v \mapsto u_v.$$

Then, T is well defined. Let $v, w \in AA_c(\mathbb{R}, \mathcal{H})$. Applying Theorem 4.7.11 to $\tilde{u} = u_v$, $\tilde{f} = g(\cdot, v(\cdot))$, $\hat{u} = u_w$, and $\hat{f} = g(\cdot, w(\cdot))$, we obtain that

$$|Tv(t) - Tw(t)| \leq e^{-\alpha(t-\sigma)}|u_v(\sigma) - u_w(\sigma)| + \int_\sigma^t e^{-\alpha(t-s)}|g(s, v(s)) - g(s, w(s))| \mathrm{d}s \quad \text{for } t \geq \sigma.$$

Letting $\sigma \to -\infty$, we obtain that, for each $t \in \mathbb{R}$,

$$\begin{aligned} |Tv(t) - Tw(t)| &\leq \int_{-\infty}^t e^{-\alpha(t-s)}|g(s, v(s)) - g(s, w(s))| \mathrm{d}s \\ &\leq \frac{\mathrm{Lip}(g)}{\alpha}|v - w|_\infty. \end{aligned}$$

This means that T is a strict contraction. We deduce that the operator T has a unique fixed point that is the unique compact almost automorphic weak solution of (4.249). □

4.7.4 Hyperbolic and parabolic inclusions

4.7.4.1 A dissipative hyperbolic system

We give an existence theorem of compact almost automorphic weak solutions for the following dissipative nonlinear wave inclusion:

$$\begin{cases} \frac{\partial^2}{\partial t^2}u(t,x) - \Delta u(t,x) + \beta(\frac{\partial}{\partial t}u(t,x)) \ni \theta(t,x) & \text{for } (t,x) \in \mathbb{R} \times \Omega, \\ u(t,x) = 0 \quad \text{for } (t,x) \in \mathbb{R} \times \partial\Omega. \end{cases} \tag{4.289}$$

We assume that:

(A1) Ω is a bounded open set in $\mathbb{R}^N$ with smooth boundary $\partial\Omega$ such that $\dim(\Omega) \geq 2$.

(A2) β is a strongly maximal monotone graph in $\mathbb{R} \times \mathbb{R}$ with $0 \in \beta 0$ such that

$$|\beta^0 w| \leq C_1|w|^k + C_2 \quad \text{with } 0 \leq k \leq \frac{N+2}{N-2},$$

where $\beta^0 w = \mathrm{Proj}_{\beta(w)}(0)$ (see Theorem 4.7.4 and Remark 4.7.5).

(A3) $\theta : \mathbb{R} \times \overline{\Omega} \to \mathbb{R}$ satisfies $\frac{\partial\theta}{\partial t} \in S^2(\mathbb{R}, L^2(\Omega))$ where

$$S^2(\mathbb{R}, L^2(\Omega)) = \left\{h \in L^2_{\mathrm{loc}}(\mathbb{R}, L^2(\Omega)) : \sup_{t\in\mathbb{R}} \int_t^{t+1} |h(s)|^2_{L^2(\Omega)} \mathrm{d}s < +\infty\right\},$$

and the function $t \mapsto \theta(t, \cdot)$ is in $AA_c(\mathbb{R}, L^2(\Omega))$. That is, for any $(t'_n)_n \subset \mathbb{R}$, there exist a subsequence $(t_n)_n$ and a continuous function $\tilde{\theta} : \mathbb{R} \times \overline{\Omega} \to \mathbb{R}$ such that

$$\int_\Omega \left|\theta(t+t_n,\omega)-\tilde{\theta}(t,\omega)\right|^2 d\omega \to 0 \quad \text{as } n \to +\infty,$$

$$\int_\Omega \left|\tilde{\theta}(t-t_n,\omega)-\theta(t,\omega)\right|^2 d\omega \to 0 \quad \text{as } n \to +\infty$$

uniformly on any compact subset of $\mathbb{R}$.

Let $\mathcal{H} = H_0^1(\Omega) \times L^2(\Omega)$ be the Hilbert space endowed with the following norm:

$$|(\phi_1,\phi_2)|_{\mathcal{H}} = \left(\int_\Omega \left(|\nabla\phi_1(s)|^2 + |\phi_1(s)|^2 + |\phi_2(s)|^2\right) ds\right)^{\frac{1}{2}},$$

and B be the canonical extension of β to $L^2(\Omega)$ defined in [171, p. 53] by

$$(u,v) \in G(B) \quad \text{if and only if} \quad (u(x),v(x)) \in G(\beta) \quad \text{for almost all } x \in \Omega.$$

Let

$$\begin{cases} D(\mathcal{L}) = H_0^1(\Omega) \cap H^2(\Omega) \times H_0^1(\Omega), \\ \mathcal{L} = \begin{pmatrix} 0 & -I \\ -\Delta & 0 \end{pmatrix}, \end{cases} \qquad \begin{cases} D(\mathcal{B}) = H_0^1(\Omega) \times D(B), \\ \mathcal{B} = \begin{pmatrix} 0 & 0 \\ 0 & B \end{pmatrix}. \end{cases}$$

Lemma 4.7.18 ([171, p. 93]). *The operator $D(\mathcal{A}) = D(\mathcal{L}) \cap D(\mathcal{B})$, $\mathcal{A} = \mathcal{L} + \mathcal{B}$ is maximal monotone on $\mathcal{H} = H_0^1(\Omega) \times L^2(\Omega)$.*

Let $f : \mathbb{R} \to \mathcal{H}$ be the function defined by

$$f(t)(\omega) = \begin{pmatrix} 0 \\ \theta(t,\omega) \end{pmatrix} \quad \text{for } t \in \mathbb{R} \text{ and } \omega \in \Omega.$$

Then, by assumption **(A3)**, $f \in AA_c(\mathbb{R},\mathcal{H})$. If we take $U = \begin{pmatrix} u \\ \frac{\partial u}{\partial t} \end{pmatrix}$, then (4.289) takes the following abstract form:

$$U'(t) + \mathcal{A}U(t) \ni f(t) \quad \text{for } t \in \mathbb{R}. \tag{4.290}$$

Lemma 4.7.19 ([66, Theorem 2.1]). *Let $U(t) = \begin{pmatrix} u \\ \frac{\partial u}{\partial t} \end{pmatrix}$ be a solution which starts at $\begin{pmatrix} u_0 \\ v_0 \end{pmatrix} \in D(\mathcal{A})$. Then,*

$$\frac{\partial^2 u}{\partial t^2} \in L^\infty(\mathbb{R}^+, L^2(\Omega)) \quad \textit{and} \quad \frac{\partial u}{\partial t} \in L^\infty(\mathbb{R}^+, H_0^1(\Omega)).$$

Lemma 4.7.20 ([66]). *Let $U(t) = \begin{pmatrix} u \\ \frac{\partial u}{\partial t} \end{pmatrix}$ be a solution which starts at $\begin{pmatrix} u_0 \\ v_0 \end{pmatrix} \in D(\mathcal{A})$. Then, $U(t)$ has a relatively compact range in the energy space $\mathcal{H} = H_0^1(\Omega) \times L^2(\Omega)$.*

As a consequence, we have the following result.

Theorem 4.7.21. *The differential inclusion* (4.289) *has at least a weak solution in* $AA_c(\mathbb{R}, H_0^1(\Omega) \times L^2(\Omega))$.

Proof. Using Lemmas 4.7.19 and 4.7.20, any trajectory $(U(t))_{t\geq 0}$ which starts at $U_0 \in D(\mathcal{A})$ is uniformly continuous on $\mathbb{R}^+$ in $H_0^1(\Omega) \times L^2(\Omega)$ and its range over $\mathbb{R}^+$ is relatively compact. In view of Theorem 4.7.13, the differential inclusion (4.290) has at least a compact almost automorphic weak solution. □

Remark 4.7.22. The differential inclusion (4.289) has been considered in the periodic and almost periodic case in [33–35, 63, 66, 170–172].

4.7.4.2 A dissipative parabolic system

Consider the following system:

$$\begin{cases} \frac{\partial}{\partial t} w(t,x) - \Delta w(t,x) + \beta(w(t,x)) + \alpha w(t,x) \ni \gamma(w(t,x))) + h(t,x) \\ \quad \text{for } (t,x) \in \mathbb{R} \times \Omega, \\ w(t,x) = 0 \quad \text{for } (t,x) \in \mathbb{R} \times \partial\Omega, \end{cases} \tag{4.291}$$

where $\alpha > 0$. Assume that:

(B1) Ω is a smooth subset of $\mathbb{R}^N$ with a regular boundary $\partial\Omega$.
(B2) β is a maximal monotone graph in $\mathbb{R} \times \mathbb{R}$ with $0 \in \beta 0$.
(B3) $\gamma : \mathbb{R} \to \mathbb{R}$ is a Lipschitz function such that $\gamma(0) = 0$. Let L_γ be its Lipschitz constant.
(B4) The function $t \mapsto h(t,\cdot)$ belongs to $AA_c(\mathbb{R}, L^2(\Omega))$.

Let B be the canonical extension of β to $L^2(\Omega)$ defined in [171, p. 53] by

$$(u,v) \in G(B) \quad \text{if and only if} \quad (u(x), v(x)) \in G(\beta) \quad \text{for almost all } x \in \Omega.$$

Let A_1 be defined in $L^2(\Omega)$ by

$$\begin{cases} D(A_1) = \{u \in H^2(\Omega) \cap H_0^1(\Omega) : \beta(u) \in L^2(\Omega)\}, \\ A_1 u = -\Delta u + Bu. \end{cases}$$

It is well known from [171, p. 88] that A_1 is maximal monotone. Hence, by Remark 4.7.2, the operator

$$\begin{cases} D(\mathcal{A}) = D(A_1), \\ \mathcal{A}u = A_1 u + \alpha u \end{cases}$$

is α-strongly maximal monotone. Using the fact that $0 \in \beta 0$, we get that $0 \in \mathcal{A}0$. Take $\mathcal{H} = L^2(\Omega)$.

We consider the function $f : \mathcal{H} \to \mathcal{H}$ defined by

$$f(x)(\omega) = \gamma(x(\omega)) \quad \text{for } x \in \mathcal{H} \text{ and } \omega \in \Omega.$$

By **(B3)**, one see that f is well defined. Using **(B3)** again, we obtain that f is Lipschitz with a Lipschitz constant $L_f = L_\gamma$. Furthermore, $f \in C(\mathcal{H}, \mathcal{H})$.

Let $H : \mathbb{R} \to \mathcal{H}$ be defined by

$$H(t)(\omega) = h(t, \omega) \quad \text{for } t \in \mathbb{R} \text{ and } \omega \in \Omega.$$

Assumption **(B4)** implies that $H \in AA_c(\mathbb{R}, \mathcal{H})$.

Let $g : \mathbb{R} \times \mathcal{H} \to \mathcal{H}$ be defined by

$$g(t, x) = f(x) + H(t) \quad \text{for } t \in \mathbb{R} \text{ and } x \in \mathcal{H}.$$

We deduce that $g \in AA_c(\mathbb{R}\times\mathcal{H}, \mathcal{H})$ and g is Lipschitz with respect to the second argument with a Lipschitz constant $L_g = L_\gamma$.

If we take $u(\cdot)(x) = w(\cdot, x)$, then (4.291) takes the following abstract form:

$$u'(t) + \mathcal{A}u(t) \ni g(t, u(t)) \quad \text{for } t \in \mathbb{R},$$

in the Hilbert space $\mathcal{H}$. Now, if we suppose that $L_\gamma < \alpha$, then all the assumptions of Theorem 4.7.17 are fulfilled. Consequently, we get the following result.

Theorem 4.7.23. *The system* (4.291) *has a unique compact almost automorphic weak solution provided* $L_\gamma < \alpha$.

4.8 Almost periodic elliptic equations: Sub- and super-solutions

The method of sub- and super-solutions (alternative terms are upper and lower solutions) is a popular and powerful tool in the existence theory of boundary value problems for which a maximum principle holds. In the framework of ordinary differential equations, basic ideas can be traced back to E. Picard and O. Perron. Since that time, hundreds of papers have been published in this direction. A detailed account of the existing results in the case of ordinary differential equations can be found in [106] (see also the references therein). The method extends to elliptic and parabolic partial differential equations. Simplest results of such kind can be found, e. g., in [107, Chapter IV, Appendix] and [251, Chapter 1]. For further development and applications to real world problems, we refer to [197, 262] and the references therein.

In particular, the sub- and super-solution method provides results on the existence of solutions to periodic boundary value problems, both in one and many spatial dimensions. Therefore, it is quite natural to try to develop the method for the existence of almost periodic solutions to almost periodic differential equations. First results in the case of nonlinear ordinary differential equations have been obtained by M. Krasnosel'skii, V. Burd, and Yu. Kolesov [189, Section 10] by using the theory of monotone operators. Those authors found existence results for almost periodic solutions under certain assumptions that, in particular, guarantee the uniqueness of the solution.

In 1983, A. Pankov [245] (see also [246, Section 5.1.2]) considered almost periodic semilinear elliptic equationsof second order. Assuming the existence of a properly ordered pair of sub- and super-solutions, he was able to prove the existence of a very weak almost periodic solution in the sense of Besicovitch–Sobolev spaces. This solution is almost periodic in the classical sense under certain additional assumptions that ensure the uniqueness of the solution.

In the case of ordinary differential equations, the authors of [257] made an attempt to prove that an ordered pair of Bohr almost periodic sub- and super-solutions gives rise to the existence of a Bohr almost periodic solution without any uniqueness assumption. Unfortunately, that result is wrong. R. Ortega and M. Tarallo [243] have constructed an example of an almost periodic equation of the form

$$-u'' + cu = g(t, u)$$

that possesses an ordered pair of *constant* sub- and super-solutions, but has no almost periodic solution between them. Actually, in their example the function g is quasiperiodic in t with two independent frequencies.

Our main aim in this work is to understand what happens if there are ordered sub- and super-solutions, but the uniqueness does not hold. Basically, we prove that in the envelope of the equation under consideration there exist a residual set Ω_{aa} and a set of full measure Ω_b such that for every equation in Ω_{aa} there is an almost automorphic solution, while equations in Ω_b possess bounded almost periodic in the sense of Besicovitch solutions. We do not know whether at least one equation in the envelope has a solution that is, at the same time, almost automorphic and almost periodic in Besicovitch sense, i. e., $\Omega_{aa} \cap \Omega_b \neq \emptyset$. In addition, we weaken regularity assumptions made in [245] and show that the frequency modulus of a solution obtained is contained in the frequency modulus of the equation. Let us point out that the main ingredients of this section are monotone iteration techniques and the metrizability of appropriate Bohr compactifications.

The organization of this section is as follows. In Section 4.8.1 we sketch basic facts on Bohr almost periodic functions. Our approach is based on the notion of Bohr compactification and follows [246, 260]. A standard presentation of the theory can be found in [199]. In Section 4.8.2 we remind the notion of almost automorphy (see, e. g., [234, 237] for more details) and prove a technical result needed later on. Section 4.8.3 is devoted to a brief account of Besicovitch almost periodicity. In Section 4.8.4 we study almost periodic second order elliptic equations. The main result of the section, Theorem 4.8.10, provides sufficient conditions for the existence of almost periodic solutions, including the modulus containment property. The central part is Section 4.8.5 in which the main result, Theorem 4.8.11, is proven. In Theorem 4.8.14 we give certain sufficient conditions for the existence of constant sub- and super-solutions. In addition, we show that if each equation in the envelope has at most one bounded solution, the solution constructed in Theorem 4.8.11 is almost periodic in the sense of Bohr. In Section 4.8.6 we give an application of Theorem 4.8.11 to the pendulum equation with almost periodic forcing.

Note that, to simplify the notation, we often denote by C a *generic* positive constant.

4.8.1 Bohr compactification

Let $C_b(\mathbb{R}^n)$ denote the space of all bounded continuous functions on $\mathbb{R}^n$. Endowed with the norm

$$\|f\| = \sup_{x\in\mathbb{R}^n} |f(x)|,$$

this is a Banach space. According to the Bochner definition, a function $f \in C_b(\mathbb{R}^n)$ is *almost periodic* (shortly, a. p.) in the sense of Bohr if the family of shifts $\{f(\cdot + y)\}_{y\in\mathbb{R}^n}$ is precompact in $C_b(\mathbb{R}^n)$. The set of all a. p. functions is a closed linear subspace of $C_b(\mathbb{R}^n)$, hence, a Banach space. We denote by $\mathrm{CAP}(\mathbb{R}^n)$ the space of all almost periodic functions on $\mathbb{R}^n$.

An important property of a. p. functions is the existence of the mean value. Let

$$K_T = \{x \in \mathbb{R}^n : |x_k| \le T, k = 1, 2, \dots, k\}.$$

The *mean value* of an almost periodic function f is defined by

$$\langle f\rangle = \lim_{T\to\infty} \frac{1}{(2T)^n} \int_{a+K_T} f(x)\,dx. \tag{4.292}$$

The limit in (4.292) exists uniformly in $a \in \mathbb{R}^n$ and is independent of a.

The following statement is often considered as the main result on almost periodic functions. Let $\mathrm{Trig}(\mathbb{R}^n)$ be the space of all trigonometric polynomials, i. e., finite sums of the form

$$\sum a_j \exp(i\xi_j \cdot x),$$

where $a_j \in \mathbb{C}$, $\xi_j \in \mathbb{R}^n$, and

$$x \cdot y = \sum_{k=1}^{n} x_k y_k$$

is the standard dot product in $\mathbb{R}^n$.

Proposition 4.8.1 (Approximation theorem). *The space* $\mathrm{Trig}(\mathbb{R}^n)$ *is a dense subspace of the Banach space* $\mathrm{CAP}(\mathbb{R}^n)$.

See [246, Proposition 1.3, Chapter 1].

The *Fourier–Bohr transform* of an almost periodic function f is defined by

$$\hat{f}(\xi) = \langle f(x)\exp(-\mathrm{i}\xi\cdot x)\rangle. \tag{4.293}$$

The set

$$\sigma(f) = \{\xi \in \mathbb{R}^n : \hat{f}(\xi) \neq 0\} \tag{4.294}$$

is called the *spectrum* of an almost periodic function f. It follows immediately from Proposition 4.8.1 that, for any a. p. function f, the set $\sigma(f)$ is at most countable. The additive subgroup $\mathrm{Mod}(f)$ of $\mathbb{R}^n$ generated by $\sigma(f)$ is called the *modulus* of the function f.

Now we give a brief description of Bohr compactifications of the (additive group of) space $\mathbb{R}^n$. The standard approach uses Pontryagin's duality theory (see, e. g., [177, Chapter 6] for a detailed presentation of the theory of locally compact abelian groups, including Pontryagin's duality). Consider $\mathbb{R}^n$ as a locally compact abelian group. Its dual group, $(\mathbb{R}^n)'$, consists of all characters which are, in this case, functions of the form $\exp(\mathrm{i}\xi\cdot x)$. The correspondence $\exp(\mathrm{i}\xi\cdot x)\mapsto\xi$ is an isomorphism $(\mathbb{R}^n)'\simeq\mathbb{R}^n$. Denote by $(\mathbb{R}^n)'_d$ the group $(\mathbb{R}^n)'$ endowed with the discrete topology. We set $\mathbb{R}^n_B = ((\mathbb{R}^n)'_d)'$. This is a compact abelian group called the *Bohr compactification* of $\mathbb{R}^n$. Also we introduce the dual homomorphism

$$i_B : \mathbb{R}^n = (\mathbb{R}^n)'' \to \mathbb{R}^n_B = ((\mathbb{R}^n)'_d)'$$

to the identity homomorphism $(\mathbb{R}^n)'_d \to (\mathbb{R}^n)'$. The homomorphism i_B is injective and its image $i_B(\mathbb{R}^n)$ is a dense subgroup in $\mathbb{R}^n_B$.

In what follows, we need a more general notion of relative Bohr compactification. Let $\Gamma \subseteq (\mathbb{R}^n)'$ be an nonzero additive subgroup considered as a discrete group (later on we always suppose that $\Gamma \neq \{0\}$). The *Bohr compactification* of $\mathbb{R}^n$ *relative to* Γ is defined as $\mathbb{R}^n_{B,\Gamma} = \Gamma'$. The homomorphism

$$i_{B,\Gamma} : \mathbb{R}^n \to \mathbb{R}^n_{B,\Gamma}$$

is defined as the dual to the identity map $\Gamma \to \mathbb{R}^n$. Its image is still a dense subgroup of $\mathbb{R}^n_{B,\Gamma}$. The kernel $\ker i_{B,\Gamma}$ is a linear subspace of $\mathbb{R}^n$ orthogonal to the linear subspace of $(\mathbb{R}^n)'$ generated by Γ. If $\Gamma = (\mathbb{R}^n)'$, we return to the original Bohr compactification.

The main result on Bohr compactifications is the following

Proposition 4.8.2. *A function f on* $\mathbb{R}^n$ *is almost periodic, with* $\mathrm{Mod}(f) \subseteq \Gamma$, *if and only if f is of the form*

$$f(x) = \tilde{f}(i_{B,\Gamma}x),$$

where $\tilde{f}$ *is a (unique) continuous function on* $\mathbb{R}^n_{B,\Gamma}$.

See [246, Proposition 3.5, Chapter 1].

Let f be an a. p. function and $\Gamma \supseteq \mathrm{Mod}(f)$. We use $\tilde{f}$ as a standard notation for the function on the Bohr compactification given in Proposition 4.8.2. For any $s \in \mathbb{R}^n_{B,\Gamma}$, we set

$$f^{(s)}(x) = \tilde{f}(s + i_{B,\Gamma}x), \quad x \in \mathbb{R}^n.$$

Notice that the map $s \mapsto f^{(s)}$ is a continuous map from $\mathbb{R}^n_{B,\Gamma}$ into $\mathrm{CAP}(\mathbb{R}^n)$. The set of a. p. functions $H(f) = \{f^{(s)}\}_{s\in\mathbb{R}^n_{B,\Gamma}}$ is called the *envelope* of f. The envelope $H(f)$ is independent of the choice of $\Gamma \supseteq \mathrm{Mod}(f)$. Actually, $H(f)$ is the closure of the set of shifts $\{f(\cdot + y)\}_{y\in\mathbb{R}^n}$. In the classical literature, the later property is accepted as the definition of the envelope.

Also let us point out that, for any additive subgroup $\Gamma \subseteq \mathbb{R}^n$,

$$\mathrm{CAP}_\Gamma(\mathbb{R}^n) = \{f \in \mathrm{CAP}(\mathbb{R}^n) : \mathrm{Mod}(f) \subseteq \Gamma\}$$

is a closed linear subspace of $\mathrm{CAP}(\mathbb{R}^n)$. By Proposition 4.8.2, the operator $J_\Gamma : f \mapsto \tilde{f}$ is an isometric isomorphism from the Banach space $\mathrm{CAP}_\Gamma(\mathbb{R}^n)$ onto the Banach space $C(\mathbb{R}^n_{B,\Gamma})$.

Now we complement Proposition 4.8.2 with the following result that expresses the mean value of an almost periodic function in terms of Bohr compactification (see [246, 260]). Let $\mu = \mu_\Gamma$ be the Haar measure on $\mathbb{R}^n_{B,\Gamma}$, i. e., a unique positive translation invariant measure such that $\mu(\mathbb{R}^n_{B,\Gamma}) = 1$ (see, e. g., [177, Chapter 4]).

Proposition 4.8.3. *For every* $f \in \mathrm{CAP}_\Gamma(\mathbb{R}^n)$, *we have*

$$\langle f\rangle = \int_{\mathbb{R}^n_{B,\Gamma}} \tilde{f}(s)\, d\mu(s).$$

Also we need a refined version of Proposition 4.8.1.

Proposition 4.8.4. *Given a countable subgroup* $\Gamma \subseteq \mathbb{R}^n$, *there exists a sequence of trigonometric polynomials* $P_m(x)$ *with the following properties:*

(a) $P_m(x) \geq 0$ *for all* $x \in \mathbb{R}^n$;

(b) $\langle P_m\rangle = 1$;

(c) *For any* $f \in \mathrm{CAP}_\Gamma(\mathbb{R}^n)$, *the sequence of trigonometric polynomials*

$$f_m(x) = \langle f(y)P_m(x-y)\rangle_y = \langle f(x-y)P_m(y)\rangle_y$$

belongs to $\mathrm{CAP}_\Gamma(\mathbb{R}^n)$ *and converges to* f *in that space.*

For a proof, we refer to [246, 260]. The trigonometric polynomials P_m are called the *Bochner–Fejer kernels*, while f_m are the *Bochner–Fejer approximations* of f.

It is known (see, e. g., [177, Section 24]) that a compact abelian group is metrizable if and only if its dual group is countable. Hence, the Bohr compactification $\mathbb{R}^n_{B,\Gamma}$ is metrizable whenever the subgroup $\Gamma \subseteq \mathbb{R}^n$ is countable. In the rest of this section, we accept the following convention: the symbol Γ denotes a *countable* subgroup of $\mathbb{R}^n$ so that $\mathbb{R}^n_{B,\Gamma}$ is metrizable.

4.8.2 Almost automorphic functions

Let us remind the notion of almost automorphic function due to S. Bochner. For details, we refer to [234, 237] and the references therein. A function $f \in C_b(\mathbb{R}^n)$ is *almost automorphic* if for every sequence $y_k \in \mathbb{R}^n$ there exists a subsequence $y_{k'}$ such that the pointwise limit

$$\lim f(x + y_{k'}) = g(x) \tag{4.295}$$

exists and

$$\lim g(x - y_{k'}) = f(x) \tag{4.296}$$

pointwise. Note that the function g in (4.295) is measurable, but not necessarily continuous.

An almost automorphic function f is *uniformly almost automorphic* if the limits in (4.295) and (4.296) are uniform on compact subsets of $\mathbb{R}^n$, i. e., in the space $C(\mathbb{R}^n)$ which is a Fréchet space. Equivalently, f is uniformly almost automorphic if all functions g that appear in (4.295) are continuous (see [236]).

We denote by $\mathrm{AA}(\mathbb{R}^n)$ (respectively, $\mathrm{AA}_u(\mathbb{R}^n)$) the sets of all almost automorphic (respectively, uniformly almost automorphic) functions on $\mathbb{R}^n$. These are closed linear subspaces in $C_b(\mathbb{R}^n)$. Notice that

$$\mathrm{CAP}(\mathbb{R}^n) \subset \mathrm{AA}_u(\mathbb{R}^n) \subset \mathrm{AA}(\mathbb{R}^n)$$

and all the inclusions are strict.

Proposition 4.8.5. *Let $\tilde{u}$ be a function on $\mathbb{R}^n_{B,\Gamma}$, where $\Gamma \subset \mathbb{R}^n$ is a countable subgroup. Suppose that for all $s \in \mathbb{R}^n_{B,\Gamma}$ the function $u^{(s)}(x) = \tilde{u}(s + i_{B,\Gamma}x)$ belongs to $C_b(\mathbb{R}^n)$ and is uniformly continuous. If the map*

$$U : \mathbb{R}^n_{B,\Gamma} \ni s \mapsto u^{(s)} \in C(\mathbb{R}^n)$$

is continuous at the point $s_0 \in \mathbb{R}^n_{B,\Gamma}$, then U is continuous at each point of the orbit $s_0 + i_{B,\Gamma}\mathbb{R}^n$ and $u^{(s_0)} \in \mathrm{AA}_u(\mathbb{R}^n)$.

Proof. Suppose that $s'_m \to s'_0 = s_0 + t_0$ in $\mathbb{R}^n_{B,\Gamma}$, where $t_0 = i_{B,\Gamma}x_0$. Then $s_m = s'_m - t_0 \to s_0$ and, by continuity, $u^{(s_m)} \to u^{(s_0)}$ in the space $C(\mathbb{R}^n)$. Hence,

$$u^{(s'_m)}(\cdot) = u^{(s_m)}(\cdot + x_0) \to u^{(s_0)}(\cdot + x_0) = u^{(s'_0)}(\cdot)$$

in $C(\mathbb{R}^n)$, and the first statement of the proposition follows.

Denote by T the closure of the set

$$\{(s_0 + i_{B,\Gamma}y, u^{(s_0)}(\cdot + y)) : y \in \mathbb{R}^n\}$$

in the space $\mathbb{R}^n_{B,\Gamma} \times C(\mathbb{R}^n)$. Since the function $u^{(s_0)}$ is bounded and uniformly continuous, the Arzelà–Ascoli theorem implies that T is a compact subset of $\mathbb{R}^n_{B,\Gamma} \times C(\mathbb{R}^n)$. The projection of T on the first factor is a surjective map, while the image H of the other projection is the closure of the set $\{u^{(s_0)}(\cdot + y) : y \in \mathbb{R}^n\}$ in $C(\mathbb{R}^n)$, the so-called *hull* of $u^{(s_0)}$. We set

$$H_s = \{f \in C(\mathbb{R}^n) : (s, f) \in T\}$$

for any $s \in \mathbb{R}^n_{B,\Gamma}$. This is a nonempty closed subset of T.

First, we show that $H_{s_0} = \{u^{(s_0)}\}$. Indeed, suppose that $(s_0, f) \in T$. Then there exists $s_m = i_{B,\Gamma}x_m$ such that $s_m \to 0$ in $\mathbb{R}^n_{B,\Gamma}$ and

$$u^{(s_0)}(\cdot + x_m) = u^{(s_0+s_m)} \to f.$$

The continuity of U at s_0 implies that $f = u^{(s_0)}$.

Now let us prove that the function $u^{(s_0)}$ is uniformly almost automorphic. Since T is compact, for any sequence $t_m = i_{B,\Gamma}x_m$, $x_m \in \mathbb{R}^n$, there exists a subsequence $t_{m'}$ such that $s_0 + t_{m'} \to t_0$ and $t_0 - t_{m'} \to s_0$ in $\mathbb{R}^n_{B,\Gamma}$, and $u^{(s_0)}(\cdot + x_{m'}) \to f$ and $f(\cdot - x_{m'}) \to h$ in $C(\mathbb{R}^n)$. Since T is closed, $(t_0, f) \in T$ and $(s_0, h) \in T$. Hence, $h \in H_{s_0}$. Since $H_{s_0} = \{u^{(s_0)}\}$, we conclude that $h = u^{(s_0)}$ and, therefore, $u^{(s_0)} \in \mathrm{AA}_u(\mathbb{R}^n)$. This completes the proof. □

4.8.3 Besicovitch almost periodic functions

Let $L^p_{\mathrm{loc}}(\mathbb{R}^n)$, $1 \leq p \leq \infty$, stand for the local Lebesgue space with the exponent p. For any $f \in L^p_{\mathrm{loc}}(\mathbb{R}^n)$, $p < \infty$, we introduce the quantity

$$\|f\|_{(p)} = \limsup_{T\to\infty} \frac{1}{2T}\Big[\int_{K_T} |f(x)|^p \, dx\Big]^{1/p}. \tag{4.297}$$

Functions with a finite seminorm $\|f\|_{(p)}$ form the so-called *Marcinkiewicz space* $M^p(\mathbb{R}^n)$. It is easily seen that $M^p(\mathbb{R}^n) \subseteq M^q(\mathbb{R}^n)$ whenever $q \leq p$.

A function $f \in M^p(\mathbb{R}^n)$, $p < \infty$, is *Besicovitch almost periodic*, with the exponent p, if there is a sequence $f_n \in \mathrm{CAP}(\mathbb{R}^n)$ such that

$$\lim_{n\to\infty} \|f - f_n\|_{(p)} = 0.$$

The set of all such functions is denoted by $B^p(\mathbb{R}^n)$. Obviously, $B^p(\mathbb{R}^n) \subseteq B^q(\mathbb{R}^n)$ if $q \leq p$. It is not difficult to verify that, for any $f \in B^p(\mathbb{R}^n)$, the “lim sup” in (4.297) can be replaced by “lim”.

The spaces $M^p(\mathbb{R}^n)$ and $B^p(\mathbb{R}^n)$ are complete seminormed spaces, but *not* Banach spaces because the seminorm $\|\cdot\|_{(p)}$ has a nontrivial kernel.

For Besicovitch a. p. functions, the definition of the mean value given in (4.292) makes sense. The only difference is that, in general, the limit is not uniform with respect to a. Therefore, the notions of Fourier–Bohr transform and spectrum extend immediately to Besicovitch a. p. functions. Moreover, the spectrum is at most countable. The modulus, $\mathrm{Mod}(f)$, of a Besicovitch a. p. function f is well defined as well. Moreover, for any subgroup $\Gamma \subseteq \mathbb{R}^n$, we set

$$B^p_\Gamma(\mathbb{R}^n) = \{f \in B^p(\mathbb{R}^n) : \mathrm{Mod}(f) \subseteq \Gamma\}.$$

Obviously, this is a linear subspace of $B^p(\mathbb{R}^n)$ closed in the sense that if $f_n \in B^p_\Gamma(\mathbb{R}^n)$ and $\|f_n - f\|_{(p)} \to 0$, then $f \in B^p_\Gamma(\mathbb{R}^n)$.

By Proposition 4.8.3, the operator J_Γ initially defined on $\mathrm{CAP}_\Gamma(\mathbb{R}^n)$ extends uniquely to an isometric epimorphism

$$J_\Gamma : B^p_\Gamma(\mathbb{R}^n) \to L^p(\mathbb{R}^n_{B,\Gamma}), \quad p \in [1,\infty).$$

Its kernel consists of all functions $f \in B^p_\Gamma(\mathbb{R}^n)$ such that $\|f\|_{(p)} = 0$. However, the relation between Besicovitch a. p. functions and functions on Bohr compactifications is less straightforward than in the case of Bohr a. p. functions (Proposition 4.8.2). The following statement is a direct consequence of the Birkhoff ergodic theorem (see, e. g., [132, Section VIII.7]).

Proposition 4.8.6. *Suppose that $\tilde{f} \in L^p(\mathbb{R}^n_{B,\Gamma})$. Then there exists a measurable subset $\Omega \subseteq \mathbb{R}^n_{B,\Gamma}$ such that $\mu(\Omega) = 1$, and for all $s \in \Omega$ the function*

$$f^{(s)}(x) = \tilde{f}(s + i_{B,\Gamma}x), \quad x \in \mathbb{R}^n,$$

belongs to $B^p_\Gamma(\mathbb{R}^n)$ and

$$\langle f^{(s)} \rangle = \int_{\mathbb{R}^n_{B,\Gamma}} \tilde{f}(z)\, d\mu(z).$$

Now we notice that the Bochner–Fejer approximations introduced in Proposition 4.8.4(c), make sense for Besicovitch a. p. functions. Moreover, the following statement holds.

Proposition 4.8.7. *If $f \in B^p_\Gamma(\mathbb{R}^n)$, $p \in [1,\infty)$, and f_k is the sequence of Bochner–Fejer approximations for f, then $\|f - f_k\|_{(p)} \to 0$ as $k \to \infty$.*

See [246, Theorem 2.4, Chapter 1].

Surprisingly enough, we did not find the following simple proposition in the existing literature.

Proposition 4.8.8. *Suppose that $f \in B^1(\mathbb{R}^n) \cap L^\infty(\mathbb{R}^n)$. Then $f \in B^p(\mathbb{R}^n)$ for all $p \in [1,\infty)$.*

Proof. Let $\Gamma = \mathrm{Mod}(f)$ and f_k be the sequence of Bochner–Fejer approximations for f. Making use of the properties of Bochner–Fejer kernels listed in Proposition 4.8.4, we deduce easily that

$$\|f_k\|_{L^\infty} \le \|f\|_{L^\infty}.$$

Hence,

$$\|f - f_k\|^p_{(p)} = \langle |f - f_k|^p \rangle = \langle |f - f_k| |f - f_k|^{p-1} \rangle \le (2\|f\|_{L^\infty})^{p-1} \|f - f_k\|_{(1)}.$$

This, together with Proposition 4.8.7, implies the claim. □

4.8.4 Linear almost periodic problem

First, we introduce certain functional spaces. For a detailed account of Hölder spaces on an arbitrary, not necessarily bounded, domain, we refer to [190, Section 3.1]. In this and subsequent sections, we consider real-valued functions only. Let $\alpha \in (0,1)$. The space $C^\alpha_b(\mathbb{R}^n)$ consists of all functions $f \in C_b(\mathbb{R}^n)$ that satisfy the uniform Hölder condition with the exponent α:

$$[f]_\alpha = \sup_{x,y \in \mathbb{R}^n, x \ne y} \frac{|f(x) - f(y)|}{|x - y|^\alpha} < \infty.$$

This is a Banach space with respect to the norm

$$\|f\|_{C^\alpha_b} = \|f\| + [f]_\alpha.$$

The space

$$\mathrm{CAP}^\alpha(\mathbb{R}^n) = \mathrm{CAP}(\mathbb{R}^n) \cap C^\alpha_b(\mathbb{R}^n)$$

is a closed subspace of $C^\alpha_b(\mathbb{R}^n)$.

For any positive integer m, we denote by $C^m_b(\mathbb{R}^n)$ the space of all functions $f \in C_b(\mathbb{R}^n)$ such that all derivatives of f up to order m belong to $C_b(\mathbb{R}^n)$. This is a Banach space with respect to the norm

$$\|f\|_{C_b^m} = \sum_{k=0}^{m} \|D^k f\|,$$

where $D^k f$ is a vector that consists of all kth derivatives of f, $D^0 f = f$. Similarly, we denote by $\mathrm{CAP}^m(\mathbb{R}^n)$ the space of all a. p. functions having a. p. derivatives up to order m. This is a closed subspace of $C_b^m(\mathbb{R}^n)$.

Finally, the space $C_b^{m+\alpha}(\mathbb{R}^n)$ is the space of all functions $f \in C_b^m(\mathbb{R}^n)$ such that $D^m f \in C_b^\alpha(\mathbb{R}^m)$. Endowed with the norm

$$\|f\|_{C_b^{m+\alpha}} = \|f\|_{C_b^m} + [D^m f]_\alpha,$$

this is a Banach space. We set

$$\mathrm{CAP}^{m+\alpha}(\mathbb{R}^n) = \mathrm{CAP}^m(\mathbb{R}^n) \cap C_b^{m+\alpha}(\mathbb{R}^n).$$

This is a closed subspace of $C_b^{m+\alpha}(\mathbb{R}^n)$. Actually, it is easily seen that

$$\mathrm{CAP}^{m+\alpha}(\mathbb{R}^n) = \mathrm{CAP}(\mathbb{R}^n) \cap C_b^{m+\alpha}(\mathbb{R}^n).$$

We use the following convention:

$$C_b^0(\mathbb{R}^n) = C_b(\mathbb{R}^n)$$

and

$$\mathrm{CAP}^0(\mathbb{R}^n) = \mathrm{CAP}(\mathbb{R}^n).$$

Now we consider second-order elliptic operators of the form

$$Au(x) = -\sum_{i,j=1}^{n} a_{ij}(x)\frac{\partial^2 u(x)}{\partial x_i \partial x_j} + \sum_{i=1}^{n} b_i(x)\frac{\partial u(x)}{\partial x_i} + c(x)u(x). \tag{4.298}$$

More precisely, we assume that:

(i) The matrix (a_{ij}) of leading coefficients is symmetric and there exists a constant $\lambda_0 > 0$ such that

$$\sum_{i,j=1}^{n} a_{ij}(x)\xi_i\xi_j \geq \lambda_0|\xi|^2$$

for all $\xi = (\xi_1, \ldots, \xi_n) \in \mathbb{R}^n$ and $x \in \mathbb{R}^n$.

(ii) There exists a constant $c_0 > 0$ such that $c(x) \geq c_0$ for all $x \in \mathbb{R}^n$.

We do not exclude the case when $n = 1$. In this case, without loss of generality, we may assume that the leading coefficient is equal to 1 and the operator becomes

$$Au(x) = -u''(x) + b(x)u'(x) + c(x)u(x). \tag{4.299}$$

We start with the following

Proposition 4.8.9. *Suppose that the operator A satisfies* (i) *and* (ii), *and its coefficients belong to $C_b^\alpha(\mathbb{R}^n)$, with $\alpha \in (0,1)$ if $n > 1$ and $\alpha \in [0,1)$ if $n = 1$. Then for any $f \in C_b^\alpha(\mathbb{R}^n)$ there exists a unique solution $u \in C_b^{2+\alpha}(\mathbb{R}^n)$ of the equation*

$$Au = f. \tag{4.300}$$

Moreover,

$$\|u\| \le c_0^{-1}\|f\|, \tag{4.301}$$

$$\|u\|_{C_b^1} \le C\|f\|, \tag{4.302}$$

and

$$\|u\|_{C_b^{2+\alpha}} \le C\|f\|_{C_b^\alpha}, \tag{4.303}$$

where the constant $C > 0$ depends only on λ_0, c_0, and norms of the coefficients in $C_b^\alpha(\mathbb{R}^n)$. In addition, if $f \ge 0$, then $u \ge 0$.

In the case when $\alpha \in (0,1)$, this is a well-known result (see, e. g., [190, Theorems 4.3.1 and 4.3.2]) based on the so-called Schauder's a priori estimates. We mention that estimate (4.301) and the positivity result follow from the maximum principle (see [190, Theorem 2.9.2]). The less known estimate (4.302) follows from interior L^p estimates for elliptic equations [156, Section 9.5] and the Sobolev embedding theorem. In the case when $n = 1$ and $\alpha = 0$, the statement of the proposition is also well known and can be found, e. g., in [189].

In the rest of the section, in addition to (i), we impose the following almost periodicity assumption:

(iii) The coefficients a_{ij}, b_j and c, $i,j = 1,\dots,n$, belong to $\mathrm{CAP}^\alpha(\mathbb{R}^n)$, where $\alpha \in (0,1)$ if $n > 1$ and $\alpha \in [0,1)$ if $n = 1$.

We denote by $\mathrm{Mod}(A)$ the smallest additive subgroup of $\mathbb{R}^n$ that contains the spectra of all coefficients of A. Now let $\Gamma \subseteq \mathbb{R}^n$ be any countable subgroup that contains $\mathrm{Mod}(A)$. We introduce the *envelope* $H(A) = \{A^{(s)}\}_{s\in\mathbb{R}^n_{B,\Gamma}}$ of A by

$$A^{(s)}u(x) = -\sum_{i,j=1}^n a_{ij}^{(s)}(x)\frac{\partial^2 u(x)}{\partial x_i \partial x_j} + \sum_{i=1}^n b_i^{(s)}(x)\frac{\partial u(x)}{\partial x_i} + c^{(s)}(x)u(x).$$

As in the case of functions, the *set* $H(A)$ is independent of the choice of $\Gamma \supseteq \mathrm{Mod}(A)$, while the parametrization of the envelope does depend on Γ. It is easily seen that if A satisfies (i)–(iii), then all operators $A^{(s)}$ in the envelope satisfy the same assumptions.

The key result of the section is

Theorem 4.8.10. *Assume* (i)–(iii). *If* $f \in \mathrm{CAP}^{\alpha}(\mathbb{R}^n)$, *then equation* (4.300) *has a unique solution* $u \in \mathrm{CAP}^{2+\alpha}(\mathbb{R}^n)$. *In addition,*

$$\mathrm{Mod}(u) \subseteq \mathrm{Mod}(A) + \mathrm{Mod}(f). \tag{4.304}$$

Proof. Let $\Gamma \supseteq \mathrm{Mod}(A)$ be any countable subgroup of $\mathbb{R}^n$ such that $f \in \mathrm{CAP}_{\Gamma}(\mathbb{R}^n)$. By Proposition 4.8.9, equation (4.300) has a unique solution $u \in C_b^{2+\alpha}(\mathbb{R}^n)$.

We have to show that $u \in \mathrm{CAP}_{\Gamma}(\mathbb{R}^n)$, which implies that $u \in \mathrm{CAP}_{\Gamma}^{2+\alpha}(\mathbb{R}^n)$. With this aim, we consider the family of equations

$$A^{(s)}u_s = f^{(s)}, \quad s \in \mathbb{R}^n_{B,\Gamma}.$$

By Proposition 4.8.9, each of these equations has a unique solution $u_s \in C_b^{2+\alpha}(\mathbb{R}^n)$, with $u_0 = u$.

Notice that the map $s \mapsto u_s$ is a continuous mapping $\mathbb{R}^n_{B,\Gamma} \to C_b(\mathbb{R}^n)$. Indeed, by inequality (4.301), for any $s_0, s \in \mathbb{R}^n_{B,\Gamma}$,

$$\|u_s - u_{s_0}\| \le c_0^{-1}\|A^{(s)}(u_s - u_{s_0})\| \le c_0^{-1}(\|f^{(s)} - f^{(s_0)}\| + \|(A^{(s_0)} - A^{(s)})u_{s_0}\|).$$

Since $f \in \mathrm{CAP}_{\Gamma}(\mathbb{R}^n)$ and the coefficients of A satisfy the almost periodicity assumption (ii), the right-hand side of the last inequality tends to zero as $s \to s_0$ in $\mathbb{R}^n_{B,\Gamma}$ and the conclusion follows.

As a consequence, $\tilde{u}(s) = u_s(0)$ is a well-defined continuous function on $\mathbb{R}^n_{B,\Gamma}$. Furthermore, due to the uniqueness of bounded solution (see Proposition 4.8.9),

$$u_{s+i_{B,\Gamma}x}(y) = u_s(y + x), \quad \forall s \in \mathbb{R}^n_{B,\Gamma}, x \in \mathbb{R}^n, \text{and } y \in \mathbb{R}^n.$$

Hence,

$$u(x) = u_0(x) = \tilde{u}(i_{B,\Gamma}x),$$

and, by Proposition 4.8.2, $u \in \mathrm{CAP}_{\Gamma}(\mathbb{R}^n)$. □

4.8.5 Semilinear problem

In this section we consider semilinear equations

$$Au(x) = g(x, u(x)), \quad x \in \mathbb{R}^n, \tag{4.305}$$

where A is a second-order elliptic operator of the form (4.298). We always suppose that A satisfies assumptions (i) and (iii). Assumption (ii) is not needed, in general. In addition, we impose the following assumption on the nonlinearity g:

(iv) For any $R > 0$, the function $g(x, u)$ is almost periodic in $x \in \mathbb{R}^n$ uniformly with respect to $u \in \mathbb{R}$, with $|u| \le R$, and there exists a constant $C_R > 0$ such that

$$|g(x,u) - g(x,v)| \le C_R|u - v|$$

for all $x \in \mathbb{R}^n$ and $u, v \in \mathbb{R}$, with $|u| \le R$ and $|v| \le R$, and

$$|g(x,u) - g(y,v)| \le C_R|x - y|^\alpha$$

for all $x, y \in \mathbb{R}^n$ and $u \in \mathbb{R}$, with $|u| \le R$, provided $\alpha \ne 0$ in assumption (iii).

In particular, assumption (iv) implies that the function $g(x, u)$ can be considered as an a. p. function of $x \in \mathbb{R}^n$ with values in the Fréchet space $C(\mathbb{R})$ of continuous functions on $\mathbb{R}$ endowed with the topology of uniform convergence on compact intervals. So $\bigcup_{u\in\mathbb{R}} \sigma(g(\cdot, u))$ generates a countable subgroup $\mathrm{Mod}(g) \subseteq \mathbb{R}^n$. Moreover, for any $\Gamma \supseteq \mathrm{Mod}(g)$, there exists a unique continuous function $\tilde{g}$ on $\mathbb{R}^n_{B,\Gamma}$ such that

$$g(x,u) = \tilde{g}(i_{B,\Gamma}x, u), \quad (x,u) \in \mathbb{R}^n \times \mathbb{R}.$$

If $\Gamma \supseteq \mathrm{Mod}(g)$ is a countable subgroup of $\mathbb{R}^n$, we set

$$g^{(s)}(x,u) = \tilde{g}(s + i_{B,\Gamma}x, u), \quad (x,u) \in \mathbb{R}^n \times \mathbb{R},\ s \in \mathbb{R}^n_{B,\Gamma}.$$

Notice that all the functions $g^{(s)}$ satisfy assumption (iv).

Together with equation (4.305), we consider the following family of equations:

$$A^{(s)}u(x) = g^{(s)}(x, u(x)), \quad x \in \mathbb{R}^n, \tag{4.306}$$

where $s \in \mathbb{R}^n_{B,\Gamma}$ and $\Gamma \supseteq \mathrm{Mod}(A) + \mathrm{Mod}(g)$ will be fixed later.

A function $\overline{u} \in \mathrm{CAP}^{2+\alpha}(\mathbb{R}^n)$ (respectively, $\underline{u} \in \mathrm{CAP}^{2+\alpha}(\mathbb{R}^n)$) is called a *super-solution* (respectively, a *sub-solution*) of equation (4.305) if

$$A\overline{u}(x) \ge g(x, \overline{u}(x)) \quad (\text{respectively, } A\underline{u}(x) \le g(x, \underline{u}(x)))$$

for all $x \in \mathbb{R}^n$. Given super- and sub-solutions $\overline{u}$ and $\underline{u}$, we set

$$\Gamma = \mathrm{Mod}(A) + \mathrm{Mod}(g) + \mathrm{Mod}(\overline{u}) + \mathrm{Mod}(\underline{u}).$$

Notice that the functions $\overline{u}^{(s)}$ and $\underline{u}^{(s)}$ are super- and sub-solutions for equation (4.306) for all $s \in \mathbb{R}^n_{B,\Gamma}$.

Theorem 4.8.11. *Under assumptions* (i), (iii), *and* (iv), *suppose that there exist sub- and super-solutions for equation* (4.305) *such that* $\underline{u} \le \overline{u}$. *Then for every* $s \in \mathbb{R}^n_{B,\Gamma}$ *there exists a solution* $u_s \in C_b^{2+\alpha}(\mathbb{R}^n)$ *of equation* (4.306) *such that* $\underline{u}^{(s)} \le u_s \le \overline{u}^{(s)}$. *Furthermore,* *there exist a residual set* $\Omega_{aa} \subseteq \mathbb{R}^n_{B,\Gamma}$ *and a set* $\Omega_b \subseteq \mathbb{R}^n_{B,\Gamma}$ *of measure 1, both translation invariant and such that* u_s *is uniformly almost automorphic if* $s \in \Omega_{aa}$ *and* $u_s \in B^p_\Gamma(\mathbb{R}^n)$ *for all* $p \in [1,\infty)$ *if* $s \in \Omega_b$.

Proof. Replacing $c(x)$ by $c(x) + \theta$ and $g(x,u)$ by $g(x,u) + \theta u$, we may suppose, due to assumption (iv), that $c(x) \geq c_0 > 0$ and the nonlinearity $g(x,u)$ is increasing in $u \in [\inf \underline{u}, \sup \overline{u}]$.

Let us consider a sequence of functions u_k defined recurrently as follows. We set $u_0 = \overline{u}$. Next, $u_{k+1} \in \mathrm{CAP}^{2+\alpha}_\Gamma(\mathbb{R}^n)$ is defined as a unique solution of the equation

$$Au_{k+1}(x) = g(x, u_k(x)). \tag{4.307}$$

The sequence u_k is well defined. Indeed, if $u_k \in \mathrm{CAP}^{2+\alpha}_\Gamma(\mathbb{R}^n)$, then, by assumption (iv), $g(x, u_k(x)) \in \mathrm{CAP}^\alpha_\Gamma(\mathbb{R}^n)$. By Theorem 4.8.10, equation (4.307) has a unique solution in $\mathrm{CAP}^{2+\alpha}_\Gamma(\mathbb{R}^n)$. Moreover, since $\overline{u}$ is a super-solution, the positivity statement of Proposition 4.8.9 implies that the sequence u_k is monotone decreasing, i. e., $u_{k+1} \leq u_k$. Since $u_k \in \mathrm{CAP}^{2+\alpha}(\mathbb{R}^n)$, the function u_k extends to a unique function $\tilde{u}_k \in C(\mathbb{R}^n_{B,\Gamma})$, and the sequence of functions $\tilde{u}_k$ is monotone decreasing and bounded. Hence, the sequence $\tilde{u}_k$ converges to a measurable function $\tilde{u}$ pointwise on $\mathbb{R}^n_{B,\Gamma}$.

We are going to prove that the function

$$u^{(s)}(x) = \tilde{u}(s + i_{B,\Gamma}x)$$

is actually a $C^{2+\alpha}_b$-solution of equation (4.306). It is easily seen that, for any $s \in \mathbb{R}^n_{B,\Gamma}$, the functions

$$u^{(s)}_k(x) = \tilde{u}_k(s + i_{B,\Gamma}x)$$

satisfy

$$A^{(s)}u^{(s)}_{k+1}(x) = g^{(s)}(x, u^{(s)}_k(x)). \tag{4.308}$$

As a consequence, $u^{(s)}_k \in \mathrm{CAP}^{2+\alpha}(\mathbb{R}^n)$ for all $s \in \mathbb{R}^n_{B,\Gamma}$ and integer k.

We claim that there exists a constant $C > 0$ independent of s and k such that

$$\|u^{(s)}_k\|_{C^{2+\alpha}_b} \leq C. \tag{4.309}$$

Indeed, denoting by C a generic positive constant independent of s and k, we have that $\|u^{(s)}_k\| \leq C$. Assumption (iv) implies easily that

$$\|g^{(s)}(\cdot, u^{(s)}_k)\| \leq C.$$

Equation (4.308) and estimate (4.302) of Proposition 4.8.9 imply that

$$\|u_k^{(s)}\|_{C_b^\alpha} \le \|u_k^{(s)}\|_{C_b^1} \le C$$

(the first estimate is trivial). By assumption (iv),

$$\|g^{(s)}(\cdot, u_k^{(s)})\|_{C_b^\alpha} \le C,$$

and estimate (4.309) follows from inequality (4.303) of Proposition 4.8.9.

Suppose that $\alpha > 0$. By (4.309), the functions $u_k^{(s)}$ and their derivatives up to second order are equicontinuous. Since $u_k^{(s)} \to u^{(s)}$ pointwise, by the Arzelà–Ascoli theorem, $u^{(s)} \in C_b^{2+\alpha}(\mathbb{R}^n)$ and $u_k^{(s)}$ converges to $u^{(s)}$ uniformly on compact sets together with derivatives up to second order. Passing to the limit in equation (4.308), we obtain that $u^{(s)}$ is a solution of equation (4.306).

Now suppose that $\alpha = 0$ and $n = 1$. By (4.309), the functions $u_k^{(s)}$ and their first derivatives are equicontinuous. As above, $u^{(s)} \in C_b^1(\mathbb{R}^n)$ and $u_k^{(s)}$ converges to $u^{(s)}$ uniformly on compact sets together with first derivatives. Equation (4.308) can be expressed as follows:

$$-(u_{k+1}^{(s)})'' = -b^{(s)}(x)(u_{k+1}^{(s)})' - c^{(s)}(x)u_{k+1}^{(s)} + g^{(s)}(x, u_k^{(s)}).$$

Hence, second derivatives converge uniformly on compact sets, and $u^{(s)} \in C_b^2(\mathbb{R}^n)$ satisfies equation (4.306).

Since $\tilde{u} \in \mathbb{R}^n_{B,\Gamma}$, the existence of the set Ω_b follows from Proposition 4.8.6. On the other hand, the map $U : s \mapsto u^{(s)}$ from $\mathbb{R}^n_{B,\Gamma}$ into $C(\mathbb{R}^n)$ is of the first Baire category as a pointwise limit of continuous maps. It is well known that any map of the first Baire category from a complete metric space into a separable metric space is continuous on a residual set (see, e. g., [191, Section 31.X, Theorem 1]). Hence, the existence of the set Ω_{aa} follows from Proposition 4.8.5. □

Remark 4.8.12. Bounded solutions $u^{(s)}$ obtained in the proof of Theorem 4.8.11 are maximum solutions between $\underline{u}^{(s)}$ and $\overline{u}^{(s)}$. Starting the iteration process with $u_0 = \underline{u}$, we obtain the minimum solution.

Remark 4.8.13. The function $\tilde{u}$ constructed in the proof of Theorem 4.8.11 can be considered as a generalized solution of equation (4.305) in the sense of Sobolev–Besicovitch spaces [246].

Now we give general sufficient conditions for sub- and super-solutions to exist.

Theorem 4.8.14. *Assume* (i)–(iii). *Suppose that the nonlinearity g is of the form*

$$g(x, u) = g_1(x, u) + g_2(x, u),$$

where both g_1 and g_2 satisfy assumption (iv), *$\frac{\partial g_2}{\partial u}$ is a continuous function on $\mathbb{R}^n \times \mathbb{R}$, and*

$$|g_1(x, u)| \le C \tag{4.310}$$

and

$$\frac{\partial g_2}{\partial u}(x,u) \leq 0 \tag{4.311}$$

for all $(x,u) \in \mathbb{R}^n \times \mathbb{R}$. *Then the conclusion of Theorem* 4.8.11 *holds with*

$$\Gamma = \mathrm{Mod}(A) + \mathrm{Mod}(g).$$

Proof. (a) *Reduction to the case when* $g_2 = 0$. If $u \in C_b^{2+\alpha}(\mathbb{R}^n)$ is a solution of equation (4.305), then

$$Au(x) + h(x)u = g_1(x, u(x)),$$

where

$$h(x) = \int_0^1 \frac{\partial g_2}{\partial u}(x, tu(x))\, dt.$$

It is easy to see that $h(x) \geq 0$ and, by estimate (4.301) of Proposition 4.8.9,

$$\|u\| \leq C_1, \tag{4.312}$$

where $C_1 > 0$ is independent of the solution. Modifying g_2 outside the region $|u| \geq 2C_1$, we may assume that $g = g_1$ is a bounded function.

(b) *The case when* $g = g_1$ *is bounded.* Let

$$\theta = \sup\left\{\left|\frac{\partial g}{\partial u}(x,u)\right| : x \in \mathbb{R}^n, |u| \leq C_1\right\},$$

where C_1 is the constant in the estimate (4.312). Obviously, we may assume that $C_1 \geq C$. Equation (4.305) is equivalent to the equation

$$Au + \theta u = g(x,u) + \theta u$$

and the function $g(x,u)+\theta u$ is increasing in the region $|u| \leq C_1$. We define $\overline{u} \in \mathrm{CAP}_\Gamma^{2+\alpha}(\mathbb{R}^n)$ as a unique nonnegative solution of equation $Au = C$ which exists by Proposition 4.8.9 and set $\underline{u} = -\overline{u}$. It is easy to verify that these are super- and sub-solutions, respectively, and we conclude by Theorem 4.8.11. □

Remark 4.8.15. If equation (4.305) possesses at most one solution between $\underline{u}$ and $\overline{u}$, then the map $U : \mathbb{R}_{B,\Gamma}^n \to C(\mathbb{R}^n)$ considered in the proof of Theorem 4.8.11 is continuous at the point $s = 0$ and, hence, the solution $u = u^{(0)}$ of equation (4.305) is uniformly almost automorphic. If the uniqueness of solution between $\inf \underline{u}$ and $\sup \overline{u}$ holds for all equations (4.306), $s \in \mathbb{R}_{B,\Gamma}^n$, then all solutions $u^{(s)}$ are almost periodic because the function $\tilde{u}(s) = u(s)(0)$ is continuous on $\mathbb{R}_{B,\Gamma}^n$.

Corollary 4.8.16. *Under the assumptions of Theorem* 4.8.11, *suppose in addition that* $c(x) \geq 0$ *on* $\mathbb{R}^n$, *while*

$$\frac{\partial g}{\partial u} \leq -\kappa < 0$$

and is uniformly continuous on the strip $\mathbb{R}^n \times [\inf \underline{u}, \sup \overline{u}]$. *Then equation* (4.305) *has a unique solution* $u \in \mathrm{CAP}_\Gamma^{2+\alpha}$ *between* $\underline{u}$ *and* $\overline{u}$.

Proof. Due to Remark 4.8.15, we have to verify the uniqueness needed there. Let us mention that $\partial g^{(s)}/\partial u$ is bounded and continuous on $\mathbb{R}^n \times [\inf \underline{u}, \sup \overline{u}]$ for all $s \in \mathbb{R}^n_{B,\Gamma}$. If u_1 and u_2 are two solutions of equation (4.306) between $\underline{u}$ and $\overline{u}$, then $v = u_1 - u_2$ satisfies

$$A^{(s)} v + h(x) v = 0,$$

where

$$h(x) = -\int_0^1 \frac{\partial g^{(s)}}{\partial u}(x, tu_1(x) + (1-t)u_2(x))\, dt \geq \kappa > 0.$$

By Proposition 4.8.9, $v = 0$ and we conclude. □

Remark 4.8.17. The statement of Corollary 4.8.16 remains valid if we replace the assumptions $c(x) \geq 0$ and $\partial g/\partial u < -\kappa$ by $c(x) \geq c_0 > 0$ and $\partial g/\partial u \leq 0$, respectively.

Finally, we mention that Theorem 4.8.11 covers the case when the coefficients and sub- and super-solution are periodic. In that case, it implies the existence of a periodic solution – a statement well known in the literature. Indeed, in the periodic case, $\mathbb{R}^n_{B,\Gamma}$ is a torus. Hence, the sets $\Omega_a a$ and Ω_b, being translation invariant, coincide with the whole of the torus.

4.8.6 Almost periodically forced pendulum

As an application of our main result, consider the pendulum equation

$$u'' + cu' + a \sin u = h(t), \tag{4.313}$$

with an almost periodic forcing term $h(t)$. Here $a > 0$ and the damping coefficient $c \geq 0$ so that the undamped case is allowed. The envelope of equation (4.313) is

$$u'' + cu' + a \sin u = h^{(s)}(t), \quad s \in \mathbb{R}_{B,\Gamma}, \tag{4.314}$$

where $\Gamma = \mathrm{Mod}(h)$.

Theorem 4.8.18. *Let $h \in \mathrm{CAP}(\mathbb{R})$ and $\Gamma = \mathrm{Mod}(h)$.*

(a) *If $\|h\| = a$, then for every $s \in \mathbb{R}_{B,\Gamma}$ there exists a solution $u_s \in C_b^2(\mathbb{R})$ of equation (4.314) such that*

$$\frac{\pi}{2} \le u_s(t) \le \frac{3\pi}{2} \quad \forall t \in \mathbb{R}.$$

Furthermore, there exist a residual subset $\Omega_{aa} \subseteq \mathbb{R}_{B,\Gamma}$ and a measurable subset $\Omega_b \subseteq \mathbb{R}_{B,\Gamma}$ of measure 1, both translation invariant and such that u_s is uniformly almost automorphic if $s \in \Omega_{aa}$ and $u_s \in B_\Gamma^p(\mathbb{R}^n)$ for all $p \in [1,\infty)$ if $s \in \Omega_b$.

(b) *If $\|h\| < a$, then equation (4.313) has a unique solution $u \in \mathrm{CAP}^2(\mathbb{R})$ such that*

$$\frac{\pi}{2} < u(t) < \frac{3\pi}{2} \quad \forall t \in \mathbb{R}.$$

Moreover, $\mathrm{Mod}(u) \subseteq \Gamma$.

Proof. (a) It is easy to verify that $\underline{u} = \pi/2$ and $\overline{u} = 3\pi/2$ are sub- and super-solutions, and the result follows from Theorem 4.8.11.

(b) If $\delta > 0$ is sufficiently small, then $\underline{u} = \pi/2+\delta$ and $\overline{u} = 3\pi/2-\delta$ are sub- and super-solutions. Since the derivative $(a \sin u)' = a \cos u$ is strictly negative on $[\pi/2+\delta, 3\pi/2-\delta]$, Corollary 4.8.16 applies and we conclude. □

Let us point out that if $h(t)$ is periodic and $\|h\| \le a$, then there is a periodic solution between $\pi/2$ and $3\pi/2$. The uniqueness of such solution takes place whenever $\|h\| < a$.

The existence of a bounded solution under the assumptions of Theorem 4.8.18(a), as well as the almost periodicity of a unique bounded solution in case (b), is obtained in [155, 217] (see also [216]). In [155] the sub- and super-solution approach based on the Schauder fixed point theorem is used, while the proofs of [217] make use of an early result of Z. Opial which can be considered as a simple version of the sub- and super-solution method. In our approach, we employ relative Bohr compactifications together with monotone iteration techniques. This permits us to obtain extra information about the almost automorphy and Besicovitch almost periodicity of solutions to the equations in the envelope of (4.313), as well as the modulus containment property. According to a remark of Mawhin [217], all these results can be considered as an improvement of a result of [67] which provides the existence of a weak Besicovitch almost periodic solution to equation (4.313) (cf. Remark 4.8.13) by means of certain variational techniques. It is interesting that in [67] the same interval $[\pi/2, 3\pi/2]$ appears.

5 Abstract delay equations in Banach spaces

In this chapter, we examine the methods used to establish the existence, uniqueness, and the relationship between boundedness, almost periodicity, and almost automorphic dynamics of solutions. We specifically extend Massera- and Bohr–Neugebauer-type results for various partial functional differential equations, including those with finite delay, infinite delay, and neutral type, as well as equations with nonautonomous perturbations.

In the first four sections, under compactness assumptions, we use the variation of constants formula in the phase space, along with spectral decomposition, to show that the dynamics of the complex system can be studied through an ordinary differential equation. This is known as the reduction principle. Additionally, under the hypothesis of exponential dichotomy, uniqueness results are also obtained.

In the last two sections, we once again apply the reduction principle, but in a different sense. Specifically, we show that the dynamics of bounded solutions for nonautonomous partial functional differential equations can be reduced to a linear discrete dynamical system.

5.1 Partial functional differential equations with finite delay

In [57], under some compactness assumptions on the semigroup generated by the differential operator A on some functional Banach spaces, we proved that the following delay differential equation:

$$\frac{\mathrm{d}}{\mathrm{d}t}x(t) = Ax(t) + L(x_t) + f(t) \quad \text{for } t \in \mathbb{R}, \tag{5.1}$$

has the Bohr–Neugebauer property, that is, all bounded solutions on $\mathbb{R}$ have the same oscillating behavior as the forcing f. The compactness assumption we used finds its applications in several classes of reaction diffusion partial differential equations. Our result is achieved by showing that the partial functional differential equation (5.1) can be partially reduced to a finite-dimensional ordinary differential equation via a spectral decomposition of the phase space.

We assume that A is a linear operator on a Banach space X. We assume that the domain $D(A)$ is not necessarily dense and A satisfies the well-known Hille–Yosida condition, namely, we suppose

(H0) There exist $\bar{M} \geq 1$ and $\omega \in \mathbb{R}$ such that $(\omega, +\infty) \subset \rho(A)$ and

$$|R(\lambda, A)^n| \leq \frac{\bar{M}}{(\lambda - \omega)^n} \quad \text{for } n \in \mathbb{N} \text{ and } \lambda > \omega,$$

where $\rho(A)$ is the resolvent set of A and $R(\lambda, A) := (\lambda I - A)^{-1}$ for $\lambda \in \rho(A)$.

https://doi.org/10.1515/9783111684710-005

Let $C := C([-r,0],X)$ be the space of continuous functions from $[-r,0]$ to X endowed with the uniform norm topology. For every $t \in \mathbb{R}$, the history function $u_t \in C$ is defined by

$$u_t(\theta) := u(t+\theta) \quad \text{for } \theta \in [-r,0].$$

In what follows, L is a bounded linear operator from C to X and the forcing function f is Stepanov almost automorphic from $\mathbb{R}$ to X which is not necessarily bounded.

5.1.1 Variation of constants formula and reduction principle

In the sequel, we assume that the operator A satisfies the Hille–Yosida condition **(H0)**. To equation (5.1), we associate the following Cauchy problem:

$$\begin{cases} \frac{\mathrm{d}}{\mathrm{d}t}u(t) = Au(t) + L(u_t) + f(t) & \text{for } t \geq \sigma, \\ u_\sigma = \varphi \in C. \end{cases} \tag{5.2}$$

Definition 5.1.1 ([7]). A continuous function $u : [-r+\sigma, +\infty) \to X$ is called an integral solution of equation (5.2) if
(i) $\int_\sigma^t u(s)\mathrm{d}s \in D(A)$ for $t \geq \sigma$,
(ii) $u(t) = \varphi(0) + A\int_\sigma^t u(s)\mathrm{d}s + \int_\sigma^t [L(u_s) + f(s)]\mathrm{d}s$ for $t \geq \sigma$,
(iii) $u_\sigma = \varphi$.

If u is an integral solution of equation (5.2), then, from the continuity of u, we have $u(t) \in \overline{D(A)}$ for all $t \geq \sigma$. In particular, $\varphi(0) \in \overline{D(A)}$.

Let us introduce the part A_0 of the operator A in $\overline{D(A)}$ defined by

$$\begin{cases} D(A_0) = \{x \in D(A) : Ax \in \overline{D(A)}\}, \\ A_0 x = Ax \quad \text{for } x \in D(A_0). \end{cases}$$

Lemma 5.1.2 ([7]). *Assume that* **(H0)** *holds. Then, A_0 generates a strongly continuous semigroup $(T_0(t))_{t\geq 0}$ on $\overline{D(A)}$.*

For the existence of integral solutions, one has the following result:

Theorem 5.1.3 ([7]). *Assume that* **(H0)** *holds. Then, for all $\varphi \in C$ such that $\varphi(0) \in \overline{D(A)}$, equation* (5.2) *has a unique integral solution u on $[-r+\sigma, +\infty)$. Moreover, u is given by*

$$\begin{cases} u(t) = T_0(t-\sigma)\varphi(0) + \lim_{\lambda\to+\infty} \int_\sigma^t T_0(t-s)B_\lambda[L(u_s) + f(s)]\mathrm{d}s & \text{for } t \geq \sigma, \\ u_\sigma = \varphi, \end{cases}$$

where $B_\lambda = \lambda R(\lambda, A)$ for $\lambda > \omega$.

In the sequel of this chapter, for simplicity, integral solutions are called solutions; $u(\cdot,\sigma,\varphi,f)$ denotes the solution of equation (5.2). The phase space C_0 of equation (5.2) is given by

$$C_0 = \{\varphi \in C : \varphi(0) \in \overline{D(A)}\}.$$

For each $t \geq 0$, we define the linear operator $U(t)$ on C_0 by

$$U(t)\varphi = u_t(\cdot,0,\varphi,0),$$

where $u(\cdot,0,\varphi,0)$ is the solution of the following homogeneous equation:

$$\begin{cases} \frac{\mathrm{d}}{\mathrm{d}t}u(t) = Au(t) + L(v_t) & \text{for } t \geq 0, \\ u_0 = \varphi. \end{cases}$$

We have the following result:

Proposition 5.1.4 ([7]). *Assume that* **(H0)** *holds. Then* $(U(t))_{t\geq 0}$ *is a strongly continuous semigroup on* C_0. *Moreover, the operator* $\mathcal{A}_U$ *defined on* C_0 *by*

$$\begin{cases} D(\mathcal{A}_U) = \{\varphi \in C^1([-r,0],X) : \varphi(0) \in D(A), \varphi'(0) \in \overline{D(A)} \text{ and } \varphi'(0) = A\varphi(0) + L(\varphi)\}, \\ \mathcal{A}_U\varphi = \varphi' \end{cases}$$

is the infinitesimal generator of $(U(t))_{t\geq 0}$ *on* C_0.

In order to give the variation of constants formula associated to equation (5.2), we need to recall some notations and results which are taken from [7]. Let $\langle X_0\rangle$ be the space defined by

$$\langle X_0\rangle = \{X_0 y : y \in X\},$$

where the function $X_0 y$ is given, for each $y \in X$ by

$$(X_0 y)(\theta) = \begin{cases} 0 & \text{if } \theta \in [-r,0), \\ y & \text{if } \theta = 0. \end{cases}$$

The space $C_0 \oplus \langle X_0\rangle$ equipped with the norm $|\varphi + X_0 y| = |\varphi| + |y|$, for $(\varphi, y) \in C_0 \times X$, is a Banach space. Consider the extension $\widetilde{\mathcal{A}_U}$ of the operator $\mathcal{A}_U$ on $C_0 \oplus \langle X_0\rangle$ defined by

$$\begin{cases} D(\widetilde{\mathcal{A}_U}) = \{\varphi \in C^1([-r,0],X) : \varphi(0) \in D(A) \text{ and } \varphi'(0) \in \overline{D(A)}\}, \\ \widetilde{\mathcal{A}_U}\varphi = \varphi' + X_0(A\varphi(0) + L(\varphi) - \varphi'(0)). \end{cases}$$

Lemma 5.1.5 ([7, Theorem 13]). *Assume that* **(H0)** *holds. Then* $\widetilde{\mathcal{A}_U}$ *satisfies the Hille–Yosida condition on* $C_0 \oplus \langle X_0\rangle$: *there exist* $\widetilde{M} \geq 0$, $\widetilde{\omega} \in \mathbb{R}$ *such that* $(\widetilde{\omega}, +\infty) \subset \rho(\widetilde{\mathcal{A}_U})$ *and*

$$|R(\lambda, \widetilde{\mathcal{A}_U})^n| \leq \frac{\widetilde{M}}{(\lambda - \widetilde{\omega})^n} \quad \text{for } n \in \mathbb{N} \text{ and } \lambda > \widetilde{\omega}.$$

Theorem 5.1.6 ([7, Theorem 16]). *Assume that* **(H0)** *holds. Then, for all* $\varphi \in C_0$, *the solution* $u(\cdot, \sigma, \varphi, f)$ *of equation* (5.2) *is given by the following variation of constants formula:*

$$u_t(., \sigma, \varphi, f) = U(t - \sigma)\varphi + \lim_{n \to +\infty} \int_{\sigma}^{t} U(t - s)\widetilde{B}_n(X_0 f(s))ds \quad \text{for } t \geq \sigma,$$

where $\widetilde{B}_n = nR(n, \widetilde{\mathcal{A}_U})$ *for* $n > \widetilde{\omega}$.

The following assumption plays a crucial role to get the reduction principle:
(H1) The operator $T_0(t)$ is compact on $\overline{D(A)}$ for every $t > 0$.

We have the following fundamental result on the semigroup $(U(t))_{t\geq 0}$:

Theorem 5.1.7 ([7, Lemma 10]). *Assume that* **(H0)** *and* **(H1)** *hold. Then the operator* $U(t)$ *is compact for* $t > r$.

As a consequence, from the compactness property of the operator $U(t)$, we have that the spectrum $\sigma(A_U)$ is the point spectrum. Moreover, we have the following spectral decomposition result:

Theorem 5.1.8 ([136]). *The phase space* C_0 *is decomposed as follows:*

$$C_0 = S \oplus V, \tag{5.3}$$

where S is U-invariant and there are positive constants a and N such that

$$|U(t)\varphi| \leq Ne^{-at}|\varphi| \quad \text{for each } \varphi \in S \text{ and } t \geq 0.$$

Further, V is a finite-dimensional space and the restriction of U to V is a group.

In the sequel, $U^s(t)$ and $U^v(t)$ denote the restriction of $U(t)$ on S and V, respectively, which correspond to the above decomposition. Let $d := \dim(V)$ with a vector basis $\Phi = \{\varphi_1, \dots, \varphi_d\}$. Then, there exist d-elements $\{\psi_1, \dots, \psi_d\}$ in C_0^* such that

$$\begin{cases} \langle \psi_i, \varphi_j \rangle = \delta_{ij}, \\ \langle \psi_i, \varphi \rangle = 0 \quad \text{for all } \varphi \in S \text{ and } i \in \{1, \dots, d\}, \end{cases} \tag{5.4}$$

where $\langle \cdot, \cdot \rangle$ denotes the duality pairing between C_0^* and C_0, and

$$\delta_{ij} = \begin{cases} 1 & \text{if } i = j, \\ 0 & \text{if } i \neq j. \end{cases}$$

Letting $\Psi = \mathrm{col}\{\psi_1, \dots \psi_2\}$, $\langle \Psi, \Phi \rangle$ is a $(d \times d)$-matrix, where the (i, j)-component is $\langle \psi_i, \varphi_j \rangle$.

Denote by Π^s and Π^v the projections on S and V, respectively. For each $\varphi \in C_0$, we have

$$\Pi^v\varphi = \Phi\langle\Psi, \varphi\rangle.$$

In fact, for $\varphi \in C_0$, we have $\varphi = \Pi^s\varphi + \Pi^v\varphi$ with $\Pi^v\varphi = \sum_{i=1}^d a_i\varphi_i$ and $a_i \in \mathbb{R}$. By (5.4), we conclude that

$$a_i = \langle\psi_i, \varphi\rangle.$$

Hence

$$\Pi^v\varphi = \sum_{i=1}^{d}\langle\psi_i, \varphi\rangle\varphi_i = \Phi\langle\Psi, \varphi\rangle.$$

Since $(U^v(t))_{t\geq 0}$ is a group on V, there exists a $(d \times d)$-matrix G such that

$$U^v(t)\Phi = \Phi e^{tG} \quad \text{for } t \in \mathbb{R}.$$

For $n, n_0 \in \mathbb{N}$ such that $n \geq n_0 \geq \widetilde{\omega}$ and $i \in \{1, \dots, d\}$, we define the linear operator $x_{i,n}^*$ by

$$x_{i,n}^*(a) = \langle\psi_i, \widetilde{B}_n X_0 a\rangle \quad \text{for } a \in X.$$

Since $|\widetilde{B}_n| \leq \frac{n}{n-\widetilde{\omega}}\widetilde{M}$ for any $n \geq n_0$, $x_{i,n}^*$ is a bounded linear operator from X to $\mathbb{R}$ such that

$$|x_{i,n}^*| \leq \frac{n}{n-\widetilde{\omega}}\widetilde{M}|\psi_i| \quad \text{for any } n \geq n_0.$$

Define the d-column vector $x_n^* = \operatorname{col}(x_{1,n}^*, \dots, x_{d,n}^*)$. Then one can see that

$$\langle x_n^*, a\rangle = \langle\Psi, \widetilde{B}_n X_0 a\rangle \quad \text{for } a \in X,$$

with

$$\langle x_n^*, a\rangle_i = \langle\psi_i, \widetilde{B}_n X_0 a\rangle \quad \text{for } i = 1, \dots, d \text{ and } a \in X.$$

Consequently, we have

$$\sup_{n\geq n_0}|x_n^*| < +\infty,$$

which implies that $(x_n^*)_{n\geq n_0}$ is a bounded sequence in $\mathcal{L}(X, \mathbb{R}^d)$. We recall the following important results:

Theorem 5.1.9 ([146]). *There exists $x^* \in \mathcal{L}(X, \mathbb{R}^d)$ such that $(x_n^*)_{n\geq n_0}$ converges weakly to x^* in the sense that*

$$\langle x_n^*, x\rangle \to \langle x^*, x\rangle \quad \text{as } n \to +\infty \text{ for all } x \in X.$$

For the proof, we need the following fundamental theorem.

Theorem 5.1.1 ([289, p. 776]). *Let Y be any separable Banach space and $(z_n^*)_{n\in\mathbb{N}}$ any bounded sequence in Y^*. Then there exists a subsequence $(z_{n_k}^*)_{k\in\mathbb{N}}$ of $(z_n^*)_{n\in\mathbb{N}}$ which converges weakly in Y^* in the sense that there exists $z^* \in Y^*$ such that*

$$\langle z_{n_k}^*, x\rangle \to \langle z^*, x\rangle, \quad \text{as } n \to \infty, \text{ for all } x \in Y.$$

Proof. Let Z_0 be any closed separable subspace of X. Since $(x_n^*)_{n\geq n_0}$ is a bounded sequence, by Theorem 5.1.1 we get that the sequence $(x_n^*)_{n\geq n_0}$ has a subsequence $(x_{n_k}^*)_{k\in\mathbb{N}}$ which converges weakly to some $x_{Z_0}^*$ in Z_0. We claim that the whole sequence $(x_n^*)_{n\geq n_0}$ converges weakly to $x_{Z_0}^*$ in Z_0. In fact, we proceed by contradiction and suppose that there exists a subsequence $(x_{n_p}^*)_{p\in\mathbb{N}}$ of $(x_n^*)_{n\geq n_0}$ which converges weakly to some $\widetilde{x}_{Z_0}^*$ with $\widetilde{x}_{Z_0}^* \neq x_{Z_0}^*$. Let $u_t(\cdot, \sigma, \varphi, f)$ denote the solution of equation (5.1). Then

$$\Pi^\nu u_t(\cdot, \sigma, 0, f) = \lim_{n\to+\infty} \int_\sigma^t \mathcal{U}^\nu(t-\xi)\Pi^\nu(\widetilde{B}_n X_0 f(\xi))d\xi$$

and

$$\Pi^\nu(\widetilde{B}_n X_0 f(\xi)) = \Phi\langle \Psi, \widetilde{B}_n X_0 f(\xi)\rangle = \Phi\langle x_n^*, f(\xi)\rangle.$$

It follows that

$$\begin{aligned}\Pi^\nu u_t(\cdot, \sigma, 0, f) &= \lim_{n\to+\infty} \Phi \int_\sigma^t e^{(t-\xi)G}\langle \Psi, \widetilde{B}_n X_0 f(\xi)\rangle d\xi,\\ &= \lim_{n\to+\infty} \Phi \int_\sigma^t e^{(t-\xi)G}\langle x_n^*, f(\xi)\rangle d\xi.\end{aligned}$$

For any $a \in Z_0$, set $f(\cdot) = a$, then

$$\lim_{k\to+\infty} \int_\sigma^t e^{(t-\xi)G}\langle x_{n_k}^*, a\rangle d\xi = \lim_{p\to+\infty} \int_\sigma^t e^{(t-\xi)G}\langle x_{n_p}^*, a\rangle d\xi \quad \text{for } a \in Z_0,$$

which implies that

$$\int_\sigma^t e^{(t-\xi)G}\langle x_{Z_0}^*, a\rangle d\xi = \int_\sigma^t e^{(t-\xi)G}\langle \widetilde{x}_{Z_0}^*, a\rangle d\xi \quad \text{for } a \in Z_0.$$

Consequently, $x^*_{Z_0} \equiv \tilde{x}^*_{Z_0}$, which gives a contradiction. We conclude that the whole sequence $(x^*_n)_{n\geq n_0}$ converges weakly to $x^*_{Z_0}$ in Z_0. Let Z_1 be another closed separable subspace of X. Then, by using the same argument as above, we get that $(x^*_n)_{n\geq n_0}$ converges weakly to $x^*_{Z_1}$ in Z_1. Since $Z_0 \cap Z_1$ is a closed separable subspace of X, we get that $x^*_{Z_1} \equiv x^*_{Z_0}$ in $Z_0 \cap Z_1$. For any $x \in X$, we define x^* by

$$\langle x^*, x\rangle = \langle x^*_Z, x\rangle,$$

where Z is any closed separable subspace of X such that $x \in Z$. Then x^* is well defined on X and x^* is a bounded linear from X to $\mathbb{R}^d$ such that

$$|x^*| \leq \sup_{n\geq n_0}|x^*_n| < \infty,$$

and $(x^*_n)_{n\geq n_0}$ converges weakly to x^* in X. □

As a consequence, we conclude that

Corollary 5.1.2. *For any continuous function $h : \mathbb{R} \to X$, we have*

$$\lim_{n\to+\infty} \int_\sigma^t \mathcal{U}^\nu(t-\xi)\Pi^\nu(\widetilde{B}_n X_0 h(\xi))\mathrm{d}\xi = \Phi \int_\sigma^t e^{(t-\xi)G}\langle x^*, h(\xi)\rangle \mathrm{d}\xi \quad \textit{for all } t,\sigma \in \mathbb{R}.$$

Theorem 5.1.10 ([146]). *Assume that* **(H0)** *and* **(H1)** *hold, f is continuous, and let u be a solution of equation* (5.1) *on $\mathbb{R}$. Then the function z defined by $z(t) := \langle \Psi, u_t\rangle$ is a solution of the ordinary differential equation*

$$\frac{\mathrm{d}}{\mathrm{d}t} z(t) = Gz(t) + \langle x^*, f(t)\rangle \quad \textit{for } t \in \mathbb{R}. \tag{5.5}$$

Conversely, if z is a solution of equation (5.5) *on $\mathbb{R}$ and if, in addition, f is bounded on $\mathbb{R}$, then the function u given by*

$$u(t) := \left[\Phi z(t) + \lim_{n\to+\infty} \int_{-\infty}^t U^s(t-s)\Pi^s(\widetilde{B}_n X_0 f(s))\mathrm{d}s\right](0) \quad \textit{for } t \in \mathbb{R},$$

is a solution of equation (5.1) *on $\mathbb{R}$.*

Proof. Let u be a solution of equation (5.2) on $\mathbb{R}$. Then

$$u_t = \Pi^s u_t + \Pi^\nu u_t \quad \text{for all } t \in \mathbb{R}$$

and

$$\Pi^\nu u_t = \mathcal{U}^\nu(t-\sigma)\Pi^\nu u_\sigma + \lim_{n\to+\infty} \int_\sigma^t \mathcal{U}^\nu(t-\xi)\Pi^\nu(\widetilde{B}_n X_0 f(\xi))\mathrm{d}\xi \quad \text{for } t,\sigma \in \mathbb{R}.$$

Since $\Pi^v u_t = \Phi\langle\Psi, u_t\rangle$ and by Corollary 5.1.2, we get that

$$\begin{aligned}\Phi\langle\Psi, u_t\rangle &= \mathcal{U}^v(t-\sigma)\Phi\langle\Psi, u_\sigma\rangle + \Phi\int_\sigma^t e^{(t-\xi)G}\langle x^*, f(\xi)\rangle d\xi \\ &= \Phi e^{(t-\sigma)G}\langle\Psi, u_\sigma\rangle + \Phi\int_\sigma^t e^{(t-\xi)G}\langle x^*, f(\xi)\rangle d\xi \quad \text{for } t,\sigma \in \mathbb{R}.\end{aligned}$$

Let $z(t) = \langle\Psi, u_t\rangle$. Then

$$z(t) = e^{(t-\sigma)G}z(\sigma) + \int_\sigma^t e^{(t-\xi)G}\langle x^*, f(\xi)\rangle d\xi \quad \text{for } t,\sigma \in \mathbb{R}.$$

Consequently, z is a solution of the ordinary differential equation (3.43) on $\mathbb{R}$.

Conversely, assume that f is bounded on $\mathbb{R}$, then $\int_{-\infty}^t \mathcal{U}^s(t-\xi)\Pi^s(\widetilde{B}_n X_0 f(\xi))d\xi$ is well defined on $\mathbb{R}$. Let z be a solution of (3.43) on $\mathbb{R}$ and v be defined by

$$v(t) = \Phi z(t) + \lim_{n\to+\infty}\int_{-\infty}^t \mathcal{U}^s(t-\xi)\Pi^s(\widetilde{B}_n X_0 f(\xi))d\xi \quad \text{for } t \in \mathbb{R}.$$

Since

$$z(t) = e^{(t-\sigma)G}z(\sigma) + \int_\sigma^t e^{(t-\xi)G}\langle x^*, f(\xi)\rangle d\xi \quad \text{for } t,\sigma \in \mathbb{R},$$

using Corollary 5.1.2, the function v_1 given by

$$v_1(t) = \Phi z(t) \quad \text{for } t \in \mathbb{R}$$

satisfies

$$v_1(t) = \mathcal{U}^v(t-\sigma)v_1(\sigma) + \lim_{n\to+\infty}\int_\sigma^t \mathcal{U}^v(t-\xi)\Pi^v(\widetilde{B}_n X_0 f(\xi))d\xi, \quad \text{for } t,\sigma \in \mathbb{R}.$$

Moreover, the function v_2 given by

$$v_2(t) = \lim_{n\to+\infty}\int_{-\infty}^t \mathcal{U}^s(t-\xi)\Pi^s(\widetilde{B}_n X_0 f(\xi))d\xi \quad \text{for } t \in \mathbb{R}$$

satisfies

$$v_2(t) = \mathcal{U}^s(t-\sigma)v_2(\sigma) + \lim_{n\to+\infty}\int_\sigma^t \mathcal{U}^s(t-\xi)\Pi^s(\widetilde{B}_n X_0 f(\xi))d\xi \quad \text{for all } t \geq \sigma.$$

Then, for all $t \geq \sigma$ with $t, \sigma \in \mathbb{R}$, one has

$$\begin{aligned}\mathcal{U}(t-\sigma)v(\sigma) &= \mathcal{U}^{v}(t-\sigma)v_1(\sigma) + \mathcal{U}^{s}(t-\sigma)v_2(\sigma),\\ &= v_1(t) - \lim_{n\to+\infty}\int_{\sigma}^{t}\mathcal{U}^{v}(t-\xi)\Pi^{v}(\widetilde{B}_n X_0 f(\xi))d\xi\\ &\quad + v_2(t) - \lim_{n\to+\infty}\int_{\sigma}^{t}\mathcal{U}^{s}(t-\xi)\Pi^{s}(\widetilde{B}_n X_0 f(\xi))d\xi\\ &= v(t) - \lim_{n\to+\infty}\int_{\sigma}^{t}\mathcal{U}(t-\xi)(\widetilde{B}_n X_0 f(\xi))d\xi.\end{aligned}$$

Therefore

$$v(t) = \mathcal{U}(t-\sigma)v(\sigma) + \lim_{n\to+\infty}\int_{\sigma}^{t}\mathcal{U}(t-\xi)(\widetilde{B}_n X_0 f(\xi))\xi \quad \text{for } t \geq \sigma.$$

By Theorem 3.3.3, we obtain that the function u defined by $u(t) = v(t)(0)$ is a solution of equation (5.2) on $\mathbb{R}$. □

We get the same result as in the above theorem by weakening the boundedness assumption of the forcing term f. Namely, f is only required to be bounded by mean values.

Theorem 5.1.11. *Assume that* **(H0)** *and* **(H1)** *hold, f is locally integrable, and let u be a solution of equation* (5.1) *on $\mathbb{R}$. Then the function z defined by $z(t) := \langle \Psi, u_t \rangle$ is given by*

$$z(t) = e^{tG}z(0) + \int_{0}^{t} e^{(t-s)G}\langle x^*, f(s)\rangle ds \quad \textit{for } t \in \mathbb{R}. \tag{5.6}$$

Conversely, if z satisfies equation (5.6) *on $\mathbb{R}$ and if, in addition, f satisfies*

$$\sup_{t\in\mathbb{R}}\int_{t}^{t+1}|f(s)|ds < \infty,$$

then the function u given by

$$u(t) := \left[\Phi z(t) + \lim_{n\to\infty}\int_{-\infty}^{t} U^{s}(t-s)\Pi^{s}(\widetilde{B}_n X_0 f(s))ds\right](0) \quad \textit{for } t \in \mathbb{R},$$

is a solution of equation (5.1) *on $\mathbb{R}$.*

Proof. The proof is similar to that of Theorem 5.1.10. We only have to prove that the limit

$$\lim_{n\to+\infty}\int_{-\infty}^{t} U^s(t-s)\Pi^s(\widetilde{B}_n X_0 f(s))ds$$

exists in C_0. For $t\in\mathbb{R}$ and for n sufficiently large, we have

$$\begin{aligned}\left|\int_{-\infty}^{t} U^s(t-s)\Pi^s(\widetilde{B}_n X_0 f(s))ds\right| &\le 2\widetilde{M}N|\Pi^s|\int_{-\infty}^{t} e^{-a(t-s)}|f(s)|ds\\ &\le 2\widetilde{M}N|\Pi^s|\sum_{k=1}^{+\infty}\left(\int_{t-k}^{t-k+1} e^{-a(t-s)}|f(s)|ds\right)\\ &\le 2\widetilde{M}N|\Pi^s|\sum_{k=1}^{+\infty}\left(e^{-a(k-1)}\int_{t-k}^{t-k+1}|f(s)|ds\right)\le K,\end{aligned}$$

where $K = 2\widetilde{M}N|\Pi^s|\sup_{t\in\mathbb{R}}(\int_t^{t+1}|f(s)|ds)\frac{1}{1-e^{-a}}$. Let

$$H(n,t',t) := U^s(t-t')\Pi^s(\widetilde{B}_n X_0 f(t')) \quad \text{for } n\in\mathbb{N} \text{ and } t'\le t.$$

For n and m sufficiently large and $\sigma\le t$, we have

$$\begin{aligned}\left|\int_{-\infty}^{t} H(n,s,t)ds-\int_{-\infty}^{t} H(m,s,t)ds\right| &\le \left|\int_{-\infty}^{\sigma} H(n,s,t)ds\right|+\left|\int_{-\infty}^{\sigma} H(m,s,t)ds\right|\\ &\quad+\left|\int_{\sigma}^{t} H(n,s,t)ds-\int_{\sigma}^{t} H(m,s,t)ds\right|\\ &\le 2Ke^{-a(t-\sigma)}+\left|\int_{\sigma}^{t} H(n,s,t)ds-\int_{\sigma}^{t} H(m,s,t)ds\right|.\end{aligned}$$

Since $\lim_{n\to+\infty}\int_\sigma^t H(n,s,t)ds$ exists, it follows that

$$\limsup_{n,m\to+\infty}\left|\int_{-\infty}^{t} H(n,s,t)ds-\int_{-\infty}^{t} H(m,s,t)ds\right|\le 2Ke^{-a(t-\sigma)}.$$

Letting $\sigma\to-\infty$, we get

$$\limsup_{n,m\to+\infty}\left|\int_{-\infty}^{t} H(n,s,t)ds-\int_{-\infty}^{t} H(m,s,t)ds\right|=0.$$

Thus, due to the completeness of the phase space C_0, we deduce that the limit

$$\lim_{n\to+\infty}\int_{-\infty}^{t} H(n,s,t)ds$$

exists in C_0. □

5.1.2 Almost automorphic forcing term

In the following, we assume that
(H2) f is an almost automorphic function.

Theorem 5.1.3. *Assume that* **(H0)**, **(H1)**, *and* **(H2)** *hold. If equation* (5.2) *has a bounded solution on* $\mathbb{R}^+$, *then it has an almost automorphic solution.*

Proof. Let u be a bounded solution of equation (5.1) on $\mathbb{R}^+$. By Theorem 3.3.10, the function $z(t) = \langle\Psi, u_t\rangle$, for $t \geq 0$, is a solution of the ordinary differential equation (3.43) and z is bounded on $\mathbb{R}^+$. Moreover, the function

$$\varrho(t) = \langle x^*, f(t)\rangle \quad \text{for } t \in \mathbb{R}$$

is almost automorphic from $\mathbb{R}$ to $\mathbb{R}^d$. By Theorem 3.1.3, we get that the reduced system (3.43) has an almost automorphic solution $\tilde{z}$. Consequently, $\Phi\tilde{z}(\cdot)$ is an almost automorphic function on $\mathbb{R}$. By Theorem 3.3.10, the function $u(t) = v(t)(0)$, where

$$v(t) = \Phi\tilde{z}(t) + \lim_{n\to+\infty}\int_{-\infty}^{t} \mathcal{U}^s(t-\xi)\Pi^s(\widetilde{B}_n X_0 f(\xi))d\xi \quad \text{for } t \in \mathbb{R},$$

is a solution of equation (5.1) on $\mathbb{R}$. We claim that v is almost automorphic. In fact, consider the function y defined by

$$y(t) = \lim_{n\to+\infty}\int_{-\infty}^{t} \mathcal{U}^s(t-\xi)\Pi^s(\widetilde{B}_n X_0 f(\xi))d\xi \quad \text{for } t \in \mathbb{R}.$$

Since f is almost automorphic, for any sequence of real numbers $(a'_p)_{p\geq 0}$ there exists a subsequence $(a_p)_{p\geq 0}$ of $(a'_p)_{p\geq 0}$ such that

$$\lim_{p\to\infty} f(t+a_p) = h(t) \quad \text{for all } t \in \mathbb{R}$$

and

$$\lim_{p\to\infty} h(t-a_p) = f(t) \quad \text{for all } t \in \mathbb{R}.$$

Now

$$y(t+a_p) = \lim_{n\to+\infty} \int_{-\infty}^{t+a_p} \mathcal{U}^s(t+a_p-\xi)\Pi^s(\widetilde{B}_n X_0 f(\xi))d\xi \quad \text{for } t \in \mathbb{R},$$

which gives

$$y(t+a_p) = \lim_{n\to+\infty} \int_{-\infty}^{t} \mathcal{U}^s(t-\xi)\Pi^s(\widetilde{B}_n X_0 f(\xi+a_p))d\xi \quad \text{for } t \in \mathbb{R}.$$

By the Lebesgue's dominated convergence theorem, we get that

$$y(t+a_p) \to w(t) \quad \text{as } p \to \infty,$$

where w is given by

$$w(t) = \lim_{n\to+\infty} \int_{-\infty}^{t} \mathcal{U}^s(t-\xi)\Pi^s(\widetilde{B}_n X_0 h(\xi))d\xi \quad \text{for } t \in \mathbb{R}.$$

Using the same arguments as above, we prove that

$$w(t-a_p) \to \lim_{n\to+\infty} \int_{-\infty}^{t} \mathcal{U}^s(t-\xi)\Pi^s(\widetilde{B}_n X_0 f(\xi))d\xi \quad \text{as } p \to \infty,$$

which implies that y is almost automorphic. Consequently, v is an almost automorphic solution of equation (5.1). □

5.1.3 Stepanov almost automorphic forcing term

The goal of this section is to investigate the almost automorphic aspect of bounded solutions of equation (5.1).

Since almost automorphic functions have relatively compact ranges, it is appropriate to investigate first when a bounded solution of equation (5.1) has a relatively compact range for an input function $f \in BS^p(\mathbb{R}, X)$. We distinguish two cases: $p > 1$ and $p = 1$.

5.1.3.1 The case $p > 1$

Lemma 5.1.12. *Assume that* **(H0)** *and* **(H1)** *hold and* $f \in BS^p(\mathbb{R}, X)$ *with* $p > 1$. *Then, every bounded solution of equation* (5.1) *on* $\mathbb{R}^+$ *(resp. on* $\mathbb{R}$*) has a relatively compact range.*

Proof. Let x be a solution of equation (5.1) which is bounded on $\mathbb{R}^+$ and $\overline{x} := \sup_{t\geq 0}|x_t|$. Let $M_0 \geq 1$ and $\omega_0 \in \mathbb{R}$ be such that $|T(t)| \leq M_0 e^{\omega_0 t}$ for all $t \geq 0$. For $0 < \varepsilon \leq 1$ and $t \geq 1$, we have

$$x(t) = T_0(\varepsilon)x(t-\varepsilon) + \lim_{\lambda\to\infty}\int_{t-\varepsilon}^{t} T_0(t-s)B_\lambda[L(x_s)+f(s)]\mathrm{d}s.$$

Since $T_0(\varepsilon)$ is compact, there exists a compact subset K_ε of X such that $T_0(\varepsilon)x(t-\varepsilon) \in K_\varepsilon$ for all $t \geq 1$. Moreover, for each $t \geq 1$, we have

$$\begin{aligned}\left|\lim_{\lambda\to\infty}\int_{t-\varepsilon}^{t} T_0(t-s)B_\lambda[L(x_s)+f(s)]\mathrm{d}s\right| &\leq M_0\overline{M}|L|\overline{x}\int_{t-\varepsilon}^{t} e^{\omega_0(t-s)}\mathrm{d}s + M_0\overline{M}\int_{t-\varepsilon}^{t} e^{\omega_0(t-s)}|f(s)|\mathrm{d}s\\ &\leq M_0\overline{M}|L|\overline{x}\int_0^{\varepsilon} e^{\omega_0 s}\mathrm{d}s\\ &\quad + M_0\overline{M}\left(\int_{t-\varepsilon}^{t} e^{q\omega_0(t-s)}\mathrm{d}s\right)^{\frac{1}{q}}\left(\int_{t-\varepsilon}^{t}|f(s)|^p\mathrm{d}s\right)^{\frac{1}{p}}\\ &\leq \delta(\varepsilon),\end{aligned}$$

where $\delta(\varepsilon) := M_0\overline{M}|L|\overline{x}\int_0^{\varepsilon} e^{\omega_0 s}\mathrm{d}s + M_0\overline{M}|f|_{BS^p}(\int_0^{\varepsilon} e^{q\omega_0 s}\mathrm{d}s)^{\frac{1}{q}}$. Then for all $t \geq 1$ we have

$$\lim_{\lambda\to\infty}\int_{t-\varepsilon}^{t} T_0(t-s)B_\lambda[L(x_s)+f(s)]\mathrm{d}s \in \overline{B}(0,\delta(\varepsilon)).$$

It follows that

$$\{x(t) : t \geq 1\} \subset K_\varepsilon + \overline{B}(0,\delta(\varepsilon)).$$

Taking the Kuratowski's measure of noncompactness, we obtain that

$$\alpha(\{x(t) : t \geq 1\}) \leq 2\delta(\varepsilon).$$

The above inequality holds for all $0 < \varepsilon \leq 1$, thus, by letting ε to 0, we deduce that

$$\alpha(\{x(t) : t \geq 1\}) = 0.$$

Thus $\{x(t) : t \geq 1\}$ is relatively compact. Since $x(\cdot)$ is continuous, we conclude that $\{x(t) : t \geq 0\}$ is relatively compact. The same proof can be used for solutions bounded on the whole real line. □

Lemma 5.1.13. *Assume that* **(H0)** *and* **(H1)** *hold and* $f \in BS^p(\mathbb{R}, X)$ *with* $p > 1$. *Then, every bounded solution of equation* (5.1) *on* $\mathbb{R}^+$ *(resp. on* $\mathbb{R}$*) is uniformly continuous.*

Proof. Let x be a solution of equation (5.1) which is bounded on $\mathbb{R}^+$ and let $\overline{x} := \sup_{t\geq 0}|x_t|$. Let $M_0 \geq 1$ and $\omega_0 \in \mathbb{R}$ be such that $|T(t)| \leq M_0 e^{\omega_0 t}$ for all $t \geq 0$. For $0 \leq t' \leq t$, we have

$$x(t) = T_0(t-t')x(t') + \lim_{\lambda\to\infty}\int_{t'}^{t} T_0(t-s)B_\lambda[L(x_s) + f(s)]\mathrm{d}s.$$

Let $K = \overline{\{x(t) : t \geq 0\}}$, which is a compact subset of X by Lemma 5.1.12. It follows that

$$\begin{aligned}
|x(t)-x(t')| &\leq |T_0(t-t')x(t') - x(t')| + M_0\overline{M}|L|\overline{x}\int_{t'}^{t} e^{\omega_0(t-s)}\mathrm{d}s + M_0\overline{M}\int_{t'}^{t} e^{\omega_0(t-s)}|f(s)|\mathrm{d}s\\
&= \sup_{y\in K}|T_0(t-t')y-y| + M_0\overline{M}|L|\overline{x}\int_0^{t-t'} e^{\omega_0 s}\mathrm{d}s + M_0\overline{M}\int_0^{t-t'} e^{\omega_0 s}|f(t-s)|\mathrm{d}s\\
&\leq \sup_{y\in K}|T_0(t-t')y-y| + M_0\overline{M}|L|\overline{x}\int_0^{t-t'} e^{\omega_0 s}\mathrm{d}s\\
&\quad + M_0\overline{M}\left(\int_0^{t-t'} e^{q\omega_0 s}\mathrm{d}s\right)^{\frac{1}{q}}\left(\int_0^{t-t'}|f(t-s)|^p\mathrm{d}s\right)^{\frac{1}{p}}\\
&= \sup_{y\in K}|T_0(t-t')y-y| + M_0\overline{M}|L|\overline{x}\int_0^{t-t'} e^{\omega_0 s}\mathrm{d}s\\
&\quad + M_0\overline{M}\left(\int_0^{t-t'} e^{q\omega_0 s}\mathrm{d}s\right)^{\frac{1}{q}}\left(\int_{t'}^{t}|f(s)|^p\mathrm{d}s\right)^{\frac{1}{p}}.
\end{aligned}$$

If $|t-t'|$ is small enough, then, since $f \in BS^p(\mathbb{R}, X)$, we have

$$\begin{aligned}
|x(t)-x(t')| &\leq \sup_{y\in K}|T_0(t-t')y-y| + M_0\overline{M}|L|\overline{x}\int_0^{t-t'} e^{\omega_0 s}\mathrm{d}s\\
&\quad + M_0\overline{M}\left(\int_0^{t-t'} e^{q\omega_0 s}\mathrm{d}s\right)^{\frac{1}{q}}\left(\int_{t'}^{t'+1}|f(s)|^p\mathrm{d}s\right)^{\frac{1}{p}}\\
&\leq \sup_{y\in K}|T_0(t-t')y-y| + M_0\overline{M}|L|\overline{x}\int_0^{t-t'} e^{\omega_0 s}\mathrm{d}s + M_0\overline{M}|f|_{BS^p}\left(\int_0^{t-t'} e^{q\omega_0 s}\mathrm{d}s\right)^{\frac{1}{q}}.
\end{aligned}$$

The Banach–Steinhaus theorem ensures that $\sup_{y\in K}|T_0(t-t')y-y| \to 0$ as $|t-t'| \to 0$. Thus x is uniformly continuous. □

Corollary 5.1.14. *Assume that* **(H0)** *and* **(H1)** *hold and* $f \in BS^p(\mathbb{R}, X)$ *with* $p > 1$. *If* $t \mapsto x(t)$ *is a bounded solution of equation* (5.1) *on* $\mathbb{R}^+$ *(resp. on* $\mathbb{R}$*), then the history function* $t \mapsto x_t$ *has a relatively compact range on* $\mathbb{R}^+$ *(resp. on* $\mathbb{R}$*).*

Proof. From Lemma 5.1.13, the solution $t \mapsto x(t)$ is uniformly continuous. Thus the family of functions $\theta \mapsto x_t(\theta)$ indexed by t is equicontinuous. Since for each $\theta \in [-r, 0]$ the set $\{x_t(\theta) : t \geq 0\}$ is relatively compact (by Lemma 5.1.12), we get the compactness of $\overline{\{x_t : t \geq 0\}}$ in the space $C([-r, 0], X)$ by applying Arzelà–Ascoli theorem. □

The following result follows immediately from Lemma 5.1.13.

Corollary 5.1.15. *Assume that* **(H0)** *and* **(H1)** *hold and* $f \in BS^p(\mathbb{R}, X)$ *with* $p > 1$. *If* $t \mapsto x(t)$ *is a bounded solution of equation* (5.1) *on* $\mathbb{R}^+$ *(resp. on* $\mathbb{R}$*), then the history function* $t \mapsto x_t$ *is uniformly continuous on* $\mathbb{R}^+$ *(resp. on* $\mathbb{R}$*).*

5.1.3.2 The case $p = 1$

To reproduce the above results for $f \in BS^p(\mathbb{R}, X)$ with $p = 1$, one needs to assume more than the boundedness in the Stepanov norm.

In what follows, let $M_1 \geq 1$ and $\omega_1 \in \mathbb{R}$ be such that $|U(t)| \leq M_1 e^{\omega_1 t}$ for all $t \geq 0$.

Lemma 5.1.16. *Assume that* **(H0)** *and* **(H1)** *hold and* $f \in SAA^1(\mathbb{R}, X)$. *If* $t \mapsto x(t)$ *is a bounded solution of equation* (5.1) *on* $\mathbb{R}^+$ *(resp. on* $\mathbb{R}$*), then the history function* $t \mapsto x_t$ *has a relatively compact range on* $\mathbb{R}^+$ *(resp. on* $\mathbb{R}$*).*

Proof. Let x be a solution of equation (5.1) which is bounded on $\mathbb{R}^+$. Consider the following decomposition:

$$\{x_t : t \geq 0\} = \{x_t : t \geq 2r\} \cup \{x_t : 0 \leq t < 2r\}. \tag{5.7}$$

Recall that the operator $U(t)$ is compact whenever $t > r$, where r is the delay (see Theorem 5.1.7). Let $(t_n)_n$ be a sequence of real numbers such that $t_n \geq 2r$ for all $n \in \mathbb{N}$. Then we have

$$\begin{aligned} x_{t_n} &= U(2r)x_{t_n-2r} + \lim_{m\to\infty} \int_{t_n-2r}^{t_n} U(t_n - s)\widetilde{B}_m(X_0 f(s))ds \\ &= U(2r)x_{t_n-2r} + \lim_{m\to\infty} \int_0^{2r} U(s)\widetilde{B}_m(X_0 f(t_n - s))ds. \end{aligned}$$

Since the operator $U(2r)$ is compact and $f \in SAA^1(\mathbb{R}, X)$, there exist a subsequence $(t'_n)_n \subset (t_n)_n$ and a function $\widetilde{f} \in L^1_{loc}(\mathbb{R}, X)$ such that $U(2r)x_{t'_n-2r}$ converges to some $\varphi_1 \in C_0$ and, for all $t \in \mathbb{R}$,

$$\int_t^{t+1} |f(t'_n + s) - \widetilde{f}(s)|ds \to 0 \quad \text{as } n \to \infty.$$

Let $\varphi_2 := \lim_{m\to\infty} \int_0^{2r} U(s)\widetilde{B}_m(X_0\widetilde{f}(-s))ds$. Then $x_{t_n'} \to \varphi_1 + \varphi_2$ as $n \to \infty$. In fact,

$$\begin{aligned}\left|\lim_{m\to\infty}\int_0^{2r} U(s)\widetilde{B}_m(X_0 f(t_n' - s))ds - \varphi_2\right| &\le M_1\widetilde{M}\int_0^{2r} e^{\omega_1 s}|f(t_n' - s) - \widetilde{f}(-s)|ds \\ &\le M_1\widetilde{M}e^{|\omega_1|2r}\int_0^{2r}|f(t_n' - s) - \widetilde{f}(-s)|ds \\ &\le M_1\widetilde{M}e^{|\omega_1|2r}\sum_{k=0}^{[2r]+1}\int_{-k-1}^{-k}|f(t_n' + s) - \widetilde{f}(s)|ds \to 0,\end{aligned}$$

as $n \to \infty$. Therefore, the set $\{x_t : t \ge 2r\}$ is relatively compact. The relative compactness of $\{x_t : 0 \le t < 2r\}$ follows from the continuity of the history function $t \mapsto x_t$. The proof is complete. □

Lemma 5.1.17. *Assume that* **(H0)** *and* **(H1)** *hold and* $f \in SAA^1(\mathbb{R}, X)$. *If* $t \mapsto x(t)$ *is a bounded solution of equation* (5.1) *on* $\mathbb{R}^+$ *(resp. on* $\mathbb{R}$*), then the history function* $t \mapsto x_t$ *is uniformly continuous on* $\mathbb{R}^+$ *(resp. on* $\mathbb{R}$*).*

Proof. If $t \mapsto x_t$ is not uniformly continuous, then there exist $\varepsilon > 0$ and two real sequences $(s_n)_n$ and $(h_n)_n$ such that $\lim_{n\to\infty} h_n = 0$ and

$$|x_{s_n+h_n} - x_{s_n}| > \varepsilon \quad \text{for all } n \in \mathbb{N}. \tag{5.8}$$

For all $n \in \mathbb{N}$ such that $h_n \ge 0$, we have

$$\begin{aligned}x_{s_n+h_n} &= U(h_n)x_{s_n} + \lim_{m\to\infty}\int_{s_n}^{s_n+h_n} U(s_n + h_n - s)(\widetilde{B}_m(X_0 f(s)))ds \\ &= U(h_n)x_{s_n} + \lim_{m\to\infty}\int_0^{h_n} U(h_n - s)(\widetilde{B}_m(X_0 f(s + s_n)))ds.\end{aligned}$$

Let $K := \overline{\{x_t : t \in \mathbb{R}\}}$. Then we have

$$\begin{aligned}|x_{s_n+h_n} - x_{s_n}| &\le |U(h_n)x_{s_n} - x_{s_n}| + \widetilde{M}\int_0^{h_n}|U(h_n - s)||f(s + s_n)|ds \\ &\le \sup_{\phi\in K}|U(h_n)\phi - \phi| + \widetilde{M}M_1\int_0^{h_n} e^{\omega_1(h_n-s)}|f(s + s_n)|ds \\ &\le \sup_{\phi\in K}|U(h_n)\phi - \phi| + \widetilde{M}M_1 e^{|\omega_1|h_n}\int_0^{h_n}|f(s + s_n)|ds.\end{aligned}$$

On the other hand, for all $n \in \mathbb{N}$ such that $h_n < 0$, we have

$$x_{s_n} = U(-h_n)x_{s_n+h_n} + \lim_{m\to\infty} \int_{s_n+h_n}^{s_n} U(s_n - s)(\widetilde{B}_m(X_0 f(s)))ds$$
$$= U(-h_n)x_{s_n+h_n} + \lim_{m\to\infty} \int_{h_n}^{0} U(-s)(\widetilde{B}_m(X_0 f(s + s_n)))ds.$$

Therefore,

$$|x_{s_n} - x_{s_n+h_n}| \le |U(-h_n)x_{s_n+h_n} - x_{s_n+h_n}| + \widetilde{M} \int_{h_n}^{0} |U(-s)||f(s + s_n)|ds$$
$$\le \sup_{\phi\in K}|U(-h_n)\phi - \phi| + \widetilde{M}M_1 \int_{h_n}^{0} e^{-\omega_1 s}|f(s + s_n)|ds$$
$$\le \sup_{\phi\in K}|U(-h_n)\phi - \phi| + \widetilde{M}M_1 e^{-|\omega_1|h_n} \int_{h_n}^{0} |f(s + s_n)|ds.$$

We deduce that, for all $n \in \mathbb{N}$,

$$|x_{s_n+h_n} - x_{s_n}| \le \sup_{\phi\in K}|U(|h_n|)\phi - \phi| + \widetilde{M}M_1 e^{|\omega_1 h_n|} \int_{-|h_n|}^{|h_n|} |f(s + s_n)|ds. \tag{5.9}$$

From Lemma 5.1.16, K is a compact subset of C. It follows by the Banach–Steinhaus theorem that

$$\sup_{\phi\in K}|U(|h_n|)\phi - \phi| \to 0 \quad \text{as } n \to \infty.$$

On the other hand, from the S^1-almost automorphy of f, there exist a subsequence $(s'_n)_n \subset (s_n)_n$ and a function $\widehat{f} \in L^1_{loc}(\mathbb{R}, X)$ such that, for each $t \in \mathbb{R}$,

$$\int_{t}^{t+1} |f(s + s'_n) - \widehat{f}(s)|ds \to 0 \quad \text{as } n \to \infty. \tag{5.10}$$

Let $(h'_n)_n$ be the corresponding subsequence of $(h_n)_n$. We can assume that $|h_n| \le 1$ for all $n \in \mathbb{N}$. Then, we have

$$\begin{aligned}\int_{-|h'_n|}^{|h'_n|} |f(s+s'_n)|ds &\le \int_{-|h'_n|}^{|h'_n|} |f(s+s'_n)-\hat{f}(s)|ds + \int_{-|h'_n|}^{|h'_n|} |\hat{f}(s)|ds \\ &\le \int_{-1}^{1} |f(s+s'_n)-\hat{f}(s)|ds + \int_{-|h'_n|}^{|h'_n|} |\hat{f}(s)|ds \\ &\le \int_{-1}^{0} |f(s+s'_n)-\hat{f}(s)|ds + \int_{0}^{1} |f(s+s'_n)-\hat{f}(s)|ds + \int_{-|h'_n|}^{|h'_n|} |\hat{f}(s)|ds.\end{aligned}$$

It follows from (5.10) and Lebesgue's dominated convergence theorem that

$$\int_{-|h'_n|}^{|h'_n|} |f(s+s'_n)|ds \to 0 \quad \text{as } n \to \infty.$$

Therefore, we deduce from (5.9) that

$$|x_{s'_n+h'_n} - x_{s'_n}| \to 0 \quad \text{as } n \to \infty,$$

which contradicts (5.8). We conclude that $t \mapsto x_t$ must be uniformly continuous. □

The following result shows that to get a bounded solution on $\mathbb{R}$, one only needs to have a bounded solution on $\mathbb{R}^+$.

Theorem 5.1.18. *Assume that* **(H0)** *and* **(H1)** *hold and* $f \in SAA^1(\mathbb{R}, X)$. *Then, if equation* (5.1) *has an integral solution which is bounded on* $\mathbb{R}^+$, *then it has an integral solution which is bounded on* $\mathbb{R}$.

Proof. Let x be a solution of equation (5.1) which is bounded on $\mathbb{R}^+$. From Lemmas 5.1.16 and 5.1.17, x is compact and uniformly continuous. Let $(t_n)_n$ be a sequence of real numbers such that $\lim_{n\to\infty} t_n = \infty$. If $t \in [-1,1]$, then for sufficiently large n the sequence of functions $x_n : t \mapsto x(t+t_n)$ is well defined and is equicontinuous. It follows by Arzelà–Ascoli theorem that there exist a function y and a subsequence $(t_n^1)_n$ such that

$$x(t+t_n^1) \to y(t) \quad \text{as } n \to \infty$$

uniformly on $[-1,1]$. Using the same argument, we deduce that for each $N \in \mathbb{N}^*$ there exists a subsequence $(t_n^N)_n \subset (t_n^{N-1})_n \subset \cdots \subset (t_n^1)_n \subset (t_n)_n$ such that

$$x(t+t_n^N) \to y(t) \quad \text{as } n \to \infty$$

uniformly on $[-N,N]$. Let $(t'_n)_n := (t_n^n)_n$ be the Cantor's diagonal sequence. It is clear that

$$x(t+t'_n) \to y(t) \quad \text{as } n \to \infty$$

uniformly on each compact subset of $\mathbb{R}$. Since $f \in SAA^1(\mathbb{R}, X)$, there exist a subsequence $(t_n'')_n \subset (t_n')_n$ and a function $\tilde{f} \in L^1_{\mathrm{loc}}(\mathbb{R}, X)$ such that

$$\int_t^{t+1} |f(t_n'' + s) - \tilde{f}(s)| \mathrm{d}s \to 0 \tag{5.11}$$

and

$$\int_t^{t+1} |\tilde{f}(t_n'' - s) - f(s)| \mathrm{d}s \to 0 \tag{5.12}$$

as $n \to \infty$. For each $t \geq s$ and for $n \in \mathbb{N}$ sufficiently large, we have

$$x(t + t_n'') = T_0(t - s)x(s + t_n'') + \lim_{\lambda \to \infty} \int_s^t T_0(t - \sigma)B_\lambda[L(x_{\sigma + t_n''}) + f(\sigma + t_n'')]\mathrm{d}\sigma. \tag{5.13}$$

Taking the limit as $n \to \infty$ in (5.13), using (5.11) and the fact that $x_{\sigma + t_n''} \to y_\sigma$, we get, for each $t \geq s$,

$$y(t) = T_0(t - s)y(s) + \lim_{\lambda \to \infty} \int_s^t T_0(t - \sigma)B_\lambda[L(y_\sigma) + \tilde{f}(\sigma)]\mathrm{d}\sigma.$$

We observe that y is also compact and uniformly continuous. Using (5.12) and applying the above argument to the returning sequence $(-t_n'')_n$, we obtain a solution z of equation (5.1) which is bounded on $\mathbb{R}$. □

Now we use the reduction principle in Theorem 5.1.11 to extend Bohr–Neugebauer-type theorem to the partial functional differential equation (5.1).

The following theorem shows the nature of all bounded solutions of equation (5.1) on $\mathbb{R}$.

Theorem 5.1.19. *Assume that* **(H0)** *and* **(H1)** *hold and $f : \mathbb{R} \to X$ is S^1-almost automorphic. Then, every bounded integral solution of equation* (5.1) *on $\mathbb{R}$ is compact almost automorphic.*

Proof of Theorem 5.1.19. Let x be a bounded solution of equation (5.1) on $\mathbb{R}$. Using the spectral decomposition (5.3), we have, for each $t \in \mathbb{R}$,

$$x_t = \Pi^v x_t + \Pi^s x_t. \tag{5.14}$$

On the one hand, we have, for $t \geq \sigma$,

$$\Pi^s x_t = U^s(t - \sigma)\Pi^s x_\sigma + \lim_{n \to \infty} \int_\sigma^t U^s(t - s)\Pi^s(\widetilde{B}_n(X_0 f(s)))\mathrm{d}s. \tag{5.15}$$

Since $t \mapsto x_t$ is bounded on $\mathbb{R}$ and $U(t)$ is exponentially stable in S, letting $\sigma \to -\infty$ in (5.15), we obtain that, for all $t \in \mathbb{R}$,

$$\Pi^s x_t = \lim_{n\to\infty} \int_{-\infty}^{t} U^s(t-s)\Pi^s(\widetilde{B}_n(X_0 f(s)))ds.$$

In fact, for each fixed $t \in \mathbb{R}$, we have, for all $\sigma \leq t$,

$$|U^s(t-\sigma)\Pi^s x_\sigma| \leq e^{-\alpha(t-\sigma)}|\Pi^s| \sup_{s\in\mathbb{R}} |x_s| \to 0 \quad \text{as } \sigma \to -\infty$$

and

$$\begin{aligned}
&\left|\lim_{n\to\infty} \int_{\sigma}^{t} U^s(t-s)\Pi^s(\widetilde{B}_n(X_0 f(s)))ds - \lim_{n\to\infty} \int_{-\infty}^{t} U^s(t-s)\Pi^s(\widetilde{B}_n(X_0 f(s)))ds\right| \\
&= \left|\lim_{n\to\infty} \int_{-\infty}^{\sigma} U^s(t-s)\Pi^s(\widetilde{B}_n(X_0 f(s)))ds\right| \\
&\leq \widetilde{M}N|\Pi^s| \sum_{k=1}^{\infty} \int_{\sigma-k}^{\sigma-k+1} e^{-\alpha(t-s)}|f(s)|ds \\
&\leq \widetilde{M}N|\Pi^s||f|_{BS^1} \frac{e^{-\alpha(t-\sigma)}}{1-e^{-\alpha}} \to 0 \quad \text{as } \sigma \to -\infty.
\end{aligned}$$

Using the same approach as in the proof (case 2) of Theorem 3.1.1, one can prove that the function

$$t \mapsto \Pi^s x_t = \lim_{n\to\infty} \int_{-\infty}^{t} U^s(t-s)\Pi^s(\widetilde{B}_n(X_0 f(s)))ds$$

is almost automorphic.

On the other hand, for each $t \in \mathbb{R}$,

$$\Pi^v x_t = \Phi\langle\Psi, x_t\rangle = \sum_{i=1}^{d} \langle\psi_i, x_t\rangle\varphi_i. \tag{5.16}$$

From Theorem 5.1.11, the function $z(t) = \langle\Psi, x_t\rangle$ is an integral solution of the following differential equation:

$$z'(t) = Gz(t) + \langle x^*, f(t)\rangle \quad \text{for } t \in \mathbb{R}.$$

Moreover, the function $t \mapsto \langle\Psi, x_t\rangle$ is bounded on $\mathbb{R}$ and the function $t \mapsto \langle x^*, f(t)\rangle$ is S^1-almost automorphic. It follows by Theorem 3.1.1 that $t \mapsto \langle\Psi, x_t\rangle$ is almost automorphic.

We deduce from (5.16) that the function $t \mapsto \Pi^{\nu} x_t$ is almost automorphic. The almost automorphy of $t \mapsto x_t$ follows from (5.14).

From Lemma 5.1.17, the solution $t \mapsto x_t$ is uniformly continuous. The compact almost automorphy of x follows from Proposition 2.3.6. □

The following result is a consequence of Theorems 5.1.18 and 5.1.19. It shows that Theorem 5.1.3 holds even if we assume that f is S^1-almost automorphic, which is a weaker condition than the almost automorphy. Moreover, this result gives more than almost automorphy.

Corollary 5.1.20. *Assume that* **(H0)** *and* **(H1)** *hold and $f : \mathbb{R} \to X$ is S^1-almost automorphic. If equation* (5.1) *has a bounded solution on $\mathbb{R}^+$, then it has a compact almost automorphic solution.*

5.1.4 Applications

5.1.4.1 Lotka–Volterra equation

In order to apply the previous results, we consider the model of Lotka–Volterra with diffusion which is taken from [272]

$$\begin{cases} \frac{\partial}{\partial t} v(t,x) = \frac{\partial^2}{\partial x^2} v(t,x) + \int_{-r}^{0} G(\theta) v(t+\theta, x) d\theta + h(t,x) \\ \quad \text{for } t \geq 0 \text{ and } x \in [0,\pi], \\ u(t,x) = 0 \quad \text{for } x = 0, \pi \text{ and } t \geq 0, \\ u(\theta, x) = \varphi(\theta, x) \quad \text{for } \theta \in [-r, 0] \text{ and } x \in [0,\pi], \end{cases} \tag{5.17}$$

where $G : [-r, 0] \to \mathbb{R}$, $\varphi : [-r, 0] \times [0,\pi] \to \mathbb{R}$ and $h : \mathbb{R} \times [0,\pi] \to \mathbb{R}$ are continuous functions.

Let $X = C([0,\pi]; \mathbb{R})$ be the space of continuous functions from $[0,\pi]$ to $\mathbb{R}$ endowed with the uniform norm topology. Define the operator $A : D(A) \subset X \to X$ by

$$\begin{cases} D(A) = \{y \in C^2([0,\pi]; \mathbb{R}) : y(0) = y(\pi) = 0\}, \\ Ay = y''. \end{cases}$$

Lemma 5.1.4 ([113, Proposition 14.6, pp. 319–320]).

$$(0, +\infty) \subset \rho(A) \quad \textit{and} \quad |(\lambda - A)^{-1}| \leq \frac{1}{\lambda} \quad \textit{for } \lambda > 0.$$

Moreover,

$$\overline{D(A)} = \{y \in X : y(0) = y(\pi) = 0\}.$$

This lemma implies that condition **(H0)** is satisfied.

We introduce $L : C \to X$ by

$$L(\phi)(x) = \int_{-r}^{0} G(\theta)\phi(\theta)(x)d\theta \quad \text{for } x \in [0,\pi] \text{ and } \phi \in C.$$

The function $f : \mathbb{R} \to X$ is defined by

$$f(t)(x) = h(t,x) \quad \text{for } t \in \mathbb{R} \text{ and } x \in [0,\pi].$$

Operator L is bounded linear from C to X and, by continuity of h, f is a continuous function from $\mathbb{R}$ to X. Equation (5.17) then takes the abstract form (5.1).

Let A_0 be the part of A in $\overline{D(A)}$. Then, A_0 is given by

$$\begin{cases} D(A_0) = \{y \in C^2([0,\pi];\mathbb{R}) : y(0) = y(\pi) = y''(0) = y''(\pi) = 0\}, \\ A_0 y = Ay \quad \text{for } y \in D(A_0). \end{cases}$$

It is well known [136, Example 1.4.34, p. 123] that A_0 generates a strongly continuous compact semigroup $(T_0(t))_{t\geq 0}$ on $\overline{D(A)}$ and

$$|T_0(t)| \leq e^{-t} \quad \text{for } t \geq 0.$$

In order to study the existence of almost automorphic solution of equation (5.1), we suppose that

(H3) h is almost automorphic in t uniformly for $x \in [0,\pi]$, which means that there exists a measurable function $g : \mathbb{R} \times [0,\pi] \to \mathbb{R}$ such that

$$\lim_{n\to\infty} h(t+s_n, x) = g(t,x) \quad \text{exists for all } t \text{ in } \mathbb{R} \text{ uniformly in } x \in [0,\pi]$$

and

$$\lim_{n\to\infty} g(t-s_n, x) = h(t,x) \quad \text{for all } t \text{ in } \mathbb{R} \text{ uniformly in } x \in [0,\pi].$$

Moreover, we suppose that

(H4) there exists a constant $\beta \in (0,1)$ such that

$$\int_{-r}^{0} |G(\theta)| d\theta \leq (1-\beta).$$

Proposition 5.1.5. *Assume that* **(H3)** *and* **(H4)** *hold. Then equation* (5.1) *has a bounded solution. Consequently, equation* (5.1) *has an almost automorphic solution.*

Proof. The first goal is to prove that equation (5.1) has a bounded solution on $\mathbb{R}^+$. Let $\rho = (1 + \frac{|f|_\infty}{\beta})$, where $|f|_\infty = \sup_{s\in\mathbb{R}} |f(s)|$. Consider $\varphi \in C_0$ such that $|\varphi|_C < \rho$. We claim that

$$|u(t)| \le \rho \quad \text{for all } t \ge 0. \tag{5.18}$$

We proceed by contradiction. Let t_0 be the first time such that (5.18) is not true. Then

$$t_0 = \inf\{t > 0 : |u(t)| > \rho\}.$$

By continuity of u, one has

$$|u(t_0)| = \rho,$$

and there exists a positive constant $\varepsilon > 0$ such that

$$|u(t)| > \rho \quad \text{for } t \in (t_0, t_0 + \varepsilon).$$

We have

$$u(t_0) = T_0(t_0)\varphi(0) + \lim_{\lambda\to+\infty} \int_0^{t_0} T_0(t_0 - s)B_\lambda[L(u_s) + f(s)]ds,$$

which implies that

$$|u(t_0)| \le e^{-t_0}\rho + \int_0^{t_0} e^{-(t_0-s)}\left[\int_{-r}^{0} |G(\theta)||u(s+\theta)|d\theta + |f|_\infty\right]ds.$$

Since $|u(t)| \le \rho$, for $t \le t_0$, one gets

$$|u(t)| \le \rho \quad \text{for } t \in [-r, t_0].$$

Therefore

$$|u(t_0)| \le e^{-t_0}\rho + (1 - e^{-t_0})\left[\int_{-r}^{0} |G(\theta)|d\theta\rho + |f|_\infty\right].$$

Condition **(H4)** implies that

$$|u(t_0)| \le e^{-t_0}\rho + (1 - e^{-t_0})[(1-\beta)\rho + |f|_\infty]$$

and

$$|u(t_0)| \le e^{-t_0}\rho + (1 - e^{-t_0})\rho + (1 - e^{-t_0})[-\beta\rho + |f|_\infty].$$

Consequently, we obtain that

$$|u(t_0)| \leq \rho - (1 - e^{-t_0})\beta < \rho,$$

and, by continuity of u, there exists a positive ε_0 such that

$$|u(t)| < \rho \quad \text{for } t \in (t_0, t_0 + \varepsilon_0),$$

which gives a contradiction, so we deduce that equation (5.1) has a bounded solution u on $\mathbb{R}^+$. Finally, by Theorem 5.1.3, we get that equation (5.1) has an almost automorphic solution. □

5.1.4.2 Reaction–diffusion model with delay

To apply our results, we consider the following reaction–diffusion equation with delay in a bounded open subset Ω of $\mathbb{R}^n$ with a smooth boundary $\partial\Omega$:

$$\begin{cases} \frac{\partial}{\partial t}v(t,x) = \Delta v(t,x) + av(t,x) + \int_{t-r}^{t} h(s-t)v(s,x)ds + F(t)\psi(x) \\ \quad \text{for } t \in \mathbb{R} \text{ and } x \in \Omega, \\ v(t,x) = 0 \quad \text{on } \mathbb{R} \times \partial\Omega, \end{cases} \tag{5.19}$$

where $a \in \mathbb{R}$, $h : [-r, 0] \to \mathbb{R}$ and $\psi : \overline{\Omega} \to \mathbb{R}$ are continuous functions. The function $F : \mathbb{R} \to \mathbb{R}$ is given by

$$F(t) = \sum_{n\geq 1} F_n(t),$$

where F_n are defined for every integer $n \geq 1$ by

$$F_n(t) = \sum_{k\in P_n} H(n^2(t-k)),$$

with $P_n = 3^n(2\mathbb{Z}+1) = \{3^n(2k+1), k \in \mathbb{Z}\}$ and $H \in C_0^\infty(\mathbb{R},\mathbb{R})$, with support in $(-\frac{1}{2}, \frac{1}{2})$ such that

$$H \geq 0, \quad H(0) = 1, \quad \text{and} \quad \int_{\frac{-1}{2}}^{\frac{1}{2}} H(s)ds = 1.$$

The function F is not almost automorphic, since it is not bounded. However, $F \in C^\infty(\mathbb{R},\mathbb{R}) \cap AA^1(\mathbb{R},\mathbb{R})$ (see [264]).

To rewrite equation (5.19) in the abstract form (5.1), we introduce the space $X := C(\overline{\Omega})$ of continuous functions from $\overline{\Omega}$ to $\mathbb{R}$ endowed with the uniform norm topology and define the operator $A : D(A) \subset X \to X$ by

$$\begin{cases} D(A) = \{u \in C_0(\Omega) \cap H_0^1(\Omega) : \Delta u \in C(\overline{\Omega})\}, \\ Au = \Delta u. \end{cases}$$

where Δ is the Laplace operator. Let us denote by λ_1 the smallest eigenvalue of $-\Delta$ in $H_0^1(\Omega)$ ($\lambda_1 > 0$ since Ω is bounded and smooth). The operator A satisfies the Hille–Yosida condition **(H0)** on X and

$$\overline{D(A)} = \{u \in X : u|_{\partial\Omega} = 0\} \neq X.$$

Let A_0 be the part of the operator A in $\overline{D(A)}$. Then A_0 is given by

$$\begin{cases} D(A_0) = \{u \in C_0(\Omega) \cap H_0^1(\Omega) : \Delta u \in C_0(\Omega)\}, \\ A_0 u = \Delta u. \end{cases}$$

Lemma 5.1.21 ([97]). *The linear operator A_0 generates a compact C_0-semigroup $(T_0(t))_{t\geq 0}$ on $\overline{D(A)}$ such that, for each $t \geq 0$,*

$$|T_0(t)| \leq \exp\left(\frac{\lambda_1 |\Omega|^{\frac{2}{n}}}{4\pi}\right) e^{-\lambda_1 t}. \tag{5.20}$$

Let $L : C \to X$ be the operator defined by

$$L(\varphi)(\xi) = a\varphi(0)(\xi) + \int_{-r}^{0} h(s)\varphi(s)(\xi)\mathrm{d}s \quad \text{for } \xi \in \overline{\Omega} \text{ and } \varphi \in C,$$

and $f : \mathbb{R} \to X$ be given by

$$f(t)(\xi) = F(t)\psi(\xi) \quad \text{for } \xi \in \overline{\Omega} \text{ and } t \in \mathbb{R}.$$

Then, L is a bounded linear operator from C to X, and $f \in AA^1(\mathbb{R}, X)$. Equation (5.19) takes the following abstract form:

$$\frac{\mathrm{d}}{\mathrm{d}t}u(t) = Au(t) + L(u_t) + f(t) \quad \text{for } t \in \mathbb{R}. \tag{5.21}$$

Theorem 5.1.22. *Assume that*

$$|a| + \int_{-r}^{0} |h(s)|\mathrm{d}s < \lambda_1 \exp\left[-\lambda_1\left(\frac{|\Omega|^{\frac{2}{n}}}{2\pi} + r\right)\right]. \tag{5.22}$$

Then equation (5.21) *has a unique compact almost automorphic solution that is globally attractive.*

For the proof of Theorem 5.1.22, we need the following lemma.

Lemma 5.1.23 ([104]). *If*

$$x(t) \le h(t) + \int_{t_0}^{t} k(s)x(s)ds \quad for\ t \in [t_0, \tau),$$

where all the functions involved are continuous and nonnegative on $[t_0, \tau)$, $\tau \le \infty$, *and* $k(t) \ge 0$, *then* x *satisfies*

$$x(t) \le h(t) + \int_{t_0}^{t} h(s)k(s)e^{\int_s^t k(u)du}ds \quad for\ t \in [t_0, \tau).$$

Proof of Theorem 5.1.22. From Theorem 5.1.3, for any initial data $\varphi \in C_0$, equation (5.21) has a solution x given by

$$x(t) = T_0(t)\varphi(0) + \lim_{\lambda \to \infty} \int_0^t T_0(t-s)B_\lambda[L(x_s) + f(s)]ds \quad \text{for } t \ge 0,$$

where $B_\lambda = \lambda R(\lambda, A)$. Let $M := \exp(\frac{\lambda_1 |\Omega|^{\frac{2}{n}}}{4\pi})$. Then we have $(-\lambda_1, \infty) \subset \rho(A)$ and

$$|R(\lambda, A)| \le \frac{M}{\lambda + \lambda_1} \quad \text{for } \lambda > -\lambda_1. \tag{5.23}$$

Therefore, by (5.23) and (5.20), we get

$$|x(t)| \le Me^{-\lambda_1 t}|\varphi| + \int_0^t M^2 e^{-\lambda_1(t-s)}[|L||x_s| + |f(s)|]ds.$$

It follows that

$$e^{\lambda_1 t}|x(t)| \le M|\varphi| + M^2 \int_0^t e^{\lambda_1 s}[|L||x_s| + |f(s)|]ds \quad \text{for } t \ge 0. \tag{5.24}$$

Let $\theta \in [-r, 0]$ and $t \ge 0$. If $t + \theta < 0$ then

$$e^{\lambda_1 t}|x(t+\theta)| = e^{\lambda_1 t}|\varphi(t+\theta)| \le e^{\lambda_1 r}|\varphi| \le Me^{\lambda_1 r}|\varphi|.$$

If $t + \theta \ge 0$, then from (5.24) we have

$$e^{\lambda_1(t+\theta)}|x(t+\theta)| \le M|\varphi| + M^2 \int_0^{t+\theta} e^{\lambda_1 s}|f(s)|ds + M^2|L| \int_0^{t+\theta} e^{\lambda_1 s}|x_s|ds.$$

Since $-\theta \leq r$, we then obtain

$$e^{\lambda_1 t}|x(t+\theta)| \leq Me^{\lambda_1 r}|\varphi| + M^2 e^{\lambda_1 r}\int_0^t e^{\lambda_1 s}|f(s)|ds + M^2|L|e^{\lambda_1 r}\int_0^t e^{\lambda_1 s}|x_s|ds.$$

Thus, for each $t \geq 0$,

$$\begin{aligned}
e^{\lambda_1 t}|x_t| &\leq \sup_{-r\leq\theta\leq 0} e^{\lambda_1 t}|x(t+\theta)| \\
&\leq Me^{\lambda_1 r}|\varphi| + M^2 e^{\lambda_1 r}\int_0^t e^{\lambda_1 s}|f(s)|ds + M^2|L|e^{\lambda_1 r}\int_0^t e^{\lambda_1 s}|x_s|ds \\
&\leq Me^{\lambda_1 r}|\varphi| + M^2 e^{\lambda_1 r}\sum_{k=0}^{[t]}\int_k^{k+1} e^{\lambda_1 s}|f(s)|ds + M^2|L|e^{\lambda_1 r}\int_0^t e^{\lambda_1 s}|x_s|ds \\
&\leq Me^{\lambda_1 r}|\varphi| + M^2 e^{\lambda_1 r}|f|_{\mathrm{BS}^1}\sum_{k=0}^{[t]} e^{\lambda_1(k+1)} + M^2|L|e^{\lambda_1 r}\int_0^t e^{\lambda_1 s}|x_s|ds \\
&\leq Me^{\lambda_1 r}|\varphi| + M_2(e^{\lambda_1(t+1)} - 1) + M^2|L|e^{\lambda_1 r}\int_0^t e^{\lambda_1 s}|x_s|ds,
\end{aligned}$$

where $M_2 = \frac{M^2 e^{\lambda_1(r+1)}|f|_{\mathrm{BS}^1}}{e^{\lambda_1}-1}$. By the generalized Gronwall's inequality in Lemma 5.1.23, we obtain, for each $t \geq 0$,

$$\begin{aligned}
e^{\lambda_1 t}|x_t| &\leq Me^{\lambda_1 r}|\varphi| + M_2(e^{\lambda_1(t+1)} - 1) \\
&\quad + M^2|L|e^{\lambda_1 r}\int_0^t (Me^{\lambda_1 r}|\varphi| + M_2(e^{\lambda_1(s+1)} - 1))e^{(M^2|L|e^{\lambda_1 r})(t-s)}ds.
\end{aligned}$$

But from (5.22) we have $|L| \leq |a| + \int_{-r}^0 |h(s)|ds < \frac{\lambda_1}{M^2 e^{r\lambda_1}}$, thus

$$\lambda_1 - M^2|L|e^{r\lambda_1} > 0. \tag{5.25}$$

It follows that, for each $t \geq 0$,

$$|x_t| \leq M_2 e^{\lambda_1} + M|\varphi|e^{r\lambda_1} + \frac{M^2 M_2|L|e^{\lambda_1(r+1)}}{\lambda_1 - M^2|L|e^{r\lambda_1}}.$$

This shows that x is a bounded solution of equation (5.21) on $\mathbb{R}^+$. Using Corollary 5.1.20, we deduce that equation (5.21) has a compact almost automorphic solution y.

Let z be another solution. Then

$$y(t) - z(t) = T_0(t)(y(0) - z(0)) + \lim_{\lambda\to\infty} \int_0^t T_0(t-s)B_\lambda[L(y_s - z_s)]\mathrm{d}s \quad \text{for } t \geq 0.$$

Using the same computations as above, we have, for $t \geq 0$,

$$e^{\lambda_1 t}|y_t - z_t| \leq M e^{\lambda_1 r}|y_0 - z_0| + M^2 \int_0^t |L| e^{\lambda_1 r} e^{\lambda_1 s}|y_s - z_s|\mathrm{d}s.$$

Now using the classical Gronwall's lemma, we obtain, for $t \geq 0$,

$$|y_t - z_t| \leq M e^{\lambda_1 r}|y_0 - z_0| e^{(M^2|L|e^{\lambda_1 r} - \lambda_1)t}.$$

Thus from (5.25), we deduce that $|y_t - z_t| \to 0$ as $t \to \infty$, that is, y is globally attractive.

We claim that y is the unique solution of (5.21) which is bounded on the whole real line. In fact, if w is another solution which is bounded on $\mathbb{R}$, then for all $t, \sigma \in \mathbb{R}$ with $\sigma \leq t$ we can show that

$$|y_t - w_t| \leq M e^{\lambda_1 r}|y_\sigma - w_\sigma| e^{(M^2|L|e^{\lambda_1 r} - \lambda_1)(t-\sigma)}.$$

Letting $\sigma \to -\infty$, we deduce that $y_t = w_t$ for all $t \in \mathbb{R}$. □

5.2 Partial functional differential equations with infinite delay

In this section, we consider the following partial functional differential equation:

$$\frac{\mathrm{d}}{\mathrm{d}t}u(t) = Au(t) + L(u_t) + f(t) \quad \text{for } t \in \mathbb{R}, \tag{5.26}$$

where A is a linear operator on a Banach space X. We assume that the domain $D(A)$ is not necessarily dense and A satisfies the well-known Hille–Yosida condition, namely, we suppose that there exist $\overline{M} \geq 1$ and $\omega \in \mathbb{R}$ such that $(\omega, \infty) \subset \rho(A)$ and

$$|R(\lambda, A)^n| \leq \frac{\overline{M}}{(\lambda - \omega)^n} \quad \text{for } n \in \mathbb{N} \text{ and } \lambda > \omega,$$

where $\rho(A)$ is the resolvent set of A and $R(\lambda, A) := (\lambda I - A)^{-1}$ for $\lambda \in \rho(A)$.

Operator L is bounded linear from $\mathcal{B}$ to X, where $\mathcal{B}$ is a normed linear space of functions mapping $(-\infty, 0]$ into X satisfying the fundamental axioms introduced by Hale and Kato in [165]. For every $t \in \mathbb{R}$, the history function $u_t \in \mathcal{B}$ is defined by

$$u_t(\theta) := u(t + \theta) \quad \text{for } \theta \leq 0.$$

The input function f is Stepanov almost automorphic from $\mathbb{R}$ to X.

We are interested in studying the nature of all bounded solutions of equation (5.26) on $\mathbb{R}$. We are also concerned with the problem of finding sufficient conditions for the existence of an almost automorphic solution of equation (5.26).

In this work, we prove the Bohr–Neugebauer property for equation (5.26) when A generates an immediately compact C_0-semigroup on a Banach space and the phase space $\mathcal{B}$ has the uniform fading memory property. More specifically, we prove that all bounded solutions of equation (5.26) on $\mathbb{R}$ are compact almost automorphic, even if the input term f is only Stepanov almost automorphic.

5.2.1 Phase space, integral solutions, and the variation of constants formula

The choice of the phase space $\mathcal{B}$ plays an important role in the qualitative analysis of partial functional differential equations with infinite delay. In fact, the choice of $\mathcal{B}$ affects some properties of solutions. In this work, we employ an axiomatic definition of the phase space $\mathcal{B}$ which has first been introduced by Hale and Kato [165]. We assume that $(\mathcal{B}, \|\cdot\|_{\mathcal{B}})$ is a normed space of functions mapping $(-\infty, 0]$ into a Banach space $(X, \|\cdot\|)$ and satisfying the following fundamental axioms:

(A) There exist a positive constant N and functions $K, \widetilde{K} : [0,\infty) \to [0,\infty)$, with K continuous and $\widetilde{K}$ locally bounded, such that if a function $x : (-\infty, a] \to X$ is continuous on $[\sigma, a]$ with $x_\sigma \in \mathcal{B}$, for some $\sigma < a$, then for all $t \in [\sigma, a]$:
  - (i) $x_t \in \mathcal{B}$;
  - (ii) $t \mapsto x_t$ is continuous with respect to $\|\cdot\|_{\mathcal{B}}$ on $[\sigma, a]$;
  - (iii) $N\|x(t)\| \le \|x_t\|_{\mathcal{B}} \le K(t-\sigma)\sup_{\sigma\le s\le t}\|x(s)\| + \widetilde{K}(t-\sigma)\|x_\sigma\|_{\mathcal{B}}$.

(B) $\mathcal{B}$ is a Banach space.

(C) If $(\phi_n)_{n\ge0}$ is a Cauchy sequence in $\mathcal{B}$ which converges compactly to a function ϕ, then $\phi \in \mathcal{B}$ and $\|\phi_n - \phi\|_{\mathcal{B}} \to 0$ as $n \to \infty$.

As a consequence of Axiom **(A)**, we deduce the following result.

Lemma 5.2.1 ([181]). *Let $C_{00}((-\infty, 0], X)$ be the space of continuous functions mapping $(-\infty, 0]$ into X with compact supports. Then $C_{00}((-\infty, 0], X) \subset \mathcal{B}$. In addition, for $a < 0$, we have*

$$\|\phi\|_{\mathcal{B}} \le K(-a)\sup_{\theta\le0}\|\phi(\theta)\|,$$

for any $\phi \in C_{00}((-\infty, 0], X)$ with the support included in $[a, 0]$.

In addition, we consider

(C2) If a uniformly bounded sequence $(\phi_n)_{n\ge0}$ in $C_{00}((-\infty, 0], X)$ converges to a function ϕ compactly on $(-\infty, 0]$, then $\phi \in \mathcal{B}$ and $\|\phi_n - \phi\|_{\mathcal{B}} \to 0$ as $n \to \infty$.

Let $(S_0(t))_{t\geq 0}$ be the strongly continuous semigroup defined on the subspace

$$\mathcal{B}_0 = \{\phi \in \mathcal{B} : \phi(0) = 0\}$$

by

$$(S_0(t)\phi)(\theta) = \begin{cases} \phi(t+\theta) & \text{if } t+\theta \leq 0, \\ 0 & \text{if } t+\theta > 0. \end{cases}$$

Definition 5.2.2 ([181]). Assume that the space $\mathcal{B}$ satisfies Axioms **(A)**, **(B)**, and **(C2)**. Then $\mathcal{B}$ is said to be a fading memory space if, for all $\phi \in \mathcal{B}_0$,

$$S_0(t)\phi \underset{t\to\infty}{\longrightarrow} 0 \quad \text{in } \mathcal{B}.$$

Moreover, $\mathcal{B}$ is said to be a uniform fading memory space, if

$$S_0(t) \underset{t\to\infty}{\longrightarrow} 0 \quad \text{in the uniform operator topology.}$$

The following results give some properties of fading and uniform fading memory spaces.

Proposition 5.2.3 ([181, p. 190]). *The following statements hold:*
(i) *If $\mathcal{B}$ is a fading memory space, then the functions $K(\cdot)$ and $\widetilde{K}(\cdot)$ in Axiom* **(A)** *can be chosen to be constants.*
(ii) *If $\mathcal{B}$ is a uniform fading memory space, then the functions $K(\cdot)$ and $\widetilde{K}(\cdot)$ can be chosen such that $K(\cdot)$ is constant and $\widetilde{K}(t) \to 0$ as $t \to \infty$.*

In the sequel, we assume that the operator A satisfies the Hille–Yosida condition. Consider the following Cauchy problem:

$$\begin{cases} \frac{\mathrm{d}}{\mathrm{d}t}x(t) = Ax(t) + L(x_t) + f(t) & \text{for } t \geq \sigma, \\ x_\sigma = \phi \in \mathcal{B}. \end{cases} \tag{5.27}$$

Definition 5.2.4 ([9]). Let $\phi \in \mathcal{B}$. A function $x : \mathbb{R} \to X$ is called an integral solution of equation (5.27) on $\mathbb{R}$ if the following conditions hold:
(i) x is continuous on $[\sigma, \infty)$,
(ii) $x_\sigma = \phi$,
(iii) $\int_\sigma^t x(s)\mathrm{d}s \in D(A)$ for $t \geq \sigma$,
(iv) $x(t) = \phi(0) + A\int_\sigma^t x(s)\mathrm{d}s + \int_\sigma^t [L(x_s) + f(s)]\mathrm{d}s$ for $t \geq \sigma$.

If x is an integral solution of equation (5.27), then from the continuity of x, we have $x(t) \in \overline{D(A)}$ for all $t \geq \sigma$. In particular, $\phi(0) \in \overline{D(A)}$. Let us introduce the part A_0 of the operator A in $\overline{D(A)}$ defined by

$$\begin{cases} D(A_0) := \{x \in D(A) : Ax \in \overline{D(A)}\}, \\ A_0 x := Ax \quad \text{for } x \in D(A_0). \end{cases}$$

Lemma 5.2.5 ([265]). *The operator A_0 generates a strongly continuous semigroup $(T_0(t))_{t\geq 0}$ on $\overline{D(A)}$.*

For the existence of integral solutions, one has the following result:

Theorem 5.2.6 ([56, 265]). *Assume that $\mathcal{B}$ satisfies* (**A**) *and* (**B**). *Then for all $\phi \in \mathcal{B}$ such that $\phi(0) \in \overline{D(A)}$, equation* (5.27) *has a unique integral solution x on $\mathbb{R}$. Moreover, x is given by*

$$\begin{cases} x(t) = T_0(t-\sigma)\phi(0) + \lim_{\lambda\to\infty} \int_\sigma^t T_0(t-s)B_\lambda[L(x_s) + f(s)]\mathrm{d}s & \textit{for } t \geq \sigma, \\ x_\sigma = \phi, \end{cases}$$

where $B_\lambda := \lambda R(\lambda, A)$ for $\lambda > \omega$.

In the sequel of this work, for simplicity, integral solutions are called solutions. The phase space $\mathcal{B}_A$ of equation (5.27) is given by

$$\mathcal{B}_A := \{\phi \in \mathcal{B} : \phi(0) \in \overline{D(A)}\}.$$

For each $t \geq 0$, $V(t)$ is the bounded linear operator defined on $\mathcal{B}_A$ by

$$V(t)\phi = y_t,$$

where y is the solution of the homogeneous equation

$$\begin{cases} \frac{\mathrm{d}}{\mathrm{d}t} y(t) = Ay(t) + L(y_t) \quad \text{for } t \geq 0, \\ y_0 = \phi. \end{cases}$$

We have the following result.

Proposition 5.2.7 ([1, Proposition 2]). *$(V(t))_{t\geq 0}$ is a strongly continuous semigroup on $\mathcal{B}_A$. Moreover, $(V(t))_{t\geq 0}$ satisfies the following translation property:*

$$(V(t)\phi)(\theta) = \begin{cases} V(t+\theta)\phi(0) & \textit{for } t+\theta \geq 0, \\ \phi(t+\theta) & \textit{for } t+\theta \leq 0. \end{cases}$$

Let $\mathcal{A}_V$ denote the infinitesimal generator of the semigroup $(V(t))_{t\geq 0}$ on $\mathcal{B}_A$.

We introduce the following sequence of linear operators Θ^n mapping X into $\mathcal{B}_A$ defined for $n > \omega$ and $y \in X$ by

$$(\Theta^n y)(\theta) = \begin{cases} (n\theta + 1)B_n y & \text{for } -\frac{1}{n} \leq \theta \leq 0, \\ 0 & \text{for } \theta < -\frac{1}{n}, \end{cases}$$

where B_n is the bounded operator defined for sufficiently large n by $B_n := nR(n, A)$. For $y \in X$, the function $\Theta^n y$ belongs to $C_{00}((-\infty, 0], X)$ with the support included in $[-1, 0]$. By Lemma 5.2.1, we deduce that $\Theta^n y \in \mathcal{B}$ and

$$\|\Theta^n y\|_{\mathcal{B}} \leq \widetilde{N} K(1)\|y\|, \tag{5.28}$$

where $\widetilde{N} := \sup_{n>\omega} \|B_n\|$. In addition, we have, for each $y \in X$,

$$(\Theta^n y)(0) = B_n y = nR(n, A)y \in D(A).$$

It follows that $\Theta^n y \in \mathcal{B}_A$.

Now we give the variation of constants formula for equation (5.27) established in [9].

Theorem 5.2.8 ([9]). *For all $\phi \in \mathcal{B}_A$, the solution x of equation* (5.27) *satisfies the following variation of constants formula:*

$$x_t = V(t)\phi + \lim_{n\to\infty} \int_{\sigma}^{t} V(t-s)\Theta^n f(s)ds \quad \textit{for } t \geq \sigma. \tag{5.29}$$

5.2.2 Spectral decomposition of the phase space and reduction principle

The spectral decomposition of the phase space provides a powerful tool to analyze the asymptotic behavior of solutions. We know that each $\lambda \in \sigma(\mathcal{A}_V)$ with $\operatorname{Re}\lambda > \omega_{\text{ess}}(V)$ is an isolated eigenvalue of the operator $\mathcal{A}_V$. Let $\rho > \omega_{\text{ess}}(V)$ and consider the set

$$\Sigma_\rho := \{\lambda \in \sigma(\mathcal{A}_V) : \operatorname{Re}\lambda \geq \rho\}. \tag{5.30}$$

From [136, Corollary 2.11, Chapter IV and Theorem 3.1, Chapter V], the set Σ_ρ is finite, and we have the following decomposition of the phase space $\mathcal{B}_A$:

$$\mathcal{B}_A = U_\rho \oplus S_\rho, \tag{5.31}$$

where U_ρ and S_ρ are closed subspaces of $\mathcal{B}_A$ which are invariant under $(V(t))_{t\geq 0}$. The subspace U_ρ is finite-dimensional. For every sufficiently small $\varepsilon > 0$, there exists $C_\varepsilon > 0$ such that

$$\|V(t)\phi\|_{\mathcal{B}} \leq C_\varepsilon e^{(\rho-\varepsilon)t}\|\phi\|_{\mathcal{B}} \quad \text{for } t \geq 0 \text{ and } \phi \in S_\rho. \tag{5.32}$$

In what follows, $V^{U_\rho}(t)$ and $V^{S_\rho}(t)$ denote the restrictions of $V(t)$ on U_ρ and S_ρ, respectively. Also Π^{U_ρ} and Π^{S_ρ} denote the projections on U_ρ and S_ρ, respectively, and $(V^{U_\rho}(t))_{t\in\mathbb{R}}$ is a group of operators.

We have the following result.

Theorem 5.2.9 ([56]). *If $\mathcal{B}$ is a uniform fading memory space and $(T_0(t))_{t\geq 0}$ is immediately compact then $\omega_{\text{ess}}(V) < 0$.*

As a consequence of Theorem 5.2.9, for $\rho = 0$, we have the following spectral decomposition:

$$\mathcal{B}_A = U_0 \oplus S_0, \tag{5.33}$$

and there exist two positive constants a and N such that

$$\|V(t)\phi\|_{\mathcal{B}} \leq Ne^{-at}\|\phi\|_{\mathcal{B}} \quad \text{for each } \phi \in S_0 \text{ and } t \geq 0.$$

Let $d := \dim(U_0)$ with a vector basis $\Phi = \{\varphi_1, \ldots, \varphi_d\}$. Then, there exist d-elements $\{\psi_1, \ldots, \psi_d\}$ in $\mathcal{B}_A^*$ such that

$$\begin{cases} \langle \psi_i, \varphi_j \rangle = \delta_{ij}, \\ \langle \psi_i, \varphi \rangle = 0 \quad \text{for all } \varphi \in S_0 \text{ and } i \in \{1, \ldots, d\}, \end{cases} \tag{5.34}$$

where $\langle \cdot, \cdot \rangle$ denotes the duality pairing between $\mathcal{B}_A^*$ and $\mathcal{B}_A$, and

$$\delta_{ij} = \begin{cases} 1 & \text{if } i = j, \\ 0 & \text{if } i \neq j. \end{cases}$$

Letting $\Psi = \text{col}\{\psi_1, \ldots \psi_2\}$, $\langle \Psi, \Phi \rangle$ is a $(d \times d)$-matrix, where the (i,j)-component is $\langle \psi_i, \varphi_j \rangle$.

Denote by Π^{S_0} and Π^{U_0} the projections on S_0 and U_0, respectively. For each $\varphi \in \mathcal{B}_A$, we have

$$\Pi^{U_0}\varphi = \Phi\langle \Psi, \varphi \rangle.$$

In fact, for $\varphi \in \mathcal{B}_A$, we have $\varphi = \Pi^{S_0}\varphi + \Pi^{U_0}\varphi$ with $\Pi^{U_0}\varphi = \sum_{i=1}^{d} a_i\varphi_i$ and $a_i \in \mathbb{R}$. By (5.34), we conclude that

$$a_i = \langle \psi_i, \varphi \rangle.$$

Hence

$$\Pi^{U_0}\varphi = \sum_{i=1}^{d} \langle \psi_i, \varphi \rangle \varphi_i = \Phi\langle \Psi, \varphi \rangle.$$

Since $(V^{U_0}(t))_{t\geq 0}$ is a group on U_0, there exists a $(d \times d)$-matrix G such that

$$V^{U_0}(t)\Phi = \Phi e^{tG} \quad \text{for } t \in \mathbb{R}.$$

For $n, n_0 \in \mathbb{N}$ such that $n \geq n_0 \geq \widetilde{\omega}$ and $i \in \{1, \dots, d\}$, we define the linear operator $x^*_{i,n}$ by

$$x^*_{i,n}(a) = \langle \psi_i, \Theta^n a \rangle \quad \text{for } a \in X.$$

Since $|\widetilde{B}_n| \leq \frac{n}{n-\widetilde{\omega}} \widetilde{M}$ for any $n \geq n_0$, one gets that $x^*_{i,n}$ is a bounded linear operator from X to $\mathbb{R}$ such that

$$|x^*_{i,n}| \leq \frac{n}{n - \widetilde{\omega}} \widetilde{M} |\psi_i| \quad \text{for any } n \geq n_0.$$

Define the d-column vector $x^*_n = \mathrm{col}(x^*_{1,n}, \dots, x^*_{d,n})$. Then one can see that

$$\langle x^*_n, a \rangle = \langle \Psi, \widetilde{B}_n X_0 a \rangle \quad \text{for } a \in X,$$

with

$$\langle x^*_n, a \rangle_i = \langle \psi_i, \widetilde{B}_n X_0 a \rangle \quad \text{for } i = 1, \dots, d \text{ and } a \in X.$$

Consequently, we have

$$\sup_{n \geq n_0} |x^*_n| < \infty,$$

which implies that $(x^*_n)_{n \geq n_0}$ is a bounded sequence in $\mathcal{L}(X, \mathbb{R}^d)$. We recall the following important results:

Theorem 5.2.10 ([8]). *There exists $x^* \in \mathcal{L}(X, \mathbb{R}^d)$ such that $(x^*_n)_{n \geq n_0}$ converges weakly to x^* in the sense that*

$$\langle x^*_n, x \rangle \to \langle x^*, x \rangle \quad \text{as } n \to \infty \text{ for all } x \in X.$$

Theorem 5.2.11 ([8]). *Assume that $(T_0(t))_{t \geq 0}$ is immediately compact, f is continuous, and let u be a solution of equation (5.26) on $\mathbb{R}$. Then the function z defined by $z(t) := \langle \Psi, u_t \rangle$ is a solution of the ordinary differential equation*

$$\frac{\mathrm{d}}{\mathrm{d}t} z(t) = G z(t) + \langle x^*, f(t) \rangle \quad \text{for } t \in \mathbb{R}. \tag{5.35}$$

Conversely, if z is a solution of equation (5.35) on $\mathbb{R}$ and if, in addition, f is bounded on $\mathbb{R}$, then the function u given by

$$u(t) := \left[\Phi z(t) + \lim_{n \to \infty} \int_{-\infty}^{t} V^{S_0}(t-s) \Pi^{S_0}(\Theta^n f(s)) \mathrm{d}s \right](0) \quad \text{for } t \in \mathbb{R}$$

is a solution of equation (5.26) on $\mathbb{R}$.

The following result relaxes the assumptions of Theorem 5.2.11.

Theorem 5.2.12. *Assume that* $(T_0(t))_{t\geq 0}$ *is immediately compact,* f *is locally integrable, and let* u *be a solution of equation* (5.26) *on* $\mathbb{R}$. *Then the function* z *defined by* $z(t) := \langle \Psi, u_t \rangle$ *is given by*

$$z(t) = e^{tG} z(0) + \int_0^t e^{(t-s)G} \langle x^*, f(s) \rangle ds \quad \text{for } t \in \mathbb{R}. \tag{5.36}$$

Conversely, if z *satisfies equation* (5.36) *on* $\mathbb{R}$ *and if, in addition,* f *satisfies*

$$\sup_{t\in\mathbb{R}} \int_t^{t+1} |f(s)| ds < \infty,$$

then the function u *given by*

$$u(t) := \left[\Phi z(t) + \lim_{n\to\infty} \int_{-\infty}^{t} V^{S_0}(t-s) \Pi^{S_0}(\Theta^n f(s)) ds \right](0) \quad \text{for } t \in \mathbb{R}$$

is a solution of equation (5.26) *on* $\mathbb{R}$.

Proof. The proof is similar to that of Theorem 5.2.11. We only have to prove that the limit

$$\lim_{n\to\infty} \int_{-\infty}^{t} V^{S_0}(t-s) \Pi^{S_0}(\Theta^n f(s)) ds$$

exists in $\mathcal{B}_A$. For $t \in \mathbb{R}$ and for n sufficiently large, we have

$$\begin{aligned}
\left\| \int_{-\infty}^{t} V^{S_0}(t-s) \Pi^{S_0}(\Theta^n f(s)) ds \right\|_{\mathcal{B}} &\leq 2\widetilde{M}N |\Pi^{S_0}| \int_{-\infty}^{t} e^{-a(t-s)} |f(s)| ds \\
&\leq 2\widetilde{M}N |\Pi^{S_0}| \sum_{k=1}^{\infty} \left(\int_{t-k}^{t-k+1} e^{-a(t-s)} |f(s)| ds \right) \\
&\leq 2\widetilde{M}N |\Pi^{S_0}| \sum_{k=1}^{\infty} \left(e^{-a(k-1)} \int_{t-k}^{t-k+1} |f(s)| ds \right) \leq K,
\end{aligned}$$

where $K = 2\widetilde{M}N |\Pi^{S_0}| \sup_{t\in\mathbb{R}} (\int_t^{t+1} |f(s)| ds) \frac{1}{1-e^{-a}}$. Let

$$H(n,s,t) := V^{S_0}(t-s) \Pi^{S_0}(\Theta^n f(s)) \quad \text{for } n \in \mathbb{N} \text{ and } s \leq t.$$

For n and m sufficiently large and $\sigma \le t$, we have

$$\begin{aligned}\left\| \int_{-\infty}^{t} H(n,s,t)ds - \int_{-\infty}^{t} H(m,s,t)ds \right\|_{\mathcal{B}} &\le \left\| \int_{-\infty}^{\sigma} H(n,s,t)ds \right\|_{\mathcal{B}} + \left\| \int_{-\infty}^{\sigma} H(m,s,t)ds \right\|_{\mathcal{B}} \\ &\quad + \left\| \int_{\sigma}^{t} H(n,s,t)ds - \int_{\sigma}^{t} H(m,s,t)ds \right\|_{\mathcal{B}} \\ &\le 2Ke^{-\alpha(t-\sigma)} + \left\| \int_{\sigma}^{t} H(n,s,t)ds - \int_{\sigma}^{t} H(m,s,t)ds \right\|_{\mathcal{B}}.\end{aligned}$$

Since $\lim_{n\to\infty} \int_{\sigma}^{t} H(n,s,t)ds$ exists, it follows that

$$\limsup_{n,m\to\infty} \left\| \int_{-\infty}^{t} H(n,s,t)ds - \int_{-\infty}^{t} H(m,s,t)ds \right\|_{\mathcal{B}} \le 2Ke^{-\alpha(t-\sigma)}.$$

Letting $\sigma \to -\infty$, we get

$$\limsup_{n,m\to\infty} \left\| \int_{-\infty}^{t} H(n,s,t)ds - \int_{-\infty}^{t} H(m,s,t)ds \right\|_{\mathcal{B}} = 0.$$

Thus by the completeness of the phase space $\mathcal{B}_A$, we deduce that the limit

$$\lim_{n\to\infty} \int_{-\infty}^{t} H(n,s,t)ds$$

exists in $\mathcal{B}_A$. □

5.2.3 Solutions having relatively compact ranges

The goal of this section is to investigate when a bounded solution of equation (5.26) has a relatively compact range for an input function $f \in BS^p(\mathbb{R}, X)$. We distinguish two cases, $p > 1$ and $p = 1$.

5.2.3.1 Case $p > 1$

Lemma 5.2.13. *Assume that $(T_0(t))_{t\ge 0}$ is immediately compact, $\mathcal{B}$ has the fading memory property, and $f \in BS^p(\mathbb{R}, X)$ with $p > 1$. Then, every bounded solution of equation (5.26) on $\mathbb{R}^+$ has a relatively compact range.*

Proof. Let x be a bounded solution of equation (5.26) on $\mathbb{R}^+$. Then by Axiom (**A**)(iii), we have

$$\|x_t\|_{\mathcal{B}} \leq K(t) \sup_{0\leq s\leq t} \|x(s)\| + \widetilde{K}(t)\|x_0\|_{\mathcal{B}}.$$

From Proposition 5.2.3, $K(t)$ and $\widetilde{K}(t)$ can be chosen as constant functions. It follows that

$$\overline{x} := \sup_{t\geq 0} \|x_t\|_{\mathcal{B}} < \infty.$$

Let $M_0 \geq 1$ and $\omega_0 \in \mathbb{R}$ be such that $|T_0(t)| \leq M_0 e^{\omega_0 t}$ for all $t \geq 0$. For $0 < \varepsilon \leq 1$ and $t \geq 1$, we have

$$x(t) = T_0(\varepsilon)x(t-\varepsilon) + \lim_{\lambda\to\infty} \int_{t-\varepsilon}^{t} T_0(t-s)B_\lambda[L(x_s) + f(s)]\mathrm{d}s.$$

Since $T_0(\varepsilon)$ is compact, there exists a compact subset K_ε of X such that $T_0(\varepsilon)x(t-\varepsilon) \in K_\varepsilon$ for all $t \geq 1$. Moreover, for each $t \geq 1$ we have

$$\begin{aligned}
\left|\lim_{\lambda\to\infty} \int_{t-\varepsilon}^{t} T_0(t-s)B_\lambda[L(x_s) + f(s)]\mathrm{d}s\right| &\leq M_0\overline{M}|L|\overline{x} \int_{t-\varepsilon}^{t} e^{\omega_0(t-s)}\mathrm{d}s + M_0\overline{M} \int_{t-\varepsilon}^{t} e^{\omega_0(t-s)}|f(s)|\mathrm{d}s \\
&\leq M_0\overline{M}|L|\overline{x} \int_0^{\varepsilon} e^{\omega_0 s}\mathrm{d}s \\
&\quad + M_0\overline{M}\left(\int_{t-\varepsilon}^{t} e^{q\omega_0(t-s)}\mathrm{d}s\right)^{\frac{1}{q}}\left(\int_{t-\varepsilon}^{t} |f(s)|^p\mathrm{d}s\right)^{\frac{1}{p}} \\
&\leq \delta(\varepsilon),
\end{aligned}$$

where $\delta(\varepsilon) := M_0\overline{M}|L|\overline{x}\int_0^{\varepsilon} e^{\omega_0 s}\mathrm{d}s + M_0\overline{M}|f|_{BS^p}(\int_0^{\varepsilon} e^{q\omega_0 s}\mathrm{d}s)^{\frac{1}{q}}$. Then for all $t \geq 1$, we have

$$\lim_{\lambda\to\infty} \int_{t-\varepsilon}^{t} T_0(t-s)B_\lambda[L(x_s) + f(s)]\mathrm{d}s \in \overline{B}(0, \delta(\varepsilon)).$$

It follows that

$$\{x(t) : t \geq 1\} \subset K_\varepsilon + \overline{B}(0, \delta(\varepsilon)).$$

Taking the Kuratowski's measure of noncompactness, we obtain that

$$\alpha(\{x(t) : t \geq 1\}) \leq 2\delta(\varepsilon).$$

The above inequality holds for all $0 < \varepsilon \le 1$, thus by letting ε to 0, we deduce that

$$\alpha(\{x(t) : t \ge 1\}) = 0.$$

Thus $\{x(t) : t \ge 1\}$ is relatively compact. Since $x(\cdot)$ is continuous, we conclude that $\{x(t) : t \ge 0\}$ is relatively compact. □

Lemma 5.2.14. *Assume that $(T_0(t))_{t\ge0}$ is immediately compact, $\mathcal{B}$ has the fading memory property, and $f \in BS^p(\mathbb{R}, X)$ with $p > 1$. Then, every bounded solution of equation (5.26) on $\mathbb{R}^+$ is uniformly continuous.*

Lemma 5.2.15 ([41, Theorem 4.1.2]). *Let Y be a normed space and $(T_n)_n$ be a sequence of bounded linear operators on Y such that $\sup_n \|T_n\| < \infty$. If D is a dense subset of Y and if, for each $y \in D$,*

$$T_n y \to Ty \quad \text{as } n \to \infty,$$

for some bounded linear operator T, then, for every compact set K of Y,

$$\sup_{y\in K} \|T_n y - Ty\| \to 0 \quad \text{as } n \to \infty.$$

Proof of Lemma 5.2.14. Let x be a solution of equation (5.26) which is bounded on $\mathbb{R}^+$ and let $\overline{x} := \sup_{t\ge0} |x_t|$. Let $M_0 \ge 1$ and $\omega_0 \in \mathbb{R}$ such that $|T(t)| \le M_0 e^{\omega_0 t}$ for all $t \ge 0$. Let $(t_n)_n$ and $(s_n)_n$ be two real sequences such that $t_n - s_n \to 0$ as $n \to \infty$. Assume without loss of generality that $t_n \ge s_n$ for all $n \in \mathbb{N}$. Then we have

$$x(t_n) = T_0(t_n - s_n)x(s_n) + \lim_{\lambda\to\infty} \int_{s_n}^{t_n} T_0(t_n - s)B_\lambda[L(x_s) + f(s)]ds.$$

Let $K = \overline{\{x(t) : t \ge 0\}}$, which is a compact subset of X by Lemma 5.2.13. It follows that

$$\begin{aligned}
|x(t_n) - x(s_n)| &\le |T_0(t_n - s_n)x(s_n) - x(s_n)| + M_0\overline{M}|L|\overline{x} \int_{s_n}^{t_n} e^{\omega_0(t_n-s)}ds \\
&\quad + M_0\overline{M} \int_{s_n}^{t_n} e^{\omega_0(t_n-s)}|f(s)|ds \\
&\le \sup_{y\in K}|T_0(t_n - s_n)y - y| + M_0\overline{M}|L|\overline{x} \int_0^{t_n-s_n} e^{\omega_0 s}ds \\
&\quad + M_0\overline{M} \int_0^{t_n-s_n} e^{\omega_0 s}|f(t_n - s)|ds
\end{aligned}$$

$$
\begin{aligned}
&\le \sup_{y\in K}\big|T_0(t_n-s_n)y-y\big| + M_0\overline{M}|L|\overline{x}\int_0^{t_n-s_n} e^{\omega_0 s}\,\mathrm{d}s \\
&\quad + M_0\overline{M}\left(\int_0^{t_n-s_n} e^{q\omega_0 s}\,\mathrm{d}s\right)^{\frac{1}{q}}\left(\int_0^{t_n-s_n} |f(t_n-s)|^p\,\mathrm{d}s\right)^{\frac{1}{p}} \\
&= \sup_{y\in K}\big|T_0(t_n-s_n)y-y\big| + M_0\overline{M}|L|\overline{x}\int_0^{t_n-s_n} e^{\omega_0 s}\,\mathrm{d}s \\
&\quad + M_0\overline{M}\left(\int_0^{t_n-s_n} e^{q\omega_0 s}\,\mathrm{d}s\right)^{\frac{1}{q}}\left(\int_{s_n}^{t_n} |f(s)|^p\,\mathrm{d}s\right)^{\frac{1}{p}}.
\end{aligned}
$$

We can assume that $0 \le t_n - s_n \le 1$ for all $n \in \mathbb{N}$. Since $f \in \mathrm{BS}^p(\mathbb{R}, X)$, we have

$$
\begin{aligned}
\big|x(t_n)-x(s_n)\big| &\le \sup_{y\in K}\big|T_0(t_n-s_n)y-y\big| + M_0\overline{M}|L|\overline{x}\int_0^{t_n-s_n} e^{\omega_0 s}\,\mathrm{d}s \\
&\quad + M_0\overline{M}\left(\int_0^{t_n-s_n} e^{q\omega_0 s}\,\mathrm{d}s\right)^{\frac{1}{q}}\left(\int_{s_n}^{s_n+1} |f(s)|^p\,\mathrm{d}s\right)^{\frac{1}{p}} \\
&\le \sup_{y\in K}\big|T_0(t_n-s_n)y-y\big| + M_0\overline{M}|L|\overline{x}\int_0^{t_n-s_n} e^{\omega_0 s}\,\mathrm{d}s \\
&\quad + M_0\overline{M}|f|_{\mathrm{BS}^p}\left(\int_0^{t_n-s_n} e^{q\omega_0 s}\,\mathrm{d}s\right)^{\frac{1}{q}}.
\end{aligned}
$$

The semigroup $(T_0(t))_{t\ge 0}$ being strongly continuous, we have, for each $y \in K$, $T_0(t_n - s_n)y \to y$ as $n \to \infty$. This implies that $\sup_n \|T_0(t_n - s_n)y\| < \infty$ for each $y \in X$ and thus by the Banach–Steinhaus theorem $\sup_n \|T_0(t_n - s_n)\| < \infty$. It follows from Lemma 5.2.15 that $\sup_{y\in K} |T_0(t_n - s_n)y - y| \to 0$ as $n \to \infty$. Thus x is uniformly continuous. □

Corollary 5.2.16. *Assume that $(T_0(t))_{t\ge 0}$ is immediately compact, $\mathcal{B}$ has the fading memory property, and $f \in \mathrm{BS}^p(\mathbb{R}, X)$ with $p > 1$. If $t \mapsto x(t)$ is a bounded solution of equation* (5.26) *on $\mathbb{R}^+$, then the history function $t \mapsto x_t$ has a relatively compact range on $\mathbb{R}^+$.*

Proof. The result is a consequence of Lemmas 5.2.13, 5.2.14 and [225, Corollary 4.1]. □

5.2.3.2 Case $p = 1$

To reproduce the above results for $f \in \mathrm{BS}^p(\mathbb{R}, X)$ with $p = 1$, one needs to assume more than the boundedness in the Stepanov norm. In addition, we will need to assume that the phase space $\mathcal{B}$ has a property stronger than the fading memory property, which is the uniform fading memory property. Consider the following hypothesis:

(H) $f \in L^p_{\text{loc}}(\mathbb{R}, X)$ and for every sequence of real numbers $(s_n)_n$ there exist a subsequence $(s'_n)_n \subset (s_n)_n$ and a function $g \in L^p_{\text{loc}}(\mathbb{R}, X)$ such that, for each $t \in \mathbb{R}$,

$$\left(\int_t^{t+1} |f(s + s'_n) - g(s)|^p ds \right)^{\frac{1}{p}} \to 0 \quad \text{as } n \to \infty. \tag{5.37}$$

Remark. Note that if f is S^p-almost automorphic then f satisfies **(H)**. However, the converse is not true. In fact, let $f : \mathbb{R} \to \mathbb{R}$ be a continuous nonconstant function such that $\lim_{t \to \pm\infty} f(t) = l$ exists. Assume without loss of generality that $p = 1$ and let $(s_n)_n$ be a sequence of real numbers. If $(s_n)_n$ is bounded, then it has a convergent subsequence $(s'_n)_n$, say, $\lim_{n \to \infty} s'_n = s_0$. In this case we have, for all $t \in \mathbb{R}$,

$$f(t + s'_n) \to f(t + s_0) =: g(t)$$

as $n \to \infty$. Now if $(s_n)_n$ is not bounded then it has a subsequence $(s'_n)_n$ such that $\lim_{n \to \infty} s'_n = \pm\infty$. In this case we have, for all $t \in \mathbb{R}$,

$$f(t + s'_n) \to l =: g(t).$$

This shows that f satisfies **(H)**. On the other hand, the function f is not S^1-almost automorphic. In fact, if we assume that f is S^1-almost automorphic, we choose a sequence $(s_n)_n$ such that $\lim_{n \to \infty} s_n = \infty$. Thus by the S^1-almost automorphy, there exist a subsequence $(s'_n)_n \subset (s_n)_n$ and a function $g \in L^1_{\text{loc}}(\mathbb{R}, X)$ such that, for each $t \in \mathbb{R}$,

$$\int_t^{t+1} |f(s + s'_n) - g(s)| ds \to 0 \tag{5.38}$$

and

$$\int_t^{t+1} |g(s - s'_n) - f(s)| ds \to 0 \tag{5.39}$$

as $n \to \infty$. For each $t \in \mathbb{R}$, we have

$$\int_t^{t+1} |g(s) - l| ds \le \int_t^{t+1} |g(s) - f(s + s'_n)| ds + \int_t^{t+1} |f(s + s'_n) - l| ds.$$

Letting $n \to \infty$ in the above inequality, we get that $g(t) = l$ for almost all $t \in \mathbb{R}$. Therefore, for each $t \in \mathbb{R}$,

$$\begin{aligned}\int_t^{t+1}|l-f(s)|\mathrm{d}s &= \int_{t-s_n'}^{t+1-s_n'}|l-f(\theta+s_n')|\mathrm{d}\theta\\ &= \int_{t-s_n'}^{t+1-s_n'}|g(\theta)-f(\theta+s_n')|\mathrm{d}\theta\\ &= \int_t^{t+1}|g(s-s_n')-f(s)|\mathrm{d}s.\end{aligned}$$

Thus from (5.39) we get $f(t) = l$ for almost all $t \in \mathbb{R}$ and thus, by continuity, $f(t) = l$ for all $t \in \mathbb{R}$, which is a contradiction because f is assumed to be nonconstant.

In what follows, let $M_1 \geq 1$ and $\omega_1 \in \mathbb{R}$ be such that $|V(t)| \leq M_1 e^{\omega_1 t}$ for all $t \geq 0$.

Lemma 5.2.17. *Assume that $(T_0(t))_{t\geq 0}$ is immediately compact, $\mathcal{B}$ has the uniform fading memory property, and f satisfies* (**H**). *If $t \mapsto x(t)$ is a bounded solution of equation* (5.26) *on $\mathbb{R}^+$, then the history function $t \mapsto x_t$ has a relatively compact range on $\mathbb{R}^+$.*

Proof. Let x be a bounded solution of equation (5.26) on $\mathbb{R}$. Using the spectral decomposition (5.33), we have, for each $t \in \mathbb{R}$,

$$x_t = \Pi^{U_0}x_t + \Pi^{S_0}x_t. \tag{5.40}$$

On the one hand, $\{\Pi^{U_0}x_t : t \geq 0\}$ is relatively compact since it is bounded in a finite-dimensional subspace U_0.

On the other hand, we have, for $t \geq 0$,

$$\Pi^{S_0}x_t = V^{S_0}(t)\Pi^{S_0}x_0 + \lim_{n\to\infty}\int_0^t V^{S_0}(t-s)\Pi^{S_0}(\widetilde{B}_n(X_0 f(s)))\mathrm{d}s. \tag{5.41}$$

Let $(t_n)_n$ be a sequence of nonnegative real numbers. If $(t_n)_n$ is bounded then it has a subsequence $(t_n')_n$ which converges to some $t_0 \geq 0$, and thus $\Pi^{S_0}x_{t_n'} \to \Pi^{S_0}x_{t_0}$. Now if $(t_n)_n$ is unbounded then it has a subsequence $(t_n')_n$ such that $t_n' \to \infty$. It follows that

$$\begin{aligned}\Pi^{S_0}x_{t_n'} &= V^{S_0}(t_n')\Pi^{S_0}x_0 + \lim_{m\to\infty}\int_0^{t_n'} V^{S_0}(t_n'-s)\Pi^{S_0}(\widetilde{B}_m(X_0 f(s)))\mathrm{d}s\\ &= V^{S_0}(t_n')\Pi^{S_0}x_0 + \lim_{m\to\infty}\int_{-t_n'}^{0} V^{S_0}(-s)\Pi^{S_0}(\widetilde{B}_m(X_0 f(s+t_n')))\mathrm{d}s.\end{aligned}$$

It is clear that $V^{S_0}(t_n')\Pi^{S_0}x_0 \to 0$ as $n \to \infty$. We claim that

$$\lim_{m\to\infty}\int_{-t'_n}^{0} V^{S_0}(-s)\Pi^{S_0}(\widetilde{B}_m(X_0 f(s+t'_n)))\mathrm{d}s \to \lim_{m\to\infty}\int_{-\infty}^{0} V^{S_0}(-s)\Pi^{S_0}(\widetilde{B}_m(X_0\widetilde{f}(s)))\mathrm{d}s$$

as $n\to\infty$. In fact,

$$\begin{aligned}
&\left\|\lim_{m\to\infty}\int_{-t'_n}^{0} V^{S_0}(-s)\Pi^{S_0}(\widetilde{B}_m(X_0 f(s+t'_n)))\mathrm{d}s - \lim_{m\to\infty}\int_{-\infty}^{0} V^{S_0}(-s)\Pi^{S_0}(\widetilde{B}_m(X_0\widetilde{f}(s)))\mathrm{d}s\right\|_{\mathcal{B}}\\
&= \left\|\lim_{m\to\infty}\int_{-t'_n}^{0} V^{S_0}(-s)\Pi^{S_0}(\widetilde{B}_m(X_0 f(s+t'_n)-\widetilde{f}(s)))\mathrm{d}s\right.\\
&\qquad \left. - \lim_{m\to\infty}\int_{-\infty}^{-t'_n} V^{S_0}(-s)\Pi^{S_0}(\widetilde{B}_m(X_0\widetilde{f}(s)))\mathrm{d}s\right\|\\
&\le \left\|\lim_{m\to\infty}\int_{-t'_n}^{0} V^{S_0}(-s)\Pi^{S_0}(\widetilde{B}_m(X_0 f(s+t'_n)-\widetilde{f}(s)))\mathrm{d}s\right\|_{\mathcal{B}}\\
&\qquad + \left\|\lim_{m\to\infty}\int_{-\infty}^{-t'_n} V^{S_0}(-s)\Pi^{S_0}(\widetilde{B}_m(X_0\widetilde{f}(s)))\mathrm{d}s\right\|_{\mathcal{B}}\\
&\le \widetilde{M}N|\Pi^{S_0}|\int_{-t'_n}^{0} e^{as}\|f(s+t'_n)-\widetilde{f}(s)\|\mathrm{d}s + \widetilde{M}N|\Pi^{S_0}|\int_{-\infty}^{-t'_n} e^{as}\|\widetilde{f}(s)\|\mathrm{d}s.
\end{aligned}$$

We can prove that both $\int_{-t'_n}^{0} e^{as}\|f(s+t'_n)-\widetilde{f}(s)\|\mathrm{d}s$ and $\int_{-\infty}^{-t'_n} e^{as}\|\widetilde{f}(s)\|\mathrm{d}s$ converges to 0 as $n\to\infty$. Therefore $\{\Pi^{S_0}x_t : t\ge 0\}$ is relatively compact and thus $\{x_t : t\ge 0\}$ is also relatively compact. □

Lemma 5.2.18. *Assume that $(T_0(t))_{t\ge 0}$ is immediately compact, $\mathcal{B}$ has the uniform fading memory property, and f satisfies* (**H**). *If $t\mapsto x(t)$ is a bounded solution of equation* (5.26) *on $\mathbb{R}^+$, then the history function $t\mapsto x_t$ is uniformly continuous on $\mathbb{R}^+$.*

Proof. If $t\mapsto x_t$ is not uniformly continuous on $\mathbb{R}^+$, then there exist $\varepsilon>0$ and two nonnegative real sequences $(s_n)_n$ and $(t_n)_n$ such that $t_n - s_n \to 0$ and

$$\|x_{t_n} - x_{s_n}\|_{\mathcal{B}} > \varepsilon \quad \text{for all } n\in\mathbb{N}.$$

Let $h_n := t_n - s_n$ for all $n\in\mathbb{N}$. Then $h_n\to 0$ and

$$\|x_{s_n+h_n} - x_{s_n}\|_{\mathcal{B}} > \varepsilon \quad \text{for all } n\in\mathbb{N}. \tag{5.42}$$

For all $n\in\mathbb{N}$ such that $h_n\ge 0$, we have

$$\begin{aligned} x_{s_n+h_n} &= V(h_n)x_{s_n} + \lim_{m\to\infty} \int_{s_n}^{s_n+h_n} V(s_n + h_n - s)(\widetilde{B}_m(X_0 f(s)))\mathrm{d}s \\ &= V(h_n)x_{s_n} + \lim_{m\to\infty} \int_0^{h_n} V(h_n - s)(\widetilde{B}_m(X_0 f(s + s_n)))\mathrm{d}s. \end{aligned}$$

Let $K := \overline{\{x_t : t \geq 0\}}$. Then we have

$$\begin{aligned} \|x_{s_n+h_n} - x_{s_n}\|_{\mathcal{B}} &\leq \|V(h_n)x_{s_n} - x_{s_n}\|_{\mathcal{B}} + \widetilde{M} \int_0^{h_n} |V(h_n - s)||f(s + s_n)|\mathrm{d}s \\ &\leq \sup_{\phi\in K}\|V(h_n)\phi - \phi\|_{\mathcal{B}} + \widetilde{M}M_1 \int_0^{h_n} e^{\omega_1(h_n-s)}|f(s + s_n)|\mathrm{d}s \\ &\leq \sup_{\phi\in K}\|V(h_n)\phi - \phi\|_{\mathcal{B}} + \widetilde{M}M_1 e^{|\omega_1|h_n} \int_0^{h_n} |f(s + s_n)|\mathrm{d}s. \end{aligned}$$

In the other hand, for all $n \in \mathbb{N}$ such that $h_n < 0$, we have

$$\begin{aligned} x_{s_n} &= V(-h_n)x_{s_n+h_n} + \lim_{m\to\infty} \int_{s_n+h_n}^{s_n} V(s_n - s)(\widetilde{B}_m(X_0 f(s)))\mathrm{d}s \\ &= V(-h_n)x_{s_n+h_n} + \lim_{m\to\infty} \int_{h_n}^{0} V(-s)(\widetilde{B}_m(X_0 f(s + s_n)))\mathrm{d}s. \end{aligned}$$

Therefore,

$$\begin{aligned} \|x_{s_n} - x_{s_n+h_n}\|_{\mathcal{B}} &\leq \|V(-h_n)x_{s_n+h_n} - x_{s_n+h_n}\|_{\mathcal{B}} + \widetilde{M} \int_{h_n}^{0} |V(-s)||f(s + s_n)|\mathrm{d}s \\ &\leq \sup_{\phi\in K}\|V(-h_n)\phi - \phi\|_{\mathcal{B}} + \widetilde{M}M_1 \int_{h_n}^{0} e^{-\omega_1 s}|f(s + s_n)|\mathrm{d}s \\ &\leq \sup_{\phi\in K}\|V(-h_n)\phi - \phi\|_{\mathcal{B}} + \widetilde{M}M_1 e^{-|\omega_1|h_n} \int_{h_n}^{0} |f(s + s_n)|\mathrm{d}s. \end{aligned}$$

We deduce that, for all $n \in \mathbb{N}$,

$$\|x_{s_n+h_n} - x_{s_n}\|_{\mathcal{B}} \leq \sup_{\phi\in K}\|V(|h_n|)\phi - \phi\|_{\mathcal{B}} + \widetilde{M}M_1 e^{|\omega_1 h_n|} \int_{-|h_n|}^{|h_n|} |f(s + s_n)|\mathrm{d}s. \tag{5.43}$$

From Lemma 5.2.17, K is a compact subset of $\mathcal{B}$. It follows by the Banach–Steinhaus theorem and Lemma 5.2.15 that

$$\sup_{\phi\in K}\|V(|h_n|)\phi-\phi\|_{\mathcal{B}}\to 0\quad\text{as } n\to\infty.$$

On the other hand, since f satisfies **(H)**, there exist a subsequence $(s'_n)_n\subset(s_n)_n$ and a function $\widehat{f}\in L^1_{\mathrm{loc}}(\mathbb{R},X)$ such that, for each $t\in\mathbb{R}$,

$$\int_t^{t+1}|f(s+s'_n)-\widehat{f}(s)|\mathrm{d}s\to 0\quad\text{as } n\to\infty. \tag{5.44}$$

Let $(h'_n)_n$ be the corresponding subsequence of $(h_n)_n$. We can assume that $|h_n|\le 1$ for all $n\in\mathbb{N}$. Then, we have

$$\begin{aligned}\int_{-|h'_n|}^{|h'_n|}|f(s+s'_n)|\mathrm{d}s &\le \int_{-|h'_n|}^{|h'_n|}|f(s+s'_n)-\widehat{f}(s)|\mathrm{d}s+\int_{-|h'_n|}^{|h'_n|}|\widehat{f}(s)|\mathrm{d}s\\ &\le \int_{-1}^{1}|f(s+s'_n)-\widehat{f}(s)|\mathrm{d}s+\int_{-|h'_n|}^{|h'_n|}|\widehat{f}(s)|\mathrm{d}s\\ &\le \int_{-1}^{0}|f(s+s'_n)-\widehat{f}(s)|\mathrm{d}s+\int_{0}^{1}|f(s+s'_n)-\widehat{f}(s)|\mathrm{d}s+\int_{-|h'_n|}^{|h'_n|}|\widehat{f}(s)|\mathrm{d}s.\end{aligned}$$

It follows from (5.44) and Lebesgue's dominated convergence theorem that

$$\int_{-|h'_n|}^{|h'_n|}|f(s+s'_n)|\mathrm{d}s\to 0\quad\text{as } n\to\infty.$$

Therefore, we deduce from (5.43) that

$$\|x_{s'_n+h'_n}-x_{s'_n}\|_{\mathcal{B}}\to 0\quad\text{as } n\to\infty,$$

which contradicts (5.42). We conclude that $t\mapsto x_t$ must be uniformly continuous. □

The following result shows that to get a bounded solution on $\mathbb{R}$, one only needs to have a bounded solution on $\mathbb{R}^+$.

Theorem 5.2.19. *Assume that $(T_0(t))_{t\ge0}$ is immediately compact, $\mathcal{B}$ has the uniform fading memory property, and $f\in \mathrm{SAA}^1(\mathbb{R},X)$. Then, if equation* (5.26) *has a solution which is bounded on $\mathbb{R}^+$, it has a solution which is bounded on $\mathbb{R}$.*

Proof. Let x be a solution of equation (5.26) which is bounded on $\mathbb{R}^+$. From Lemmas 5.2.17 and 5.2.18, x has a relatively compact range and is uniformly continuous. Let $(t_n)_n$ be a sequence of real numbers such that $\lim_{n\to\infty} t_n = \infty$. If $t \in [-1,1]$ then for sufficiently large n the sequence of functions $x_n : t \mapsto x(t+t_n)$ is well defined and equicontinuous. It follows by Arzelà–Ascoli theorem that there exist a function y and a subsequence $(t_n^1)_n \subset (t_n)_n$ such that

$$x(t+t_n^1) \to y(t) \quad \text{as } n \to \infty$$

uniformly on $[-1,1]$. Using a diagonal procedure, one can construct a subsequence $(t_n')_n$ such that

$$x(t+t_n') \to y(t) \quad \text{as } n \to \infty$$

uniformly on each compact subset of $\mathbb{R}$. Since $f \in SAA^1(\mathbb{R},X)$, there exist a subsequence $(t_n'')_n \subset (t_n')_n$ and a function $\tilde{f} \in L^1_{loc}(\mathbb{R},X)$ such that

$$\int_t^{t+1} |f(t_n''+s) - \tilde{f}(s)| ds \to 0 \tag{5.45}$$

and

$$\int_t^{t+1} |\tilde{f}(t_n''-s) - f(s)| ds \to 0 \tag{5.46}$$

as $n \to \infty$. For each $t \geq s$ and for $n \in \mathbb{N}$ sufficiently large, we have

$$x(t+t_n'') = T_0(t-s)x(s+t_n'') + \lim_{\lambda\to\infty} \int_s^t T_0(t-\sigma)B_\lambda[L(x_{\sigma+t_n''}) + f(\sigma+t_n'')]d\sigma. \tag{5.47}$$

Taking the limit as $n \to \infty$ in (5.47) using (5.45) and using the fact that $x_{\sigma+t_n''} \to y_\sigma$, we get, for each $t \geq s$,

$$y(t) = T_0(t-s)y(s) + \lim_{\lambda\to\infty} \int_s^t T_0(t-\sigma)B_\lambda[L(y_\sigma) + \tilde{f}(\sigma)]d\sigma.$$

We observe that y has also a relatively compact range and is uniformly continuous. Using (5.46) and applying the above argument to the returning sequence $(-t_n'')_n$, we get a solution z of equation (5.26) which is bounded on $\mathbb{R}$. □

5.2.4 Bohr–Neugebauer property

Now we use the reduction principle in Theorem 5.2.12 to extend Bohr–Neugebauer-type theorem to the partial functional differential equation (5.26).

Theorem 5.2.20. *Assume that $(T_0(t))_{t\geq 0}$ is immediately compact, $\mathcal{B}$ has the uniform fading memory property, and $f : \mathbb{R} \to X$ is S^1-almost automorphic. Then equation* (5.26) *possesses the Bohr–Neugebauer property. More specifically, every bounded solution of equation* (5.26) *on $\mathbb{R}$ is compact almost automorphic.*

Proof. Let x be a bounded solution of equation (5.26) on $\mathbb{R}$. Using the spectral decomposition (5.33), we have, for each $t \in \mathbb{R}$,

$$x_t = \Pi^{U_0} x_t + \Pi^{S_0} x_t. \tag{5.48}$$

On the one hand, we have, for $t \geq \sigma$,

$$\Pi^{S_0} x_t = V^{S_0}(t-\sigma)\Pi^{S_0} x_\sigma + \lim_{n\to\infty} \int_\sigma^t V^{S_0}(t-s)\Pi^{S_0}(\widetilde{B}_n(X_0 f(s)))ds. \tag{5.49}$$

Since $t \mapsto x_t$ is bounded on $\mathbb{R}$ and $V(t)$ is exponentially stable in S_0, letting $\sigma \to -\infty$ in (5.49), we obtain that, for all $t \in \mathbb{R}$,

$$\Pi^{S_0} x_t = \lim_{n\to\infty} \int_{-\infty}^t V^{S_0}(t-s)\Pi^{S_0}(\widetilde{B}_n(X_0 f(s)))ds.$$

Using the same approach as in the proof of [137, Theorem 30], one can prove that the function

$$t \mapsto \Pi^{S_0} x_t = \lim_{n\to\infty} \int_{-\infty}^t V^{S_0}(t-s)\Pi^{S_0}(\widetilde{B}_n(X_0 f(s)))ds$$

is almost automorphic.

On the other hand, for each $t \in \mathbb{R}$,

$$\Pi^{U_0} x_t = \Phi\langle\Psi, x_t\rangle = \sum_{i=1}^d \langle\psi_i, x_t\rangle\varphi_i. \tag{5.50}$$

From Theorem 5.2.12, the function $z(t) = \langle\Psi, x_t\rangle$ is an integral solution of the following differential equation:

$$z'(t) = Gz(t) + \langle x^*, f(t)\rangle \quad \text{for } t \in \mathbb{R}.$$

Moreover, the function $t \mapsto \langle \Psi, x_t \rangle$ is bounded on $\mathbb{R}$ and the function $t \mapsto \langle x^*, f(t) \rangle$ is S^1-almost automorphic. It follows by Theorem 3.1.1 that $t \mapsto \langle \Psi, x_t \rangle$ is almost automorphic. We deduce from (5.50) that the function $t \mapsto \Pi^{U_0} x_t$ is almost automorphic. The almost automorphy of $t \mapsto x_t$ follows from (5.48).

Since $K := \overline{\{x_t : t \in \mathbb{R}\}}$ is compact, using the same argument as in the proof of Lemma 5.2.18 on the whole real line, we can prove that the solution $t \mapsto x_t$ is uniformly continuous. The compact almost automorphy of x follows from Lemma 2.3.6. □

The following result is a consequence of Theorems 5.2.19 and 5.2.20.

Corollary 5.2.21. *Assume that $(T_0(t))_{t \geq 0}$ is immediately compact, $\mathcal{B}$ has the uniform fading memory property, and $f : \mathbb{R} \to X$ is S^1-almost automorphic. If equation* (5.26) *has a bounded solution on $\mathbb{R}^+$, then it has a compact almost automorphic solution.*

5.2.5 Application

To apply our results, we consider the following reaction–diffusion equation with delay in a bounded open subset Ω of $\mathbb{R}^n$ with a smooth boundary $\partial\Omega$:

$$\begin{cases} \frac{\partial}{\partial t} v(t,x) = \Delta v(t,x) + av(t,x) + \int_{-\infty}^{t} h(s-t) v(s,x) \mathrm{d}s + F(t)\psi(x) \\ \quad \text{for } t \in \mathbb{R} \text{ and } x \in \Omega, \\ v(t,x) = 0 \quad \text{on } \mathbb{R} \times \partial\Omega, \end{cases} \tag{5.51}$$

where $a \in \mathbb{R}$, $h : [-\infty, 0] \to \mathbb{R}$ and $\psi : \overline{\Omega} \to \mathbb{R}$ are continuous functions. The function $F : \mathbb{R} \to \mathbb{R}$ is given by

$$F(t) = \sum_{n \geq 1} F_n(t),$$

where F_n are defined for every integer $n \geq 1$ by

$$F_n(t) = \sum_{k \in P_n} H(n^2(t-k)),$$

with $P_n = 3^n(2\mathbb{Z}+1) = \{3^n(2k+1), k \in \mathbb{Z}\}$ and $H \in C_0^\infty(\mathbb{R}, \mathbb{R})$, with support in $(-\frac{1}{2}, \frac{1}{2})$ such that

$$H \geq 0, \quad H(0) = 1, \quad \text{and} \quad \int_{\frac{-1}{2}}^{\frac{1}{2}} H(s) \mathrm{d}s = 1.$$

The function F is not almost automorphic, since it is not bounded. However, $F \in C^\infty(\mathbb{R}, \mathbb{R}) \cap SAA^1(\mathbb{R}, \mathbb{R})$ (see [264]).

To rewrite equation (5.51) in the abstract form (5.26), we introduce the space $X := C(\overline{\Omega})$ of continuous functions from $\overline{\Omega}$ to $\mathbb{R}$ endowed with the uniform norm topology and define the operator $A : D(A) \subset X \to X$ by

$$\begin{cases} D(A) = \{u \in C_0(\Omega) \cap H_0^1(\Omega) : \Delta u \in C(\overline{\Omega})\}, \\ Au = \Delta u, \end{cases}$$

where Δ is the Laplace operator. Let us denote by λ_1 the smallest eigenvalue of $-\Delta$ in $H_0^1(\Omega)$ ($\lambda_1 > 0$ since Ω is bounded and smooth). The operator A satisfies the Hille–Yosida condition on X and

$$\overline{D(A)} = \{u \in X : u|_{\partial\Omega} = 0\} \neq X.$$

Let A_0 be the part of the operator A in $\overline{D(A)}$. Then A_0 is given by

$$\begin{cases} D(A_0) = \{u \in C_0(\Omega) \cap H_0^1(\Omega) : \Delta u \in C_0(\Omega)\}, \\ A_0 u = \Delta u. \end{cases}$$

Lemma 5.2.22 ([97]). *The linear operator A_0 generates a compact C_0-semigroup $(T_0(t))_{t\geq 0}$ on $\overline{D(A)}$ such that, for each $t \geq 0$,*

$$|T_0(t)| \leq \exp\left(\frac{\lambda_1 |\Omega|^{\frac{2}{n}}}{4\pi}\right) e^{-\lambda_1 t}. \tag{5.52}$$

Consider the phase space C_γ defined by

$$C_\gamma := \left\{\varphi \in C((-\infty, 0], X) : \lim_{\theta \to -\infty} e^{\gamma\theta} \|\varphi(\theta)\| = 0\right\},$$

equipped with the following norm:

$$\|\varphi\|_\gamma = \sup_{\theta \leq 0} e^{\gamma\theta} \|\varphi(\theta)\| \quad \text{for } \varphi \in C_\gamma,$$

where γ is a real number. From [181, Theorem 3.7], the phase space C_γ satisfies Axioms **(A)**, **(B)**, and **(C)**. Moreover, the space C_γ has the uniform fading memory property when $\gamma > 0$.

In what follows, we assume that $\gamma > 0$ and

$$\int_{-\infty}^{0} e^{-\gamma\theta} |h(\theta)| \, d\theta < \infty.$$

Let $L : C_\gamma \to X$ be the linear operator defined by

$$L(\varphi)(\xi) = a\varphi(0)(\xi) + \int_{-\infty}^{0} h(\theta)\varphi(\theta)(\xi)\mathrm{d}\theta \quad \text{for } \xi \in \overline{\Omega} \text{ and } \varphi \in C_\gamma,$$

and $f : \mathbb{R} \to X$ be given by

$$f(t)(\xi) = F(t)\psi(\xi) \quad \text{for } \xi \in \overline{\Omega} \text{ and } t \in \mathbb{R}.$$

Then, L is a bounded linear operator from C_γ to X, and $f \in \mathrm{SAA}^1(\mathbb{R}, X)$. Equation (5.51) takes the following abstract form:

$$\frac{\mathrm{d}}{\mathrm{d}t}u(t) = Au(t) + L(u_t) + f(t) \quad \text{for } t \in \mathbb{R}. \tag{5.53}$$

Theorem 5.2.23. *Assume that $\gamma \geq \lambda_1$ and*

$$|a| + \int_{-\infty}^{0} e^{-\gamma\theta}|h(\theta)|\mathrm{d}\theta < \lambda_1 \exp\left(\frac{-\lambda_1|\Omega|^{\frac{2}{n}}}{2\pi}\right). \tag{5.54}$$

Then equation (5.53) *has a unique compact almost automorphic solution that is globally attractive.*

Proof of Theorem 5.2.23. From Theorem 5.2.6, for any initial data $\varphi \in \mathcal{B}_A$, equation (5.53) has a solution x given by

$$x(t) = T_0(t)\varphi(0) + \lim_{\lambda\to\infty} \int_0^t T_0(t-s)B_\lambda[L(x_s) + f(s)]\mathrm{d}s \quad \text{for } t \geq 0,$$

where $B_\lambda = \lambda R(\lambda, A)$. Let $M := \exp(\frac{\lambda_1|\Omega|^{\frac{2}{n}}}{4\pi})$. Then, $(-\lambda_1, \infty) \subset \rho(A)$ and

$$|R(\lambda, A)| \leq \frac{M}{\lambda + \lambda_1} \quad \text{for } \lambda > -\lambda_1. \tag{5.55}$$

Therefore by (5.55) and (5.52), we get

$$|x(t)| \leq Me^{-\lambda_1 t}|\varphi|_\gamma + \int_0^t M^2 e^{-\lambda_1(t-s)}[|L||x_s|_\gamma + |f(s)|]\mathrm{d}s.$$

It follows that

$$e^{\lambda_1 t}|x(t)| \leq M|\varphi|_\gamma + M^2 \int_0^t e^{\lambda_1 s}[|L||x_s|_\gamma + |f(s)|]\mathrm{d}s \quad \text{for } t \geq 0. \tag{5.56}$$

Let $\theta \leq 0$ and $t \geq 0$. If $t + \theta < 0$, then

$$\begin{aligned}
e^{\lambda_1 t}e^{\gamma\theta}|x(t+\theta)| &= e^{\lambda_1 t}e^{\gamma\theta}|\varphi(t+\theta)| \\
&= e^{\lambda_1 t}e^{-\gamma t}e^{\gamma\theta}e^{\gamma t}|\varphi(t+\theta)| \\
&= e^{\lambda_1 t}e^{-\gamma t}e^{\gamma(t+\theta)}|\varphi(t+\theta)| \\
&\le e^{(\lambda_1-\gamma)t}|\varphi|_\gamma \\
&\le |\varphi|_\gamma \le M|\varphi|_\gamma.
\end{aligned}$$

If $t+\theta \ge 0$, then from (5.56) we have

$$e^{\lambda_1(t+\theta)}|x(t+\theta)| \le M|\varphi|_\gamma + M^2\int_0^{t+\theta} e^{\lambda_1 s}|f(s)|ds + M^2|L|\int_0^{t+\theta} e^{\lambda_1 s}|x_s|_\gamma ds,$$

$$e^{\lambda_1 t}|x(t+\theta)| \le e^{-\lambda_1\theta}M|\varphi|_\gamma + e^{-\lambda_1\theta}M^2\int_0^{t+\theta} e^{\lambda_1 s}|f(s)|ds + e^{-\lambda_1\theta}M^2|L|\int_0^{t+\theta} e^{\lambda_1 s}|x_s|_\gamma ds,$$

$$\begin{aligned}
e^{\lambda_1 t}e^{\gamma\theta}|x(t+\theta)| &\le e^{(\gamma-\lambda_1)\theta}M|\varphi|_\gamma + e^{(\gamma-\lambda_1)\theta}M^2\int_0^{t+\theta} e^{\lambda_1 s}|f(s)|ds + e^{(\gamma-\lambda_1)\theta}M^2|L|\int_0^{t+\theta} e^{\lambda_1 s}|x_s|_\gamma ds \\
&\le M|\varphi|_\gamma + M^2\int_0^{t+\theta} e^{\lambda_1 s}|f(s)|ds + M^2|L|\int_0^{t+\theta} e^{\lambda_1 s}|x_s|_\gamma ds.
\end{aligned}$$

Thus, for each $t \ge 0$,

$$\begin{aligned}
e^{\lambda_1 t}|x_t|_\gamma &\le \sup_{\theta\le 0} e^{\lambda_1 t}e^{\gamma\theta}|x(t+\theta)| \\
&\le M|\varphi|_\gamma + M^2\int_0^t e^{\lambda_1 s}|f(s)|ds + M^2|L|\int_0^t e^{\lambda_1 s}|x_s|_\gamma ds \\
&\le M|\varphi|_\gamma + M^2\sum_{k=0}^{[t]}\int_k^{k+1} e^{\lambda_1 s}|f(s)|ds + M^2|L|\int_0^t e^{\lambda_1 s}|x_s|_\gamma ds \\
&\le M|\varphi|_\gamma + M^2|f|_{\mathrm{BS}^1}\sum_{k=0}^{[t]} e^{\lambda_1(k+1)} + M^2|L|\int_0^t e^{\lambda_1 s}|x_s|_\gamma ds \\
&\le M|\varphi|_\gamma + M_2(e^{\lambda_1(t+1)}-1) + M^2|L|\int_0^t e^{\lambda_1 s}|x_s|_\gamma ds,
\end{aligned}$$

where $M_2 = \frac{M^2 e^{\lambda_1}|f|_{\mathrm{BS}^1}}{e^{\lambda_1}-1}$. By the generalized Gronwall's inequality in Lemma 5.1.23, we obtain, for each $t \ge 0$,

$$e^{\lambda_1 t}|x_t|_\gamma \le Me^{\lambda_1 r}|\varphi|_\gamma + M_2(e^{\lambda_1(t+1)}-1) + M^2|L|\int_0^t (M|\varphi|_\gamma + M_2(e^{\lambda_1(s+1)}-1))e^{(M^2|L|)(t-s)}ds.$$

But from (5.54) we have $|L| \leq |a| + \int_{-\infty}^{0} e^{-\gamma\theta}|h(\theta)|d\theta < \frac{\lambda_1}{M^2}$, thus

$$\lambda_1 - M^2|L| > 0. \tag{5.57}$$

It follows that, for each $t \geq 0$,

$$|x_t|_\gamma \leq M_2 e^{\lambda_1} + M|\varphi|_\gamma + \frac{M^2 M_2 |L| e^{\lambda_1}}{\lambda_1 - M^2|L|}.$$

This shows that x is a bounded solution of equation (5.53) on $\mathbb{R}^+$. Using Corollary 5.2.21, we deduce that equation (5.53) has a compact almost automorphic solution y.

Let z be another solution. Then

$$y(t) - z(t) = T_0(t)(y(0) - z(0)) + \lim_{\lambda\to\infty} \int_0^t T_0(t-s)B_\lambda[L(y_s - z_s)]ds \quad \text{for } t \geq 0.$$

Using the same computations as above we have, for $t \geq 0$,

$$e^{\lambda_1 t}|y_t - z_t|_\gamma \leq M|y_0 - z_0|_\gamma + M^2 \int_0^t |L|e^{\lambda_1 s}|y_s - z_s|_\gamma ds.$$

Now using the classical Gronwall's lemma, we obtain, for $t \geq 0$,

$$|y_t - z_t|_\gamma \leq M|y_0 - z_0|_\gamma e^{(M^2|L| - \lambda_1)t}.$$

Thus from (5.57), we deduce that $|y_t - z_t|_\gamma \to 0$ as $t \to \infty$, that is, y is globally attractive.

We claim that y is the unique solution of (5.53) which is bounded on the whole real line. In fact, if w is another solution which is bounded on $\mathbb{R}$, then for all $t, \sigma \in \mathbb{R}$ with $\sigma \leq t$ we can show that

$$|y_t - w_t|_\gamma \leq M|y_\sigma - w_\sigma|_\gamma e^{(M^2|L| - \lambda_1)(t-\sigma)}.$$

Letting $\sigma \to -\infty$, we deduce that $y_t = w_t$ for all $t \in \mathbb{R}$. □

5.3 Neutral partial functional differential equations under exponential dichotomy condition

The aim of this section is to study the existence and uniqueness of a μ-pseudo almost periodic integral solution for the following neutral partial functional differential equation:

$$\frac{d}{dt}\mathcal{D}(u_t) = A\mathcal{D}(u_t) + L(u_t) + f(t) \quad \text{for } t \in \mathbb{R}, \tag{5.58}$$

where A is a linear operator on a Banach space X satisfying the well-known Hille–Yosida condition, the domain is not necessarily dense, namely, we suppose that:

(H0) There exist $\bar{M} \geq 1$ and $\omega \in \mathbb{R}$ such that $(\omega, \infty) \subset \rho(A)$ and

$$|R(\lambda, A)^n| \leq \frac{\bar{M}}{(\lambda - \omega)^n} \quad \text{for } n \in \mathbb{N} \text{ and } \lambda > \omega,$$

where $\rho(A)$ is the resolvent set of A and $R(\lambda, A) := (\lambda I - A)^{-1}$ for $\lambda \in \rho(A)$.

Operator $\mathcal{D} : C \to X$ is bounded and linear, where $C := C([-r, 0], X)$, with $r > 0$, is the space of continuous functions from $[-r, 0]$ to X endowed with the uniform norm topology. For the well posedness of equation (5.58), we assume that $\mathcal{D}$ has the following form:

$$\mathcal{D}(\varphi) = \varphi(0) - \int_{-r}^{0} [\mathrm{d}\eta(\theta)]\varphi(\theta) \quad \text{for } \varphi \in C,$$

for a mapping $\eta : [-r, 0] \to \mathcal{L}(X)$ of bounded variation and nonatomic at zero, which means that there exists a continuous nondecreasing function $\delta : [0, r] \to [0, \infty)$ such that $\delta(0) = 0$ and

$$\left|\int_{-s}^{0} [\mathrm{d}\eta(\theta)]\varphi(\theta)\right| \leq \delta(s) \sup_{-r \leq \theta \leq 0} |\varphi(\theta)| \quad \text{for } \varphi \in C \text{ and } s \in [0, r],$$

where $\mathcal{L}(X)$ is the space of bounded linear operators from X to X. For every $t \in \mathbb{R}$, the history function $u_t \in C$ is defined by

$$u_t(\theta) = u(t + \theta) \quad \text{for } \theta \in [-r, 0].$$

Operator L is bounded and linear from C to X and the input function f is Stepanov μ-pseudo almost periodic from $\mathbb{R}$ to X.

5.3.1 Variation of constants formula and spectral decomposition

In the sequel, we assume that the operator A satisfies the Hille–Yosida condition **(H0)**. To equation (5.58), we associate the following Cauchy problem:

$$\begin{cases} \frac{\mathrm{d}}{\mathrm{d}t}\mathcal{D}(u_t) = A\mathcal{D}(u_t) + L(u_t) + f(t) & \text{for } t \geq \sigma, \\ u_\sigma = \varphi \in C. \end{cases} \tag{5.59}$$

Definition 5.3.1 ([7]). A continuous function $u : [-r + \sigma, \infty) \to X$ is called an integral solution of (5.59) if

(i) $\int_\sigma^t \mathcal{D}(u_s)ds \in D(A)$ for $t \geq \sigma$,
(ii) $\mathcal{D}(u_t) = \mathcal{D}(\varphi) + A\int_\sigma^t \mathcal{D}(u_s)ds + \int_\sigma^t [L(u_s) + f(s)]ds$ for $t \geq \sigma$,
(iii) $u_\sigma = \varphi$.

If u is an integral solution of (5.59), then from the continuity of u, we have $\mathcal{D}(u_t) \in \overline{D(A)}$, for all $t \geq \sigma$. In particular, $\mathcal{D}(\varphi) \in \overline{D(A)}$.

Let us introduce the part A_0 of the operator A in $\overline{D(A)}$ defined by

$$\begin{cases} D(A_0) := \{x \in D(A) : Ax \in \overline{D(A)}\}, \\ A_0x := Ax \quad \text{for } x \in D(A_0). \end{cases}$$

Lemma 5.3.2 ([108]). *Assume that* **(H0)** *holds. Then A_0 generates a strongly continuous semigroup $(T_0(t))_{t\geq 0}$ on $\overline{D(A)}$.*

For the existence of the integral solutions, one has the following result.

Theorem 5.3.3 ([7]). *Assume that* **(H0)** *holds. Then for all $\varphi \in C$ such that $\mathcal{D}(\varphi) \in \overline{D(A)}$, equation* (5.59) *has a unique integral solution u on $[-r+\sigma, \infty)$. Moreover, u satisfies*

$$\mathcal{D}(u_t) = T_0(t-\sigma)\mathcal{D}(\varphi) + \lim_{\lambda\to\infty} \int_\sigma^t T_0(t-s)B_\lambda[L(u_s) + f(s)]ds \quad \text{for } t \geq \sigma,$$

where $B_\lambda := \lambda R(\lambda, A)$ for $\lambda > \omega$.

In the sequel, $u(\cdot, \sigma, \varphi, f)$ denotes the integral solution of (5.59). The phase space C_0 of equation (5.59) is given by

$$C_0 = \{\varphi \in C : \mathcal{D}(\varphi) \in \overline{D(A)}\}.$$

For each $t \geq 0$, we define the linear operator $T(t)$ on C_0 by

$$T(t)\varphi = u_t(\cdot, 0, \varphi, 0),$$

where $u(\cdot, 0, \varphi, 0)$ is the integral solution of the homogeneous equation

$$\begin{cases} \frac{d}{dt}\mathcal{D}(u_t) = A\mathcal{D}(u_t) + L(u_t) \quad \text{for } t \geq 0, \\ u_0 = \varphi. \end{cases} \tag{5.60}$$

We have the following result.

Proposition 5.3.4 ([7]). *Assume that* **(H0)** *holds. Then $(T(t))_{t\geq 0}$ is a strongly continuous semigroup on C_0. Moreover, the operator $\mathcal{A}$ defined on C_0 by*

$$\begin{cases} D(\mathcal{A}) = \{\varphi \in C^1([-r,0],X) : \mathcal{D}(\varphi) \in D(A), \mathcal{D}(\varphi') \in \overline{D(A)} \text{ and } \mathcal{D}(\varphi') = A\mathcal{D}(\varphi) + L(\varphi)\}, \\ \mathcal{A}\varphi = \varphi' \end{cases}$$

is the infinitesimal generator of $(T(t))_{t\geq 0}$ *on* C_0.

To determine the asymptotic behavior of the semigroup $(T(t))_{t\geq 0}$, we need to introduce some preliminary results. In neutral systems, many fundamental properties depend essentially on the choice of the difference operator $\mathcal{D}$. Here we suppose that $\mathcal{D}$ is stable in the sense given in the literature; more details can be found in [164, 166, 272].

Definition 5.3.5 ([164, 166]). The operator $\mathcal{D}$ is said to be stable if there exist positive constants η and μ such that the solution of the following homogeneous difference equation:

$$\begin{cases} \mathcal{D}(u_t) = 0 & \text{for } t \geq 0, \\ u_0 = \varphi, \end{cases}$$

where $\varphi \in \{\psi \in C : \mathcal{D}(\psi) = 0\}$, satisfies

$$|u_t(\cdot,\varphi)| \leq \mu e^{-\eta t}|\varphi| \quad \text{for } t \geq 0.$$

Example. The operator $\mathcal{D}$ defined by

$$\mathcal{D}(\varphi) = \varphi(0) - q\varphi(-r) \quad \text{for } \varphi \in C,$$

is stable if and only if $|q| < 1$.

In the following, we suppose that

(H1) The semigroup $(T_0(t))_{t\geq 0}$ is compact on $\overline{D(A)}$ whenever $t > 0$;

(H2) The operator $\mathcal{D}$ is stable.

Then, we have the following fundamental result on the semigroup $(T(t))_{t\geq 0}$.

Theorem 5.3.6 ([7, Lemma 10]). *Assume that* **(H0)–(H2)** *hold. Then the semigroup* $(T(t))_{t\geq 0}$ *is decomposed on* C_0 *as follows:*

$$T(t) = T_1(t) + T_2(t) \quad \textit{for } t \geq 0,$$

where $(T_1(t))_{t\geq 0}$ *is an exponentially stable semigroup on* C_0, *which means that there are positive constants* γ_0 *and* N_0 *such that*

$$|T_1(t)\varphi| \leq N_0 e^{-\gamma_0 t}|\varphi| \quad \textit{for } t \geq 0 \textit{ and } \varphi \in C_0.$$

Moreover $T_2(t)$ *is compact for every* $t > 0$.

Consequently from Theorem 5.3.6, we obtain the following interesting result that will be used for the spectral decomposition.

Corollary 5.3.7 ([7]). *Assume that* **(H0)–(H2)** *hold. Then* $\omega_{\text{ess}}(T) < 0$.

Definition 5.3.8 ([7]). The semigroup $(T(t))_{t\geq 0}$ is said to have an exponential dichotomy if

$$\sigma(\mathcal{A}) \cap i\mathbb{R} = \emptyset.$$

Since **(H0)**, **(H1)**, and **(H2)** hold, by Corollary 5.3.7 we have $\omega_{\text{ess}}(T) < 0$. Consequently, we get the following result on the spectral decomposition of C_0.

Theorem 5.3.9 ([7]). *Assume that* **(H0)–(H2)** *hold. If the semigroup* $(T(t))_{t\geq 0}$ *has an exponential dichotomy, then the space* C_0 *is decomposed as a direct sum* $C_0 = \mathcal{S} \oplus \mathcal{U}$ *of two* $T(t)$ *invariant closed subspaces* $\mathcal{S}$ *and* $\mathcal{U}$ *such that the restricted semigroup on* $\mathcal{U}$ *is a group and there exist positive constants* M *and* c *such that*

$$\begin{cases} |T(t)\varphi| \leq Me^{-ct}|\varphi| & \text{for } t \geq 0 \text{ and } \varphi \in \mathcal{S}, \\ |T(t)\varphi| \leq Me^{ct}|\varphi| & \text{for } t \leq 0 \text{ and } \varphi \in \mathcal{U}. \end{cases}$$

Subspaces $\mathcal{S}$ *and* $\mathcal{U}$ *are respectively called the stable and unstable spaces.*

To give a variation of constants formula associated to equation (5.59), we need to extend the semigroup $(T(t))_{t\geq 0}$ to the space $C_0 \oplus \langle X_0 \rangle$, where $\langle X_0 \rangle$ is the space defined by

$$\langle X_0 \rangle = \{X_0 y : y \in X\},$$

and the function $X_0 y$ is given, for $y \in X$, by

$$(X_0 y)(\theta) = \begin{cases} 0 & \text{if } \theta \in [-r, 0), \\ y & \text{if } \theta = 0. \end{cases}$$

The space $C_0 \oplus \langle X_0 \rangle$ equipped with the norm $|\varphi + X_0 y| = |\varphi| + |y|$, for $(\varphi, y) \in C_0 \times X$, is a Banach space. Consider the extension $\widetilde{\mathcal{A}}$ of the operator $\mathcal{A}$ on $C_0 \oplus \langle X_0 \rangle$ defined by

$$\begin{cases} D(\widetilde{\mathcal{A}}) = \{\varphi \in C^1([-r, 0], X) : \mathcal{D}(\varphi) \in D(A) \text{ and } \mathcal{D}(\varphi') \in \overline{D(A)}\}, \\ \widetilde{\mathcal{A}}\varphi = \varphi' + X_0(A\mathcal{D}(\varphi) + L(\varphi) - \mathcal{D}(\varphi')). \end{cases}$$

To compute the resolvent operator $R(\lambda, \widetilde{\mathcal{A}}) = (\lambda - \widetilde{\mathcal{A}})^{-1}$, we introduce the following assumption:

(H3) $\mathcal{D}(e^{\lambda \cdot} y) \in D(A)$, for all $y \in D(A)$ and all complex λ, where $e^{\lambda \cdot} y \in C$ is defined by

$$(e^{\lambda \cdot} y)(\theta) = e^{\lambda\theta} y \quad \text{for } \theta \in [-r, 0].$$

Lemma 5.3.10 ([7, Theorem 13]). *Assume that* **(H0)** *and* **(H3)** *hold. Then $\widetilde{\mathcal{A}}$ satisfies the Hille–Yosida condition on $C_0 \oplus \langle X_0 \rangle$: there exist $\widetilde{M} \geq 0$ and $\widetilde{\omega} \in \mathbb{R}$ such that $(\widetilde{\omega}, \infty) \subset \rho(\widetilde{\mathcal{A}})$ and*

$$|R(\lambda, \widetilde{\mathcal{A}})^n| \leq \frac{\widetilde{M}}{(\lambda - \widetilde{\omega})^n} \quad \text{for } n \in \mathbb{N} \text{ and } \lambda > \widetilde{\omega}.$$

Now, we can state the variation of constants formula associated with (5.59).

Theorem 5.3.11 ([7, Theorem 16]). *Assume that* **(H0)** *and* **(H3)** *hold. Then, for all $\varphi \in C_0$, the integral solution $u(\cdot, \sigma, \varphi, f)$ of* (5.59) *is given by the following variation of constants formula:*

$$u_t(., \sigma, \varphi, f) = T(t - \sigma)\varphi + \lim_{n\to\infty} \int_\sigma^t T(t - s)(\widetilde{B}_n X_0 f(s))ds \quad \text{for } t \geq \sigma,$$

where $\widetilde{B}_n X_0 y = nR(n, \widetilde{\mathcal{A}})(X_0 y)$ for $n > \widetilde{\omega}$ and $y \in X$.

Theorem 5.3.12 ([7]). *Assume that* **(H0)–(H3)** *hold and the semigroup $(T(t))_{t\geq 0}$ has an exponential dichotomy. If f is bounded on $\mathbb{R}$, then equation* (5.58) *has a unique bounded integral solution on $\mathbb{R}$ which is given by*

$$u_t = \lim_{n\to\infty} \int_{-\infty}^t T^s(t - \tau)\Pi^s(\widetilde{B}_n X_0 f(\tau))\, d\tau + \lim_{n\to\infty} \int_{\infty}^t T^u(t - \tau)\Pi^u(\widetilde{B}_n X_0 f(\tau))\, d\tau \quad \text{for } t \in \mathbb{R},$$

where Π^s and Π^u are the projections of C onto the stable and unstable subspaces, T^s and T^u are the restrictions of $T(t)$ on $\mathcal{S}$ and $\mathcal{U}$, respectively.

5.3.2 μ-Pseudo almost periodic integral solutions

In this section, we study the almost periodicity of the integral solution of equation (5.58) perturbed by an ergodic term. We give a result which shows the existence and uniqueness of a μ-pseudo almost periodic integral solution if the input function f is μ-pseudo almost periodic in Stepanov's sense.

The following theorem is similar to [144, Theorem 3.14], but here we show the existence and uniqueness of a bounded integral solution even if the forcing term f belongs to $\mathrm{BS}^1(\mathbb{R}, X)$ instead of $\mathrm{BC}(\mathbb{R}, X)$.

Theorem 5.3.13. *Assume that* **(H0)–(H3)** *hold and the semigroup $(T(t))_{t\geq 0}$ has an exponential dichotomy. If $f \in \mathrm{BS}^1(\mathbb{R}, X)$. Then equation* (5.58) *has a unique bounded integral solution on $\mathbb{R}$ which is given by*

$$u_t = \lim_{n\to\infty} \int_{-\infty}^t T^s(t - \tau)\Pi^s(\widetilde{B}_n X_0 f(\tau))\, d\tau - \lim_{n\to\infty} \int_t^{\infty} T^u(t - \tau)\Pi^u(\widetilde{B}_n X_0 f(\tau))\, d\tau. \tag{5.61}$$

Proof. Let us first prove that the limits in (5.61) exist in the phase space C. For n sufficiently large, we have $|\widetilde{B}_n| \leq \frac{n\widetilde{M}}{n-\widetilde{\omega}} \leq 2\widetilde{M}$, it follows for $t \in \mathbb{R}$ that

$$\begin{aligned}\int_{-\infty}^{t} |T^s(t-\tau)\Pi^s(\widetilde{B}_n X_0 f(\tau))|\,d\tau &\leq 2\widetilde{M}M|\Pi^s| \int_{-\infty}^{t} e^{-c(t-\tau)}|f(\tau)|\,d\tau \\ &\leq 2\widetilde{M}M|\Pi^s| \sum_{k=1}^{\infty}\left(\int_{t-k}^{t-k+1} e^{-c(t-\tau)}|f(\tau)|\,d\tau\right) \\ &\leq 2\widetilde{M}M|\Pi^s| \sum_{k=1}^{\infty}\left(e^{-c(k-1)}\int_{t-k}^{t-k+1} |f(\tau)|d\tau\right) \\ &\leq K,\end{aligned} \tag{5.62}$$

where $K := 2\widetilde{M}M|\Pi^s||f|_{BS^1}\frac{1}{1-e^{-c}}$. Let

$$H(n,\tau,t) := T^s(t-\tau)\Pi^s(\widetilde{B}_n X_0 f(\tau)) \quad \text{for } n \in \mathbb{N} \text{ and } \tau \leq t.$$

For n sufficiently large and $\sigma \leq t$, we have

$$\begin{aligned}\left|\int_{-\infty}^{\sigma} H(n,\tau,t)d\tau\right| &\leq 2\widetilde{M}M|\Pi^s| \sum_{k=1}^{\infty}\left(\int_{\sigma-k}^{\sigma-k+1} e^{-c(t-\tau)}|f(\tau)|\,d\tau\right) \\ &\leq 2\widetilde{M}M|\Pi^s| \sum_{k=1}^{\infty}\left(e^{-c(t-\sigma+k-1)}\int_{\sigma-k}^{\sigma-k+1} |f(\tau)|d\tau\right) \\ &\leq 2\widetilde{M}M|\Pi^s||f|_{BS^1}e^{-c(t-\sigma)}\sum_{k=1}^{\infty} e^{-c(k-1)} = Ke^{-c(t-\sigma)}.\end{aligned}$$

It follows that, for n and m sufficiently large and $\sigma \leq t$,

$$\begin{aligned}\left|\int_{-\infty}^{t} H(n,\tau,t)d\tau - \int_{-\infty}^{t} H(m,\tau,t)d\tau\right| &\leq \left|\int_{-\infty}^{\sigma} H(n,\tau,t)d\tau\right| + \left|\int_{-\infty}^{\sigma} H(m,\tau,t)d\tau\right| \\ &\quad + \left|\int_{\sigma}^{t} H(n,\tau,t)d\tau - \int_{\sigma}^{t} H(m,\tau,t)d\tau\right| \\ &\leq 2Ke^{-c(t-\sigma)} + \left|\int_{\sigma}^{t} H(n,\tau,t)d\tau - \int_{\sigma}^{t} H(m,\tau,t)d\tau\right|.\end{aligned}$$

Since $\lim_{n\to\infty}\int_{\sigma}^{t} H(n,\tau,t)d\tau$ exists, then

$$\limsup_{n,m\to\infty}\left|\int_{-\infty}^{t} H(n,\tau,t)\mathrm{d}\tau - \int_{-\infty}^{t} H(m,\tau,t)\mathrm{d}\tau\right| \leq 2Ke^{-\alpha(t-\sigma)}.$$

Letting $\sigma \to -\infty$, we get

$$\limsup_{n,m\to\infty}\left|\int_{-\infty}^{t} H(n,\tau,t)\mathrm{d}\tau - \int_{-\infty}^{t} H(m,\tau,t)\mathrm{d}\tau\right| = 0.$$

Thus by the completeness of the phase space C, we deduce that the limit

$$\lim_{n\to\infty}\int_{-\infty}^{t} H(n,\tau,t)\mathrm{d}\tau = \lim_{n\to\infty}\int_{-\infty}^{t} T^s(t-\tau)\Pi^s(\widetilde{B}_n X_0 f(\tau))\,\mathrm{d}\tau$$

exists. In addition, one can see from (5.62) that the function

$$\xi_1 : t \mapsto \lim_{n\to\infty}\int_{-\infty}^{t} T^s(t-\tau)\Pi^s(\widetilde{B}_n X_0 f(\tau))\,\mathrm{d}\tau$$

is bounded on $\mathbb{R}$. Similarly, we can show that the function

$$\xi_2 : t \mapsto \lim_{n\to\infty}\int_{t}^{\infty} T^u(t-\tau)\Pi^u(\widetilde{B}_n X_0 f(\tau))\,\mathrm{d}\tau$$

is well defined and bounded on $\mathbb{R}$. Using the same argument as in [3, Theorem 5.9], the integral solution u given by the formula

$$u_t = \lim_{n\to\infty}\int_{-\infty}^{t} T^s(t-\tau)\Pi^s(\widetilde{B}_n X_0 f(\tau))\,\mathrm{d}\tau - \lim_{n\to\infty}\int_{t}^{\infty} T^u(t-\tau)\Pi^u(\widetilde{B}_n X_0 f(\tau))\,\mathrm{d}\tau$$

is the only bounded integral solution of equation (5.58) on $\mathbb{R}$. □

Now we study the existence and uniqueness of a Bohr almost periodic integral solution to equation (5.58), if the input function f is Stepanov almost periodic.

Theorem 5.3.14. *Assume that* **(H0)–(H3)** *hold and the semigroup* $(T(t))_{t\geq 0}$ *has an exponential dichotomy. If f is S^1-almost periodic, then equation* (5.58) *has a unique integral solution which is almost periodic* (*in Bohr's sense*).

Proof. From Theorem 5.3.13, equation (5.58) has a unique bounded integral solution on $\mathbb{R}$ which is given by

$$\begin{aligned} u_t &= \lim_{n\to\infty} \int_{-\infty}^{t} T^s(t-\tau)\Pi^s(\widetilde{B}_n X_0 f(\tau))\,d\tau + \lim_{n\to\infty} \int_{\infty}^{t} T^u(t-\tau)\Pi^u(\widetilde{B}_n X_0 f(\tau))\,d\tau \\ &= X(t) + Y(t), \end{aligned} \tag{5.63}$$

with

$$\begin{aligned} X(t) &:= \lim_{n\to\infty} \int_{-\infty}^{t} T^s(t-\tau)\Pi^s(\widetilde{B}_n X_0 f(\tau))\,d\tau, \\ Y(t) &:= \lim_{n\to\infty} \int_{\infty}^{t} T^u(t-\tau)\Pi^u(\widetilde{B}_n X_0 f(\tau))\,d\tau. \end{aligned} \tag{5.64}$$

We will show that both of the above terms are almost periodic. Consider, for each integer $k \geq 1$,

$$X_k(t) := \lim_{n\to\infty} \int_{t-k}^{t-k+1} T^s(t-\tau)\Pi^s(\widetilde{B}_n X_0 f(\tau))\,d\tau \quad \text{for } t \in \mathbb{R}.$$

We will show that $X_k(t) \in AP(\mathbb{R}, X)$, for each $k \geq 1$. Note that

$$X_k(t) = \lim_{n\to\infty} \int_{k-1}^{k} T^s(\tau)\Pi^s(\widetilde{B}_n X_0 f(t-\tau))\,d\tau.$$

First, the function $t \mapsto X_k(t)$ is continuous on $\mathbb{R}$. In fact, since $f \in L^1_{loc}(\mathbb{R}, X)$, we have, for each $t \in \mathbb{R}$,

$$\begin{aligned} |X_k(t+h) - X_k(t)| &= \left| \lim_{n\to\infty} \int_{k-1}^{k} T^s(\tau)\Pi^s(\widetilde{B}_n X_0(f(t+h-\tau) - f(t-\tau)))\,d\tau \right| \\ &\leq \widetilde{M}M|\Pi^s| \int_{k-1}^{k} e^{-c\tau}|f(t+h-\tau) - f(t-\tau)|\,d\tau \\ &\leq \widetilde{M}M|\Pi^s|e^{-c(k-1)} \int_{k-1}^{k} |f(t+h-\tau) - f(t-\tau)|\,d\tau \\ &= \widetilde{M}M|\Pi^s|e^{-c(k-1)} \int_{t-k}^{t-k+1} |f(\tau+h) - f(\tau)|\,d\tau \to 0 \quad \text{as } h \to 0. \end{aligned}$$

Let $(s'_m)_{m\in\mathbb{N}}$ be a sequence of real numbers. Since $f \in SAP^1(\mathbb{R}, X)$, there exist a subsequence $(s_m)_{m\in\mathbb{N}}$ and a function $g \in L^1_{loc}(\mathbb{R}, X)$ such that

$$\sup_{t\in\mathbb{R}}\int_t^{t+1}|f(s_m+s)-g(s)|\,\mathrm{d}s \to 0 \quad \text{as } m\to\infty.$$

Consider the function $Z_k(t) := \lim_{n\to\infty}\int_{k-1}^{k} T^s(\tau)\Pi^s(\widetilde{B}_n X_0 g(t-\tau))\,\mathrm{d}\tau$. Then we have

$$\begin{aligned}
|X_k(t+s_m)-Z_k(t)| &= \left|\lim_{n\to\infty}\int_{k-1}^{k} T^s(\tau)\Pi^s(\widetilde{B}_n X_0(f(t+s_m-\tau)-g(t-\tau)))\,\mathrm{d}\tau\right| \\
&\le \widetilde{M}M|\Pi^s|\int_{k-1}^{k} e^{-c\tau}|f(t+s_m-\tau)-g(t-\tau)|\,\mathrm{d}\tau \\
&\le \widetilde{M}M|\Pi^s|e^{-c(k-1)}\int_{k-1}^{k}|f(t+s_m-\tau)-g(t-\tau)|\,\mathrm{d}\tau \\
&= \widetilde{M}M|\Pi^s|e^{-c(k-1)}\int_{t-k}^{t-k+1}|f(\tau+s_m)-g(\tau)|\,\mathrm{d}\tau \\
&\le \widetilde{M}M|\Pi^s|e^{-c(k-1)}\sup_{s\in\mathbb{R}}\int_s^{s+1}|f(s_m+\tau)-g(\tau)|\mathrm{d}\tau.
\end{aligned}$$

It follows that $\sup_{t\in\mathbb{R}}|X_k(t+s_m)-Z_k(t)|\to 0$ as $m\to\infty$. Therefore $X_k(t)\in AP(\mathbb{R},X)$ for each $k\ge 1$. In addition, for each $t\in\mathbb{R}$ and $k\ge 1$,

$$\begin{aligned}
|X_k(t)| &\le \widetilde{M}M|\Pi^s|\int_{k-1}^{k} e^{-c\tau}|f(t-\tau)|\,\mathrm{d}\tau \\
&\le \widetilde{M}M|\Pi^s|e^{-c(k-1)}\int_{k-1}^{k}|f(t-\tau)|\,\mathrm{d}\tau \\
&= \widetilde{M}M|\Pi^s|e^{-c(k-1)}\int_{t-k}^{t-k+1}|f(\tau)|\,\mathrm{d}\tau \\
&\le \widetilde{M}M|\Pi^s|\|f\|_{BS^1}e^{-c(k-1)}.
\end{aligned}$$

We deduce from the well-known Weierstrass M-test that the series $\sum_{k=1}^{\infty}X_k(t)$ is uniformly convergent on $\mathbb{R}$. Let

$$H(n,\tau,t) := T^s(t-\tau)\Pi^s(\widetilde{B}_n X_0 f(\tau)) \quad \text{for } n\in\mathbb{N} \text{ and } \tau\le t.$$

We claim that $X(t)=\sum_{k=1}^{\infty}X_k(t)$, where $X(\cdot)$ is the function defined by (5.64). In fact,

$$\begin{aligned}\left|\sum_{k=1}^{N} X_k(t) - X(t)\right| &= \left|\sum_{k=1}^{N} \lim_{n\to\infty} \int_{t-k}^{t-k+1} H(n,\tau,t)\,d\tau - \lim_{n\to\infty}\int_{-\infty}^{t} H(n,\tau,t)\,d\tau\right| \\ &= \left|\lim_{n\to\infty} \sum_{k=N+1}^{\infty} \int_{t-k}^{t-k+1} H(n,\tau,t)\,d\tau\right| \\ &\le \widetilde{M}M|\Pi^s| \sum_{k=N+1}^{\infty} \int_{t-k}^{t-k+1} e^{-c(t-\tau)}|f(\tau)|\,d\tau \\ &\le \widetilde{M}M|\Pi^s||f|_{BS^1} \sum_{k=N+1}^{\infty} e^{-c(k-1)} \to 0 \quad \text{as } N\to\infty.\end{aligned}$$

Since the convergence of the series $\sum_{k=1}^{\infty} X_k(t)$ is uniform, we deduce that $X(t) \in AP(\mathbb{R}, X)$. Similarly, we show that $Y(t) \in AP(\mathbb{R}, X)$. The integral solution given by (5.63) is then almost periodic. □

In the next theorem, we show that to have a μ-pseudo almost periodic integral solution, we only need f to be μ-pseudo almost periodic in Stepanov's sense.

Theorem 5.3.15. *Assume that* **(H0)–(H4)** *hold and the semigroup* $(T(t))_{t\ge0}$ *has an exponential dichotomy. If* $f \in PAP^1(\mathbb{R}, X, \mu)$, *then equation* (5.58) *has a unique integral solution which is* μ*-pseudo almost periodic.*

Proof. From Theorem 5.3.13, equation (5.58) has a unique bounded integral solution on $\mathbb{R}$ which is given by

$$\begin{aligned}u_t &= \lim_{n\to\infty}\int_{-\infty}^{t} T^s(t-\tau)\Pi^s(\widetilde{B}_n X_0 f(\tau))\,d\tau + \lim_{n\to\infty}\int_{\infty}^{t} T^u(t-\tau)\Pi^u(\widetilde{B}_n X_0 f(\tau))\,d\tau \\ &= X(t) + Y(t),\end{aligned} \tag{5.65}$$

with

$$X(t) := \lim_{n\to\infty}\int_{-\infty}^{t} T^s(t-\tau)\Pi^s(\widetilde{B}_n X_0 f(\tau))\,d\tau,$$

$$Y(t) := \lim_{n\to\infty}\int_{\infty}^{t} T^u(t-\tau)\Pi^u(\widetilde{B}_n X_0 f(\tau))\,d\tau.$$

We will show that both of the above terms are μ-pseudo almost periodic. Let $f = f_1 + f_2$, with $f_1 \in SAP^1(\mathbb{R}, X)$ and $f_2 \in \mathcal{E}^1(\mathbb{R}, X, \mu)$. Then

$$\begin{aligned}X(t) &= \lim_{n\to\infty}\int_{-\infty}^{t} T^s(t-\tau)\Pi^s(\widetilde{B}_n X_0 f_1(\tau))\,d\tau + \lim_{n\to\infty}\int_{-\infty}^{t} T^s(t-\tau)\Pi^s(\widetilde{B}_n X_0 f_2(\tau))\,d\tau \\ &= X^1(t) + X^2(t),\end{aligned}$$

with

$$X^1(t) := \lim_{n\to\infty} \int_{-\infty}^{t} T^s(t-\tau)\Pi^s(\widetilde{B}_n X_0 f_1(\tau))\,d\tau,$$
$$X^2(t) := \lim_{n\to\infty} \int_{-\infty}^{t} T^s(t-\tau)\Pi^s(\widetilde{B}_n X_0 f_2(\tau))\,d\tau. \tag{5.66}$$

Using the proof of Theorem 5.3.14, we have $X^1(t) \in AP(\mathbb{R}, X)$. Let us show that $X^2(t) \in \mathcal{E}(\mathbb{R}, X, \mu)$. Consider, for each $k \geq 1$,

$$X_k^2(t) := \lim_{n\to\infty} \int_{t-k}^{t-k+1} T^s(t-\tau)\Pi^s(\widetilde{B}_n X_0 f_2(\tau))\,d\tau \quad \text{for } t \in \mathbb{R}.$$

We have

$$\begin{aligned}|X_k^2(t)| &\leq \widetilde{M}M|\Pi^s| \int_{k-1}^{k} e^{-c\tau}|f_2(t-\tau)|\,d\tau \\ &\leq \widetilde{M}M|\Pi^s|e^{-c(k-1)} \int_{k-1}^{k} |f_2(t-\tau)|\,d\tau \\ &= \widetilde{M}M|\Pi^s|e^{-c(k-1)} \int_{t}^{t+1} |f_2(\tau-k)|\,d\tau.\end{aligned}$$

It follows that

$$\frac{1}{\mu[-T,T]} \int_{-T}^{T} |X_k^2(t)|d\mu(t) \leq \widetilde{M}M|\Pi^s|e^{-c(k-1)} \frac{1}{\mu[-T,T]} \int_{-T}^{T}\int_{t}^{t+1} |f_2(\tau-k)|\,d\tau d\mu(t).$$

Since μ satisfies **(H4)**, we get that $\mathcal{E}^1(\mathbb{R}, X, \mu)$ is invariant with respect to translation (Proposition 2.3.25), hence it follows that $X_k^2 \in \mathcal{E}(\mathbb{R}, X, \mu)$. By the same argument in the proof of Theorem 5.3.14 and using Proposition 2.3.23, we conclude that $X^2(t) = \sum_{k=1}^{\infty} X_k^2(t) \in \mathcal{E}(\mathbb{R}, X, \mu)$ as a uniform limit of ergodic functions. Therefore $X(t) \in PAP(\mathbb{R}, X, \mu)$. Employing the same approach, we can show that $Y(t) \in PAP(\mathbb{R}, X, \mu)$. The unique integral solution of equation (5.58) is then μ-pseudo almost periodic. □

5.3.3 Applications

Neutral partial differential system

To apply the abstract results of the previous section, we consider the following model proposed in [273]:

$$\begin{cases} \frac{\partial}{\partial t}[w(t,\xi)-qw(t-r,\xi)] = \frac{\partial^2}{\partial \xi^2}[w(t,\xi)-qw(t-r,\xi)] \\ \qquad + \int_{-r}^{0} \gamma(\theta)w(t+\theta,\xi)d\theta + a(t)\psi(\xi) \\ \qquad \text{for } t \in \mathbb{R} \text{ and } \xi \in [0,\pi], \\ w(t,\xi)-qw(t-r,\xi) = 0 \quad \text{for } \xi = 0,\pi \text{ and } t \in \mathbb{R}, \end{cases} \tag{5.67}$$

where $\gamma : [-r,0] \to \mathbb{R}$, $\psi : [0,\pi] \to \mathbb{R}$ are continuous functions and $q \in (0,1)$. The function $a : \mathbb{R} \to \mathbb{R}$ is given by $a(t) = a_1(t) + a_2(t)$, where

$$a_1(t) = \sum_{n\geq 0} b_n(t)$$

and b_n are defined for every integer $n \geq 1$ by

$$b_n(t) = \sum_{k\in P_n} H(n^2(t-k)),$$

with $P_n = 3^n(2\mathbb{Z}+1)$ and $H \in C_0^\infty(\mathbb{R},\mathbb{R})$, with support in $(-\frac{1}{2},\frac{1}{2})$ such that

$$H \geq 0, \quad H(0) = 1, \quad \text{and} \quad \int_{-\frac{1}{2}}^{\frac{1}{2}} H(s)ds = 1.$$

The function a_2 is given by

$$a_2(t) = \begin{cases} k & \text{if } k \leq t \leq k + \frac{1}{k^3} \text{ with } k \in \mathbb{N}^*, \\ 0 & \text{otherwise.} \end{cases}$$

To rewrite equation (5.67) in the abstract form (5.58), we introduce $X = C([0,\pi],\mathbb{R})$ the space of continuous functions from $[0,\pi]$ to $\mathbb{R}$ endowed with the uniform norm topology and define the operator $A : D(A) \subset X \to X$ by

$$\begin{cases} D(A) = \{y \in C^2([0,\pi],\mathbb{R}) : y(0) = y(\pi) = 0\}, \\ Ay = y''. \end{cases} \tag{5.68}$$

Lemma 5.3.16 ([136]). *The operator A satisfies the Hille–Yosida condition on the space X, namely, $(0,\infty) \subset \rho(A)$ and*

$$|(\lambda I - A)^{-1}| \leq \frac{1}{\lambda} \quad \textit{for } \lambda > 0.$$

This lemma implies that condition **(H0)** is satisfied. Let A_0 be the part of the operator A in $\overline{D(A)}$. Then A_0 is given by

$$\begin{cases} D(A_0) = \{y \in C^2([0,\pi];\mathbb{R}) : y(0) = y(\pi) = y''(0) = y''(\pi) = 0\}, \\ A_0 y = y'' \quad \text{for } y \in D(A_0). \end{cases}$$

It is well known that A_0 generates a strongly continuous compact semigroup $(T_0(t))_{t\geq 0}$ on $\overline{D(A)}$. This implies that **(H1)** holds. On the other hand, we can see that

$$\overline{D(A)} = \{y \in X : y(0) = y(\pi) = 0\}.$$

Let us introduce the bounded linear operator $\mathcal{D} : C([-r,0],X) \to X$ defined by

$$\mathcal{D}(\varphi) = \varphi(0) - q\varphi(-r).$$

Since $0 < q < 1$, then $\mathcal{D}$ is stable and condition **(H2)** holds. Moreover, by the definitions of the operators A and $\mathcal{D}$, it follows that condition **(H3)** is satisfied. Let $L : C \to X$ be the operator defined by

$$L(\varphi)(\xi) = \int_{-r}^{0} \gamma(\theta)\varphi(\theta)(\xi)\mathrm{d}\theta \quad \text{for } \xi \in [0,\pi] \text{ and } \varphi \in C.$$

Then L is a bounded linear operator from C to X. Let $f : \mathbb{R} \to X$ be defined by

$$f(t)(\xi) = a(t)\psi(\xi) \quad \text{for } t \in \mathbb{R} \text{ and } \xi \in [0,\pi],$$

and let $u(t) = w(t,\cdot)$ for $t \in \mathbb{R}$. Then equation (5.67) takes the abstract form

$$\frac{\mathrm{d}}{\mathrm{d}t}\mathcal{D}(u_t) = A\mathcal{D}(u_t) + L(u_t) + f(t) \quad \text{for } t \in \mathbb{R}. \tag{5.69}$$

Proposition 5.3.17. *Assume the above conditions and that*

$$\int_{-r}^{0} |\gamma(\theta)|\mathrm{d}\theta < 1 - q. \tag{5.70}$$

Then equation (5.69) *has a unique pseudo almost periodic integral solution.*

Proof. Condition (5.70) implies that the semigroup solution of the following homogeneous linear equation:

$$\begin{cases} \frac{\mathrm{d}}{\mathrm{d}t}\mathcal{D}(v_t) = A\mathcal{D}(v_t) + L(v_t) \quad \text{for } t \geq 0, \\ v_0 = \varphi \end{cases}$$

has an exponential dichotomy (see [144, Proposition 5.2]). On the other hand, we have

$$f(t) = a_1(t)\psi + a_2(t)\psi,$$

with $\psi \in C([0,\pi],\mathbb{R})$ being the function in equation (5.67). We have $a_1 \in C^\infty(\mathbb{R},\mathbb{R})$ and $a_1 \notin AP(\mathbb{R},\mathbb{R})$ because it is not bounded, but $a_1 \in SAP^1(\mathbb{R},\mathbb{R})$ (see [264]). Moreover, $a_2(\cdot)$ is not λ-ergodic since it is not bounded, where λ is the Lebesgue measure. However, we have

$$
\begin{aligned}
\frac{1}{2r}\int_{-r}^{r}\int_{t}^{t+1}|a_2(s)|\,\mathrm{d}s\,\mathrm{d}t &= \frac{1}{2r}\int_{-r}^{r}\int_{0}^{1}|a_2(t+s)|\,\mathrm{d}s\,\mathrm{d}t \\
&= \frac{1}{2r}\int_{0}^{1}\int_{-r+s}^{r+s}|a_2(t)|\,\mathrm{d}t\,\mathrm{d}s \\
&\leq \frac{1}{2r}\int_{0}^{1}\int_{0}^{\infty}|a_2(t)|\,\mathrm{d}t\,\mathrm{d}s \\
&= \frac{1}{2r}\sum_{k=1}^{\infty}\int_{k}^{k+\frac{1}{k^3}}|a_2(t)|\,\mathrm{d}t \\
&= \frac{1}{2r}\sum_{k=1}^{\infty}\frac{1}{k^2} \to 0, \quad \text{as } r \to \infty.
\end{aligned}
$$

Thus $a_2 \in \mathcal{E}^1(\mathbb{R},\mathbb{R},\lambda)$. Therefore $f \in PAP^1(\mathbb{R},X,\lambda)$. By Theorem 5.3.15, we deduce that equation (5.69) has a unique pseudo almost periodic integral solution. □

Heat equation with discrete delays

Now we consider the following model with discrete delay:

$$
\begin{cases}
\frac{\partial}{\partial t}v(t,\xi) = \frac{\partial^2}{\partial \xi^2}v(t,\xi) + av(t,\xi) + bv(t-r,\xi) + a(t)\psi(\xi) \\
\quad \text{for } t \in \mathbb{R} \text{ and } \xi \in [0,\pi], \\
v(t,0) = v(t,\pi) = 0 \quad \text{for } t \in \mathbb{R},
\end{cases}
\tag{5.71}
$$

where $a, b \in \mathbb{R}$ and the functions $a(\cdot)$, $\psi(\cdot)$ are defined as in (5.67). The system (5.71) takes the abstract form

$$
\frac{\mathrm{d}}{\mathrm{d}t}u(t) = Au(t) + L(u_t) + f(t) \quad \text{for } t \in \mathbb{R},
\tag{5.72}
$$

where

$$
L(\varphi)(\xi) := a\varphi(0)(\xi) + b\varphi(-r)(\xi) \quad \text{for } \xi \in [0,\pi] \text{ and } \varphi \in C,
$$

and A is the same operator defined by (5.68). Equation (5.72) is a special case of equation (5.58) where the operator $\mathcal{D}$ is defined for each $\varphi \in C$ by

$$\mathcal{D}(\varphi) = \varphi(0).$$

To ensure the existence and uniqueness of an almost periodic integral solution of (5.71), we give a characterization to the exponential dichotomy of the semigroup solution $(T(t))_{t\geq 0}$ of the following homogeneous system:

$$\begin{cases} \frac{d}{dt}v(t) = Av(t) + L(v_t) & \text{for } t \geq 0, \\ v_0 = \varphi. \end{cases} \tag{5.73}$$

In order to study the distribution of the spectrum of A_T, the generator of $(T(t))_{t\geq 0}$, around the imaginary axis, let us define the following sets which contain needed information about the coefficients a, b, and r in the system (5.71):

$$I_1(a,b,r) = \{n \in \mathbb{N}^* : a - |b| \leq n^2 \leq a + |b| \text{ such that } n^2 = a + b\cos(r\sqrt{b^2 - (n^2 - a)^2})\}$$

and

$$I_2(a,b,r) = \{n \in \mathbb{N}^* : a - |b| \leq n^2 \leq a + |b| \text{ such that}$$
$$\sqrt{b^2 - (n^2 - a)^2} = -b\sin(r\sqrt{b^2 - (n^2 - a)^2})\}.$$

Proposition 5.3.18. *$\sigma(A_T) \cap i\mathbb{R} = \emptyset$ if and only if $I_1(a,b,r) \cap I_2(a,b,r) = \emptyset$.*

Proof. Let $\lambda \in \sigma(A_T)$, then from [146], there exists $x \in D(A)$, $x \neq 0$ such that $\Delta(\lambda)x = 0$, which implies that

$$\lambda x - Ax - ax - be^{-r\lambda}x = 0$$

and

$$\lambda - a - be^{-r\lambda} \in \sigma_p(A).$$

On the other hand, the point spectrum $\sigma_p(A)$ of A is given by

$$\sigma_p(A) = \{-n^2 : n \in \mathbb{N}^*\}.$$

Consequently, $\lambda \in \sigma(A_T)$ if and only if

$$\lambda = -n^2 + a + be^{-r\lambda} \quad \text{for some } n \geq 1.$$

By taking the real and imaginary parts in the above formula, we obtain

$$\begin{cases} \mathrm{Re}(\lambda) = a - n^2 + be^{-r\,\mathrm{Re}(\lambda)}\cos(r\,\mathrm{Im}(\lambda)), \\ \mathrm{Im}(\lambda) = -be^{-r\,\mathrm{Re}(\lambda)}\sin(r\,\mathrm{Im}(\lambda)). \end{cases} \tag{5.74}$$

Suppose that $\sigma(A_T) \cap i\mathbb{R} \neq \emptyset$, and let $\lambda \in \sigma(A_T)$ be such that $\mathrm{Re}(\lambda) = 0$. Then there exists $n \in \mathbb{N}^*$ such that

$$\begin{cases} 0 = a - n^2 + b\cos(r\,\mathrm{Im}(\lambda)), \\ \mathrm{Im}(\lambda) = -b\sin(r\,\mathrm{Im}(\lambda)). \end{cases} \tag{5.75}$$

It follows that

$$(n^2 - a)^2 + \mathrm{Im}(\lambda)^2 = b^2$$

and

$$\mathrm{Im}(\lambda) = \pm\sqrt{b^2 - (n^2 - a)^2}.$$

We substitute into (5.75) and deduce that

$$\begin{cases} n^2 = a + b\cos(r\sqrt{b^2 - (n^2 - a)^2}), \\ \sqrt{b^2 - (n^2 - a)^2} = -b\sin(r\sqrt{b^2 - (n^2 - a)^2}). \end{cases}$$

Therefore $I_1(a,b,r) \cap I_2(a,b,r) \neq \emptyset$.

Now for the other implication, suppose that $I_1(a,b,r) \cap I_2(a,b,r) \neq \emptyset$, and let $n \in I_1(a,b,r) \cap I_2(a,b,r)$. Then by (5.74), we have

$$\lambda := \pm i\sqrt{b^2 - (n^2 - a)^2} \in \sigma(A_T),$$

which means that $\sigma(A_T) \cap i\mathbb{R} \neq \emptyset$. □

Remark. If $\sqrt{a+b} \in \mathbb{N}^*$, then $\sqrt{a+b} \in I_1(a,b,r) \cap I_2(a,b,r)$.

Corollary 5.3.19. *Under the condition $I_1(a,b,r) \cap I_2(a,b,r) = \emptyset$, equation (5.72) has a unique pseudo almost periodic integral solution.*

Heat equation with distributed delay

Now, we consider the following model with continuous delay:

$$\begin{cases} \frac{\partial}{\partial t}v(t,\xi) = \frac{\partial^2}{\partial \xi^2}v(t,\xi) + a\int_{-r}^{0} v(t+\theta,\xi)\mathrm{d}\theta + a(t)\psi(\xi) \\ \quad \text{for } t \in \mathbb{R} \text{ and } \xi \in [0,\pi], \\ v(t,0) = v(t,\pi) = 0 \quad \text{for } t \in \mathbb{R}, \end{cases} \tag{5.76}$$

where $a, b \in \mathbb{R}$ and the functions $a(\cdot)$, $\psi(\cdot)$ are defined as (5.67). The system (5.76) takes the abstract form

$$\frac{\mathrm{d}}{\mathrm{d}t}u(t) = Au(t) + L(u_t) + f(t) \quad \text{for } t \in \mathbb{R}, \tag{5.77}$$

where

$$L(\varphi)(\xi) := a\int_{-r}^{0}\varphi(\theta)(\xi)\mathrm{d}\theta \quad \text{for } \xi \in [0,\pi] \text{ and } \varphi \in C$$

and A is the same operator defined by (5.68). Let A_T be the generator of $(T(t))_{t\geq 0}$, the semigroup solution of the following homogeneous system:

$$\begin{cases} \frac{\mathrm{d}}{\mathrm{d}t}v(t) = Av(t) + L(v_t) & \text{for } t \geq 0,\\ v_0 = \varphi. \end{cases}$$

Consider the following sets:

$$I_1(a,r) := \{n \in \mathbb{N}^* : n^4 < -2a \text{ such that } n^2\sqrt{-2a-n^4} - a\sin(r\sqrt{-2a-n^4}) = 0\},$$
$$I_2(a,r) := \{n \in \mathbb{N}^* : n^4 < -2a \text{ such that } a + n^4 + a\cos(r\sqrt{-2a-n^4}) = 0\},$$

and

$$S := \{n^2 : n \in \mathbb{N}^*\}.$$

Proposition 5.3.20. $\sigma(A_T) \cap i\mathbb{R} = \emptyset$ *if and only if*

$$I_1(a,r) \cap I_2(a,r) = \emptyset \quad \text{and} \quad ar \notin S. \tag{5.78}$$

Proof. Let $\lambda \in \sigma(A_T)$, then from [146], there exists $x \in D(A)$, $x \neq 0$ such that $\Delta(\lambda)x = 0$, which implies that

$$Ax = \left(\lambda - a\int_{-r}^{0} e^{\lambda\theta}\mathrm{d}\theta\right)x.$$

Consequently, $\lambda \in \sigma(A_T)$ if and only if

$$\lambda = -n^2 + a\int_{-r}^{0} e^{\lambda\theta}\mathrm{d}\theta \quad \text{for some } n \geq 1.$$

By taking the real and imaginary parts in the above formula, we obtain

$$\begin{cases} \mathrm{Re}(\lambda) = -n^2 + a\int_{-r}^{0} e^{\mathrm{Re}(\lambda)\theta}\cos(\mathrm{Im}(\lambda)\theta)\mathrm{d}\theta,\\ \mathrm{Im}(\lambda) = a\int_{-r}^{0} e^{\mathrm{Re}(\lambda)\theta}\sin(\mathrm{Im}(\lambda)\theta)\mathrm{d}\theta. \end{cases}$$

Let us show that $\sigma(A_T) \cap i\mathbb{R} \neq \emptyset$ if and only if $I_1(a,r) \cap I_2(a,r) \neq \emptyset$ or $ar \in S$. Suppose that $\sigma(A_T) \cap i\mathbb{R} \neq \emptyset$, and let $\lambda \in \sigma(A_T)$ be such that $\mathrm{Re}(\lambda) = 0$. Then there exists $n \in \mathbb{N}^*$ such that

$$\begin{cases} 0 = -n^2 + a\int_{-r}^{0}\cos(\mathrm{Im}(\lambda)\theta)d\theta, \\ \mathrm{Im}(\lambda) = a\int_{-r}^{0}\sin(\mathrm{Im}(\lambda)\theta)d\theta. \end{cases} \tag{5.79}$$

If $\mathrm{Im}(\lambda) = 0$, then by the first equation of (5.79), $n^2 = a\,r$ and $ar \in S$.
If $\mathrm{Im}(\lambda) \neq 0$, then from (5.79) we have

$$\begin{cases} n^2 = a\frac{\sin(r\,\mathrm{Im}(\lambda))}{\mathrm{Im}(\lambda)}, \\ \mathrm{Im}(\lambda) = a\frac{\cos(r\,\mathrm{Im}(\lambda))-1}{\mathrm{Im}(\lambda)}, \end{cases} \tag{5.80}$$

and

$$\mathrm{Im}(\lambda)^2 + 2a + n^4 = 0.$$

It follows that $\mathrm{Im}(\lambda) = \pm\sqrt{-2a - n^4}$. We substitute into (5.80) and obtain

$$\begin{cases} n^2\sqrt{-2a - n^4} - a\sin(r\sqrt{-2a - n^4}) = 0, \\ a + n^4 + a\cos(r\sqrt{-2a - n^4}) = 0. \end{cases}$$

Therefore $I_1(a, r) \cap I_2(a, r) \neq \emptyset$.

Now for the other implication, suppose that $ar \in S$. Let $n \in \mathbb{N}^*$ such that $n^2 = ar$. Then by equation (5.79), we can see that $0 \in \sigma(A_T)$. Now suppose that $I_1(a, r) \cap I_2(a, r) \neq \emptyset$, and let $n \in I_1(a, r) \cap I_2(a, r)$. Since $n^4 \neq -2a$, we have

$$\begin{cases} n^2 = a\frac{\sin(r\sqrt{-2a-n^4})}{\sqrt{-2a-n^4}}, \\ \sqrt{-2a - n^4} = a\frac{\cos(r\sqrt{-2a-n^4})-1}{\sqrt{-2a-n^4}}, \end{cases}$$

and

$$\begin{cases} n^2 = a\int_{-r}^{0}\cos(\sqrt{-2a - n^4}\theta)d\theta, \\ \sqrt{-2a - n^4} = a\int_{-r}^{0}\sin(\sqrt{-2a - n^4}\theta)d\theta. \end{cases}$$

Therefore by equation (5.79), we deduce that $\pm i\sqrt{-2a - n^4} \in \sigma(A_T)$. □

We note that if $a > 0$, then $I_1(a, r) \cap I_2(a, r) = \emptyset$. Thus, a sufficient condition to the exponential dichotomy of $(T(t))_{t\geq 0}$ which is easier to check is given by the following proposition.

Proposition 5.3.21. *If $a > 0$ and $ar \notin S$, then $\sigma(A_T) \cap i\mathbb{R} = \emptyset$.*

Corollary 5.3.22. *Let a and r satisfy the condition* (5.78). *Then equation* (5.77) *has a unique pseudo almost periodic integral solution.*

5.4 Bohr–Neugebauer property for neutral partial functional differential equations

The purpose of this section is to study the behavior of bounded integral solutions of the following class of partial neutral functional differential equations:

$$\frac{\mathrm{d}}{\mathrm{d}t}\mathcal{D}u_t = A\mathcal{D}u_t + L(u_t) + f(t) \quad \text{for } t \in \mathbb{R}, \tag{5.81}$$

when the forcing term $f : \mathbb{R} \to X$ is Stepanov almost periodic and the operator A is not necessarily densely defined but satisfies the following Hille–Yosida condition:
(H0) There exist $M \geq 1$ and $\omega \in \mathbb{R}$ such that $(\omega, +\infty) \subset \rho(A)$ and

$$|R(\lambda, A)^n| \leq \frac{M}{(\lambda - \omega)^n} \quad \text{for } n \in \mathbb{N} \text{ and } \lambda > \omega,$$

where $\rho(A)$ is the resolvent set of A and $R(\lambda, A) := (\lambda I - A)^{-1}$ for $\lambda \in \rho(A)$.

Operator L is bounded and linear from C to X, where $C := C([-r, 0], X)$ denotes the space of continuous functions from $[-r, 0]$ to X endowed with the uniform norm topology. For every $t \in \mathbb{R}$, the history function $u_t \in C$ is defined by

$$u_t(\theta) := u(t + \theta) \quad \text{for } \theta \in [-r, 0].$$

In this section, we investigate the nature of bounded integral solutions of equation (5.81). More specifically, we prove, under a compactness condition, that all bounded integral solutions of equation (5.81) on $\mathbb{R}$ are almost periodic, when the forcing term f is only S^1-almost periodic. Moreover, we show that the existence of a bounded integral solution on $\mathbb{R}^+$ is sufficient to guarantee the existence of an almost periodic integral solution.

The principal working tools in this chapter are the variation of the constants formula and a reduction principle similar to that developed in [7].

5.4.1 Variation of constants formula and the reduction principle

In the sequel, we assume that the operator A satisfies the Hille–Yosida condition **(H0)**. To equation (5.81), we associate the following Cauchy problem:

$$\begin{cases} \frac{\mathrm{d}}{\mathrm{d}t}\mathcal{D}u_t = A\mathcal{D}u_t + L(u_t) + f(t) & \text{for } t \geq \sigma, \\ u_\sigma = \varphi \in C. \end{cases} \tag{5.82}$$

Definition 5.4.1 ([7]). A continuous function $u : [-r + \sigma, +\infty) \to X$ is called an *integral solution* of equation (5.82) if
(i) $\int_\sigma^t \mathcal{D}u_s \mathrm{d}s \in D(A)$ for $t \geq \sigma$,

(ii) $\mathcal{D}u_t = \mathcal{D}\varphi + A\int_\sigma^t \mathcal{D}u_s ds + \int_\sigma^t [L(u_s) + f(s)]ds$ for $t \geq \sigma$,
(iii) $u_\sigma = \varphi$.

Let us introduce the part A_0 of the operator A in $\overline{D(A)}$ defined by

$$\begin{cases} D(A_0) = \{x \in D(A) : Ax \in \overline{D(A)}\}, \\ A_0 x = Ax \quad \text{for } x \in D(A_0). \end{cases}$$

We have the following lemma.

Lemma 5.4.2 ([265]). *Assume that* **(H0)** *holds. Then, A_0 generates a strongly continuous semigroup $(T_0(t))_{t\geq 0}$ on $\overline{D(A)}$.*

For the existence of integral solutions, one has the following result.

Theorem 5.4.3 ([5]). *Assume that* **(H0)** *holds. Then, for all $\varphi \in C$ such that $\mathcal{D}\varphi \in \overline{D(A)}$, equation* (5.82) *has a unique integral solution u on $[-r + \sigma, +\infty)$. Moreover, u is given by*

$$\begin{cases} \mathcal{D}u_t = T_0(t-\sigma)\mathcal{D}\varphi + \lim_{\lambda\to+\infty} \int_\sigma^t T_0(t-s)B_\lambda[L(u_s) + f(s)]ds \quad \text{for } t \geq \sigma, \\ u_\sigma = \varphi, \end{cases}$$

where $B_\lambda = \lambda R(\lambda, A)$ for $\lambda > \omega$.

In the sequel of this chapter, for simplicity, integral solutions are called solutions. Also $u(\cdot, \sigma, \varphi, f)$ denotes the solution of equation (5.82). The phase space C_0 of equation (5.82) is given by

$$C_0 = \{\varphi \in C : \mathcal{D}\varphi \in \overline{D(A)}\}.$$

For each $t \geq 0$, we define the linear operator $U(t)$ on C_0 by

$$U(t)\varphi = u_t(\cdot, 0, \varphi, 0),$$

where $u(\cdot, 0, \varphi, 0)$ is the solution of the following homogeneous equation:

$$\begin{cases} \frac{d}{dt}u(t) = Au(t) + L(v_t) \quad \text{for } t \geq 0, \\ u_0 = \varphi. \end{cases}$$

We have the following result:

Proposition 5.4.4 ([7]). *Assume that* **(H0)** *holds. Then, $(U(t))_{t\geq 0}$ is a strongly continuous semigroup on C_0. Moreover, the operator $\mathcal{A}_U$ defined on C_0 by*

$$\begin{cases} D(\mathcal{A}_U) = \{\varphi \in C^1([-r, 0], X) : \mathcal{D}\varphi \in D(A),\ \mathcal{D}\varphi' \in \overline{D(A)} \text{ and } \mathcal{D}\varphi' = A\mathcal{D}\varphi + L(\varphi)\}, \\ \mathcal{A}_U\varphi = \varphi' \end{cases}$$

is the infinitesimal generator of $(U(t))_{t\geq 0}$ on C_0.

In order to give the variation of constants formula associated to equation (5.82), we need to extend the semigroup $(U(t))_{t\geq 0}$ to the space $C_0 \oplus \langle X_0 \rangle$, where $\langle X_0 \rangle$ is the space defined by

$$\langle X_0 \rangle = \{X_0 y : y \in X\}$$

and the function $X_0 y$ is given, for each $y \in X$, by

$$(X_0 y)(\theta) = \begin{cases} 0 & \text{if } \theta \in [-r, 0), \\ y & \text{if } \theta = 0. \end{cases}$$

The space $C_0 \oplus \langle X_0 \rangle$ equipped with the norm $|\varphi + X_0 y| = |\varphi| + |y|$, for $(\varphi, y) \in C_0 \times X$, is a Banach space. Consider the extension $\widetilde{\mathcal{A}_U}$ of the operator $\mathcal{A}_U$ on $C_0 \oplus \langle X_0 \rangle$ defined by

$$\begin{cases} D(\widetilde{\mathcal{A}_U}) = \{\varphi \in C^1([-r, 0], X) : \mathcal{D}\varphi \in D(A) \text{ and } \mathcal{D}\varphi' \in \overline{D(A)}\}, \\ \widetilde{\mathcal{A}_U}\varphi = \varphi' + X_0(A\mathcal{D}\varphi + L(\varphi) - \mathcal{D}\varphi'). \end{cases}$$

In order to compute the resolvent operator $R(\lambda, \widetilde{\mathcal{A}_U})$, we need to make the following assumption:
(H1) $\mathcal{D}(e^{\lambda \cdot} c) \in D(A)$, for all $c \in D(A)$ and all complex λ, where $e^{\lambda \cdot} c \in C$ is defined by

$$(e^{\lambda \cdot} c)(\theta) = e^{\lambda\theta} c \quad \text{for } \theta \in [-r, 0].$$

Lemma 5.4.5 ([7, Theorem 13]). *Assume that* **(H0)** *and* **(H1)** *hold. Then,* $\widetilde{\mathcal{A}_U}$ *satisfies the Hille–Yosida condition on* $C_0 \oplus \langle X_0 \rangle$*, that is, there exist* $\widetilde{M} \geq 0$ *and* $\widetilde{\omega} \in \mathbb{R}$ *such that* $(\widetilde{\omega}, +\infty) \subset \rho(\widetilde{\mathcal{A}_U})$ *and*

$$|R(\lambda, \widetilde{\mathcal{A}_U})^n| \leq \frac{\widetilde{M}}{(\lambda - \widetilde{\omega})^n} \quad \textit{for } n \in \mathbb{N} \textit{ and } \lambda > \widetilde{\omega}.$$

Now, we can state the variation of constants formula associated to equation (5.82).

Theorem 5.4.6 ([7, Theorem 16]). *Assume that* **(H0)** *and* **(H1)** *hold. Then, for all* $\varphi \in C_0$*, the solution* $u(\cdot, \sigma, \varphi, f)$ *of equation* (5.82) *is given by the following variation of constants formula:*

$$u_t(\cdot, \sigma, \varphi, f) = U(t - \sigma)\varphi + \lim_{n \to +\infty} \int_\sigma^t U(t - s)\widetilde{B}_n(X_0 f(s))ds \quad \textit{for } t \geq \sigma,$$

where $\widetilde{B}_n = nR(n, \widetilde{\mathcal{A}_U})$ *for* $n > \widetilde{\omega}$.

The following definition is essential to describe the asymptotic behavior of the semigroup $(U(t))_{t\geq 0}$.

Definition 5.4.7 ([166, Definition 3.1, p. 275]). The operator $\mathcal{D}$ is said to be stable if there exist positive constants η and μ such that the solution of the homogeneous difference equation

$$\begin{cases} \mathcal{D}u_t = 0 & \text{for } t \geq 0, \\ u_0 = \phi, \end{cases}$$

where $\phi \in \{\psi \in C : \mathcal{D}\psi = 0\}$, satisfies

$$|u_t(\cdot,\phi)| \leq \mu e^{-\eta t}|\phi| \quad \text{for } t \geq 0.$$

Example 5.4.8 ([166, 273]). The operator $\mathcal{D}$ defined by

$$\mathcal{D}\phi = \phi(0) - q\phi(-r)$$

is stable if and only if $|q| < 1$.

The following assumptions play a crucial role to get the reduction principle:

(H2) The operator $T_0(t)$ is compact on $\overline{D(A)}$ for every $t > 0$.

(H3) The operator $\mathcal{D}$ is stable.

We have the following fundamental result on the semigroup $(U(t))_{t\geq 0}$.

Theorem 5.4.9 ([7, Lemma 10]). *Assume that* **(H0)**, **(H2)**, *and* **(H3)** *hold. Then, the semigroup* $(U(t))_{t\geq 0}$ *is decomposed on* C_0 *as follows:*

$$U(t) = U_1(t) + U_2(t) \quad \textit{for } t \geq 0,$$

where $(U_1(t))_{t\geq 0}$ *is an exponentially stable semigroup on* C_0, *which means that there are positive constants* a_0 *and* N_0 *such that*

$$|U_1(t)\phi| \leq N_0 e^{-a_0 t}|\phi| \quad \textit{for } t \geq 0 \textit{ and } \phi \in C_0,$$

and $U_2(t)$ *is compact for every* $t > 0$.

By Theorem 5.4.9, we have, for $t > 0$,

$$|U(t)|_\alpha \leq |U_1(t)|_\alpha + |U_2(t)|_\alpha \leq |U_1(t)| \leq N_0 e^{-a_0 t}.$$

We deduce that $\omega_{\text{ess}}(U) \leq -a_0 < 0$. Consequently, by [136, Theorem 5.3.7, p. 333], we get the following spectral decomposition.

Theorem 5.4.10. *Assume that* **(H0)**, **(H2)**, *and* **(H3)** *hold. Then,* C_0 *is decomposed as follows:*

$$C_0 = S \oplus V, \tag{5.83}$$

where S is U-invariant and there are positive constants a and N such that

$$|U(t)\varphi| \leq Ne^{-at}|\varphi| \quad \textit{for each } \varphi \in S \textit{ and } t \geq 0.$$

Furthermore, V is a finite-dimensional space and the restriction of the semigroup $(U(t))_{t\geq 0}$ *to V becomes a group.*

In the sequel, $U^s(t)$ and $U^v(t)$ denote the restrictions of $U(t)$ on S and V, respectively, which correspond to the above decomposition. Let $d := \dim(V)$ with a vector basis $\Phi = \{\varphi_1, \dots, \varphi_d\}$. Then, there exist d-elements $\{\psi_1, \dots, \psi_d\}$ in C_0^* such that

$$\begin{cases} \langle \psi_i, \varphi_j \rangle = \delta_{ij}, \\ \langle \psi_i, \varphi \rangle = 0 \quad \text{for all } \varphi \in S \text{ and } i \in \{1, \dots, d\}, \end{cases} \tag{5.84}$$

where $\langle \cdot, \cdot \rangle$ denotes the duality pairing between C_0^* and C_0, and

$$\delta_{ij} = \begin{cases} 1 & \text{if } i = j, \\ 0 & \text{if } i \neq j. \end{cases}$$

Letting $\Psi = \text{col}\{\psi_1, \dots \psi_2\}$, $\langle \Psi, \Phi \rangle$ is a $(d \times d)$-matrix, where the (i,j)-component is $\langle \psi_i, \varphi_j \rangle$. Denote by Π^s and Π^v the projections on S and V, respectively. For each $\varphi \in C_0$, we have

$$\Pi^v \varphi = \Phi \langle \Psi, \varphi \rangle.$$

In fact, for $\varphi \in C_0$, we have $\varphi = \Pi^s \varphi + \Pi^v \varphi$ with $\Pi^v \varphi = \sum_{i=1}^d a_i \varphi_i$ and $a_i \in \mathbb{R}$. By (5.84), we conclude that

$$a_i = \langle \psi_i, \varphi \rangle.$$

Hence,

$$\Pi^v \varphi = \sum_{i=1}^d \langle \psi_i, \varphi \rangle \varphi_i = \Phi \langle \Psi, \varphi \rangle.$$

Because $(U^v(t))_{t\geq 0}$ is a group on V, there exists a $(d \times d)$-matrix G such that

$$U^v(t)\Phi = \Phi e^{tG} \quad \text{for } t \in \mathbb{R}.$$

For $n, n_0 \in \mathbb{N}$ such that $n \geq n_0 \geq \widetilde{\omega}$ and $i \in \{1, \dots, d\}$, we define the linear operator $x_{i,n}^*$ by

$$x_{i,n}^*(a) = \langle \psi_i, \widetilde{B}_n X_0 a \rangle \quad \text{for } a \in X.$$

Because $|\widetilde{B}_n| \le \frac{n}{n-\widetilde{\omega}}\widetilde{M}$ for any $n \ge n_0$, $x^*_{i,n}$ is a bounded linear operator from X to $\mathbb{R}$ such that

$$|x^*_{i,n}| \le \frac{n}{n-\widetilde{\omega}}\widetilde{M}|\psi_i| \quad \text{for any } n \ge n_0.$$

Define the d-column vector $x^*_n = \operatorname{col}(x^*_{1,n}, \dots, x^*_{d,n})$. Then, one can see that

$$\langle x^*_n, a\rangle = \langle \Psi, \widetilde{B}_n X_0 a\rangle \quad \text{for } a \in X,$$

with

$$\langle x^*_n, a\rangle_i = \langle \psi_i, \widetilde{B}_n X_0 a\rangle \quad \text{for } i = 1, \dots, d \text{ and } a \in X.$$

Consequently, we have

$$\sup_{n \ge n_0} |x^*_n| < +\infty,$$

which implies that $(x^*_n)_{n \ge n_0}$ is a bounded sequence in $\mathcal{L}(X, \mathbb{R}^d)$. We recall the following important results.

Theorem 5.4.11 ([2]). *Assume that* (**H0**), (**H1**), (**H2**), *and* (**H3**) *hold. Then, there exists* $x^* \in \mathcal{L}(X, \mathbb{R}^d)$ *such that* $(x^*_n)_{n \ge n_0}$ *converges weakly to* x^* *in the sense that*

$$\langle x^*_n, x\rangle \to \langle x^*, x\rangle \quad \text{as } n \to +\infty \text{ for all } x \in X.$$

Theorem 5.4.12 ([2]). *Assume that* (**H0**), (**H1**), (**H2**), *and* (**H3**) *hold,* f *is continuous, and let* u *be a solution of equation* (5.81) *on* $\mathbb{R}$. *Then, the function* z *defined by* $z(t) := \langle \Psi, u_t\rangle$ *is a solution of the ordinary differential equation*

$$\frac{\mathrm{d}}{\mathrm{d}t} z(t) = Gz(t) + \langle x^*, f(t)\rangle \quad \text{for } t \in \mathbb{R}. \tag{5.85}$$

Conversely, if z *is a solution of equation* (5.85) *on* $\mathbb{R}$ *and if, in addition,* f *is bounded on* $\mathbb{R}$, *then the function* u *given by*

$$u(t) := \left[\Phi z(t) + \lim_{n \to +\infty} \int_{-\infty}^{t} U^s(t-s)\Pi^s(\widetilde{B}_n X_0 f(s))\mathrm{d}s\right](0) \quad \text{for } t \in \mathbb{R},$$

is a solution of equation (5.81) *on* $\mathbb{R}$.

We obtain the same result in Theorem 5.4.12 if we weaken the boundedness assumption of the function f.

Theorem 5.4.13. *Assume that* (H0), (H1), (H2), *and* (H3) *hold, f is locally integrable, and let u be a solution of equation* (5.81) *on $\mathbb{R}$. Then, the function z defined by $z(t) := \langle \Psi, u_t \rangle$ is given by*

$$z(t) = e^{tG} z(0) + \int_0^t e^{(t-s)G} \langle x^*, f(s) \rangle ds \quad \text{for } t \in \mathbb{R}. \tag{5.86}$$

Conversely, if z satisfies equation (5.86) *on $\mathbb{R}$ and if, in addition, f satisfies*

$$\sup_{t \in \mathbb{R}} \int_t^{t+1} |f(s)| ds < +\infty,$$

then the function u given by

$$u(t) := \left[\Phi z(t) + \lim_{n \to \infty} \int_{-\infty}^t U^s(t-s) \Pi^s (\widetilde{B}_n X_0 f(s)) ds \right](0) \quad \text{for } t \in \mathbb{R},$$

is a solution of equation (5.81) *on $\mathbb{R}$.*

Proof. The proof is similar to that given for Theorem 5.4.12. We only have to prove that the limit

$$\lim_{n \to +\infty} \int_{-\infty}^t U^s(t-s) \Pi^s (\widetilde{B}_n X_0 f(s)) ds$$

exists in C_0. For $t \in \mathbb{R}$ and for n sufficiently large, we have

$$\begin{aligned} \left| \int_{-\infty}^t U^s(t-s) \Pi^s (\widetilde{B}_n X_0 f(s)) ds \right| &\le 2\widetilde{M} N |\Pi^s| \int_{-\infty}^t e^{-a(t-s)} |f(s)| ds \\ &\le 2\widetilde{M} N |\Pi^s| \sum_{k=1}^{+\infty} \left(\int_{t-k}^{t-k+1} e^{-a(t-s)} |f(s)| ds \right) \\ &\le 2\widetilde{M} N |\Pi^s| \sum_{k=1}^{+\infty} \left(e^{-a(k-1)} \int_{t-k}^{t-k+1} |f(s)| ds \right) \le K, \end{aligned}$$

where $K = 2\widetilde{M} N |\Pi^s| \sup_{t \in \mathbb{R}} (\int_t^{t+1} |f(s)| ds) \frac{1}{1-e^{-a}}$. Let

$$H(n, s, t) := U^s(t-s) \Pi^s (\widetilde{B}_n X_0 f(s)) \quad \text{for } n \in \mathbb{N} \text{ and } s \le t.$$

For n and m sufficiently large and $\sigma \le t$, we have

$$\begin{aligned}\left|\int_{-\infty}^{t} H(n,s,t)ds - \int_{-\infty}^{t} H(m,s,t)ds\right| &\le \left|\int_{-\infty}^{\sigma} H(n,s,t)ds\right| + \left|\int_{-\infty}^{\sigma} H(m,s,t)ds\right| \\ &\quad + \left|\int_{\sigma}^{t} H(n,s,t)ds - \int_{\sigma}^{t} H(m,s,t)ds\right| \\ &\le 2Ke^{-c(t-\sigma)} + \left|\int_{\sigma}^{t} H(n,s,t)ds - \int_{\sigma}^{t} H(m,s,t)ds\right|.\end{aligned}$$

Because $\lim_{n\to+\infty} \int_{\sigma}^{t} H(n,s,t)ds$ exists, it follows that

$$\limsup_{n,m\to+\infty}\left|\int_{-\infty}^{t} H(n,s,t)ds - \int_{-\infty}^{t} H(m,s,t)ds\right| \le 2Ke^{-c(t-\sigma)}.$$

Letting $\sigma \to -\infty$, we get

$$\limsup_{n,m\to+\infty}\left|\int_{-\infty}^{t} H(n,s,t)ds - \int_{-\infty}^{t} H(m,s,t)ds\right| = 0.$$

Thus, from the completeness of the phase space C_0, we deduce that the limit

$$\lim_{n\to+\infty}\int_{-\infty}^{t} H(n,s,t)ds$$

exists in C_0. □

5.4.2 Behavior of bounded solutions of equation (5.81)

In this section, we investigate the nature of bounded integral solutions of the neutral functional differential equation (5.81).

The following theorem is of Bohr–Neugebauer type.

Theorem 5.4.14. *Assume that* **(H0)**–**(H3)** *hold and* $f : \mathbb{R} \to X$ *is* S^1*-almost periodic. Then, every bounded integral solution of equation* (5.81) *on* $\mathbb{R}$ *is almost periodic.*

Proof. Let x be a bounded integral solution of equation (5.81) on $\mathbb{R}$. Using the spectral decomposition (5.83), we have, for each $t \in \mathbb{R}$,

$$x_t = \Pi^{\nu} x_t + \Pi^{s} x_t. \tag{5.87}$$

On the one hand, we have, for $t \geq \sigma$,

$$\Pi^s x_t = U^s(t-\sigma)\Pi^s x_\sigma + \lim_{n\to+\infty} \int_\sigma^t U^s(t-s)\Pi^s(\widetilde{B}_n(X_0 f(s)))ds. \tag{5.88}$$

Since $t \mapsto x_t$ is bounded on $\mathbb{R}$ and $U(t)$ is exponentially stable in the subspace S, letting $\sigma \to -\infty$ in (5.88), we obtain, for all $t \in \mathbb{R}$,

$$\Pi^s x_t = \lim_{n\to+\infty} \int_{-\infty}^t U^s(t-s)\Pi^s(\widetilde{B}_n(X_0 f(s)))ds.$$

In fact, for each fixed $t \in \mathbb{R}$, we have, for all $\sigma \leq t$,

$$|U^s(t-\sigma)\Pi^s x_\sigma| \leq e^{-a(t-\sigma)}|\Pi^s| \sup_{s\in\mathbb{R}} |x_s| \to 0,$$

as $\sigma \to -\infty$, and

$$\begin{aligned}
&\left|\lim_{n\to+\infty} \int_\sigma^t U^s(t-s)\Pi^s(\widetilde{B}_n(X_0 f(s)))ds - \lim_{n\to+\infty} \int_{-\infty}^t U^s(t-s)\Pi^s(\widetilde{B}_n(X_0 f(s)))ds\right| \\
&= \left|\lim_{n\to+\infty} \int_{-\infty}^\sigma U^s(t-s)\Pi^s(\widetilde{B}_n(X_0 f(s)))ds\right| \\
&\leq \widetilde{M}N|\Pi^s| \sum_{k=1}^{+\infty} \int_{\sigma-k}^{\sigma-k+1} e^{-a(t-s)}|f(s)|ds \\
&\leq \widetilde{M}N|\Pi^s||f|_{BS^1} \frac{e^{-a(t-\sigma)}}{1-e^{-a}} \to 0, \quad \text{as } \sigma \to -\infty.
\end{aligned}$$

Using a proof similar to that in [137, Theorem 30], one can prove that $t \mapsto \Pi^s x_t$ is almost periodic.

On the other hand, for each $t \in \mathbb{R}$,

$$\Pi^v x_t = \Phi\langle\Psi, x_t\rangle = \sum_{i=1}^d \langle\psi_i, x_t\rangle\varphi_i. \tag{5.89}$$

By Theorem 5.4.13, the function $t \mapsto \langle\Psi, x_t\rangle$ is a bounded solution of the integral equation (5.86) on $\mathbb{R}$. Moreover, the function $t \mapsto \langle x^*, f(t)\rangle$ is S^1-almost periodic. It follows by Theorem 3.1.1 that $t \mapsto \langle\Psi, x_t\rangle$ is almost periodic. We deduce from (5.89) that the function $t \mapsto \Pi^v x_t$ is almost periodic. The almost periodicity of $t \mapsto x_t$ follows from (5.87). □

Theorem 5.4.15. *Assume that* **(H0)–(H3)** *hold and $f : \mathbb{R} \to X$ is S^1-almost periodic. If equation* (5.81) *has a bounded integral solution on $\mathbb{R}^+$, then it has a bounded integral solution on $\mathbb{R}$.*

Proof. Let u be a bounded integral solution of equation (5.81) on $\mathbb{R}^+$. By Theorem 5.4.13, the function $t \mapsto \langle \Psi, u_t \rangle$ is a bounded solution of the integral equation (5.86) on $\mathbb{R}^+$. We deduce from Theorem 3.1.2 that the integral equation (5.86) has a bounded solution on $\mathbb{R}$. Let y be this solution. From Theorem 5.4.13, the function x defined by

$$x(t) := \left(\Phi y(t) + \lim_{n\to+\infty} \int_{-\infty}^{t} \mathcal{U}^s(t-s)\Pi^s(\widetilde{B}_n(X_0 f(s)))ds \right)(0) \quad \text{for all } t \in \mathbb{R},$$

is a solution of equation (5.81) on $\mathbb{R}$. It is clear from the proof of Theorem 5.4.13 that the function

$$t \mapsto \lim_{n\to+\infty} \int_{-\infty}^{t} \mathcal{U}^s(t-s)\Pi^s(\widetilde{B}_n(X_0 f(s)))ds$$

is bounded on $\mathbb{R}$. Therefore the solution x is bounded on $\mathbb{R}$. □

From Theorems 5.4.14 and 5.4.15, we have the following result.

Corollary 5.4.16. *Assume that* **(H0)–(H3)** *hold and* $f : \mathbb{R} \to X$ *is* S^1*-almost periodic. If equation* (5.81) *has a bounded integral solution on* $\mathbb{R}^+$*, then it has an almost periodic integral solution.*

5.4.3 Application

To apply our results, we consider the following model proposed in [273] which describes the evolution of the voltage across of a transmission line:

$$\begin{cases} \frac{\partial}{\partial t}[v(t,x) - qv(t-r,x)] = \frac{\partial^2}{\partial x^2}[v(t,x) - qv(t-r,x)] + av(t,x) + bv(t-r,x) \\ \qquad\qquad + \int_{t-r}^{t} h(s-t)v(s,x)ds + F(t)\psi(x) \\ \qquad\qquad \text{for } t \in \mathbb{R} \text{ and } x \in [0,\pi], \\ v(t,0) - qv(t-r,0) = v(t,\pi) - qv(t-r,\pi) = 0 \quad \text{for } t \in \mathbb{R}, \end{cases} \tag{5.90}$$

where $a, b \in \mathbb{R}$, $h : [-r,0] \to \mathbb{R}$, $\psi : [0,\pi] \to \mathbb{R}$ are continuous functions, and $q \in (0,1)$. The function $F : \mathbb{R} \to \mathbb{R}$ is given by

$$F(t) = \sum_{n\geq 0} F_n(t)$$

and F_n are defined for every integer $n \geq 1$ by

$$F_n(t) = \sum_{k\in P_n} H(n^2(t-k)),$$

with $P_n = 3^n(2\mathbb{Z}+1) = \{3^n(2k+1), k \in \mathbb{Z}\}$ and $H \in C_0^\infty(\mathbb{R},\mathbb{R})$, with support in $(-\frac{1}{2},\frac{1}{2})$ such that

$$H \geq 0, \quad H(0) = 1, \quad \text{and} \quad \int_{-\frac{1}{2}}^{\frac{1}{2}} H(s)ds = 1.$$

The function F is not almost periodic, since it is not bounded. However, $F \in C^\infty(\mathbb{R},\mathbb{R}) \cap SAP^1(\mathbb{R},\mathbb{R})$.

To rewrite equation (5.90) in the abstract form (5.81), we introduce $X = C([0,\pi],\mathbb{R})$, the space of continuous functions from $[0,\pi]$ to $\mathbb{R}$ endowed with the uniform norm topology, and define the operator $A : D(A) \subset X \to X$ by

$$\begin{cases} D(A) = \{y \in C^2([0,\pi],\mathbb{R}) : y(0) = y(\pi) = 0\}, \\ Ay = y''. \end{cases}$$

Lemma 5.4.17 ([113]). *The operator A satisfies the Hille–Yosida condition on the space X, namely, $(0,+\infty) \subset \rho(A)$ and*

$$|(\lambda I - A)^{-1}| \leq \frac{1}{\lambda} \quad \textit{for } \lambda > 0.$$

This lemma implies that condition **(H0)** is satisfied. We can see that

$$\overline{D(A)} = \{y \in X : y(0) = y(\pi) = 0\} \neq X.$$

Let A_0 be the part of the operator A in $\overline{D(A)}$. Then, A_0 is given by

$$\begin{cases} D(A_0) = \{y \in C^2([0,\pi],\mathbb{R}) : y(0) = y(\pi) = y''(0) = y''(\pi) = 0\}, \\ A_0 y = y'' \quad \text{for } y \in D(A_0). \end{cases}$$

By Lemma 5.4.2, the operator A_0 generates a strongly continuous semigroup $(T_0(t))_{t\geq 0}$ on $\overline{D(A)}$ which is compact on $\overline{D(A)}$. This implies that **(H2)** holds. Let us introduce the bounded linear operator $\mathcal{D} : C = C([-r,0],X) \to X$ by

$$\mathcal{D}\phi = \phi(0) - q\phi(-r).$$

Since $0 < q < 1$, $\mathcal{D}$ is stable and hypothesis **(H3)** holds. Moreover, by definitions of the operators A and $\mathcal{D}$, we can see that **(H1)** is satisfied.

If we put

$$u(t)(x) = v(t,x) \quad \text{for } t \in \mathbb{R} \text{ and } x \in [0,\pi],$$

then equation (5.90) takes the following abstract form:

$$\frac{d}{dt}\mathcal{D}u(t) = A\mathcal{D}u(t) + L(u_t) + f(t) \quad \text{for } t \in \mathbb{R}, \tag{5.91}$$

where $L : C \to X$ is defined by

$$L(\varphi)(x) = a\varphi(0)(x) + b\varphi(-r)(x) + \int_{-r}^{0} h(s)\varphi(s)(x)ds \quad \text{for } x \in [0,\pi] \text{ and } \varphi \in C,$$

and $f : \mathbb{R} \to X$ is given by

$$f(t)(x) = F(t)\psi(x) \quad \text{for } x \in [0,\pi] \text{ and } t \in \mathbb{R}.$$

It follows that L is a bounded linear operator from C to X, and $f \in SAP^1(\mathbb{R}, X)$.

Lemma 5.4.18. *Under the condition* $|a| + |b| + \int_{-r}^{0} |h(\theta)|d\theta < 1 - q$, *the semigroup solution corresponding to equation (5.91) with $f = 0$ is uniformly exponentially stable.*

Proof. The proof is similar to that of [141, Proposition 5.2]. □

Proposition 5.4.19. *Assume that* $|a| + |b| + \int_{-r}^{0} |h(\theta)|d\theta < 1 - q$. *Then, equation (5.91) has a unique globally attractive almost periodic integral solution. As a consequence, all other integral solutions are asymptotically almost periodic.*

Proof. From Lemma 5.4.18, we have $\omega_0(U) < 0$. Hence, there exists two positive constants $\widehat{M}$ and α such that $|U(t)| \le \widehat{M}e^{-\alpha t}$ for all $t \ge 0$. We claim that all solutions of equation (5.91) are bounded on $\mathbb{R}^+$. In fact, let $\phi \in C_0$ and u a solution such that $u_0 = \phi$. Then,

$$u_t = U(t)\phi + \lim_{n \to +\infty} \int_0^t U(t-s)\widetilde{B}_n(X_0 f(s))ds \quad \text{for } t \ge 0. \tag{5.92}$$

It follows for $t \ge 0$ that

$$|u_t| \le \widehat{M}e^{-\alpha t}|\phi| + \widehat{M}\widetilde{M}e^{-\alpha t}\int_0^t e^{\alpha s}|f(s)|ds$$

$$\le \widehat{M}e^{-\alpha t}|\phi| + \widehat{M}\widetilde{M}e^{-\alpha t}\sum_{k=0}^{[t]} e^{\alpha(k+1)}\int_k^{k+1}|f(s)|ds$$

$$= \widehat{M}e^{-\alpha t}|\phi| + \widehat{M}\widetilde{M}|f|_{BS^1}e^{\alpha}\frac{e^{\alpha([t]+1-t)} - e^{-\alpha t}}{e^{\alpha} - 1}$$

$$\le \widehat{M}|\phi| + \widehat{M}\widetilde{M}|f|_{BS^1}\frac{e^{2\alpha}}{e^{\alpha} - 1}.$$

We deduce from Corollary 5.4.16 the existence of an almost periodic solution z. If $\tilde{z}$ is another solution, then from the variation of constants formula (5.92), we have

$$|z_t - \widetilde{z}_t| \le |U(t)||z_0 - \widetilde{z}_0| \le \widehat{M}e^{-at}|z_0 - \widetilde{z}_0| \to 0,$$

as $t \to +\infty$. Thus, the almost periodic solution z is globally attractive. In addition, since $\widetilde{z}$ can be decomposed as follows:

$$\widetilde{z}_t = z_t + (\widetilde{z}_t - z_t),$$

the solution $\widetilde{z}$ is asymptotically almost periodic.

The almost periodic solution z is the only solution which is bounded on $\mathbb{R}$. In fact, if $\overline{z}$ is another bounded solution on $\mathbb{R}$. Then, by using the variation of constant formula in the phase space, we have for all $\sigma, t \in \mathbb{R}$ with $\sigma \le t$,

$$z_t - \overline{z}_t = U(t-\sigma)(z_\sigma - \overline{z}_\sigma).$$

It follows that, for all $\sigma, t \in \mathbb{R}$ with $\sigma \le t$,

$$|z_t - \overline{z}_t| \le \widehat{M}e^{-a(t-\sigma)}|z_\sigma - \overline{z}_\sigma|. \tag{5.93}$$

Letting $\sigma \to -\infty$ in (5.93), we get $z_t = \overline{z}_t$ for all $t \in \mathbb{R}$. □

5.5 Nonautonomous partial functional differential equations through discrete dynamical systems: finite delay case

5.5.1 Introduction

Consider the following partial functional differential equation:

$$\begin{cases} \frac{\mathrm{d}}{\mathrm{d}t}u(t) = Au(t) + L(t)u_t + f(t) & \text{for } t \ge \sigma, \\ u_\sigma = \phi \in C := C([-r,0];X), \end{cases} \tag{5.94}$$

where $(A, D(A))$ is a Hille–Yosida operator on a Banach space $(X, |\cdot|_X)$; that is, $(A, D(A))$ satisfies the following condition:

(H$_0$) $(A, D(A))$ is a Hille–Yosida operator in X with coefficients $\omega_0 \in \mathbb{R}$ and $M_0 \ge 1$.

We suppose that the condition $\overline{D(A)} = X$ is not necessary. Also $(C, |\cdot|_C)$ is the Banach space of continuous functions $\phi : [-r,0] \to X$ with the uniform norm topology while, for every $t \ge 0$, the function $u_t : [-r,0] \to X$ defined, for $\theta \in [-r,0]$, by $u_t(\theta) = u(t+\theta)$ is called the history function; $f : \mathbb{R} \to X$ is a continuous function and the family $(L(t))_{t\in\mathbb{R}}$ satisfies the following assumption:

(H$_1$) For all t, $L(t)$ is a linear bounded operator from C into X and, for each $\phi \in C$, the mapping $t \mapsto L(t)\phi$ is continuous.

The problem of existence of almost periodic and almost automorphic solutions of equation (5.94) has been investigated by many works. In particular, the authors of [2] and [146] have shown that, when $L(\cdot)$ does not depend on the time and under additional assumptions, the problem can be reduced to an ordinary differential equation in finite dimension. However, for the nonautonomous case, the situation is more complicated since even in finite-dimensional spaces the Bohr–Neugebauer- and Massera-type theorems are not always satisfied. In fact, in [185] the author has given an interesting example of an almost periodic scalar linear equation with bounded solutions, but none of them is almost periodic. For more specific details, see [21, 124, 182, 199, 226, 228, 240–242, 259]. In the same framework, in [180] and [226] the authors proposed an interesting method to solve this problem for equation (5.94) when $\overline{D(A)} = X$ and with unbounded delay. More precisely, they used a variation of constants formula in the phase space and spectral properties of the monodromy operator associated to the equation when A generated a compact C_0-semigroup to show that the problem could be reduced to a linear difference equation in $\mathbb{Z}$. Sadly, the variation of constants formula used in [180] and [226] does not hold when $\overline{D(A)} \neq X$, so we cannot use the same approach here.

Motivated by all these articles, the suggestion of this chapter is to extend the results and methods in [180] and [226] to prove Massera- and Bohr–Neugebauer-type theorems for equation (5.94) when $\overline{D(A)} \neq X$. First, we construct a new variation of constants formula adapted to our problem. Second, we study the existence of an almost periodic and almost automorphic solution for a class of linear difference equations on $\mathbb{Z}$ in a Banach space. Based on all this, we deduce the existence of an almost periodic and almost automorphic solutions of equation (5.94).

5.5.2 Integral solutions

We begin this subsection with a brief recall of some basic results related to equation (5.94), which will be used throughout this work.

Definition 5.5.1 ([5]). Let $\phi \in C$ and $u \in C([-r+\sigma, +\infty[; X)$. Then u is said to be an integral solution for equation (5.94) on $[-r+\sigma, +\infty[$ with initial data ϕ, if the following are true:
(i) $\int_\sigma^t u(s)ds \in D(A)$ for $t \geq \sigma$,
(ii) $u(t) = \phi(0) + A(\int_\sigma^t u(s)ds) + \int_\sigma^t (L(s)u_s + f(s))ds$ for $t \geq \sigma$,
(iii) $u_\sigma = \phi$.

As a result of (i), if u is an integral solution of (5.94), then $u(t) \in \overline{D(A)}$, for any $t \geq \sigma$. Particularly, $\phi(0) \in \overline{D(A)}$. Therefore, the space $C_0 := \{\phi \in C : \phi(0) \in \overline{D(A)}\}$ is the phase space of equation (5.94).

Throughout this section, an integral solution will simply be called a solution.

Let $(A_0, D(A_0))$, the part of $(A, D(A))$ on $\overline{D(A)}$. Then, $(A_0, D(A_0))$ generates a C_0-semigroup on $\overline{D(A)}$, denoted by $(T(t))_{t\geq 0}$, which satisfies

$$|T(t)| \leq M_0 e^{\omega_0 t} \quad \text{for } t \geq 0.$$

The following result is about the existence and uniqueness of the solution. It also gives a useful variation of constants formula in X.

Theorem 5.5.2 ([5]). *Assume that* $(\mathbf{H}_0)$ *and* $(\mathbf{H}_1)$ *hold. Then, for all* $\phi \in C_0$*, equation* (5.94) *has a unique global solution given by*

$$u(t) = \begin{cases} T(t-\sigma)\phi(0) + \lim_{\lambda\to+\infty} \int_\sigma^t T(t-s)B_\lambda(L(s)u_s + f(s))ds & \text{for } t \geq \sigma, \\ u_\sigma = \phi, \end{cases}$$

where $B_\lambda = \lambda \mathcal{R}(\lambda, A)$ *for* $\lambda > \omega_0$.

In the rest of this work, the solutions of equation (5.94) are indicated by $u(\cdot, \sigma, \phi, L, f)$. For $t \geq \sigma$, we define the operator $\mathbb{U}(t, \sigma)$ on C_0 by

$$\mathbb{U}(t,\sigma)\phi = u_t(\cdot, \sigma, \phi, L, 0) \quad \text{for } \phi \in C_0,$$

where $u(\cdot, \sigma, \phi, L, 0)$ is the unique solution of equation (5.94) with $f = 0$.

Next results show that the family $(\mathbb{U}(t,\sigma))_{t\geq\sigma}$ satisfies the proprieties of an evolution family which are important tools in this work.

Theorem 5.5.3. *Assume that* $(\mathbf{H}_0)$ *and* $(\mathbf{H}_1)$ *hold. Then, the family* $(\mathbb{U}(t,\sigma))_{t\geq\sigma}$ *is a strongly continuous evolutionary process on* C_0*, that is:*

(i) $\mathbb{U}(t,t) = I_{C_0}$ *for all*
(ii) $\mathbb{U}(t,\sigma)\mathbb{U}(\sigma,\eta) = \mathbb{U}(t,\eta)$ *for all* $t \geq \sigma \geq \eta$.
(iii) *For all* $t \geq \sigma$ *and* $\phi \in C_0$*, we have that*

$$|\mathbb{U}(t,\sigma)\phi|_C \leq M_1 e^{(M_2 \sup_{s\in[\sigma,t]} |L(s)|_{\mathcal{L}(C,X)} + \omega_0)(t-\sigma)} |\phi|_C,$$

where

$$M_1 := \begin{cases} M_0 & \text{if } \omega_0 \geq 0, \\ M_0 e^{-\omega_0 r} & \text{if } \omega_0 < 0, \end{cases} \qquad M_2 := \begin{cases} M_0^2 & \text{if } \omega_0 \geq 0, \\ M_0^2 e^{-\omega_0 r} & \text{if } \omega_0 < 0. \end{cases}$$

(iv) *For* $t \geq \sigma$, $\mathbb{U}(t,\sigma) \in \mathcal{L}(C_0)$.
(v) *For any fixed* $\phi \in C_0$*, the following map is continuous:*

$$\{(\alpha,\beta) \in \mathbb{R}^2, \alpha \geq \beta\} \ni (t,\sigma) \to \mathbb{U}(t,\sigma)\phi.$$

(vi) *If we suppose that* $L(t+q) = L(t)$*, for all* $t \in \mathbb{R}$*, then the evolution family* $(\mathbb{U}(t,\sigma))_{t\geq\sigma}$ *is* q*-periodic, which means that*

$$\mathbb{U}(t+q, \sigma+q) = \mathbb{U}(t, \sigma) \quad \textit{for all } t \geq \sigma.$$

Proof. Claims (i) and (ii) are outcomes of the definition and uniqueness of the solution.
Claim (iii) is only a consequence of Gronwall's lemma.

(iv) By the uniqueness of solution and (iii), we deduce that, for $t \geq \sigma$, $\mathbb{U}(t, \sigma) \in \mathcal{L}(C_0)$.

(v) Let $\phi \in C_0$. Then, for each $\sigma \in \mathbb{R}$ the function $t \to \mathbb{U}(t, \sigma)\phi$ is continuous at any $t \geq \sigma$. Also, for any $t \in \mathbb{R}$ the function $\sigma \to \mathbb{U}(t, \sigma)\phi$ is continuous on $]-\infty, t]$. Actually, for $\sigma < t$ and $h > 0$ small enough, we have

$$\begin{aligned}
|\mathbb{U}(t, \sigma+h)\phi - \mathbb{U}(t, \sigma)\phi|_C &= |\mathbb{U}(t, \sigma+h)\phi - \mathbb{U}(t, \sigma+h)\mathbb{U}(\sigma+h, \sigma)\phi|_C, \\
&\leq |\mathbb{U}(t, \sigma+h)|_{\mathcal{L}(C_0)} |\mathbb{U}(\sigma+h, \sigma)\phi - \phi|_C, \\
&\leq M_1 e^{(M_2 \sup_{s\in[\sigma,t]} |L(s)|_{\mathcal{L}(C,X)} + \omega_0)(t-\sigma-h)} |\mathbb{U}(\sigma+h, s)\phi - \phi|_C.
\end{aligned}$$

Then,

$$\lim_{h\to 0^+} |\mathbb{U}(t, \sigma+h)\phi - \mathbb{U}(t, \sigma)\phi|_C = 0.$$

Let $\sigma \leq t$ and $h > 0$. Then

$$\begin{aligned}
&|\mathbb{U}(t, \sigma-h)\phi - \mathbb{U}(t, \sigma)\phi|_C \\
&\quad = |\mathbb{U}(t, \sigma)\mathbb{U}(\sigma, \sigma-h)\phi - \mathbb{U}(t, \sigma)\phi|_C \\
&\quad \leq |\mathbb{U}(t, \sigma)|_{\mathcal{L}(C_0)} (|\mathbb{U}(\sigma+h, \sigma)\phi - \mathbb{U}(\sigma, \sigma-h)\phi|_C + |\mathbb{U}(\sigma+h, \sigma)\phi - \phi|_C).
\end{aligned}$$

Thus,

$$\lim_{h\to 0^+} |\mathbb{U}(t, \sigma-h)\phi - \mathbb{U}(t, \sigma)\phi|_C = 0.$$

Now, we claim that the function $(t, \sigma) \to \mathbb{U}(t, \sigma)\phi$ is continuous for all $t \geq \sigma$. Indeed, let $t > \sigma$ and consider a sequence $(t_n, \sigma_n)_{n\in\mathbb{N}}$ in $\mathbb{R}^2$ with $t_n \geq \sigma_n$ for all $n \in \mathbb{N}$, and satisfying

$$\lim_{n\to+\infty} (t_n, \sigma_n) = (t, \sigma).$$

Since $t > \sigma$, we then can find $t_0 \in \mathbb{R}$ such that $t > t_0 > \sigma$. Therefore, there exists $N_1 \in \mathbb{N}$ such that $(t_n, \sigma_n) \in \Delta_{t_0} := \{(\alpha, \beta) \in \mathbb{R}^2, \alpha > t_0 > \beta\}$ for all $n \geq N_1$. As a result,

$$\begin{aligned}
|\mathbb{U}(t_n, \sigma_n)\phi - \mathbb{U}(t, \sigma)\phi|_C &= |\mathbb{U}(t_n, t_0)\mathbb{U}(t_0, \sigma_n)\phi - \mathbb{U}(t, t_0)\mathbb{U}(t_0, \sigma)\phi|_C \\
&\leq |\mathbb{U}(t_n, t_0)\mathbb{U}(t_0, \sigma_n)\phi - \mathbb{U}(t_n, t_0)\mathbb{U}(t_0, \sigma)\phi|_C \\
&\quad + |\mathbb{U}(t_n, t_0)\mathbb{U}(t_0, \sigma)\phi - \mathbb{U}(t, t_0)\mathbb{U}(t_0, \sigma)\phi|_C \\
&\leq M_1 e^{(M_2 \sup_{s\in[t_0,t_n]} |L(s)|_{\mathcal{L}(C,X)} + \omega_0)(t_n-t_0)} |\mathbb{U}(t_0, \sigma_n)\phi - \mathbb{U}(t_0, \sigma)\phi|_C \\
&\quad + |\mathbb{U}(t_n, t_0)\mathbb{U}(t_0, \sigma)\phi - \mathbb{U}(t, t_0)\mathbb{U}(t_0, \sigma)\phi|_C
\end{aligned}$$

for all $n \geq N_1$. Then,

$$\lim_{n\to+\infty}|\mathbb{U}(t_n,\sigma_n)\phi - \mathbb{U}(t,\sigma)\phi|_C = 0.$$

Consequently, the function $(t,\sigma) \to \mathbb{U}(t,\sigma)\phi$ is continuous for all $t > \sigma$. Let $t \in \mathbb{R}$ and $(t_n,\sigma_n)_{n\in\mathbb{N}}$ be any sequence in $\mathbb{R}^2$ such that $t_n \geq \sigma_n$ for all $n \in \mathbb{N}$, and $(t_n,\sigma_n) \to (t,t)$ as $n \to +\infty$. Then,

$$|\mathbb{U}(t_n,\sigma_n)\phi - \mathbb{U}(t,t)\phi|_C \leq |\mathbb{U}(t_n,\sigma_n)\phi - \mathbb{U}(t_n-\sigma_n,0)\phi|_C + |\mathbb{U}(t_n-\sigma_n,0)\phi - \phi|_C. \quad (5.95)$$

Since $(t_n - \sigma_n) \to 0$ as $n \to +\infty$, we then can find $\overline{N_1} \geq 1$ such that

$$-r < -(t_n - \sigma_n) \leq 0.$$

Observe, for $n \geq \overline{N_1} \geq 1$ and $\theta \in [-r, -(t_n - \sigma_n)]$, that

$$\mathbb{U}(t_n,\sigma_n)\phi(\theta) - \mathbb{U}(t_n-\sigma_n,0)\phi(\theta) = \phi(t_n-\sigma_n+\theta) - \phi(t_n-\sigma_n+\theta) = 0.$$

Consequently, for any $n \geq \overline{N_1}$,

$$|\mathbb{U}(t_n,\sigma_n)\phi - \mathbb{U}(t_n-\sigma_n,0)\phi|_C = \sup_{\theta\in[-(t_n-\sigma_n),0]} |\mathbb{U}(t_n,\sigma_n)\phi(\theta) - \mathbb{U}(t_n-\sigma_n,0)\phi(\theta)|_X.$$

So, for each $n \geq \overline{N_1}$ there exists $\theta_n \in [-(t_n-\sigma_n),0]$ such that

$$\begin{aligned}
&\sup_{\theta\in[-(t_n-\sigma_n),0]} |\mathbb{U}(t_n,\sigma_n)\phi(\theta) - \mathbb{U}(t_n-\sigma_n,0)\phi(\theta)|_X \\
&= |\mathbb{U}(t_n,\sigma_n)\phi(\theta_n) - \mathbb{U}(t_n-\sigma_n,0)\phi(\theta_n)|_X \\
&\leq \left|\lim_{\lambda\to+\infty}\int_{\sigma_n}^{t_n+\theta_n} T(t_n+\theta_n-s)B_\lambda L(s)\mathbb{U}(s,\sigma_n)\phi ds\right|_X \\
&\quad + \left|\lim_{\lambda\to+\infty}\int_0^{t_n-\sigma_n+\theta_n} T(t_n-\sigma_n+\theta_n-s)B_\lambda L(s)\mathbb{U}(s,0)\phi ds\right|_X.
\end{aligned}$$

Then,

$$\begin{aligned}
&|\mathbb{U}(t_n,\sigma_n)\phi - \mathbb{U}(t_n-\sigma_n,0)\phi|_C \\
&\leq \int_0^{t_n-\sigma_n+\theta_n} M_0^2 e^{(t_n-\sigma_n+\theta_n-s)\omega_0}|L(s+\sigma_n)|_{\mathcal{L}(C,X)} M_1 e^{(M_2\sup_{\eta\in[\sigma_n,\sigma_n+s]}|L(\eta)|_{\mathcal{L}(C,X)}+\omega_0)s}|\phi|_C ds \\
&+ \int_0^{t_n-\sigma_n+\theta_n} M_0^2 e^{(t_n-\sigma_n+\theta_n-s)\omega_0}|L(s)|_{\mathcal{L}(C,X)} M_1 e^{(M_2\sup_{\eta\in[0,s]}|L(\eta)|_{\mathcal{L}(C,X)}+\omega_0)s}|\phi|_C ds.
\end{aligned}$$

Using the fact that $\lim_{n\to+\infty}(t_n - \sigma_n) = 0$, we obtain that $\lim_{n\to+\infty}\theta_n = 0$. Since the sequence $(\sigma_n)_{n\geq0}$ is bounded,

$$\lim_{n\to+\infty} |\mathbb{U}(t_n, \sigma_n)\phi - \mathbb{U}(t_n - \sigma_n, 0)\phi|_C = 0.$$

By (5.95) and the continuity of the map $\eta \mapsto \mathbb{U}(\eta, 0)$, we conclude that

$$\lim_{n\to+\infty} |\mathbb{U}(t_n, \sigma_n)\phi - \mathbb{U}(t, t)\phi|_C = 0.$$

Hence, the function $(t, \sigma) \to \mathbb{U}(t, \sigma)\phi$ is continuous at (t, t) for all $t \in \mathbb{R}$.

(vi) Suppose that L is q-periodic. Consider $\sigma \in \mathbb{R}$ and $\phi \in C$. Then,

$$\begin{aligned}
\mathbb{U}(t+q, \sigma+q)\phi(0) &= T(t+q-\sigma-q)\phi(0) \\
&\quad + \lim_{\lambda\to\infty} \int_{\sigma+q}^{t+q} T(t+q-s)B_\lambda L(s)\mathbb{U}(s, \sigma+q)\phi ds \\
&= T(t-\sigma)\phi(0) + \lim_{\lambda\to\infty} \int_{\sigma}^{t} T(t-s)B_\lambda L(s+q)\mathbb{U}(s+q, \sigma+q)\phi ds \\
&= T(t-\sigma)\phi(0) + \lim_{\lambda\to\infty} \int_{\sigma}^{t} T(t-s)B_\lambda L(s+q)\mathbb{U}(s+q, \sigma+q)\phi ds \\
&= T(t-\sigma)\phi(0) + \lim_{\lambda\to\infty} \int_{\sigma}^{t} T(t-s)B_\lambda L(s)\mathbb{U}(s+q, \sigma+q)\phi ds
\end{aligned}$$

for all $t \geq \sigma$. By the uniqueness of the solution, we deduce that

$$\mathbb{U}(t, \sigma)\phi = \mathbb{U}(t+q, \sigma+q)\phi \quad \text{for all } t \geq \sigma.$$

□

5.5.3 A new variation of constants formula

Inspired by some results in [9, 226], in this section we establish a representation of the solutions of equation (5.94) in terms of f and the solution process $(\mathbb{U}(t, \sigma))_{t\geq\sigma}$. We consider the following family of linear operators Υ^n mapping X into C defined, for any $n > n_0 := [\max\{\omega_0, \frac{1}{r}\}] + 1$ and $x \in X$, by

$$\Upsilon^n x(\theta) = \begin{cases} (n\theta+1)B_n x & \text{for } -\frac{1}{n} \leq \theta \leq 0, \\ 0 & \text{for } -r \leq \theta \leq -\frac{1}{n}, \end{cases}$$

where $[t]$ denotes the integer part of t. Observe that, for any $x \in X$, $\Upsilon^n x \in C$ and there exists $\tilde{M} \geq 1$ such that

$$|\Upsilon^n x|_C \leq \tilde{M}|x|_X. \tag{5.96}$$

Since, for any $x \in X$, $\Upsilon^n x(0) = B_n x \in \overline{D(A)}$, one has $\Upsilon^n x \in C_0$.

The next theorem gives a variation of constants formula of the solutions of equation (5.94) in C_0.

Theorem 5.5.4. *Let assumptions* $(\mathbf{H}_0)$ *and* $(\mathbf{H}_1)$ *be satisfied. Then,*

$$\lim_{n\to+\infty}\int_\sigma^t \mathbb{U}(t,s)\mathrm{Y}^n f(s)ds = u_t(\cdot,\sigma,0,L,f)$$

uniformly on any compact interval in $[\sigma,+\infty[$. *Moreover, for all* $\phi \in C_0$, *the solution* $u(\cdot,\sigma,\phi,L,f)$ *of equation* (5.94) *with initial condition* ϕ *satisfies the following variation of constants formula in* C_0:

$$u_t(\cdot,\sigma,\phi,L,f) = \mathbb{U}(t,\sigma)\phi + \lim_{n\to+\infty}\int_\sigma^t \mathbb{U}(t,s)\mathrm{Y}^n f(s)ds \quad \textit{for } t \geq \sigma, \tag{5.97}$$

Proof. First, we observe that

$$\begin{aligned} u_t(\cdot,\sigma,\phi,L,f) &= u_t(\cdot,\sigma,\phi,L,0) + u_t(\cdot,\sigma,0,L,f) \\ &= \mathbb{U}(t,\sigma)\phi + u_t(\cdot,\sigma,0,L,f) \end{aligned}$$

for all $t \geq \sigma$. Then, it remains to prove that

$$\lim_{n\to+\infty}\sup_{t\in E}\left|\int_\sigma^t \mathbb{U}(t,s)\mathrm{Y}^n f(s)ds - u_t(\cdot,\sigma,0,L,f)\right|_C = 0$$

for each compact interval E in $[\sigma,+\infty[$. Let start by showing this result in $[\sigma,\sigma+r]$.

Let $n > n_0$ be large enough. Letting $t \in [\sigma,\sigma+r]$, we define the following function:

$$\xi^n(\sigma,t) = \int_\sigma^t \mathbb{U}(t,\sigma)\mathrm{Y}^n f(s)ds.$$

Then, $\xi^n(\sigma,t) \in C$ and, for $\theta \in [-r,0]$, we have

$$\begin{aligned} \xi^n(\sigma,t)(\theta) &= \int_\sigma^t \mathbb{U}(t,\sigma)\mathrm{Y}^n f(s)(\theta)ds \\ &= \int_\sigma^t u(t+\theta,s,\mathrm{Y}^n f(s),L,0)ds. \end{aligned}$$

On the other hand, we have that

$$u(t+\theta, s, \Upsilon^n f(s), L, 0) = \begin{cases} T(t+\theta-s)B_n f(s) \\ \quad + \lim_{\lambda\to+\infty} \int_s^{t+\theta} T(t+\theta-\eta)B_\lambda L(\eta)(\mathbb{U}(\eta,s))\Upsilon^n f(s))d\eta & \text{for } s \le t+\theta, \\ (n(t+\theta-s)+1)B_n f(s) & \text{for } t+\theta \le s \le t+\theta+\frac{1}{n}, \\ 0 & \text{for } s \ge t+\theta+\frac{1}{n}. \end{cases}$$

It follows that

$$\xi^n(\sigma,t)(\theta) = \begin{cases} \int_{t+\theta}^{\min\{t,t+\theta+\frac{1}{n}\}} (n(t+\theta-s)+1)B_n f(s)ds \\ \quad + \int_\sigma^{t+\theta} T(t+\theta-s)B_n f(s)ds \\ \quad + \lim_{\lambda\to+\infty} \int_\sigma^{t+\theta} T(t+\theta-s)B_\lambda L(s)(\xi^n(\sigma,s))ds & \text{for } -(t-\sigma) \le \theta \le 0, \\ \int_\sigma^{\min\{t,t+\theta+\frac{1}{n}\}} (n(t+\theta-s)+1)B_n f(s)ds & \text{for } -(t-\sigma)-\frac{1}{n} \le \theta \le -(t-\sigma), \\ 0 & \text{for } -r \le \theta \le -(t-\sigma)-\frac{1}{n}. \end{cases}$$

Now, for $n > m$ large enough and $t \in [\sigma, \sigma+r]$, let

$$\chi^{n,m}(\sigma,t) := \xi^n(\sigma,t) - \xi^m(\sigma,t).$$

If $-(t-\sigma) \le \theta \le 0$, then

$$\begin{aligned} \chi^{n,m}(\sigma,t)(\theta) = & \int_{t+\theta}^{\min\{t,t+\theta+\frac{1}{n}\}} (n(t+\theta-s)+1)B_n f(s)ds \\ & - \int_{t+\theta}^{\min\{t,t+\theta+\frac{1}{m}\}} (m(t+\theta-s)+1)B_m f(s)ds \\ & + \int_\sigma^{t+\theta} T(t+\theta-s)(B_n f(s) - B_m f(s))ds \\ & + \lim_{\lambda\to+\infty} \int_\sigma^{t+\theta} T(t+\theta-s)B_\lambda L(s)(\xi^n(\sigma,s) - \xi^m(\sigma,s))ds. \end{aligned}$$

If $-(t-\sigma) - \frac{1}{m} \le \theta \le -(t-\sigma)$, we thus obtain

$$\begin{aligned} \chi^{n,m}(\sigma,t)(\theta) = & \int_\sigma^{\min\{t,t+\theta+\frac{1}{n}\}} (n(t+\theta-s)+1)B_n f(s)ds \\ & - \int_\sigma^{\min\{t,t+\theta+\frac{1}{m}\}} (m(t+\theta-s)+1)B_m f(s)ds. \end{aligned}$$

If $-(t-\sigma)-\frac{1}{n} \leq \theta \leq -(t-\sigma)-\frac{1}{m}$, one has

$$\chi^{n,m}(\sigma,t)(\theta) = \int_{\sigma}^{\min\{t,t+\theta+\frac{1}{n}\}} (n(t+\theta-s)+1)B_n f(s)\mathrm{d}s.$$

If $\theta \leq -(t-\sigma)-\frac{1}{n}$, then

$$\chi^{n,m}(\sigma,t)(\theta) = 0.$$

In addition, if $-(t-\sigma) \leq \theta \leq 0$, then

$$\begin{aligned}
|\chi^{n,m}(\sigma,t)(\theta)|_X \leq & \left| \int_{t+\theta}^{\min\{t,t+\theta+\frac{1}{n}\}} (n(t+\theta-s)+1)B_n f(s)\mathrm{d}s \right|_X \\
& + \left| \int_{\sigma}^{t+\theta} T(t+\theta-s)(B_n f(s) - B_m f(s))\mathrm{d}s \right|_X \\
& + \left| \lim_{\lambda\to\infty} \int_{\sigma}^{t+\theta} T(t+\theta-s)B_\lambda L(s)(\chi^{n,m}(\sigma,s))\mathrm{d}s \right|_X \\
& + \left| \int_{t+\theta}^{\min\{t,t+\theta+\frac{1}{m}\}} (m(t+\theta-s)+1)B_m f(s)\mathrm{d}s \right|_X .
\end{aligned}$$

Thus,

$$\begin{aligned}
|\chi^{n,m}(\sigma,t)(\theta)|_X \leq & \sup_{\sigma\leq\eta\leq\sigma+r} \left| \int_{\sigma}^{\eta} T(\eta-s)B_n f(s)\mathrm{d}s - \int_{\sigma}^{\eta} T(\eta-s)B_m f(s)\mathrm{d}s \right|_X \\
& + \int_{\sigma}^{t} M_0^2 \sup_{0\leq\eta\leq r} (e^{\omega_0\eta})\|L(s)\|_{\mathcal{L}(C,X)} \|\chi^{n,m}(\sigma,s)\|_C \mathrm{d}s \\
& + \sup_{\sigma\leq\eta\leq\sigma+r} |f(\eta)|_X \int_{t+\theta}^{t+\theta+\frac{1}{n}} |n(t+\theta-s)+1| \frac{nM_0}{n-\omega_0}\mathrm{d}s \\
& + \sup_{\sigma\leq\eta\leq\sigma+r} |f(\eta)|_X \int_{t+\theta}^{t+\theta+\frac{1}{m}} |m(t+\theta-s)+1| \frac{mM_0}{m-\omega_0}\mathrm{d}s.
\end{aligned}$$

Therefore,

$$\begin{aligned}|\chi^{n,m}(\sigma,t)(\theta)|_X &\le M_0 \sup_{\sigma\le\eta\le\sigma+r}|f(\eta)|_X\left(\frac{1}{n-\omega_0}+\frac{1}{m-\omega_0}\right)\\&\quad+\sup_{\sigma\le\eta\le\sigma+r}\left|\int_\sigma^\eta T(\eta-s)B_nf(s)\mathrm{d}s-\int_\sigma^\eta T(\eta-s)B_mf(s)\mathrm{d}s\right|_X\\&\quad+M_0^2\sup_{0\le\eta\le r}(e^{\omega_0\eta})\sup_{\sigma\le\eta\le\sigma+r}|L(\eta)|_{\mathcal{L}(C,X)}\int_\sigma^t|\chi^{n,m}(\sigma,s)|_C\mathrm{d}s.\end{aligned}$$

If $-(t-\sigma)-\frac{1}{m}\le\theta\le-(t-\sigma)$, we have that

$$\begin{aligned}|\chi^{n,m}(\sigma,t)(\theta)|_X &\le \left|\int_{t+\theta}^{\min\{t,t+\theta+\frac{1}{n}\}}(n(t+\theta-s)+1)B_nf(s)\mathrm{d}s\right|_X\\&\quad+\left|\int_{t+\theta}^{\min\{t,t+\theta+\frac{1}{m}\}}(m(t+\theta-s)+1)B_mf(s)\mathrm{d}s\right|_X\\&\le M_0\sup_{\sigma\le\eta\le\sigma+r}|f(\eta)|_X\left(\frac{1}{n-\omega_0}+\frac{1}{m-\omega_0}\right).\end{aligned}$$

Then,

$$\begin{aligned}|\chi^{n,m}(\sigma,t)(\theta)|_X &\le M_0 \sup_{\sigma\le\eta\le\sigma+r}|f(\eta)|_X\left(\frac{1}{n-\omega_0}+\frac{1}{m-\omega_0}\right)\\&\quad+\sup_{\sigma\le\eta\le\sigma+r}\left|\int_\sigma^\eta T(\eta-s)B_nf(s)\mathrm{d}s-\int_\sigma^\eta T(\eta-s)B_mf(s)\mathrm{d}s\right|_X\\&\quad+M_0^2\sup_{0\le\eta\le r}(e^{\omega_0\eta})\sup_{\sigma\le\eta\le\sigma+r}|L(\eta)|_{\mathcal{L}(C,X)}\int_\sigma^t|\chi^{n,m}(\sigma,s)|_C\mathrm{d}s.\end{aligned}$$

If $-(t-\sigma)-\frac{1}{n}\le\theta\le-(t-\sigma)-\frac{1}{m}$, then

$$\begin{aligned}|\chi^{n,m}(\sigma,t)(\theta)|_X &\le \left|\int_{t+\theta}^{\min\{t,t+\theta+\frac{1}{n}\}}(n(t+\theta-s)+1)B_nf(s)\mathrm{d}s\right|_X\\&\le M_0\sup_{\sigma\le\eta\le\sigma+r}|f(\eta)|_X\left(\frac{1}{n-\omega_0}\right).\end{aligned}$$

Therefore,

$$
\begin{aligned}
|\chi^{n,m}(\sigma,t)(\theta)|_X \le{}& M_0 \sup_{\sigma\le\eta\le\sigma+r} |f(\eta)|_X\left(\frac{1}{n-\omega_0}+\frac{1}{m-\omega_0}\right)\\
&+\sup_{\sigma\le\eta\le\sigma+r}\left|\int_\sigma^\eta T(\eta-s)B_nf(s)ds-\int_\sigma^\eta T(\eta-s)B_mf(s)ds\right|_X\\
&+M_0^2\sup_{0\le\eta\le r}(e^{\omega_0\eta})\sup_{\sigma\le\eta\le\sigma+r}|L(\eta)|_{\mathcal{L}(C,X)}\int_\sigma^t|\chi^{n,m}(\sigma,s)|_C ds.
\end{aligned}
$$

Consequently,

$$
\begin{aligned}
|\chi^{n,m}(\sigma,t)|_C \le{}& M_0 \sup_{\sigma\le\eta\le\sigma+r} |f(\eta)|_X\left(\frac{1}{n-\omega_0}+\frac{1}{m-\omega_0}\right)\\
&+\sup_{\sigma\le\eta\le\sigma+r}\left|\int_\sigma^\eta T(\eta-s)B_nf(s)ds-\int_\sigma^\eta T(\eta-s)B_mf(s)ds\right|_X\\
&+M_0^2\sup_{0\le\eta\le r}(e^{\omega_0\eta})\sup_{\sigma\le\eta\le\sigma+r}|L(\eta)|_{\mathcal{L}(C,X)}\int_\sigma^t|\chi^{n,m}(\sigma,s)|_C ds.
\end{aligned}
$$

By Gronwall's lemma, we deduce that

$$
|\chi^{n,m}(\sigma,t)|_C \le \vartheta^{n,m}(\sigma)\exp\Big(\Big(M_0^2\sup_{0\le\eta\le r}(e^{\omega_0\eta})\sup_{\sigma\le\eta\le\sigma+r}|L(\eta)|_{\mathcal{L}(C,X)}\Big)(t-\sigma)\Big),
$$

where

$$
\begin{aligned}
\vartheta^{n,m}(\sigma) ={}& M_0 \sup_{\sigma\le\eta\le\sigma+r} |f(\eta)|_X\left(\frac{1}{n-\omega_0}+\frac{1}{m-\omega_0}\right)\\
&+\sup_{\sigma\le\eta\le\sigma+r}\left|\int_\sigma^\eta T(\eta-s)B_nf(s)ds-\int_\sigma^\eta T(\eta-s)B_mf(s)ds\right|_X.
\end{aligned}
$$

Furthermore,

$$
\sup_{t\in[\sigma,\sigma+r]}|\chi^{n,m}(\sigma,t)|_C \le \vartheta^{n,m}(\sigma)\exp\Big(M_0^2 r\sup_{0\le\eta\le r}(e^{\omega_0\eta})\sup_{\sigma\le\eta\le\sigma+r}|L(\eta)|_{\mathcal{L}(C,X)}\Big).
$$

Observe that $(\int_\sigma^\eta T(\eta-s)B_nf(s)ds)_{n\ge0}$ converges uniformly on $[\sigma,\sigma+r]$ as $n\to+\infty$, thus

$$
\lim_{n,m\to+\infty}\sup_{\sigma\le\eta\le\sigma+r}\left|\int_\sigma^\eta T(\eta-s)B_nf(s)ds-\int_\sigma^\eta T(\eta-s)B_mf(s)ds\right|_X=0.
$$

Consequently, $\vartheta^{n,m}(\sigma)\to 0$ as $n,m\to+\infty$. Thus,

$$
\lim_{n,m\to+\infty}\sup_{t\in[\sigma,\sigma+r]}|\chi^{n,m}(\sigma,t)|_C=0,
$$

which means that $(\xi^n(\sigma,\cdot))_{n\in\mathbb{N}}$ is a Cauchy sequence in $C([\sigma,\sigma+r];C)$, the space of all continuous functions from $[\sigma,\sigma+r]$ into C. Therefore, there is a function $\xi(\cdot)\in C([\sigma,\sigma+r];C)$ such that

$$\lim_{n\to+\infty}\sup_{t\in[\sigma,\sigma+r]}|\xi^n(\sigma,t)-\xi(t)|_C=0.$$

Let $y:[-r+\sigma,\sigma+r]\to X$ be the function defined by

$$y(t)=\begin{cases}\lim_{\lambda\to+\infty}\int_\sigma^t T(t-s)B_\lambda(L(s)\xi(s)+f(s))ds & \text{for } \sigma\le t\le\sigma+r,\\ 0 & \text{for } \sigma-r\le t\le\sigma.\end{cases}$$

Then, for $t\in[\sigma,\sigma+r]$, we obtain

$$y_t(\theta)=\begin{cases}\lim_{\lambda\to+\infty}\int_\sigma^{t+\theta} T(t+\theta-s)B_\lambda(L(s)\xi(s)+f(s))ds & \text{for } \sigma\le t+\theta\le\sigma+r,\\ 0 & \text{for } \sigma-r\le t+\theta\le\sigma.\end{cases}$$

We claim that $y_t=\xi(t)$ for each $t\in[\sigma,\sigma+r]$. Indeed, if $\theta\in[-(t-\sigma),0]$, we obtain

$$\begin{aligned}|\xi^n(\sigma,t)(\theta)-y_t(\theta)|_X\le&\left|\int_{t+\theta}^{\min\{t,t+\theta+\frac{1}{n}\}}(n(t+\theta-s)+1)B_nf(s)ds\right|_X\\&+\left|\int_\sigma^{t+\theta}T(t+\theta-s)(B_nf(s))ds-\lim_{\lambda\to\infty}\int_\sigma^{t+\theta}T(t+\theta-s)B_\lambda f(s)ds\right|_X\\&+\left|\lim_{\lambda\to\infty}\int_\sigma^{t+\theta}T(t+\theta-s)B_\lambda L(s)(\xi^n(\sigma,s)-\xi(s))ds\right|_X.\end{aligned}$$

It follows that

$$\begin{aligned}|\xi^n(\sigma,t)(\theta)-y_t(\theta)|_X\le&\frac{M_0}{n-\omega_0}\sup_{s\in[\sigma,\sigma+r]}|f(s)|_X\\&+M_0^2\sup_{0\le\eta\le r}(e^{\omega_0\eta})\sup_{\eta\in[\sigma,\sigma+r]}|L(\eta)|_{\mathcal{L}(C,X)}\int_\sigma^{\sigma+r}|\xi^n(\sigma,s)-\xi(s))|_Cds\\&+\sup_{\eta\in[\sigma,\sigma+r]}\left|\int_\sigma^\eta T(\eta-s)B_nf(s)ds-\lim_{\lambda\to\infty}\int_\sigma^\eta T(\eta-s)B_\lambda f(s)ds\right|_X.\end{aligned}$$

If $-(t-\sigma)-\frac{1}{n}\le\theta\le-(t-\sigma)$, then

$$
\begin{aligned}
|\xi^n(\sigma,t)(\theta)-y_t(\theta)|_X &\le \left|\int_\sigma^{\min\{t,t+\theta+\frac{1}{n}\}} (n(t+\theta-s)+1)B_nf(s)ds\right|_X \\
&\le \frac{M_0}{n-\omega_0}\sup_{s\in[\sigma,\sigma+r]}|f(s)|_X \\
&\quad + M_0^2\sup_{0\le\eta\le r}(e^{\omega_0\eta})\sup_{\eta\in[\sigma,\sigma+r]}|L(\eta)|_{\mathcal{L}(C,X)}\int_\sigma^{\sigma+r}|\xi^n(\sigma,s)-\xi(s))|_C ds \\
&\quad + \sup_{\eta\in[\sigma,\sigma+r]}\left|\int_\sigma^\eta T(\eta-s)B_nf(s)ds-\lim_{\lambda\to\infty}\int_\sigma^\eta T(\eta-s)B_\lambda f(s)ds\right|_X.
\end{aligned}
$$

Consequently,

$$
\begin{aligned}
|\xi^n(\sigma,t)-y_t|_C \le \frac{M_0}{n-\omega_0}&\sup_{s\in[\sigma,\sigma+r]}|f(s)|_X \\
&+ M_0^2\sup_{0\le\eta\le r}(e^{\omega_0\eta})\sup_{\eta\in[\sigma,\sigma+r]}|L(\eta)|_{\mathcal{L}(C,X)}\int_\sigma^{\sigma+r}|\xi^n(\sigma,s)-\xi(s))|_C ds \\
&+ \sup_{\eta\in[\sigma,\sigma+r]}\left|\int_\sigma^\eta T(\eta-s)B_nf(s)ds-\lim_{\lambda\to\infty}\int_\sigma^\eta T(\eta-s)B_\lambda f(s)ds\right|_X.
\end{aligned}
$$

Therefore,

$$
\lim_{n\to+\infty}\sup_{t\in[\sigma,\sigma+r]}|\xi^n(\sigma,t)-y_t|_C = 0.
$$

Thus, $y_t = \xi(t)$ for any $t\in[\sigma,\sigma+r]$, which implies

$$
u_t(.,\sigma,0,L,f) = \lim_{n\to+\infty}\int_\sigma^t \mathbb{U}(t,\sigma)\Upsilon^n f(s)ds \quad \text{for all } t\in[\sigma,\sigma+r]. \tag{5.98}
$$

Furthermore, the above limit exists uniformly on $[\sigma,\sigma+r]$.

Let now consider the case when $t\in[\sigma+r,\sigma+2r]$. By using the same reasoning as above, we can show that, for $t\in[\sigma+r,\sigma+2r]$,

$$
u_t(\cdot,\sigma+r,0,L,f) = \lim_{n\to+\infty}\int_{\sigma+r}^t \mathbb{U}(t,\sigma)\Upsilon^n f(s)ds
$$

and this limit exists uniformly on $[\sigma+r,\sigma+2r]$. On the other hand, we observe, for $t\in[\sigma+r,\sigma+2r]$, that

$$
u_t(\cdot,\sigma,0,L,f) = \mathbb{U}(t,\sigma+r)u_{\sigma+r}(\cdot,\sigma,0,L,f)+u_t(\cdot,\sigma+r,0,L,f).
$$

In fact, let $t \in [\sigma + r, \sigma + 2r]$ and $\theta \in [-r, 0]$. Then,

$$u_t(\theta, \sigma, 0, L, f) = \lim_{\lambda\to\infty} \int_{\sigma}^{t+\theta} T(t+\theta-s)B_\lambda(L(s)u_s(\cdot, \sigma, 0, L, f) + f(s))ds.$$

If $\theta \in [-r, -(t-\sigma-r)]$, we obtain

$$\begin{aligned}
&[\mathbb{U}(t, \sigma + r)u_{\sigma+r}(\cdot, \sigma, 0, L, f)](\theta) + u_t(\theta, \sigma + r, 0, L, f) \\
&\quad = u_{\sigma+r}(t + \theta - \sigma - r, \sigma, 0, L, f) \\
&\quad = \lim_{\lambda\to\infty} \int_{\sigma}^{t+\theta} T(t+\theta-s)B_\lambda(L(s)u_s(\cdot, \sigma, 0, L, f) + f(s))ds.
\end{aligned}$$

If $\theta \in [-(t-\sigma-r), 0]$, then

$$\begin{aligned}
&[\mathbb{U}(t, \sigma + r)u_{\sigma+r}(\cdot, \sigma, 0, L, f)](\theta) + u_t(\theta, \sigma + r, 0, L, f) \\
&\quad = T(t + \theta - \sigma - r)u_{\sigma+r}(0, \sigma, 0, L, f) \\
&\quad\quad + \lim_{\lambda\to\infty} \int_{\sigma+r}^{t+\theta} T(t+\theta-s)B_\lambda L(s)\mathbb{U}(s, \sigma + r)u_{\sigma+r}(\cdot, \sigma, 0, L, f)ds \\
&\quad\quad + \lim_{\lambda\to\infty} \int_{\sigma+r}^{t+\theta} T(t+\theta-s)B_\lambda(L(s)u_s(\cdot, \sigma + r, 0, L, f) + f(s))ds.
\end{aligned}$$

Thus,

$$\begin{aligned}
&[\mathbb{U}(t, \sigma + r)u_{\sigma+r}(\cdot, \sigma, 0, L, f)](\theta) + u_t(\theta, \sigma + r, 0, L, f) \\
&\quad = \lim_{\lambda\to\infty} \int_{\sigma}^{\sigma+r} T(t+\theta-s)B_\lambda(L(s)u_s(\cdot, \sigma, 0, L, f) + f(s))ds \\
&\quad\quad + \lim_{\lambda\to\infty} \int_{\sigma+r}^{t+\theta} T(t+\theta-s)B_\lambda L(s)\mathbb{U}(s, \sigma + r)u_{\sigma+r}(\cdot, \sigma, 0, L, f)ds \\
&\quad\quad + \lim_{\lambda\to\infty} \int_{\sigma+r}^{t+\theta} T(t+\theta-s)B_\lambda(L(s)u_s(\cdot, \sigma + r, 0, L, f) + f(s))ds.
\end{aligned}$$

Therefore, for all $t \in [\sigma + r, \sigma + 2r]$,

$$\begin{aligned}
&|u_t(\cdot, \sigma, 0, L, f) - \mathbb{U}(t, \sigma + r)u_{\sigma+r}(\cdot, \sigma, 0, L, f) - u_t(\cdot, \sigma + r, 0, L, f)|_C \\
&\quad \le M(\sigma) \int_{\sigma+r}^{t} |u_s(\cdot, \sigma, 0, L, f) - \mathbb{U}(s, \sigma + r)u_{\sigma+r}(\cdot, \sigma, 0, L, f) - u_t(\cdot, \sigma + r, 0, L, f))|_C ds,
\end{aligned}$$

where

$$M(\sigma) = M_0^2 \sup_{\eta\in[0,r]} (e^{w\eta}) \sup_{\eta\in[\sigma+r,\sigma+2r]} |L(\eta)|_{\mathcal{L}(C,X)}.$$

By Gronwall's lemma, we deduce, for $t \in [\sigma + r, \sigma + 2r]$, that

$$u_t(\cdot, \sigma, 0, L, f) = \mathbb{U}(t, \sigma + r)u_{\sigma+r}(\cdot, \sigma, 0, L, f) + u_t(\cdot, \sigma + r, 0, L, f).$$

Consequently, for all $n > n_0$ large enough and $t \in [\sigma + r, \sigma + 2r]$, we have

$$\begin{aligned}
&\left| u_t(\cdot, \sigma, 0, L, f) - \int_\sigma^t \mathbb{U}(t, s)\Upsilon^n f(s)ds \right|_C \\
&\quad\le \left| \mathbb{U}(t, \sigma + r)\left\{ u_{\sigma+r}(\cdot, \sigma, 0, L, f) - \int_\sigma^{\sigma+r} \mathbb{U}(\sigma + r, s)\Upsilon^n f(s)ds \right\} \right|_C \\
&\quad\quad + \left| u_t(\cdot, \sigma + r, 0, L, f) - \int_{\sigma+r}^t \mathbb{U}(t, s)\Upsilon^n f(s)ds \right|_C .
\end{aligned}$$

Hence,

$$\lim_{n\to+\infty} \sup_{t\in[\sigma+r,\sigma+2r]} \left| u_t(\cdot, \sigma, 0, L, f) - \int_\sigma^t \mathbb{U}(t, s)\Upsilon^n f(s)ds \right|_C = 0.$$

Continuing this process, we can prove that

$$\lim_{n\to+\infty} \sup_{t\in[\sigma+jr,\sigma+(j+1)r]} \left| u_t(\cdot, \sigma, 0, L, f) - \int_\sigma^t \mathbb{U}(t, s)\Upsilon^n f(s)ds \right|_C = 0,$$

for all $j \in \mathbb{N}$. The proof is completed. □

5.5.4 Almost periodicity and almost automorphy in linear discrete dynamical systems

In this section, we study the existence of almost periodic and almost automorphic solutions to the following linear difference equation:

$$x(n + 1) = Bx(n) + c(n) \quad \text{for } n \in \mathbb{Z}, \tag{DE}$$

where B is a bounded linear operator on a Banach space X and $(c(n))_{n\in\mathbb{Z}}$ is a sequence in X. Here, we discuss the Massera result and the Bohr–Neugebauer property for the discrete dynamic system (DE), under conditions related to spectral properties of B. These conditions do not include any specific requirement on the spectral properties of the sequence $(c(n))_{n\in\mathbb{Z}}$, except that it should be almost periodic or almost automorphic. We also avoid assuming that "X does not contain any subspace isomorphic to c_0."

We have the following Bohr–Neugebauer-type results for equation (DE) when the dimension of X is finite.

Theorem 5.5.5 ([102]). *Assume that the dimension of X is finite and $(c(n))_{n\in\mathbb{Z}}$ is almost periodic. Then, every bounded solution of* (DE) *on $\mathbb{Z}$ is almost periodic.*

Theorem 5.5.6 ([37]). *Assume that the dimension of X is finite and $(c(n))_{n\in\mathbb{Z}}$ is almost automorphic. Then, every bounded solution of* (DE) *on $\mathbb{Z}$ is almost automorphic.*

The goal here is to extend the above theorems to the case when the dimension of X is infinite. To achieve this goal, we need the following spectral decomposition result.

Theorem 5.5.7. *Let $B \in \mathcal{L}(X)$ be such that $r_{\mathrm{ess}}(B) < 1$. Then, the space X is decomposed as follows:*

$$X = X_U \oplus X_S,$$

where
(1) *X_U and X_S are two closed subspaces of X, invariant under B,*
(2) *the dimension of X_U is finite,*
(3) *$\sigma(B_S) = \{\lambda \in \sigma(B) : |\lambda|_{\mathbb{C}} < 1\}$ and $\sigma(B_U) = \{\lambda \in \sigma(B) : |\lambda|_{\mathbb{C}} \geq 1\}$, where B_S is the restriction of B to X_S and B_U is the restriction of B to X_U,*
(4) $\lim_{n\to+\infty} B_S^n = 0$.

Proof. Consider the following set:

$$\sigma_1 := \{\lambda \in \sigma(B) : |\lambda|_{\mathbb{C}} \geq 1\}.$$

Since $r_{\mathrm{ess}}(B) < 1$, one has that σ_1 has finitely many elements. In fact, if not, then there exists a sequence $\{\lambda_n\}_{n\in\mathbb{N}} \subset \sigma_1$ such that $\lambda_n \neq \lambda_m$ for $n \neq m$. By virtue of the compactness of $\sigma(B)$, there exists a subsequence $\{\lambda_{n_k}\}_{k\in\mathbb{N}}$ such that

$$\lambda_{n_k} \to \lambda_0 \in \sigma_1 \quad \text{as } k \to +\infty.$$

Since $\lambda_{n_k} \neq \lambda_{n_{k'}}$ for $n_k \neq n_{k'}$, λ_0 is a limit point of $\sigma(B) \setminus \{\lambda_0\}$. Therefore,

$$\lambda_0 \in \sigma_{\mathrm{ess}}(B).$$

Consequently,

$$|\lambda_0|_{\mathbb{C}} \leq r_{\mathrm{ess}}(B) < 1,$$

which leads to a contradiction.

Let $\sigma_1 = \{\mu_1, \ldots, \mu_m\}$. Then, for each $j \in \{1, \ldots, m\}$, μ_j is a isolated eigenvalue of B and there exists $k_j \in \mathbb{N}^*$ such that

$$\dim \ker\big((\mu_j I_X - B)^{k_j}\big) < \infty.$$

Let consider for $j \in \{1, \ldots, m\}$ the following operator:

$$\mathcal{J}_j = \frac{1}{2\pi i} \int_{\gamma_j} \mathcal{R}(\mu, B) d\mu,$$

where γ_j is a positively oriented curve in $\mathbb{C}$ enclosing the isolated singularity μ_j, but no other points of $\sigma(B)$ (see Figure 5.1). Then, by Theorem 2.1.9, for all $j \in \{1, \ldots, m\}$, the operator $\mathcal{J}_j$ is a projection in Y and

$$\text{range}(\mathcal{J}_j) = \ker\big((\mu_j I_X - B)^{k_j}\big).$$

Observe that, for any $s \neq j$, if γ_j does not enclose λ_s and γ_s does not enclose λ_j, then

$$\mathcal{J}_s \mathcal{J}_j = \mathcal{J}_j \mathcal{J}_s = 0.$$

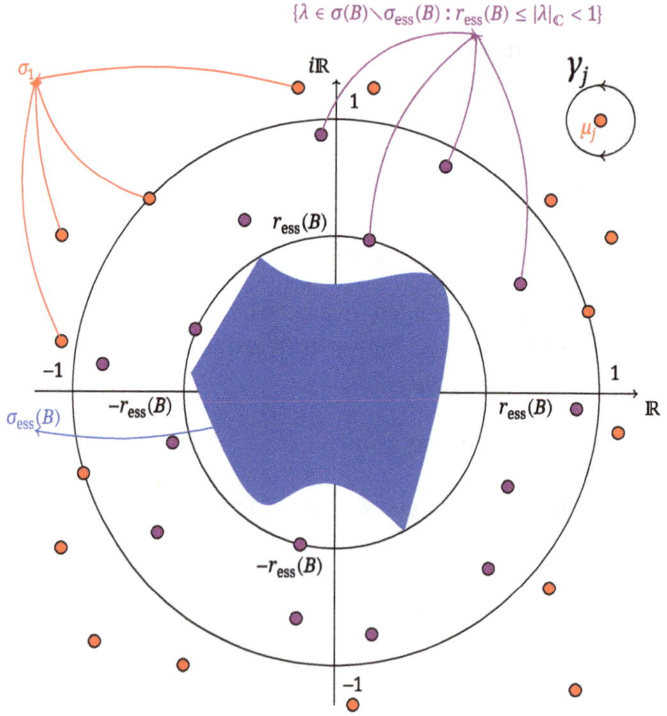

Figure 5.1: Spectrum of the operator B.

Therefore, the following operator:

$$\mathcal{J} = \sum_{j=1}^{m} \mathcal{J}_j$$

is a projection on X. Moreover,

$$\operatorname{range}(\mathcal{J}) = \bigoplus_{j=1}^{m} \ker((\mu_j I_X - B)^{k_j}) \quad \text{and} \quad \dim \operatorname{range}(\mathcal{J}) < \infty.$$

Consequently,

$$X = X_U \oplus X_S,$$

where $X_U = \operatorname{range}(\mathcal{J})$ and $X_S = \operatorname{range}(I_X - \mathcal{J})$.

Since for any $\lambda \in \rho(B)$, $\mathcal{R}(\lambda, B)B = B\mathcal{R}(\lambda, B)$, one gets

$$B\mathcal{J} = \mathcal{J}B.$$

Thus, X_U and X_S are B-invariant. Moreover, by Theorem 2.1.9, we deduce that $\sigma(B_U) = \sigma_1$ and B_S has spectrum $\sigma(B) \setminus \sigma_1$. Hence,

$$r(B_S) < 1,$$

where

$$r(B_S) := \max\{|\lambda|_{\mathbb{C}} : \lambda \in \sigma(B_S)\}.$$

As a consequence of Theorem 2.1.5, we deduce that

$$\lim_{n \to +\infty} [B_S]^n = 0.$$

The proof is finished. □

Thanks to the above theorem, we can assert that the study of the behavior of bounded solutions of equation (DE) can be reduced to that of a linear difference equation posed in a finite-dimensional space. As a result of this fact, in [226, Theorem 3.1], the authors established Massera's theorem for equation (DE) when $(c(n))_{n\geq 0}$ is almost periodic, as follows:

Theorem 5.5.8 ([226]). *Assume that $r_{\text{ess}}(B) < 1$ and $(c(n))_{n\in\mathbb{Z}}$ is almost periodic. If equation* (DE) *has a bounded solution on $\mathbb{N}$, then it has an almost periodic solution on $\mathbb{Z}$.*

Using the same approach as in the proof of [226, Theorem 3.1], we derive the following Bohr–Neugebauer theorem.

Theorem 5.5.9. *Assume that* $r_{\mathrm{ess}}(B) < 1$ *and* $(c(n))_{n\in\mathbb{Z}}$ *is almost periodic. Then, every bounded solution of* (DE) *on* $\mathbb{Z}$ *is almost periodic.*

Proof. Let $\mathcal{J}^U$ and $\mathcal{J}^S$ denote the projection operators on X_U and X_S, respectively (see Theorem (5.5.7)). For any $n \in \mathbb{Z}$, we have that

$$x(n) = x_1(n) + x_2(n),$$

where $x_1(n) := \mathcal{J}^U x(n)$ and $x_2(n) := \mathcal{J}^S x(n)$. First, we observe that $x_2(\cdot)$ satisfies the following difference equation:

$$x_2(n) = \mathcal{J}^S B \mathcal{J}^S x_2(n-1) + \mathcal{J}^S c(n-1) \quad \text{for } n \in \mathbb{Z}.$$

Then,

$$x_2(\cdot) = \mathcal{J}^S B \mathcal{J}^S \Lambda(-1) x_2(\cdot) + \mathcal{J}^S \Lambda(-1) c(\cdot),$$

where $\Lambda(-1)$ is the operator of translation defined for any sequence $b(\cdot)$ by

$$(\Lambda(-1)b)(n) = b(n-1) \quad \text{for any } n \in \mathbb{Z}.$$

Therefore,

$$(x_2(\cdot) - \mathcal{J}^S B \mathcal{J}^S \Lambda(-1)) x_2(\cdot) = \mathcal{J}^S \Lambda(-1) c(\cdot).$$

As for any $\zeta \in \mathcal{L}(X)$, $n \in \mathbb{N}^*$ and $b(\cdot) \in l^\infty(\mathbb{Z}, X)$,

$$|[\zeta\Lambda(-1)]^n b(\cdot)|_{l^\infty(\mathbb{Z},X)} = |\zeta^n b(\cdot)|_{l^\infty(\mathbb{Z},X)},$$

then

$$|[\mathcal{J}^S B \mathcal{J}^S \Lambda(-1)]^n|_{\mathcal{L}(l^\infty(\mathbb{Z},X))} \le |[\mathcal{J}^S B \mathcal{J}^S]^n|_{\mathcal{L}(X)}.$$

Since $\lim_{n\to+\infty}(\mathcal{J}^S B \mathcal{J}^S)^n = 0$,

$$\lim_{n\to+\infty} (\mathcal{J}^S B \mathcal{J}^S \Lambda(-1))^n = 0.$$

By Theorem 2.1.5, we obtain that

$$r(\mathcal{J}^S B \mathcal{J}^S \Lambda(-1)) < 1.$$

Therefore,

$$1 \in \rho(\mathcal{J}^S B \mathcal{J}^S \Lambda(-1)).$$

Consequently, for any $n \in \mathbb{Z}$,

$$x_2(n) = \mathcal{R}(1, \mathcal{J}^S B \mathcal{J}^S \Lambda(-1)) \mathcal{J}^S \Lambda(-1) c(n).$$

Then, $x_2(\cdot)$ is almost periodic. On the other hand, we have that $x_1(\cdot)$ is a bounded solution on $\mathbb{Z}$ of the difference equation

$$x_1(n) = \Psi x_1(n-1) + \mathcal{J}^U c(n-1) \quad \text{for } n \in \mathbb{Z},$$

where $\Psi = \mathcal{J}^U B \mathcal{J}^U$. Since the dimension of X_U is finite, by Theorem 5.5.5, we have that $x_1(\cdot)$ is almost periodic. Hence, $x(\cdot)$ is almost periodic. □

Continuing our analysis, we apply the spectral decomposition from Theorem 5.5.7 and employ similar reasoning as in [180, Lemma 3.3] to establish the following result.

Theorem 5.5.10. *Assume that $r_{\text{ess}}(B) < 1$ and $(c(n))_{n\in\mathbb{Z}}$ is almost automorphic. Then, every bounded solution of* (DE) *on $\mathbb{Z}$ is almost automorphic. Moreover, if equation* (DE) *has a bounded solution on $\mathbb{N}$, then it has an almost automorphic solution on $\mathbb{Z}$.*

Proof. Let $x(\cdot)$ be a bounded solution of (DE) on $\mathbb{Z}$. Then, we have that

$$x(n) = x_1(n) + x_2(n) \quad \text{for all } n \in \mathbb{Z},$$

where $x_1(n) := \mathcal{J}^U x(n)$ and $x_2(n) = \mathcal{J}^S x(n)$. By the same reasoning as in the first step of the above proof, we have that

$$x_2(n) = \mathcal{R}(1, \mathcal{J}^S B \mathcal{J}^S \Lambda(-1)) \mathcal{J}^S \Lambda(-1) c(n).$$

Because of $c(\cdot)$ is almost automorphic, we have that $x_2(\cdot)$ is also almost automorphic. Since $x_1(\cdot)$ is a bounded solution of the following difference equation:

$$x_1(n) = \Psi x_1(n-1) + \mathcal{J}^U c(n-1) \quad \text{for } n \in \mathbb{Z},$$

by Theorem 5.5.6 it is almost automorphic. Hence, $x(\cdot)$ is almost automorphic.

Assume now (DE) has a bounded solution $x(\cdot)$ on $\mathbb{N}$. Let us consider the following difference equation:

$$y(n) = \Psi y(n-1) + c_1(n-1) \quad \text{for } n \in \mathbb{Z}, \tag{5.99}$$

where $\Psi = \mathcal{J}^U B \mathcal{J}^U$ and $c_1(\cdot) = \mathcal{J}^U c(\cdot)$. Then, the sequence $y_1(\cdot) = \mathcal{J}^U x(\cdot)$ is a bounded solution of equation (5.99) on $\mathbb{N}$. Therefore, for any $m \in \mathbb{N}, y_1(m+\cdot)$ is a bounded solution on $\mathbb{N}$ of the following equation:

$$y_1(m+n) = \Psi y_1(m+n-1) + c_1(m+n-1) \quad \text{for } n \in \mathbb{Z}.$$

On the other hand, since $c(\cdot)$ is almost automorphic, there exist a subsequence $\{m_j''\}_{j\in\mathbb{N}} \subset \mathbb{N}$ and a sequence $\overline{c}_1(\cdot)$ such that

$$\lim_{j\to+\infty} c_1(n+m_j'') = \overline{c}_1(n) \quad \text{and} \quad \lim_{j\to+\infty} \overline{c}_1(n-m_j'') = c(n)$$

for all $n \in \mathbb{Z}$. Furthermore, since $(y_1(m_j''))_{j\in\mathbb{N}}$ is bounded on X_U, there exist a subsequence $\{m_j'\}_{j\in\mathbb{N}}$ of $\{m_j''\}_{j\in\mathbb{N}}$ and $\overline{y}_1(0)$ such that

$$\lim_{j\to+\infty} y_1(m_j') = \overline{y}_1(0).$$

Since $y_1(m_j'+1) = \Psi y_1(m_j') + c_1(m_j')$, one then gets

$$\lim_{j\to+\infty} y_1(m_j'+1) = \overline{y}_1(1) := \Psi\overline{y}_1(0) + \overline{c}_1(0).$$

In similar steps, we can show, for any $n \in \mathbb{N}$, that the sequence $(y_1(m_j'+n))_{j\in\mathbb{N}}$ converges to $\overline{y}_1(n)$ in X_U. Moreover, for any $n \in \mathbb{N}$,

$$\overline{y}_1(n+1) = \Psi\overline{y}_1(n) + \overline{c}_1(n).$$

We can see that

$$y_1(m_j') = \Psi y_1(m_j'-1) + c_1(m_j'-1).$$

Then,

$$y_1(m_j'-1) = \Psi^{-1}[y_1(m_j') - c_1(m_j'-1)]. \tag{5.100}$$

Note that Ψ^{-1} exists, since $0 \in \rho(\Psi)$ (see Theorem 5.5.7). Thus,

$$\overline{y}_1(-1) := \lim_{j\to+\infty} y_1(m_j'-1) = \Psi^{-1}[\overline{y}_1(0) - \overline{c}_1(-1)].$$

Continuing this process, we deduce, for any $n \in \mathbb{Z}_-$, that the sequence $(y_1(m_j'+n))_{j\in\mathbb{N}}$ converges to $\overline{y}_1(n) \in X_U$ which satisfies, for $n \in \mathbb{Z}_-$,

$$\overline{y}_1(n) = \Psi^{-1}[\overline{y}_1(n+1) - \overline{c}_1(n)].$$

Therefore, for $n \in \mathbb{Z}_-$, we have

$$\overline{y}_1(n+1) = \Psi\overline{y}_1(n) + \overline{c}_1(n).$$

As a result, we have constructed a bounded sequence $(\overline{y}_1(n))_{n\in\mathbb{Z}}$ defined, for $n \in \mathbb{Z}$, by

$$\overline{y}_1(n) = \lim_{j\to+\infty} y_1(m_j'+n),$$

which is a bounded solution of the following difference equation:

$$\overline{y}_1(n+1) = \Psi\overline{y}_1(n) + \overline{c}_1(n) \quad \text{for } n \in \mathbb{Z}.$$

Additionally, we have that the sequence $(\overline{y}_1(-m'_j))_{j\in\mathbb{N}}$ is bounded, and then there exist a subsequence $(\overline{y}_1(-m_j))_{j\in\mathbb{N}}$ and $y(0) \in X_U$ such that

$$\lim_{j\to+\infty} \overline{y}_1(-m_j) = y(0).$$

Since

$$\overline{y}_1(1-m_j) = \Psi\overline{y}_1(-m_j) + \overline{c}_1(-m_j) \quad \text{and} \quad \overline{y}_1(-1-m_j) = \Psi^{-1}[\overline{y}_1(-1-m_j) - \overline{c}_1(-m_j)],$$

one gets

$$y(1) := \lim_{j\to+\infty} \overline{y}_1(1-m_j) = \Psi y(0) + c_1(0)$$

and

$$y(-1) := \lim_{j\to+\infty} \overline{y}_1(-1-m_j) = \Psi^{-1}[y(0) + c_1(-1)].$$

Step-by-step, we can prove that, for all $n \in \mathbb{Z}$, the sequence $(\overline{y}_1(n-m_j))_{j\in\mathbb{N}}$ converges to $y(n) \in X_U$. Moreover, $(y(n))_{n\in\mathbb{Z}}$ is a bounded solution of the following difference equation:

$$y(n+1) = \Psi y(n) + c_1(n) \quad \text{for } n \in \mathbb{Z}.$$

Using Theorem 5.5.6, we deduce that y is almost automorphic. On the other hand, the sequence $z(\cdot)$, given by

$$z(n) = \mathcal{R}(1, \mathcal{J}^S B \mathcal{J}^S \Lambda(-1))\mathcal{J}^S \Lambda(-1)c(n) \quad \text{for } n \in \mathbb{Z},$$

is the unique almost automorphic solution of the following difference equation:

$$z(n+1) = \mathcal{J}^S B \mathcal{J}^S z(n) + \mathcal{J}^S c(n) \quad \text{for } n \in \mathbb{Z}.$$

Consequently, the sequence $\overline{x}(\cdot)$ defined, for $n \in \mathbb{Z}$, by

$$\overline{x}(n) = z(n) + y(n)$$

is almost automorphic solution of equation (DE), completing the proof. □

5.5.5 Almost periodic and almost automorphic solutions

In this section, we use the variation of constants formula proved in Section 5.5.3 and the results established in Section 5.5.4, to prove some interesting results concerning the criteria of the existence of almost periodic solutions and almost automorphic solutions of equation (5.94) when L is periodic in time. Without loss of generality, we suppose that L is 1-periodic in t.

Let us consider the following difference equation:

$$x(n+1) = \mathbb{U}(1,0)x(n) + b(n) \quad \text{for } n \in \mathbb{Z}, \tag{5.101}$$

where $(b(n))_{n\in\mathbb{Z}}$ is the sequence defined, for $n \in \mathbb{Z}$, by

$$b(n) = u_{n+1}(\cdot, n, 0, L, f).$$

Lemma 5.5.11. *Let $x(\cdot)$ be a solution of equation (5.101) in $\mathbb{Z}$. Then, for any $m \geq n$,*

$$x(m) = \mathbb{U}(m,n)x(n) + \lim_{k\to+\infty} \int_n^m \mathbb{U}(m,s)\Upsilon^k f(s)ds.$$

Proof. The proof of this lemma is completed by showing that, for all $n \in \mathbb{Z}$ and $j \in \mathbb{N}^*$,

$$x(n+j) = \mathbb{U}(n+j,n)x(n) + \lim_{k\to+\infty} \int_n^{n+j} \mathbb{U}(n+j,s)\Upsilon^k f(s)ds. \tag{5.102}$$

Let us prove (5.102) by induction on j. Now, for $j = 1$, we get

$$\begin{aligned} x(n+1) &= \mathbb{U}(1,0)x(n) + b(n) \\ &= \mathbb{U}(n+1,n)x(n) + \lim_{k\to+\infty} \int_n^{n+1} \mathbb{U}(n+1,s)\Upsilon^k f(s)ds. \end{aligned}$$

Thus, (5.102) is valid for $j = 1$. Next we assume that (5.102) is valid for some $j \in \mathbb{N}^*$. Let us to show it is valid for $j+1$. We observe that

$$x(n+j+1) = \mathbb{U}(1,0)x(n+j) + b(n+j).$$

Then, by the induction hypothesis, we obtain that

$$\begin{aligned}
x(n+j+1) &= \mathbb{U}(1,0)\left[\mathbb{U}(n+j,n)x(n) + \lim_{k\to+\infty}\int_n^{n+j}\mathbb{U}(n+j,s)\Upsilon^k f(s)\mathrm{d}s\right] \\
&\quad + \lim_{k\to+\infty}\int_{n+j}^{n+j+1}\mathbb{U}(n+j,s)\Upsilon^k f(s)\mathrm{d}s \\
&= \mathbb{U}(n+j+1,n)x(n) + \lim_{k\to+\infty}\int_n^{n+j}\mathbb{U}(n+j+1,s)\Upsilon^k f(s)\mathrm{d}s \\
&\quad + \lim_{k\to+\infty}\int_{n+j}^{n+j+1}\mathbb{U}(n+j+1,s)\Upsilon^k f(s)\mathrm{d}s \\
&= \mathbb{U}(n+j+1,n)x(n) + \lim_{k\to+\infty}\int_n^{n+j+1}\mathbb{U}(n+j+1,s)\Upsilon^k f(s)\mathrm{d}s.
\end{aligned}$$

The induction is complete. □

We consider the following assumption:

(H$_2$) For any $t > 0$, the operator $T(t)$ is compact.

5.5.6 Almost periodic case

In this section, we assume that:

(H$_3$) f is almost periodic.

Lemma 5.5.12. *Assume that* **(H$_0$)**, **(H$_1$)**, *and* **(H$_3$)** *hold. Then, the sequence* $(b(n))_{n\in\mathbb{Z}}$ *is almost periodic.*

Proof. For $n \in \mathbb{Z}$, we have that

$$\begin{aligned}
b(n) &= u_{n+1}(\cdot, n, 0, L, f) \\
&= \lim_{k\to+\infty}\int_n^{n+1}\mathbb{U}(n+1,s)\Upsilon^k f(s)\mathrm{d}s \\
&= \lim_{k\to+\infty}\int_0^{1}\mathbb{U}(1,s)\Upsilon^k f(s+n)\mathrm{d}s.
\end{aligned}$$

Let $\{m_j'\}_{j\in\mathbb{N}} \subset \mathbb{Z}$. Since f is almost periodic function, there exist a subsequence $\{m_j\}_{j\in\mathbb{N}}$ of $\{m_j'\}_{j\in\mathbb{N}}$ and a function $g : \mathbb{R} \to X$ such that

$$\lim_{j\to+\infty}\sup_{t\in\mathbb{R}}|f(t+m_j) - g(t)|_X = 0.$$

For $n \in \mathbb{Z}$, we put

$$\overline{b}(n) := \lim_{k\to+\infty} \int_0^1 \mathbb{U}(1,s)\Upsilon^k g(s+n)ds.$$

We claim that

$$\lim_{j\to+\infty} \sup_{n\in\mathbb{Z}} |b(n+m_j) - \overline{b}(n)|_C = 0.$$

In fact, for $n \in \mathbb{Z}$ and $j \in \mathbb{N}$, we have

$$|b(n+m_j) - \overline{b}(n)|_C \le \tilde{M} \sup_{s\in[0,1]} |\mathbb{U}(1,s)|_{\mathcal{L}(C_0)} \sup_{t\in\mathbb{R}} |f(t+m_j) - g(t)|_X,$$

where $\tilde{M}$ is the constant given in (5.96). Note that the existence of $\sup_{s\in[0,1]} |\mathbb{U}(1,s)|_{\mathcal{L}(C_0)}$ is assured by Banach–Steinhaus theorem.

Then,

$$\lim_{j\to+\infty} \sup_{n\in\mathbb{Z}} |b(n+m_j) - \overline{b}(n)|_C = 0.$$

Hence, $(b(n))_{n\in\mathbb{Z}}$ is almost periodic. □

Lemma 5.5.13. *Assume that* $(\mathbf{H}_0)$, $(\mathbf{H}_1)$, $(\mathbf{H}_2)$, *and* $(\mathbf{H}_3)$ *hold. Then, every bounded solution of equation* (5.101) *on* $\mathbb{Z}$ *is almost periodic and if equation* (5.101) *has a bounded solution on* $\mathbb{N}$, *it must have an almost periodic solution on* $\mathbb{Z}$.

Proof. In light of Theorems 5.5.8 and 5.5.9, the proof of this lemma is finished by proving that

$$r_{\text{ess}}(\mathbb{U}(1,0)) < 1,$$

where $r_{\text{ess}}(\mathbb{U}(1,0))$ is the essential spectral radius of $\mathbb{U}(1,0)$. Indeed, if $1 \ge r$, then by the assumption $(\mathbf{H}_2)$ and [6, Theorem 5.2], we deduce that $\mathbb{U}(1,0)$ is compact. Consequently,

$$r_{\text{ess}}(\mathbb{U}(1,0)) = 0 < 1.$$

If $1 < r$, then by [6, Theorem 5.2], we have that

$$\mathbb{U}(1,0) = \mathcal{Y}(1) + \mathcal{W}(1),$$

where $(\mathcal{Y}(t))_{t\ge0}$ is an exponentially stable semigroup on C_0 and $\mathcal{W}(1)$ is a compact operator. Then,

$$\begin{aligned} r_{\text{ess}}(\mathbb{U}(1,0)) &= r_{\text{ess}}(\mathcal{Y}(1) + \mathcal{W}(1)) \\ &= r_{\text{ess}}(\mathcal{Y}(1)) \\ &\le r(\mathcal{Y}(1)). \end{aligned}$$

Let $\omega(\mathcal{Y})$ be the growth bound (see [136, p. 299]) of the semigroup $(\mathcal{Y}(t))_{t\geq 0}$. By [136, Proposition 1.7, p. 299] and since $(\mathcal{Y}(t))_{t\geq 0}$ is exponentially stable, one gets

$$\omega(\mathcal{Y}) < 0.$$

Thanks to [270, Proposition 4.13, pp. 170–171], we have that

$$r(\mathcal{Y}(1)) = e^{\omega(\mathcal{Y})} < 1.$$

Hence,

$$r_{\text{ess}}(\mathbb{U}(1,0)) < 1,$$

ending the proof. □

Theorem 5.5.14. *Assume that* $(\mathbf{H}_0)$, $(\mathbf{H}_1)$, $(\mathbf{H}_2)$, *and* $(\mathbf{H}_3)$ *hold. If equation* (5.94) *has an almost periodic solution* $u(\cdot)$ *on* $\mathbb{R}$, *then* $(u_n)_{n\in\mathbb{Z}}$ *is an almost periodic solution on* $\mathbb{Z}$ *of equation* (5.101). *Conversely, if* $x(\cdot)$ *is an almost periodic solution of* (5.101) *on* $\mathbb{Z}$, *then the function* $v : \mathbb{R} \to C_0$ *defined, for* $t \in \mathbb{R}$, *by*

$$v(t) = \mathbb{U}(t,[t])x([t]) + \lim_{k\to+\infty} \int_{[t]}^{t} \mathbb{U}(t,s)\mathrm{Y}^k f(s)ds$$

is almost periodic and the function $t \mapsto v(t)(0)$ *is an almost periodic solution of equation* (5.94) *on* $\mathbb{R}$.

Proof. Define $x(\cdot)$, for $n \in \mathbb{Z}$, by $x(n) = u_n$. Then, by the formula given in Theorem 5.5.4, we get that

$$\begin{aligned} x(n+1) &= u_{n+1} \\ &= \mathbb{U}(n+1,n)u_n + \lim_{k\to+\infty} \int_{n}^{n+1} \mathbb{U}(n+1,s)\mathrm{Y}^k f(s)ds \\ &= \mathbb{U}(1,0)x(n) + b(n) \quad \text{for all } n \in \mathbb{Z}, \end{aligned}$$

which means that the sequence $x(\cdot)$ is a solution of equation (5.101) on $\mathbb{Z}$.

Since $u(\cdot)$ is bounded on $\mathbb{R}$, $t \mapsto u_t$ is also bounded on $\mathbb{R}$. As a result, we get that $x(\cdot)$ is a bounded solution of equation (5.101) in $\mathbb{Z}$. Hence, by Lemma 5.5.13, we deduce that $x(\cdot)$ is almost periodic solution of (5.101) on $\mathbb{Z}$.

Conversely, we start by showing that the function $t \mapsto u(t) := v(t)(0)$ is a solution of equation (5.94) on $\mathbb{R}$. Let $\sigma \in \mathbb{R}$ and $t \geq \sigma$. Then,

$$v(t) = \mathbb{U}(t,[t])x([t]) + \lim_{k\to+\infty} \int_{[t]}^{t} \mathbb{U}(t,s)\Upsilon^k f(s)ds.$$

By Lemma 5.5.11, we obtain that

$$v(t) = \mathbb{U}(t,[t])\left\{\mathbb{U}([t],[\sigma])x([\sigma]) + \lim_{k\to\infty} \int_{[\sigma]}^{[t]} \mathbb{U}([t],s)\Upsilon^k f(s)ds\right\} + \lim_{k\to+\infty} \int_{[t]}^{t} \mathbb{U}(t,s)\Upsilon^k f(s)ds.$$

Thus,

$$v(t) = \mathbb{U}(t,[\sigma])x([\sigma]) + \lim_{k\to\infty} \int_{[\sigma]}^{t} \mathbb{U}(t,s)\Upsilon^k f(s)ds.$$

It follows that

$$v(t) = \mathbb{U}(t,\sigma)\left\{\mathbb{U}(\sigma,[\sigma])x([\sigma]) + \lim_{k\to\infty} \int_{[\sigma]}^{\sigma} \mathbb{U}(\sigma,s)\Upsilon^k f(s)ds\right\} + \lim_{k\to\infty} \int_{\sigma}^{t} \mathbb{U}(t,s)\Upsilon^k f(s)ds.$$

Consequently,

$$v(t) = \mathbb{U}(t,\sigma)v(\sigma) + \lim_{k\to\infty} \int_{\sigma}^{t} \mathbb{U}(t,s)\Upsilon^k f(s)ds \quad \text{for } t \geq \sigma.$$

By Theorem 5.5.4, we deduce that the function u is a solution of equation (5.94) on $\mathbb{R}$.

Let $\tilde{x}$ be the almost periodic extension of x on $\mathbb{R}$. Then, the following function $\mathcal{G} : \mathbb{R} \to X \times C_0$ defined, for $t \in \mathbb{R}$, by

$$\mathcal{G}(t) = (f(t), \tilde{x}(t))$$

is almost periodic. Consider $\varepsilon > 0$ and set $\varepsilon_0 := \frac{\varepsilon}{2\tilde{N}}$, where

$$\tilde{N} := \begin{cases} \tilde{M}M_1 & \text{if } \omega_0 \leq -M_2 \sup_{\eta\in\mathbb{R}}|L(\eta)|_{\mathcal{L}(C,X)}, \\ \tilde{M}M_1 e^{M_2 \sup_{\eta\in\mathbb{R}}|L(\eta)|_{\mathcal{L}(C,X)}+\omega_0} & \text{if } \omega_0 > -M_2 \sup_{\eta\in\mathbb{R}}|L(\eta)|_{\mathcal{L}(C,X)}. \end{cases}$$

Note that, since L is 1-periodic, for each $\phi \in C$, one has $\sup_{\eta\in\mathbb{R}}|L(\eta)\phi|_X < +\infty$. Thus, by Banach–Steinhaus theorem, we deduce that

$$\sup_{\eta\in\mathbb{R}}|L(\eta)|_{\mathcal{L}(C,X)} < +\infty.$$

We have that $\mathbb{Z} \cap \mathbb{T}(\mathcal{G}, \varepsilon_0)$ is relatively dense. Then, there exists $l(\varepsilon) > 0$ such that

$$(\mathbb{Z} \cap \mathbb{T}(\mathcal{G}, \varepsilon_0)) \cap [a, a + l(\varepsilon)] \neq \emptyset \quad \text{for all } a \in \mathbb{R}.$$

Let $a \in \mathbb{R}$ and $\tau_\varepsilon \in (\mathbb{Z} \cap \mathbb{T}(\mathcal{G}, \varepsilon_0)) \cap [a, a + l(\varepsilon)]$. Then,

$$\begin{aligned}|v(t + \tau_\varepsilon) - v(t)|_C &\leq |\mathbb{U}(t + \tau_\varepsilon, [t + \tau_\varepsilon])x([t + \tau_\varepsilon]) - \mathbb{U}(t, [t])x([t])|_C \\ &\quad + \left| \lim_{k\to+\infty} \int_{[t+\tau_\varepsilon]}^{t+\tau_\varepsilon} \mathbb{U}(t + \tau_\varepsilon, \sigma)\Upsilon^k f(\sigma)d\sigma - \lim_{k\to+\infty} \int_{[t]}^{t} \mathbb{U}(t, \sigma)\Upsilon^k f(\sigma)d\sigma \right|_C\end{aligned}$$

for all $t \in \mathbb{R}$. Since $\tau_\varepsilon \in \mathbb{Z}$, for all $t \in \mathbb{R}$, one then obtains $[t + \tau_\varepsilon] = [t] + \tau_\varepsilon$. Therefore,

$$\begin{aligned}|v(t + \tau_\varepsilon) - v(t)|_C &\leq |\mathbb{U}(t + \tau_\varepsilon, [t] + \tau_\varepsilon)x([t] + \tau_\varepsilon) - \mathbb{U}(t, [t])x([t])|_C \\ &\quad + \left| \lim_{k\to+\infty} \int_{[t]+\tau_\varepsilon}^{t+\tau_\varepsilon} \mathbb{U}(t + \tau_\varepsilon, \sigma)\Upsilon^k f(\sigma)d\sigma - \lim_{k\to+\infty} \int_{[t]}^{t} \mathbb{U}(t, \sigma)\Upsilon^k f(\sigma)d\sigma \right|_C \\ &\leq |\mathbb{U}(t, [t])(x([t] + \tau_\varepsilon) - x([t])|_C \\ &\quad + \left| \lim_{k\to+\infty} \int_{[t]}^{t} \mathbb{U}(t, \sigma)\Upsilon^k (f(\sigma + \tau_\varepsilon) - f(\sigma))d\sigma \right|_C \\ &< \sup_{\sigma\in[[t],t]} |\mathbb{U}(t, \sigma)|_{\mathcal{L}(C_0)} \times \varepsilon_0 + \tilde{M} \sup_{\sigma\in[[t],t]} |\mathbb{U}(t, \sigma)|_{\mathcal{L}(C_0)} \times \varepsilon_0 \\ &< \varepsilon\end{aligned}$$

for all $t \in \mathbb{R}$. Consequently, $\tau_\varepsilon \in \mathbb{T}(v, \varepsilon) \cap [a, a + l(\varepsilon)]$. Therefore, for all $\varepsilon > 0$, $\mathbb{T}(v, \varepsilon)$ is relatively dense, which means that v is almost periodic. Hence, the function u is an almost periodic solution of equation (5.94) on $\mathbb{R}$. □

Now, we are going to state a couple of the principal results of this work, namely Massera- and Bohr–Neugebauer-type theorems.

Theorem 5.5.15. *Suppose that* $(\mathbf{H}_0)$, $(\mathbf{H}_1)$, $(\mathbf{H}_2)$, *and* $(\mathbf{H}_3)$ *hold. Then, the existence of a bounded solution on* $\mathbb{R}^+$ *of equation* (5.94) *implies the existence of an almost periodic solution on* $\mathbb{R}$. *Moreover, if* u *is a bounded solution of* (5.94) *on* $\mathbb{R}$, *then the function* $t \mapsto u_t$ *is almost periodic, hence* u *is also almost periodic.*

Proof. Let u be the bounded solution of equation (5.94) on $\mathbb{R}^+$. Then, the sequence $(u_n)_{n\in\mathbb{N}}$ is a bounded solution of difference equation (5.101) on $\mathbb{N}$. Therefore, by Lemma 5.5.13, equation (5.101) has an almost periodic solution on $\mathbb{Z}$. From Theorem 5.5.14, we deduce that (5.94) has an almost periodic solution on $\mathbb{R}$.

Let u be a bounded solution of equation (5.94) on $\mathbb{R}$. Then, $(u_n)_{n\in\mathbb{Z}}$ is a bounded solution of equation (5.101) on $\mathbb{Z}$. Thus, by Lemma 5.5.13, we deduce that $(u_n)_{n\in\mathbb{Z}}$ is almost periodic on $\mathbb{Z}$. Consider the function $v : \mathbb{R} \to C_0$ defined as in Theorem 5.5.14. Then, for $t \in \mathbb{R}$, we have that

$$\begin{aligned} v(t) &= \mathbb{U}(t,[t])u_{[t]} + \lim_{k\to+\infty} \int_{[t]}^{t} \mathbb{U}(t,s)\Upsilon^k f(s)ds \\ &= u_t. \end{aligned}$$

As a consequence of Theorem 5.5.14, we conclude that $t \mapsto u_t$ is almost periodic. Therefore, u is also almost periodic. □

Corollary 5.5.16. *Let assumptions* $(\mathbf{H}_0)$, $(\mathbf{H}_1)$, $(\mathbf{H}_2)$, *and* $(\mathbf{H}_3)$ *be satisfied. If u is an almost periodic solution of equation* (5.94) *on* $\mathbb{R}$, *then the history function $t \mapsto u_t$ is almost periodic.*

5.5.7 Almost automorphic case

Now, we replace the assumption $(\mathbf{H}_3)$ by the following:
$(\mathbf{H}_4)$ f is almost automorphic.

The following lemma is needed in the next steps.

Lemma 5.5.17. *Let $\{\psi_n\}_{n\in\mathbb{N}}$ be a sequence of measurable mappings from $\mathbb{R}$ into X such that*

$$|\psi_n(t)|_X < \tilde{\psi}(t) \quad a.e.\ t \in \mathbb{R},$$

where $\tilde{\psi} \in L^1_{\mathrm{loc}}(\mathbb{R},\mathbb{R}^+)$. If $\lim_{n\to+\infty} \psi_n(t) = \psi(t)$ for a. e. $t \in \mathbb{R}$, for some measurable function ψ, then

$$\lim_{n\to+\infty} |u_t(\cdot,0,0,L,\psi_n) - u_t(\cdot,0,0,L,\psi)|_C = 0$$

uniformly on any $[a,b] \subset \mathbb{R}^+$.

Proof. Let $0 \le a < b$ and $t \in [a,b]$. Then,

$$\begin{aligned} &u(t,0,0,L,\psi_n) - u(t,0,0,L,\psi) \\ &= \lim_{\lambda\to+\infty} \int_0^t T(t-s)B_\lambda[L(s)(u_s(\cdot,0,0,L,\psi_n) - u_s(\cdot,0,0,L,\psi))]ds \\ &\quad + \lim_{\lambda\to+\infty} \int_0^t T(t-s)B_\lambda[\psi_n(s) - \psi(s)]ds. \end{aligned}$$

Therefore,

$$|u(t,0,0,L,\psi_n) - u(t,0,0,L,\psi)|_X \le \int_0^t \tilde{A}|L(s)|_{\mathcal{L}(C,X)}|u_s(\cdot,0,L,\psi_n) - u_s(\cdot,0,L,\psi)|_C ds + \int_0^t \tilde{A}|\psi_n(s) - \psi(s)|_X ds,$$

where $\tilde{A} := M_0^2 \sup_{\eta\in[0,b]} e^{\omega_0\eta}$. Consequently, for $t \in [a,b]$, we have

$$\begin{aligned}
&|u_t(\cdot,0,0,L,\psi_n) - u_t(\cdot,0,0,L,\psi)|_C \\
&\quad\le \sup_{\sigma\in[0,t]} |u(\sigma,0,0,L,\psi_n) - u(\sigma,0,0,L,\psi)|_X \\
&\quad\le \int_0^t \tilde{A} \sup_{\eta\in[0,b]} |L(\eta)|_{\mathcal{L}(C,X)}|u_s(\cdot,0,0,L,\psi_n) - u_s(\cdot,0,0,L,\psi)|_C ds \\
&\quad\quad + \int_0^t \tilde{A}|\psi_n(s) - \psi(s)|_X ds.
\end{aligned}$$

Using Gronwall's lemma, we get

$$|u_t(\cdot,0,0,L,\psi_n) - u_t(\cdot,0,0,L,\psi)|_C \le e^{(\tilde{A}\sup_{\eta\in[0,b]}|L(\eta)|_{\mathcal{L}(C,X)})t}\left[\int_0^b \tilde{A}|\psi_n(s) - \psi(s)|_X ds\right].$$

Employing Lebesque's dominated convergence theorem, we deduce

$$\lim_{n\to+\infty} \sup_{t\in[a,b]} |u_t(\cdot,0,0,L,\psi_n) - u_t(\cdot,0,0,L,\psi)|_C = 0,$$

which completes the proof. □

Theorem 5.5.18. *Suppose that* (H_0), (H_1), *and* (H_4) *hold. If u is an almost automorphic solution of* (5.94), *then it is compact almost automorphic.*

Proof. The proof of this theorem is completed by showing that u is uniformly continuous on $\mathbb{R}$. In fact, let $t, s \in \mathbb{R}$. Assume that $t \ge s$ and $|t - s|$ is small enough. Then,

$$u(t) - u(s) = T(t-s)u(s) - u(s) + \lim_{\lambda\to+\infty} \int_s^t T(t-\eta)B_\lambda[L(\eta)u_\eta + f(\eta)]d\eta.$$

Without loss of generality, we assume that $\omega_0 \le 0$, otherwise we substitute A by $A - \delta_0 I_X$ and $L(t)$ by $\tilde{L}(t)\phi = \delta_0\phi(0) + L(t)\phi$, where $\delta_0 > 0$ is taken large enough such that $\omega_0 - \delta_0 < 0$. Then,

$$|u(t)-u(s)|_X \le \sup_{\xi\in\mathcal{T}}|T(|t-s|)\xi-\xi|_X + M_0^2\Big(\sup_{\sigma\in\mathbb{R}}|L(\sigma)|_{\mathcal{L}(C,X)}\sup_{\sigma\in\mathbb{R}}|u_\sigma|_C + \sup_{\sigma\in\mathbb{R}}|f(\sigma)|_X\Big)|t-s|,$$

where $\mathcal{T} := \{u(\eta), \eta\in\mathbb{R}\}$. Since u is almost automorphic on $\mathbb{R}$, $\mathcal{T}$ is relatively compact in X, which implies

$$\lim_{t-s\to 0}\sup_{\xi\in\mathcal{T}}|T(t-s)\xi-\xi|_X = 0.$$

Furthermore,

$$\lim_{t-s\to 0}|u(t)-u(s)|_X = 0,$$

ending the proof. □

Lemma 5.5.19. *Assume that* $(\mathbf{H}_0)$, $(\mathbf{H}_1)$, *and* $(\mathbf{H}_4)$ *hold. Then, the sequence* $(b(n))_{n\in\mathbb{Z}}$ *is almost automorphic.*

Proof. As in the proof of Lemma 5.5.12, we have that

$$b(n) = \lim_{k\to+\infty}\int_0^1 \mathbb{U}(1,s)\Upsilon^k f(s+n)ds.$$

Let $\{m_j'\}_{j\in\mathbb{N}}\subset\mathbb{Z}$. Since f is almost automorphic function, there exist a subsequence $\{m_j\}_{j\in\mathbb{N}}$ of $\{m_j'\}_{j\in\mathbb{N}}$ and a function $g:\mathbb{R}\to X$ such that

$$\lim_{j\to+\infty} f(t+m_j) = g(t) \quad\text{and}\quad \lim_{j\to+\infty} g(t-m_j) = f(t) \quad \text{for all } t\in\mathbb{R}.$$

Let $\overline{b}:\mathbb{Z}\to C_0$ be the sequence defined, for $n\in\mathbb{Z}$, by

$$\overline{b}(n) = \lim_{k\to+\infty}\int_0^1 \mathbb{U}(1,s)\Upsilon^k g(s+n)ds.$$

For $n\in\mathbb{Z}$ and $j\in\mathbb{N}$, we have that

$$|b(n+m_j)-\overline{b}(n)|_C \le \tilde{M}\sup_{s\in[0,1]}|\mathbb{U}(1,s)|_{\mathcal{L}(C_0)}\int_0^1|f(s+n+m_j)-g(s+n)|_X ds$$

and

$$|\overline{b}(n-m_j)-b(n)|_C \le \tilde{M}\sup_{s\in[0,1]}|\mathbb{U}(1,s)|_{\mathcal{L}(C_0)}\int_0^1|g(s+n-m_j)-f(s+n)|_X ds.$$

Then, by Lebesgue's dominated convergence theorem, we conclude that

$$\lim_{j\to+\infty} b(n+m_j) = \overline{b}(n) \quad \text{and} \quad \lim_{j\to+\infty} \overline{b}(n-m_j) = b(n) \quad \text{for all } n \in \mathbb{Z}.$$

Hence, $(b(n))_{n\in\mathbb{Z}}$ is almost automorphic. □

Lemma 5.5.20. *Assume that* $(\mathbf{H}_0)$, $(\mathbf{H}_1)$, $(\mathbf{H}_2)$, *and* $(\mathbf{H}_4)$ *hold. Then, if equation* (5.101) *has a bounded solution on* $\mathbb{N}$, *then it must admit an almost automorphic solution on* $\mathbb{Z}$. *Moreover, each bounded solution of equation* (5.101) *on* $\mathbb{Z}$ *is almost automorphic.*

Proof. As in the proof of Lemma 5.5.13, the assumption $(\mathbf{H}_2)$ implies that $r_{\text{ess}}(\mathbb{U}(1,0)) < 1$. Then, the result of this lemma is a consequence of Theorem 5.5.10. □

Theorem 5.5.21. *Assume that* $(\mathbf{H}_0)$, $(\mathbf{H}_1)$, $(\mathbf{H}_2)$, *and* $(\mathbf{H}_4)$ *hold. If equation* (5.94) *has an almost automorphic solution* $u(\cdot)$ *on* $\mathbb{R}$, *then* $(u_n)_{n\in\mathbb{Z}}$ *is an almost automorphic solution on* $\mathbb{Z}$ *of equation* (5.101). *Conversely, if* $x(\cdot)$ *is an almost automorphic solution of* (5.101) *on* $\mathbb{Z}$, *then the function* $v : \mathbb{R} \to C_0$ *given, for* $t \in \mathbb{R}$, *by*

$$v(t) = \mathbb{U}(t,[t])x([t]) + \lim_{k\to+\infty} \int_{[t]}^{t} \mathbb{U}(t,s)\Upsilon^k f(s)ds$$

is almost automorphic. Moreover, equation (5.94) *has a compact almost automorphic solution* u *on* $\mathbb{R}$. *More precisely, this solution is given by* $u(t) = v(t)(0)$ *for all* $t \in \mathbb{R}$.

Proof. Suppose that $u : \mathbb{R} \to X$ is the almost automorphic solution of equation (5.94) in $\mathbb{R}$. Then, the sequence $(x(n))_{n\in\mathbb{Z}} := (u_n)_{n\in\mathbb{Z}}$ is a bounded solution of (5.101) on $\mathbb{Z}$. Hence, by Lemma 5.5.20, we deduce that $x(\cdot)$ is an almost automorphic solution of (5.101).

Conversely, as in the proof of Theorem 5.5.14, we have that

$$v(t) = \mathbb{U}(t,\sigma)v(\sigma) + \lim_{k\to+\infty} \int_{\sigma}^{t} \mathbb{U}(t,s)\Upsilon^k f(s)ds \quad \forall t \geq \sigma,$$

which means that the function $t \mapsto v(t)(0)$ is a solution of (5.94) on $\mathbb{R}$.

Let $(t_n''')_{n\in\mathbb{N}}$ be any sequence in $\mathbb{R}$. Set $\varsigma_n''' = t_n''' - [t_n''']$ for $n \in \mathbb{N}$. Obviously, we have that

$$(\varsigma_n''')_{n\in\mathbb{N}} \subset [0,1].$$

Then, there exist a subsequence $(\varsigma_n'')_{n\in\mathbb{N}}$ of $(\varsigma_n''')_{n\in\mathbb{N}}$ and $\varsigma_0 \in [0,1]$ such that

$$\lim_{n\to+\infty} \varsigma_n'' = \varsigma_0.$$

Since $x(\cdot)$ is almost automorphic, there exist a subsequence $([t_n'])_{n\in\mathbb{N}}$ of $([t_n''])_{n\in\mathbb{N}}$ and a sequence $\overline{x}(\cdot)$ such that

$$\lim_{n\to+\infty} x(m+[t_n']) = \overline{x}(m) \quad \text{and} \quad \lim_{n\to+\infty} \overline{x}(m-[t_n']) = x(m) \quad \text{for all } m \in \mathbb{Z}.$$

Also, there exist a subsequence $([t_n])_{n\in\mathbb{N}}$ of $([t'_n])_{n\in\mathbb{N}}$ and a measurable function $g : \mathbb{R} \to X$ such that

$$\lim_{n\to+\infty} f(t + [t_n]) = g(t) \quad \text{and} \quad \lim_{n\to+\infty} g(t - [t_n]) = f(t) \quad \text{for all } t \in \mathbb{R}.$$

For $n \in \mathbb{N}$, we have that

$$\begin{aligned} v(t_n + t) &= \mathbb{U}(t_n + t, [t_n] + [t])v([t_n] + [t]) + \lim_{k\to+\infty} \int_{[t_n]+[t]}^{t+t_n} \mathbb{U}(t + t_n, s)\Upsilon^k f(s)ds \\ &= \mathbb{U}(\varsigma_n + t - [t], 0)x([t_n] + [t]) \\ &\quad + \lim_{k\to+\infty} \int_0^{\varsigma_n+t-[t]} \mathbb{U}(\varsigma_n + t - [t], s)\Upsilon^k f([t_n] + [t] + s)ds. \end{aligned}$$

Let us consider the following function $\overline{v} : \mathbb{R} \to C_0$ defined, for $t \in \mathbb{R}$, by

$$\overline{v}(t) = \mathbb{U}(\varsigma_0 + t - [t], 0)\overline{x}([t]) + \lim_{k\to+\infty} \int_0^{\varsigma_0+t-[t]} \mathbb{U}(\varsigma_0 + t - [t], s)\Upsilon^k g([t] + s)ds.$$

Then,

$$\begin{aligned} |v(t_n + t) - \overline{v}(t)|_C &\le |\mathbb{U}(\varsigma_n + t - [t], 0)x([t_n] + [t]) - \mathbb{U}(\varsigma_0 + t - [t], 0)\overline{x}([t])|_C \\ &\quad + |u_{\varsigma_n+t-[t]}(\cdot, 0, 0, L, f([t_n] + [t] + \cdot)) - u_{\varsigma_0+t-[t]}(\cdot, 0, 0, L, g([t] + \cdot))|_C. \end{aligned}$$

Therefore,

$$\begin{aligned} |v(t_n + t) - \overline{v}(t)|_C &\le |\mathbb{U}(\varsigma_n + t - [t], 0)\{x([t_n] + [t]) - \overline{x}([t])\}|_C \\ &\quad + |\{\mathbb{U}(\varsigma_n + t - [t], 0) - \mathbb{U}(\varsigma_0 + t - [t], 0)\}\overline{x}([t])|_C \\ &\quad + 2|u_{\varsigma_0+t-[t]}(\cdot, 0, 0, L, f([t_n] + [t] + \cdot)) - u_{\varsigma_0+t-[t]}(\cdot, 0, 0, L, g([t] + \cdot))|_C \\ &\quad + |u_{\varsigma_n+t-[t]}(\cdot, 0, 0, L, f([t_n] + [t] + \cdot)) - u_{\varsigma_n+t-[t]}(\cdot, 0, 0, L, g([t] + \cdot))|_C \\ &\quad + |u_{\varsigma_n+t-[t]}(\cdot, 0, 0, L, g([t] + \cdot)) - u_{\varsigma_0+t-[t]}(\cdot, 0, 0, L, g([t] + \cdot))|_C. \end{aligned}$$

Using Lemma 5.5.17 and since the followings functions $t \mapsto u_t$, $(t, s) \mapsto \mathbb{U}(t, s)\phi$, for $\phi \in C_0$ are continuous, we deduce that

$$\lim_{n\to+\infty} |v(t_n + t) - \overline{v}(t)|_C = 0 \quad \text{for } t \in \mathbb{R}. \tag{5.103}$$

Now, we are going to show that

$$\lim_{n\to+\infty} |\overline{v}(t - t_n) - v(t)|_C = 0 \quad \text{for any } t \in \mathbb{R}.$$

Let us first consider the case $t-[t] > \varsigma_0$. Then, there exists $N(t) \in \mathbb{N}^*$ such that $\varsigma_n \leq t-[t]$, for all $n \geq N(t)$. In this situation, we get $[t-t_n] = [t]-[t_n]$. Therefore, for any $n \geq N(t)$,

$$\begin{aligned}\overline{v}(t-t_n) &= \mathbb{U}(\varsigma_0-\varsigma_n+t-[t],0)\overline{x}([t]-[t_n])\\ &\quad+\lim_{k\to+\infty}\int_0^{\varsigma_0-\varsigma_n+t-[t]}\mathbb{U}(\varsigma_0-\varsigma_n+t-[t],s)\Upsilon^k g([t]-[t_n]+s)ds.\end{aligned}$$

We can use the same reasoning used to show (5.103) and deduce

$$\lim_{n\to+\infty}|\overline{v}(t-t_n)-v(t)|_C = 0.$$

Next, we consider the case $t-[t] < \varsigma_0$. Then, there is $N(t) \in \mathbb{N}^*$ such that $\varsigma_n \geq t-[t]$, for all $n \geq N(t)$. In this case, we have that $[t-t_n] = [t]-[t_n]-1$, for any $n \geq N(t)$. Thus, for each $n \geq N(t)$,

$$\begin{aligned}\overline{v}(t-t_n) &= \mathbb{U}(\varsigma_0-\varsigma_n+t-[t]+1,0)\overline{x}([t]-[t_n]-1)\\ &\quad+\lim_{k\to+\infty}\int_0^{\varsigma_0-\varsigma_n+t-[t]+1}\mathbb{U}(\varsigma_0-\varsigma_n+t-[t]+1,s)\Upsilon^k g([t]-[t_n]-1+s)ds.\end{aligned}$$

By Lemma 5.5.17, the strong continuity of the process $(\mathbb{U}(t,s))_{t\geq s}$, and the continuity of the function $t \mapsto u_t$, we deduce that

$$\begin{aligned}\lim_{n\to+\infty}\overline{v}(t-t_n) &= \mathbb{U}(t-[t]+1,0)x([t]-1)\\ &\quad+\lim_{k\to+\infty}\int_0^{t-[t]+1}\mathbb{U}(t-[t]+1,s)\Upsilon^k f([t]-1+s)ds\\ &= \mathbb{U}(t,[t]-1)x([t]-1)+\lim_{k\to+\infty}\int_{[t]-1}^{t}\mathbb{U}(t,s)\Upsilon^k f(s)ds\\ &= \mathbb{U}(t,[t])\left\{\mathbb{U}([t],[t]-1)x([t]-1)+\lim_{k\to+\infty}\int_{[t]-1}^{[t]}\mathbb{U}([t],s)\Upsilon^k f(s)ds\cdot\right\}\\ &\quad+\lim_{k\to+\infty}\int_{[t]}^{t}\mathbb{U}(t,s)\Upsilon^k f(s)ds\\ &= \mathbb{U}(t,[t])x([t])+\lim_{k\to+\infty}\int_{[t]}^{t}\mathbb{U}(t,s)\Upsilon^k f(s)ds\\ &= v(t).\end{aligned}$$

Now, in the latter case, if $t - [t] = \varsigma_0$, there exists a subsequence $(\overline{\varsigma_n})_{n\in\mathbb{N}}$ of $(\varsigma_n)_{n\in\mathbb{N}}$, such that, for all $n \in \mathbb{N}$,

$$\overline{\varsigma_n} \le t - [t] \quad \text{or} \quad \overline{\varsigma_n} \ge t - [t].$$

Without restriction of generality, we can assume that, for all $n \in \mathbb{N}$,

$$\varsigma_n \le t - [t] \quad \text{or} \quad \varsigma_n \ge t - [t].$$

Either way, we are in one of the previous situations. Therefore, in any case we have

$$\lim_{n\to+\infty} \overline{v}(t - t_n) = v(t) \quad \text{for all } t \in \mathbb{R}.$$

Hence, the function $v(\cdot)$ is almost automorphic. Furthermore, $u(\cdot)$ is also almost automorphic. By Theorem 5.5.18, we have that u is compact almost automorphic on $\mathbb{R}$. □

Now, we use Theorem 5.5.21 to extended the Bohr–Neugebauer and Massera property to the partial functional differential equation (5.94).

Theorem 5.5.22. *Assume that* $(\mathbf{H}_0)$, $(\mathbf{H}_1)$, $(\mathbf{H}_2)$, *and* $(\mathbf{H}_4)$ *hold. If u is a bounded solution of equation* (5.94) *on* $\mathbb{R}$, *then the history function $t \mapsto u_t$ is almost automorphic and $u(\cdot)$ is compact almost automorphic.*

Proof. First, we observe that if $(u_n)_{n\in\mathbb{Z}}$ is a bounded solution of equation (5.101) on $\mathbb{Z}$, then, by Lemma 5.5.20, we deduce that $(u_n)_{n\in\mathbb{Z}}$ is almost automorphic.

Let $v : \mathbb{R} \to C_0$ be the function defined as in Theorem 5.5.21. For $t \in \mathbb{R}$, we have

$$\begin{aligned} v(t) &= \mathbb{U}(t,[t])u_{[t]} + \lim_{k\to+\infty} \int_{[t]}^{t} \mathbb{U}(t,s)\mathrm{Y}^k f(s)ds \\ &= u_t. \end{aligned}$$

From the previous theorem, we obtain the result. □

Corollary 5.5.23. *Assume that* $(\mathbf{H}_0)$, $(\mathbf{H}_1)$, $(\mathbf{H}_2)$, *and* $(\mathbf{H}_4)$ *hold. Let u be an almost automorphic solution of equation* (5.94) *on* $\mathbb{R}$. *Then, the function $t \mapsto u_t$ is almost automorphic.*

Theorem 5.5.24. *Assume that* $(\mathbf{H}_0)$, $(\mathbf{H}_1)$, $(\mathbf{H}_2)$, *and* $(\mathbf{H}_4)$ *hold. Assume further that equation* (5.94) *admits a bounded solution on* $\mathbb{R}^+$, *then it must have a compact almost automorphic solution on* $\mathbb{R}$.

Proof. Let u be the bounded solution of equation (5.94) on $\mathbb{R}^+$. Then, obviously, we have that $(u_n)_{n\in\mathbb{N}}$ is a bounded solution of difference equation (5.101) on $\mathbb{N}$. Therefore, by Lemma 5.5.20, equation (5.101) has an almost automorphic solution on $\mathbb{Z}$. From Theorem 5.5.21, we deduce that (5.94) has a compact almost automorphic solution on $\mathbb{R}$. □

5.5.8 Application

In this section, we will apply the previously obtained results to study the existence of almost periodic and almost automorphic solutions of the following model:

$$\begin{cases} \frac{\partial}{\partial t}w(t,x) = \frac{\partial^2}{\partial x^2}w(t,x) \\ \qquad + \xi_0(t)\mu(x)w(t-r,x) + \Gamma(t,x) & \text{for } t \in \mathbb{R}^+ \text{ and } x \in [0,\pi], \\ w(t,0) = w(t,\pi) = 0 & \text{for } t \in \mathbb{R}^+, \\ w(\theta,x) = \phi(\theta)(x) & \text{for } \theta \in [-r,0] \text{ and } x \in [0,\pi], \end{cases} \tag{5.104}$$

where $\xi_0 : \mathbb{R} \to \mathbb{R}$ is a continuous and 1-periodic function, $\mu \in C([0,\pi];\mathbb{R})$, $\phi \in C([-r,0];C([0,\pi];\mathbb{R}))$, and $\Gamma : \mathbb{R} \times [0,\pi] \to \mathbb{R}$ is a continuous function satisfying one of the following assumptions:

($\mathbf{A}_1$) Γ is almost periodic in t uniformly in $x \in [0,\pi]$;
($\mathbf{A}_2$) Γ is almost automorphic in t uniformly in $x \in [0,\pi]$.

In the following, we give an example of Γ which satisfies the assumptions ($\mathbf{A}_1$) and ($\mathbf{A}_2$).

Example 5.5.25.

(1) Let

$$\beta_1(t) = \sin(pt) + \sin(qt),$$

where $\frac{p}{q} \in \mathbb{R} \setminus \mathbb{Q}$. From [152], we have that β_1 is almost periodic. Consider $\Gamma : \mathbb{R} \times [0,\pi] \to \mathbb{R}$ which is defined by

$$\Gamma(t,x) = \beta_1(t)\delta_1(x) \quad \text{for } x \in [0,\pi] \text{ and } t \in \mathbb{R},$$

where $\delta_1 \in C([0,\pi];\mathbb{R})$. The function Γ satisfies assumption ($\mathbf{A}_1$).

(2) Let $y \in C([0,\pi];\mathbb{R})$. Then, the function

$$\Gamma(t,x) = \alpha_1(t)y(x) \quad \text{for } x \in [0,\pi] \text{ and } t \in \mathbb{R},$$

where α_1 is an almost automorphic function, satisfies condition ($\mathbf{A}_2$).

Let $X = C([0,\pi];\mathbb{R})$ be equipped with the uniform norm topology and $(A, D(A))$ be the linear operator defined by

$$\begin{cases} D(A) = \{y \in C^2([0,\pi];\mathbb{R}) : y(\pi) = y(0) = 0\}, \\ Ay = y'' \quad \text{for } y \in D(A). \end{cases}$$

From [113, Proposition 14.6, pp. 319–320], we have that

$$]0, +\infty[\subset \rho(A)$$

and

$$|\mathcal{R}(\lambda, A)^n|_{\mathcal{L}(X)} \le \frac{1}{\lambda^n} \quad \text{for } n \ge 1 \text{ and } \lambda > 0.$$

The part A_0 of A in $\overline{D(A)}$ is given by

$$\begin{cases} D(A_0) = \{y \in C^2([0, \pi]; \mathbb{R}) : y(\pi) = y(0) = y''(\pi) = y''(0) = 0\}, \\ A_0 y = Ay \quad \text{for } y \in D(A_0). \end{cases}$$

From [136], we have that the operator $(A_0, D(A_0))$ generates a strongly continuous compact semigroup $(T(t))_{t\ge0}$ on $\overline{D(A)}$. Then, the assumptions $(\mathbf{H}_0)$ and $(\mathbf{H}_2)$ hold. Moreover,

$$|T(t)|_{\mathcal{L}(\overline{D(A)})} \le e^{-t} \quad \text{for all } t \ge 0.$$

Consider the mapping $L : \mathbb{R} \times C \to X$ defined by

$$L(t)(\phi)(x) = \xi_0(t)\mu(x)\phi(-r)(x) \quad \text{for } x \in [0, \pi], t \in \mathbb{R}, \text{and } \phi \in C.$$

Then, condition $(\mathbf{H}_1)$ holds. Moreover, L is 1-periodic in time. Next, we consider the function $f : \mathbb{R} \to X$, given by

$$f(t)(x) = \Gamma(t, x) \quad \text{for } x \in [0, \pi] \text{ and } t \in \mathbb{R}.$$

Let $u : \mathbb{R}^+ \to X$ be defined, for $t \in \mathbb{R}$, by

$$u(t)(x) = w(t, x) \quad \text{for } x \in [0, \pi].$$

Then, the model (5.104) is equivalent to the following equation:

$$\begin{cases} \frac{d}{dt}u(t) = Au(t) + L(t)u_t + f(t) \quad \text{for } t \ge 0, \\ u_0 = \phi \in C_0. \end{cases} \tag{5.105}$$

Theorem 5.5.26. *Assume that the conditions stated above hold. We assume further that f is bounded and*

$$\sup_{t\in[0,1]} |\xi_0(t)| \sup_{x\in[0,\pi]} |\mu(x)| < 1.$$

Then, equation (5.105) *has a bounded solution on* $\mathbb{R}^+$.

Proof. Let

$$\varpi := 1 + \frac{\sup_{t\in\mathbb{R}} |f(t)|_X}{a_0},$$

where $\alpha_0 \in]0,1[$ is such that

$$\sup_{t\in[0,1]} |\xi_0(t)| \sup_{x\in[0,\pi]} |\mu(x)| < (1-\alpha_0).$$

Let $|\varphi|_C < \varpi$. We claim that

$$|u(t,\varphi)| \le \varpi \quad \text{for all } t \ge 0.$$

Indeed, suppose that the following set $\Lambda := \{t \in \mathbb{R}^+, |u(t,\varphi)|_X > \varpi\}$ is not empty and let $\tilde{t} := \inf\Lambda$. Using the continuity of $u(\cdot,\varphi)$, we obtain

$$|u(\tilde{t},\varphi)|_X = \varpi. \tag{5.106}$$

On the other hand, we have that

$$u(\tilde{t},\varphi) = T(\tilde{t})\varphi(0) + \lim_{\lambda\to\infty} \int_0^{\tilde{t}} T(\tilde{t}-s)B_\lambda\{L(s)u_s(.,\varphi) + f(s)\}ds.$$

Then,

$$|u(\tilde{t},\varphi)|_X \le e^{-\tilde{t}}\varpi + \int_0^{\tilde{t}} e^{-(\tilde{t}-s)}\{|L(s)u_s(.,\varphi)|_X + |f(s)|_X\}ds.$$

As $-r \le s+\theta \le \tilde{t}$, for all $\theta \le 0$, and $s \in [0,\tilde{t}]$, we have that

$$|L(s)u_s(\cdot,\varphi)|_X \le \varpi \sup_{t\in[0,1]} |\xi_0(t)| \sup_{x\in[0,\pi]} |\mu(x)| \le \varpi(1-\alpha_0) \quad \text{for all } s \in [0,\tilde{t}].$$

Thus,

$$\begin{aligned} |u(\tilde{t},\varphi)|_X &\le e^{-\tilde{t}}\varpi + \varpi(1-\alpha_0)\int_0^{\tilde{t}} e^{-(\tilde{t}-s)}ds + \sup_{t\in\mathbb{R}}|f(t)|_X \int_0^{\tilde{t}} e^{-(\tilde{t}-s)}ds \\ &\le e^{-\tilde{t}}\varpi + \Big\{\varpi - \alpha_0\varpi + \sup_{t\in\mathbb{R}}|f(t)|_X\Big\}(1-e^{-\tilde{t}}). \end{aligned}$$

By definition of ϖ, we conclude that

$$-\alpha_0\varpi + \sup_{t\in\mathbb{R}}|f(t)|_X = -\alpha_0.$$

Hence,

$$\begin{aligned}|u(\tilde{t},\varphi)|_X &\leq e^{-\tilde{t}}\varpi + (\varpi - \alpha_0)(1-e^{-\tilde{t}})\\ &\leq \varpi - \alpha_0(1-e^{-\tilde{t}})\\ &< \varpi,\end{aligned}$$

which contradicts (5.106). Then,

$$|u(t,\varphi)| \leq \varpi \quad \text{for all } t \geq 0.$$

Consequently, $u(\cdot,\varphi)$ is a bounded solution of equation (5.105) on $\mathbb{R}^+$. □

Corollary 5.5.27. *Assume that* $(\mathbf{A}_1)$ *(resp.* $(\mathbf{A}_2)$*) holds and*

$$\sup_{t\in[0,1]} |\xi_0(t)| \sup_{x\in[0,\pi]} |\mu(x)| < 1.$$

Then, equation (5.105) *admits an almost periodic (resp. compact almost automorphic) solution on* $\mathbb{R}$.

Proof. Under the assumption $(\mathbf{A}_1)$ (resp. $(\mathbf{A}_2)$), we obtain that f is almost periodic (resp. almost automorphic). On the other hand, from Theorem 5.5.26, we have that equation (5.105) has a bounded solution on $\mathbb{R}^+$. As a consequence of Theorem 5.5.15 (resp. Theorem 5.5.24), we conclude the result of this corollary. □

Theorem 5.5.28. *Assume that* $(\mathbf{A}_1)$ *(resp.* $(\mathbf{A}_2)$*) holds and*

$$\sup_{t\in[0,1]} |\xi_0(t)| \sup_{x\in[0,\pi]} |\mu(x)| < e^{-r}. \tag{5.107}$$

Then, equation (5.105) *has a unique almost periodic (resp. compact almost automorphic) solution which is globally attractive.*

Proof. Observe that

$$\sup_{t\in[0,1]} |\xi_0(t)| \sup_{x\in[0,\pi]} |\mu(x)| < e^{-r} < 1.$$

Then, by Corollary 5.5.27, equation (5.105) has an almost periodic (resp. compact almost automorphic) solution u on $\mathbb{R}$. Now, we claim that u is unique. Let $\tilde{u}$ be another almost periodic (resp. compact almost automorphic) solution of equation (5.105). Then, for $t \in \mathbb{R}$ and $\sigma \leq t$, we have that

$$|u_t - \tilde{u}_t|_C = |\mathbb{U}(t,\sigma)(u_\sigma - \tilde{u}_\sigma)|_C.$$

Using Theorem 5.5.3(iii), we can assert, for $t \geq \sigma$, that

$$|\mathbb{U}(t,\sigma)\phi|_C \leq e^{-r} e^{(e^{-r}\sup_{t\in[0,1]}|\xi_0(t)|\sup_{x\in[0,\pi]}|\mu(x)|-1)(t-\sigma)} |\phi|_C \quad \text{for all } \phi \in C_0.$$

Therefore,

$$|u_t - \tilde{u}_t|_C \leq e^{-r} e^{(e^{-r} \sup_{t\in[0,1]}|\xi_0(t)| \sup_{x\in[0,\pi]}|\mu(x)|-1)(t-\sigma)} \sup_{\eta\in\mathbb{R}}|u_\eta - \tilde{u}_\eta|_C. \tag{5.108}$$

Since

$$\sup_{t\in[0,1]} |\xi_0(t)| \sup_{x\in[0,\pi]} |\mu(x)| < e^{-r},$$

one gets

$$e^{-r} \sup_{t\in[0,1]} |\xi_0(t)| \sup_{x\in[0,\pi]} |\mu(x)| - 1 < 0.$$

Consequently, letting $\sigma \to -\infty$ in (5.108), we obtain

$$u_t = \tilde{u}_t.$$

Thus, u is unique. Now, let $\tilde{u}$ be any solution of equation (5.105) on $\mathbb{R}^+$. Then, for any $t \geq 0$,

$$\begin{aligned}|u_t - \tilde{u}_t|_C &= |\mathbb{U}(t,0)(u_0 - \tilde{u}_0)|_C \\ &\leq e^{-r} e^{(e^{-r} \sup_{t\in[0,1]}|\xi_0(t)| \sup_{x\in[0,\pi]}|\mu(x)|-1)t} |u_0 - \tilde{u}_0|_C.\end{aligned}$$

Therefore,

$$\lim_{t\to+\infty} |u_t - \tilde{u}_t|_C = 0.$$

That is, u is globally attractive. □

In the following, we select appropriate parameters in the system (5.104) and make corresponding numerical simulations to illustrate the results obtained in Theorem 5.5.28. Initially, we choose parameter examples that satisfy condition (5.107), specifically:

$$r = 1, \quad \xi_0(t) = \cos(2\pi t), \quad \text{and} \quad \mu(x) = \frac{1}{e^{x+1} + e^1}.$$

Numerical tests are performed for the following arbitrary initial conditions:
- $\phi_1(\theta)(x) = (\theta^2 - 0.2)(1 - \cos(2x))^3$;
- $\phi_2(\theta)(x) = (1 - e^{\theta^2}) \times (x(\pi - x))^3$;
- $\phi_3(\theta)(x) = -10 \arctan(-\theta) \sin(2x)$.

We start with the case of almost periodicity. For this purpose, we have selected the following example for $\Gamma(t, x)$:

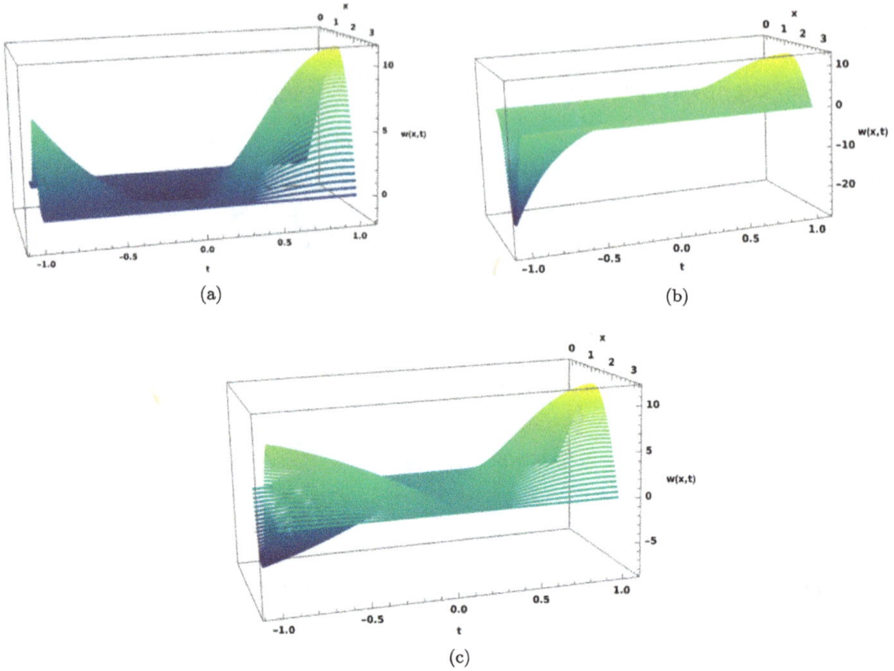

Figure 5.2: Computational solution in the time interval $[-1,1]$ corresponding to the initial condition: (a) ϕ_1, (b) ϕ_2, and (c) ϕ_3.

$$\Gamma(t,x) = (\sin(\pi t) + \sin(t))x(\pi - x)e^x.$$

The solutions corresponding to the initial conditions ϕ_1, ϕ_2, and ϕ_3 obviously cannot have the same trajectory at the beginning of time, as can be observed in the numerical results (see Figure 5.2). However, according to Theorem 5.5.28, all solutions fluctuate around the unique almost periodic solution, which implies that, from a certain point, they have the same asymptotic behavior. This fact is validated by the numerical results, as shown in Figure 5.3.

We turn our attention to the case of almost automorphy. To do this, we consider the following example of $\Gamma(t,x)$:

$$\Gamma(t,x) = \sin\left(\frac{1}{\cos t + \cos\sqrt{2}t + 2}\right)x(\pi - x)e^x.$$

Similar to the previous scenario, solutions may initiate with diverse behaviors, as illustrated in Figure 5.4. However, after a sufficiently long time, they exhibit compact almost automorphic behaviors. This is justified by the fact that the solutions are attracted by the unique compact almost automorphic solution. This phenomenon can be observed in Figure 5.5, which shows the effectiveness of the theoretical result obtained in Theorem 5.5.28.

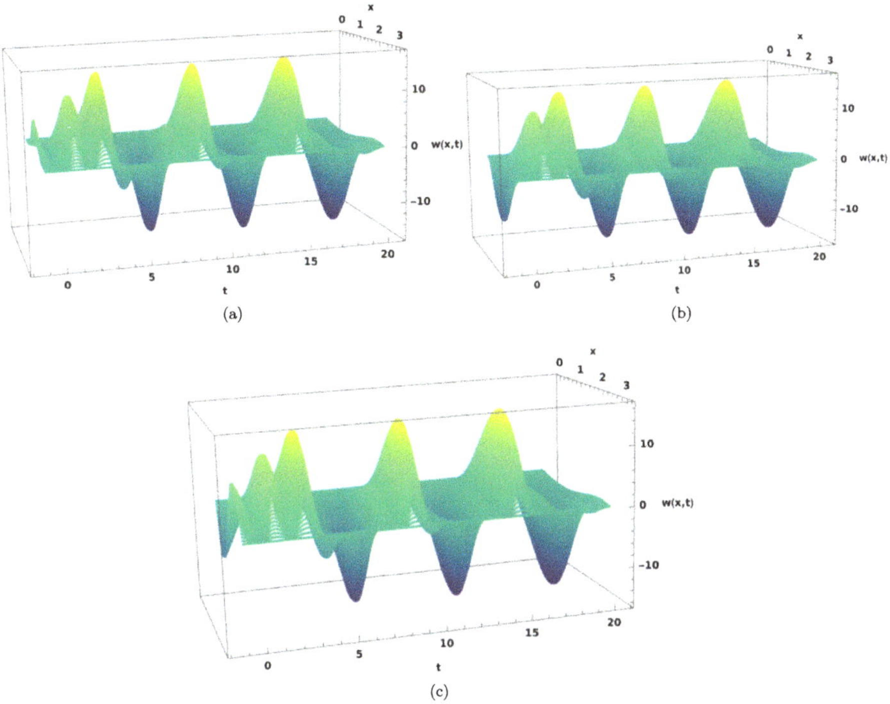

Figure 5.3: Long-time behavior of $w(t,x)$ associated to the initial condition: (a) ϕ_1, (b) ϕ_2, and (c) ϕ_3.

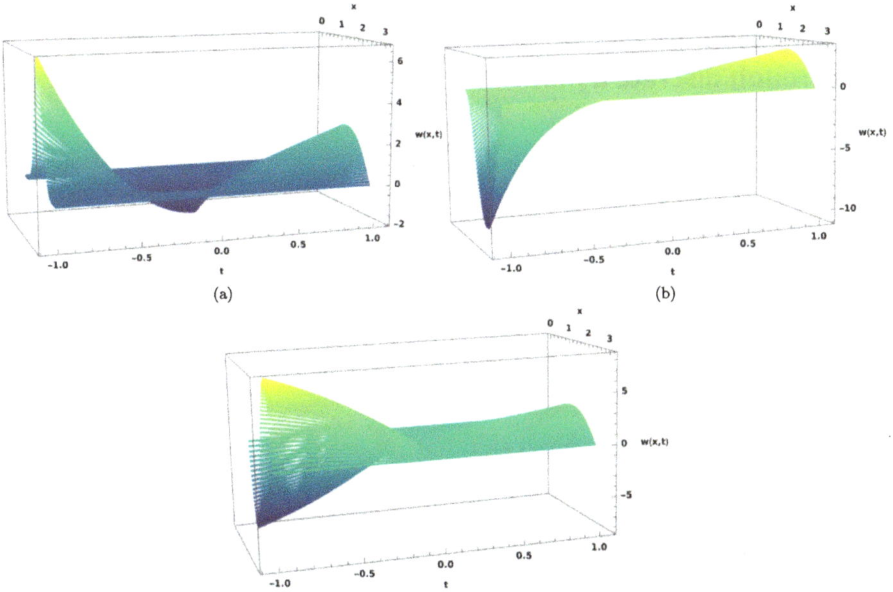

Figure 5.4: Function $w(t,x)$ in the time interval $[-1,1]$ with the initial condition: (a) ϕ_1, (b) ϕ_2, and (c) ϕ_3.

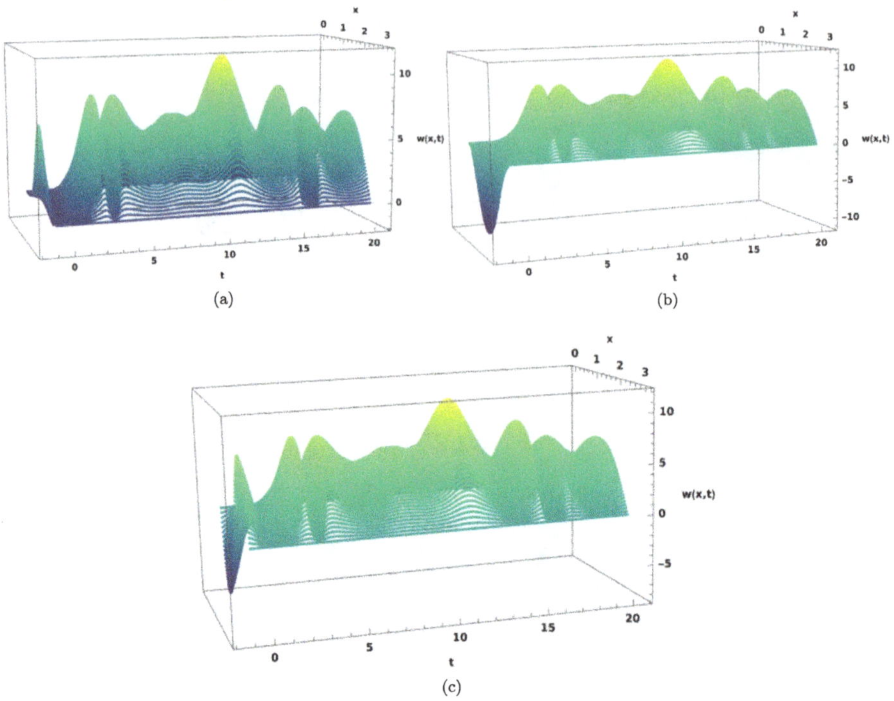

Figure 5.5: Long-time behavior of $w(t,x)$ associated to the initial condition: (a) ϕ_1, (b) ϕ_2, and (c) ϕ_3.

5.6 Nonautonomous partial functional differential equations through discrete dynamical systems: infinite delay case

5.6.1 Introduction

In this section, the main aim is to study the existence of almost periodic and almost automorphic integral solutions for the following nonautonomous linear partial functional differential equation with infinite delay:

$$\begin{cases} \frac{\mathrm{d}}{\mathrm{d}t}u(t) = Au(t) + L(t)u_t + f(t) & \text{for } t \geq 0, \\ u_0 = \phi \in \mathbb{B}, \end{cases} \tag{5.109}$$

where $(A, D(A))$ is a linear operator on a Banach space $(X, |\cdot|_X)$, with the condition $\overline{D(A)} = X$ being not necessary, and satisfies the following assumption:

($\mathbf{H}_0$) There exist $M_0 \geq 1$ and $\omega_0 \in \mathbb{R}$ such that $]\omega_0, +\infty[\subset \rho(A)$ and

$$|\mathcal{R}(\lambda, A)^n|_{\mathcal{L}(X)} \leq \frac{M_0}{(\lambda - \omega)^n} \quad \text{for } n \in \mathbb{N} \text{ and } \lambda > \omega,$$

where $\mathcal{R}(\lambda, A) = (\lambda I_X - A)^{-1}$ while $\mathbb{B}$ is a normed linear space of mappings from $]-\infty, 0]$ into X satisfying some fundamental axioms that will be introduced later. For $t \geq \sigma$, the mapping $u_t :]-\infty, 0] \to X$ is defined for $\theta \leq 0$ by $u_t(\theta) = u(t + \theta)$, which is called the history function, $f : \mathbb{R} \to X$ is a continuous function and $(L(t))_{t\in\mathbb{R}} \subset \mathcal{L}(\mathbb{B}, X)$ is such that the following hypothesis is satisfied:

(H$_1$) For each $\phi \in \mathbb{B}$, the mapping $t \to L(t)\phi$ is continuous.

In [180] and [226], the authors discussed the problem of the existence of almost periodic and almost automorphic solutions of equation (5.109) when $\overline{D(A)} = X$. Precisely, they used the variation of constants formula of the mild solutions in the phase space to prove that the problem is reduced to a linear difference equation. In the case when $\overline{D(A)} \neq X$, the variation of constants formula used in [180] and [226] is no longer applicable. In this chapter, we make a new representation of integral solutions of equation (5.109) in the phase space appropriate to our system. Next, we use this formula to discuss the existence of the almost periodic and almost automorphic integral solutions for equation (5.109), when the semigroup generated by the part of A on $\overline{D(A)}$ is compact and $\mathbb{B}$ is a uniform fading memory space.

5.6.2 Phase space and integral solutions

In this section, we will recall some basic concepts and results which will be used in the sequel of this work.

The space $(\mathbb{B}, |\cdot|_{\mathbb{B}})$ is a normed linear space formed by functions mapping $]-\infty, 0]$ into X and assumed to satisfy the following axioms:

(A) if $x :]-\infty, a] \to X$, $a \in \mathbb{R}$, is continuous on $[\sigma, a]$ with $x_\sigma \in \mathbb{B}$, for some $\sigma < a$, then the following assertions hold:
  (i) $x_t \in \mathbb{B}$ for all $t \geq \sigma$;
  (ii) $t \mapsto x_t$ is continuous on $[\sigma, a]$;
  (iii) $N|x(t)|_X \leq |x_t|_{\mathbb{B}} \leq K(t - \sigma) \sup_{\sigma\leq s\leq t}|x(s)|_X + M(t - \sigma)|x_\sigma|_{\mathbb{B}}$,
  where $N > 0$, $K : \mathbb{R}^+ \to \mathbb{R}^+$ is a continuous function and $M : \mathbb{R}^+ \to \mathbb{R}^+$ is a locally bounded function, both independent of x.

(B) $\mathbb{B}$ is a Banach space.

Lemma 5.6.1 ([181]). *Assume that the space $\mathbb{B}$ satisfies the axiom* (**A**). *Then, the following assertions hold:*

(i) *For all $\phi \in \mathbb{B}$, we have*

$$|\phi(0)|_X \leq \frac{1}{N}|\phi|_{\mathbb{B}}.$$

(ii) *The space of all continuous functions from* $]-\infty, 0]$ *into* X *with compact support, which is denoted by* $C_{00}(]-\infty, 0]; X)$, *is included in* $\mathbb{B}$. *Moreover, for every* $\phi \in C_{00}(]-\infty, 0]; X)$ *with the support in* $[a, 0]$, *we have*

$$|\phi|_{\mathbb{B}} \leq K(-a) \sup_{\theta \leq 0} |\phi(\theta)|_X.$$

Example 5.6.2 ([181]). Let the following functional spaces be endowed with the uniform norm:
- $\mathrm{BUC}(]-\infty, 0]; X)$,
- $\{\phi \in \mathrm{BC}(\mathbb{R}^-; X) : \lim_{\theta \to -\infty} \phi(\theta) \text{ exists in } X\}$,
- $\{\phi \in \mathrm{BC}(\mathbb{R}^-; X) : \lim_{\theta \to -\infty} \phi(\theta) = 0_X\}$.

They satisfy the axioms **(A)** and **(B)**.

Example 5.6.3 ([181]). Let $r, \gamma > 0$. Define $C_r \times L^p_\gamma$ as the space all measurable functions $\phi :]-\infty, 0] \to X$ which are continuous on $[-r, 0]$ such that $\theta \mapsto e^{\gamma\theta} |\phi(\theta)|^p_X$ is integrable on $]-\infty, -r]$ endowed with following norm:

$$|\phi|_{\mathbb{B}} := \sup_{\theta \in [-r, 0]} |\phi(\theta)|_X + \left(\int_{-\infty}^{-r} e^{\gamma\theta} |\phi(\theta)|^p_X \, d\theta \right)^{\frac{1}{p}}.$$

The space $C_r \times L^p_\gamma$ satisfies the axioms **(A)** and **(B)**.

In the rest of this chapter, we suppose that **(A)** and **(B)** hold on $\mathbb{B}$.

Definition 5.6.4 ([1]). Let $\phi \in \mathbb{B}$. We say that a function $u : \mathbb{R} \to X$ is an integral solution of equation (5.109) in $\mathbb{R}$ if the following statements hold:
(i) u is continuous on $[\sigma, +\infty[$,
(ii) $\int_\sigma^t u(s) ds \in D(A)$ for $t \geq \sigma$,
(iii) $u(t) = \phi(0) + A(\int_\sigma^t u(s) ds) + \int_\sigma^t (L(s)u_s + f(s)) ds$ for $t \geq \sigma$,
(iv) $u_\sigma = \phi$.

Remark 5.6.5. Note that $\phi(0) \in \overline{D(A)}$ is a necessary condition for an integral solution to exist. Indeed, (i) and (ii) imply that

$$u(t) = \lim_{h \to 0^+} \int_t^{t+h} u(s) ds \in \overline{D(A)} \quad \text{for all } t \geq \sigma.$$

In particular,

$$\phi(0) = \phi(\sigma - \sigma) = u(\sigma) \in \overline{D(A)}.$$

Hence, the phase space associated with equation (5.109) is the following:

$$\mathbb{B}_A := \{\phi \in \mathbb{B} : \phi(0) \in \overline{D(A)}\}.$$

Let $(A_0, D(A_0))$ be the part of $(A_0, D(A_0))$ on $\overline{D(A)}$. Then, $(A_0, D(A_0))$ generates a strongly continuous semigroup $(T(t))_{t\geq 0}$ on $\overline{D(A)}$. Moreover,

$$|T(t)|_{\mathcal{L}(\overline{D(A)})} \leq M_0 e^{\omega_0 t} \quad \text{for } t \geq 0.$$

The theorem stated below establishes the existence and uniqueness of an integral solution in X, along with a practical variation of constants formula.

Theorem 5.6.6 ([1]). *Let conditions* $(\mathbf{H}_0)$ *and* $(\mathbf{H}_1)$ *be satisfied. Then, for each* $\phi \in \mathbb{B}_A$, *equation* (5.109) *has a unique global integral solution* u *which satisfies the following formula:*

$$u(t) = T(t-\sigma)\phi(0) + \lim_{\lambda\to+\infty} \int_\sigma^t T(t-s)B_\lambda(L(s)u_s + f(s))ds \quad \textit{for } t \geq \sigma,$$

where $B_\lambda = \lambda\mathcal{R}(\lambda, A)$ *for* $\lambda > \omega_0$.

In the next part of this work, the integral solutions of equation (5.109) are denoted by $u(\cdot, \sigma, \phi, L, f)$. For each $t \geq \sigma$, we define the operator $\mathbb{U}(t,\sigma)$, for $\phi \in \mathbb{B}_A$, by

$$\mathbb{U}(t,\sigma)\phi = u_t(\cdot, \sigma, \phi, L, 0).$$

As an important result (see the following theorem), the family $(\mathbb{U}(t,\sigma))_{t\geq\sigma}$ is an evolution family on the phase space $\mathbb{B}_A$. This plays a crucial role in constructing proofs in this chapter.

Without loss of generality, we can assume that $\omega_0 > 0$.

Theorem 5.6.7. *Suppose that* $(\mathbf{H}_0)$ *and* $(\mathbf{H}_1)$ *are true. Then, the family* $(\mathbb{U}(t,\sigma))_{t\geq\sigma}$ *is a strongly continuous evolutionary process on* $\mathbb{B}_A$. *More precisely, the family* $(\mathbb{U}(t,\sigma))_{t\geq\sigma}$ *satisfies the properties below:*

(i) $\mathbb{U}(t,t) = I_{\mathbb{B}_A}$ *for all* $t \in \mathbb{R}$.

(ii) $\mathbb{U}(t,\sigma)\mathbb{U}(\sigma,r) = \mathbb{U}(t,r)$ *for all* $t \geq \sigma \geq r$.

(iii) *For all* $\phi \in \mathbb{B}$ *and* $t \geq \sigma$,

$$|\mathbb{U}(t,\sigma)\phi|_{\mathbb{B}} \leq m_1(t,\sigma)e^{m_2(t,\sigma)(t-\sigma)}|\phi|_{\mathbb{B}},$$

where

$$m_1(t,\sigma) := \frac{M_0}{N}e^{\omega_0(t-\sigma)} \sup_{\eta\in[0,t-\sigma]} K(\eta) + \sup_{\eta\in[0,t-\sigma]} M(\eta),$$

$$m_2(t,\sigma) := M_0^2 e^{\omega_0(t-\sigma)} \sup_{\eta\in[0,t-\sigma]} K(\eta) \sup_{\eta\in[\sigma,t]} |L(\eta)|_{\mathcal{L}(\mathbb{B},X)}.$$

(iv) *For any* $\phi \in \mathbb{B}$, *the mapping* $(t,\sigma) \to \mathbb{U}(t,\sigma)\phi$ *is continuous at any* $(t,\sigma) \in \{(\alpha,\beta) : \alpha \geq \beta\}$,

(v) *If L is q-periodic in time, then*

$$\mathbb{U}(t+q,\sigma+q)=\mathbb{U}(t,\sigma) \quad \text{for all } t \geq \sigma.$$

That is, $(\mathbb{U}(t,\sigma))_{t\geq\sigma}$ is q-periodic.

Proof. Claims (i) and (ii) are due to the definition and uniqueness of the integral solution.

(iii) By axiom **(A)**, we can observe, for $t \geq \sigma$ and $\phi \in \mathbb{B}$, that

$$|\mathbb{U}(\xi,\sigma)\phi|_{\mathbb{B}} \leq K(\xi-\sigma) \sup_{\eta\in[\sigma,\xi]} |\mathbb{U}(\eta,\sigma)\phi(0)|_X + M(\xi-\sigma)|\phi|_{\mathbb{B}} \quad \text{for } \xi \in [\sigma,t].$$

On the other hand,

$$|\mathbb{U}(\eta,\sigma)\phi(0)|_X \leq M_0 e^{\omega_0(\eta-\sigma)}|\phi(0)|_X + \int_\sigma^\eta M_0^2 e^{\omega_0(\eta-s)}|L(s)|_{\mathcal{L}(\mathbb{B},X)}|\mathbb{U}(s,\sigma)\phi|_{\mathbb{B}}ds,$$

for $\eta \in [\sigma,\xi]$. Therefore,

$$\sup_{\eta\in[\sigma,\xi]} |\mathbb{U}(\eta,\sigma)\phi(0)|_X \leq \frac{M_0}{N} e^{(t-\sigma)\omega_0}|\phi|_{\mathbb{B}} + \int_\sigma^\xi M_0^2 e^{(t-\sigma)\omega_0} \sup_{\eta\in[\sigma,t]} |L(\eta)|_{\mathcal{L}(\mathbb{B},X)}|\mathbb{U}(s,\sigma)\phi|_{\mathbb{B}}ds.$$

Consequently,

$$\begin{aligned}|\mathbb{U}(\xi,\sigma)\phi|_{\mathbb{B}} \leq {} & \left(\frac{M_0}{N} e^{(t-\sigma)\omega_0} \sup_{\eta\in[0,t-\sigma]} K(\eta) + \sup_{\eta\in[0,t-\sigma]} M(\eta)\right)|\phi|_{\mathbb{B}} \\ & + \sup_{\eta\in[0,t-\sigma]} K(\eta) \int_\sigma^\xi M_0^2 e^{(t-\sigma)\omega_0} \sup_{\eta\in[\sigma,t]} |L(\eta)|_{\mathcal{L}(\mathbb{B},X)}|\mathbb{U}(s,\sigma)\phi|_{\mathbb{B}}ds.\end{aligned}$$

As a result of Gronwall's lemma, we deduce that

$$|\mathbb{U}(\xi,\sigma)\phi|_{\mathbb{B}} \leq m_1(t,\sigma)e^{m_2(t,\sigma)(\xi-\sigma)}|\phi|_{\mathbb{B}} \quad \text{for } \xi \in [\sigma,t].$$

(v) Let us start by showing, for any $\phi \in \mathbb{B}$ and $t \in \mathbb{R}$, that the mapping $\sigma \to \mathbb{U}(t,\sigma)\phi$ is continuous at any $\sigma \leq t$. In fact, let $\sigma < t$ and $h > 0$ be small enough, then

$$\mathbb{U}(t,\sigma+h)\phi - \mathbb{U}(t,\sigma)\phi = \mathbb{U}(t,\sigma+h)(\phi - \mathbb{U}(\sigma+h,\sigma)\phi).$$

Therefore,

$$|\mathbb{U}(t,\sigma+h)\phi - \mathbb{U}(t,\sigma)\phi|_{\mathbb{B}} \leq m_1(t,\sigma+h)e^{m_2(t,\sigma+h)h}|\phi - \mathbb{U}(\sigma+h,\sigma)\phi|_{\mathbb{B}}.$$

By the definition of $m_1(\cdot)$ and $m_2(\cdot)$, we can see that

$$m_1(t,\sigma+h) \le m_1(t,\sigma) \quad \text{and} \quad m_2(t,\sigma+h) \le m_2(t,\sigma).$$

Then,

$$\left|\mathbb{U}(t,\sigma+h)\phi - \mathbb{U}(t,\sigma)\phi\right|_{\mathbb{B}} \le m_1(t,\sigma)e^{m_2(t,\sigma)h}\left|\phi - \mathbb{U}(\sigma+h,\sigma)\phi\right|_{\mathbb{B}}.$$

Thus,

$$\lim_{h\to 0^+}\left|\mathbb{U}(t,\sigma+h)\phi - \mathbb{U}(t,\sigma)\phi\right|_{\mathbb{B}} = 0.$$

Letting $\sigma \le t$ and $h > 0$ be small enough, we have

$$\begin{aligned}
&\left|\mathbb{U}(t,\sigma-h)\phi - \mathbb{U}(t,\sigma)\phi\right|_{\mathbb{B}} \\
&\quad \le \left|\mathbb{U}(t,\sigma)\right|_{\mathcal{L}(\mathbb{B}_A)}\left|(\mathbb{U}(\sigma,\sigma-h)\phi - \phi)\right|_{\mathbb{B}} \\
&\quad \le \left|\mathbb{U}(t,\sigma)\right|_{\mathcal{L}(\mathbb{B}_A)}\left(\left|\mathbb{U}(\sigma,\sigma-h)\phi - \mathbb{U}(\sigma+h,\sigma)\phi\right|_{\mathbb{B}} + \left|\mathbb{U}(\sigma,\sigma-h)\phi - \phi\right|_{\mathbb{B}}\right).
\end{aligned}$$

On the other hand, by the definition and due to the continuity of the integral solution, we can affirm that

$$\mathbb{U}(\sigma,\sigma-h)\phi - \mathbb{U}(\sigma+h,\sigma)\phi \in C_{00}(]-\infty,0];X).$$

Therefore, by Lemma 5.6.1, we obtain

$$\left|\mathbb{U}(\sigma,\sigma-h)\phi - \mathbb{U}(\sigma+h,\sigma)\phi\right|_{\mathbb{B}} \le K(-h)\sup_{\theta\in[-h,0]}\left|\mathbb{U}(\sigma,\sigma-h)\phi(\theta) - \mathbb{U}(\sigma+h,\sigma)\phi(\theta)\right|_X.$$

Consequently,

$$\lim_{h\to 0^+}\left|\mathbb{U}(\sigma,\sigma-h)\phi - \mathbb{U}(\sigma+h,\sigma)\phi\right|_{\mathbb{B}} = 0.$$

Then,

$$\lim_{h\to 0^+}\left|\mathbb{U}(t,\sigma-h)\phi - \mathbb{U}(t,\sigma)\phi\right|_{\mathbb{B}} = 0.$$

Thus, the mapping $\sigma \to \mathbb{U}(t,\sigma)\phi$ is continuous at any $\sigma \le t$.

Now, we are going to show, for each $\phi \in \mathbb{B}$, that the mapping $(t,\sigma) \to \mathbb{U}(t,\sigma)\phi$ is continuous. First, we start with the case of $t > \sigma$. Then there exists $t_0 \in \mathbb{R}$ such that $\sigma < t_0 < t$. For $(t',\sigma') \in \mathcal{O}_{t_0} := \{(\alpha,\beta) \in \mathbb{R} : \beta < t_0 < \alpha\}$, we have

$$\begin{aligned}
&\left|\mathbb{U}(t',\sigma')\phi - \mathbb{U}(t,\sigma)\phi\right|_{\mathbb{B}} \\
&\quad \le \left|\mathbb{U}(t',\sigma')\phi - \mathbb{U}(t',\sigma)\phi\right|_{\mathbb{B}} + \left|\mathbb{U}(t',\sigma)\phi - \mathbb{U}(t,\sigma)\phi\right|_{\mathbb{B}} \\
&\quad \le \left|\mathbb{U}(t',t_0)\right|_{\mathcal{L}(\mathbb{B}_A)}\left|\mathbb{U}(t_0,\sigma')\phi - \mathbb{U}(t_0,\sigma)\phi\right|_{\mathbb{B}} + \left|\mathbb{U}(t',\sigma)\phi - \mathbb{U}(t,\sigma)\phi\right|_{\mathbb{B}} \\
&\quad \le m_1(t',t_0)e^{m_2(t',t_0)(t'-t_0))}\left|\mathbb{U}(t_0,\sigma')\phi - \mathbb{U}(t_0,\sigma)\phi\right|_{\mathbb{B}} + \left|\mathbb{U}(t',\sigma)\phi - \mathbb{U}(t,\sigma)\phi\right|_{\mathbb{B}}.
\end{aligned}$$

Consequently,

$$\lim_{(t',\sigma')\to(t,\sigma)} |\mathbb{U}(t',\sigma')\phi - \mathbb{U}(t,\sigma)\phi|_{\mathbb{B}} = 0.$$

Let us consider now the case $t = \sigma$. Let $(t_n, \sigma_n)_{n\in\mathbb{N}} \in \mathbb{R}^2$ be such that $t_n \geq \sigma_n$ for all $n \in \mathbb{N}$ and

$$\lim_{n\to+\infty}(t_n, \sigma_n) = (t, t).$$

Then,

$$|\mathbb{U}(t_n,\sigma_n)\phi - \mathbb{U}(t,t)\phi|_{\mathbb{B}} \leq |\mathbb{U}(t_n,\sigma_n)\phi - \mathbb{U}(t_n - \sigma_n, 0)\phi|_{\mathbb{B}} + |\mathbb{U}(t_n - \sigma_n, 0)\phi - \phi|_{\mathbb{B}}.$$

For all $n \in \mathbb{N}$, the mapping $\theta \to \mathbb{U}(t_n,\sigma_n)\phi(\theta) - \mathbb{U}(t_n - \sigma_n, 0)\phi(\theta)$ is in $C_{00}(]-\infty, 0]; X)$. Therefore,

$$|\mathbb{U}(t_n,\sigma_n)\phi - \mathbb{U}(t_n - \sigma_n, 0)\phi|_{\mathbb{B}} \leq K(t_n - \sigma_n) \sup_{\theta\in[-(t_n-\sigma_n),0]} |\mathbb{U}(t_n,\sigma_n)\phi(\theta) - \mathbb{U}(t_n - \sigma_n, 0)\phi(\theta)|_X.$$

Since $\lim_{n\to+\infty}(t_n, \sigma_n) = (t, t)$, one has $\lim_{n\to+\infty} t_n - \sigma_n = 0$. Consequently,

$$\lim_{n\to+\infty} |\mathbb{U}(t_n,\sigma_n)\phi - \mathbb{U}(t_n - \sigma_n, 0)\phi|_{\mathbb{B}} = 0.$$

Then, we conclude that

$$\lim_{n\to+\infty} |\mathbb{U}(t_n,\sigma_n)\phi - \phi|_{\mathbb{B}} = 0,$$

completing the proof of (v).

Claim (v) is a consequence of the uniqueness of the integral solution. □

We end this section with the following lemma, which will be needed to show several main results of this section.

Lemma 5.6.8. *Assume that* $(\mathbf{H}_0)$ *and* $(\mathbf{H}_1)$ *hold. Let* $(\zeta_n)_{n\in\mathbb{N}} \in L^1_{loc}(\mathbb{R}^+, X)$ *be such that*

$$|\zeta_n(t)|_X < \tilde{\zeta}(t) \quad \text{for a. e. } t \in \mathbb{R}^+,$$

and for some $\tilde{\zeta} \in L^1_{loc}(\mathbb{R}^+, X)$*. Suppose that* $\lim_{n\to+\infty} \zeta_n(t) = \zeta(t)$ *for a. e.* $t \in \mathbb{R}^+$*, for a measurable function* ζ*, then*

$$\lim_{n\to+\infty} \sup_{t\in[0,b]} |u_t(\cdot, 0, 0, L, \zeta_n) - u_t(\cdot, 0, 0, L, \zeta)|_{\mathbb{B}_A} = 0$$

for any $[0, b] \subset \mathbb{R}^+$.

Proof. Let $t \in [0, b]$. Then,

$$
\begin{aligned}
&|u(t,0,0,L,\zeta_n) - u(t,0,0,L,\zeta)|_X \\
&\quad \le \int_0^t M_0^2 e^{\omega_0(t-s)} |L(s)|_{\mathcal{L}(\mathbb{B},X)} |u_s(\cdot,0,L,\zeta_n) - u_s(\cdot,0,L,\zeta)|_{\mathbb{B}} \mathrm{d}s \\
&\quad\quad + \int_0^t M_0^2 e^{\omega_0(t-s)} |\zeta_n(s) - \zeta(s)|_X \mathrm{d}s.
\end{aligned}
$$

On the other hand, we have that

$$
|u_t(\cdot,0,0,L,\zeta_n) - u_t(\cdot,0,0,L,\zeta)|_{\mathbb{B}} \le K(t) \sup_{s\in[0,t]} |u(s,0,0,L,\zeta_n) - u(s,0,0,L,\zeta)|_X.
$$

Therefore,

$$
\begin{aligned}
&|u_t(\cdot,0,0,L,\zeta_n) - u_t(\cdot,0,0,L,\zeta)|_{\mathbb{B}} \\
&\quad \le \int_0^t \tilde{K} \sup_{r\in[0,b]} |L(r)|_{\mathcal{L}(\mathbb{B},X)} |u_s(\cdot,0,0,L,\zeta_n) - u_s(\cdot,0,0,L,\zeta)|_{\mathbb{B}} \mathrm{d}s \\
&\quad\quad + \int_0^t \tilde{K} |\zeta_n(s) - \zeta(s)|_X \mathrm{d}s,
\end{aligned}
$$

where $\tilde{K} := \sup_{\eta\in[0,b]} M_0^2 K(\eta) e^{\omega_0 \eta}$. Using Gronwall's lemma, we obtain that

$$
\sup_{t\in[0,b]} |u_t(\cdot,0,0,L,\zeta_n) - u_t(\cdot,0,0,L,\zeta)|_{\mathbb{B}} \le e^{\tilde{K}b \sup_{r\in[0,b]} |L(r)|_{\mathcal{L}(\mathbb{B},X)}} \left(\int_0^b \tilde{K} |\zeta_n(s) - \zeta(s)|_X \mathrm{d}s \right).
$$

As a consequence of Lebesque's dominated convergence theorem, we have the result of this lemma. □

5.6.3 A new representation of integral solutions

In this subsection, we will give a proof of the first main result of this section. It is about a representation of integral solutions of equation (5.109) in terms of evolutionary process $(\mathbb{U}(t,s))_{t\ge s}$. To reach this goal, we consider, for $n > \omega_0$, the linear operator $\Upsilon^n : X \to \mathbb{B}$ given, for $x \in X$, by

$$
\Upsilon^n x(\theta) = \begin{cases} (n\theta + 1) B_n x & \text{for } -\frac{1}{n} \le \theta \le 0, \\ 0 & \text{for } \theta \le -\frac{1}{n}. \end{cases}
$$

Note that, for any $x \in X$, $\Upsilon^n x \in C_{00}(\mathbb{R}^-; X)$, thus by Lemma 5.6.1, we get $\Upsilon^n x \in \mathbb{B}$ and

$$|\Upsilon^n x|_{\mathbb{B}} \leq K(-1) \sup_{\theta \leq 0} |\Upsilon^n x(\theta)|_X.$$

Therefore, there exists $\tilde{M} > 0$ such that

$$|\Upsilon^n x|_{\mathbb{B}} \leq \tilde{M} K(-1) |x|_X.$$

Since, for any $x \in X$, $\Upsilon^n x(0) = B_n x \in \overline{D(A)}$, thus $\Upsilon^n x \in \mathbb{B}_A$.

Theorem 5.6.9. *Assume that* $(\mathbf{H}_0)$ *and* $(\mathbf{H}_1)$ *hold. Then, for all* $t \geq \sigma$,

$$\lim_{n \to +\infty} \int_\sigma^t \mathbb{U}(t,s) \Upsilon^n f(s) ds = u_t(\cdot, \sigma, 0, L, f).$$

Moreover, the above limit exists uniformly on any compact interval in $[\sigma, +\infty[$.

Proof. Let $a > \sigma$, $t \in [\sigma, a]$, and $n > \omega_0$. Then, the integral

$$\int_\sigma^t \mathbb{U}(t,s) \Upsilon^n f(s) ds$$

is the limit in $\mathbb{B}$-norm of the following Riemann sum:

$$\Pi_n^N(\sigma, t) := \frac{t - \sigma}{N} \sum_{k=0}^{N-1} \mathbb{U}(t, \eta_k) \Upsilon^n f(\eta_k),$$

where $\eta_k = \sigma + \frac{(t-\sigma)k}{N}$. We can see, for $\theta \leq 0$, that

$$\begin{aligned}
\Pi_n^N(\sigma, t)(\theta) &= \frac{t - \sigma}{N} \sum_{k=0}^{N-1} \mathbb{U}(t, \eta_k) \Upsilon^n f(\eta_k)(\theta) \\
&= \frac{t - \sigma}{N} \sum_{k=0}^{N-1} u_t(\theta, \eta_k, \Upsilon^n f(\eta_k), L, 0) \\
&= \frac{t - \sigma}{N} \sum_{k=0}^{N-1} u(t + \theta, \eta_k, \Upsilon^n f(\eta_k), L, 0).
\end{aligned}$$

Then, for all $\theta \leq 0$, $\Pi_n^N(\sigma, t)(\theta)$ is the Riemann sum in X-norm of the following integral:

$$\Psi^n(\sigma, t)(\theta) := \int_\sigma^t u(t + \theta, s, \Upsilon^n f(s), L, 0) ds.$$

In addition, by Heine's theorem, we conclude the uniform continuity of the function $(\theta, s) \to u(t+\theta, s, \Upsilon^n f(s), L, 0)$ on $[-(t-\sigma)-1, 0] \times [\sigma, t]$. Consequently,

$$\sup_{\theta \in [-(t-\sigma)-1,0]} \left|\Psi^n(\sigma,t)(\theta) - \Pi_n^N(\sigma,t)(\theta)\right|_X \to 0 \quad \text{as } N \to +\infty.$$

Observe that $\Psi^n(\sigma,t), (\Pi_n^N(\sigma,t))_{N\geq 1} \in C_{00}(\mathbb{R}^-; X)$ with the support included in $[-(t-\sigma)-1, 0]$. As a result of Lemma 5.6.1, we have that

$$\left|\Psi^n(\sigma,t) - \Pi_n^N(\sigma,t)\right|_{\mathbb{B}} \leq K(t-\sigma+1) \sup_{\theta \in [-(t-\sigma)-1,0]} \left|\Psi^n(\sigma,t)(\theta) - \Pi_n^N(\sigma,t)(\theta)\right|_X.$$

Then, $(\Pi_n^N(\sigma,t))_{N\geq 1}$ converges to $\Psi^n(\sigma,t)$ in $\mathbb{B}$-norm. Therefore,

$$\int_\sigma^t \mathbb{U}(t,s)\Upsilon^n f(s)ds = \Psi^n(\sigma,t).$$

By the definition of an integral solution, we can assert, for $s \leq t+\theta$, that

$$u(t+\theta, s, \Upsilon^n f(s), L, 0) = T(t+\theta-s)B_n f(s) + \lim_{\lambda\to+\infty} \int_s^{t+\theta} T(t+\theta-\eta)B_\lambda L(\eta)(\mathbb{U}(\eta,s))\Upsilon^n f(s))d\eta.$$

We have also that

$$u(t+\theta, s, \Upsilon^n f(s), L, 0) = \begin{cases} (n(t+\theta-s)+1)B_n f(s) & \text{for } t+\theta \leq s \leq t+\theta+\frac{1}{n}, \\ 0 & \text{for } s \geq t+\theta+\frac{1}{n}. \end{cases}$$

Therefore,

$$\Psi^n(\sigma,t)(\theta)$$
$$= \begin{cases} \int_{t+\theta}^{\min\{t,t+\theta+\frac{1}{n}\}} (n(t+\theta-s)+1)B_n f(s)ds \\ \quad + \int_\sigma^{t+\theta} T(t+\theta-s)B_n f(s)ds \\ \quad + \lim_{\lambda\to+\infty} \int_\sigma^{t+\theta} T(t+\theta-s)B_\lambda L(s)(\Psi^n(\sigma,s))ds & \text{for } -(t-\sigma) \leq \theta \leq 0, \\ \int_\sigma^{\min\{t,t+\theta+\frac{1}{n}\}} (n(t+\theta-s)+1)B_n f(s)ds & \text{for } -(t-\sigma)-\frac{1}{n} \leq \theta \leq -(t-\sigma), \\ 0 & \text{for } \theta \leq -(t-\sigma)-\frac{1}{n}. \end{cases}$$

Let $n > m > \omega_0$. If $-(t-\sigma) \leq \theta \leq 0$, we get

$$\begin{aligned}[\Psi^n(\sigma,t)-\Psi^m(\sigma,t)](\theta) &= \int\limits_{t+\theta}^{\min\{t,t+\theta+\frac{1}{n}\}} (n(t+\theta-s)+1)B_nf(s)\mathrm{d}s \\ &\quad - \int\limits_{t+\theta}^{\min\{t,t+\theta+\frac{1}{m}\}} (m(t+\theta-s)+1)B_mf(s)\mathrm{d}s \\ &\quad + \int\limits_{\sigma}^{t+\theta} T(t+\theta-s)(B_nf(s)-B_mf(s))\mathrm{d}s \\ &\quad + \lim_{\lambda\to+\infty}\int\limits_{\sigma}^{t+\theta} T(t+\theta-s)B_\lambda L(s)(\Psi^n(\sigma,s)-\Psi^m(\sigma,s))\mathrm{d}s.\end{aligned}$$

If $-(t-\sigma)-\frac{1}{m}\le\theta\le-(t-\sigma)$, we obtain

$$\begin{aligned}[\Psi^n(\sigma,t)-\Psi^m(\sigma,t)](\theta) &= \int\limits_{\sigma}^{\min\{t,t+\theta+\frac{1}{n}\}} (n(t+\theta-s)+1)B_nf(s)\mathrm{d}s \\ &\quad - \int\limits_{\sigma}^{\min\{t,t+\theta+\frac{1}{m}\}} (m(t+\theta-s)+1)B_mf(s)\mathrm{d}s.\end{aligned}$$

If $-(t-\sigma)-\frac{1}{n}\le\theta\le-(t-\sigma)-\frac{1}{m}$, then

$$[\Psi^n(\sigma,t)-\Psi^m(\sigma,t)](\theta) = \int\limits_{\sigma}^{\min\{t,t+\theta+\frac{1}{n}\}} (n(t+\theta-s)+1)B_nf(s)\mathrm{d}s.$$

If $\theta\le-(t-\sigma)-\frac{1}{n}$, then $[\Psi^n(\sigma,t)-\Psi^m(\sigma,t)](\theta)=0$.

Consequently, for $n>m>\omega_0$, we have that

$$\Psi^n(\sigma,t)-\Psi^m(\sigma,t)\in C_{00}(\mathbb{R}^-,X)$$

and

$$\operatorname{supp}(\Psi^n(\sigma,t)-\Psi^m(\sigma,t))\subset[-(a-\sigma)-1,0].$$

Therefore,

$$|\Psi^n(\sigma,t)-\Psi^m(\sigma,t)|_{\mathbb{B}}\le K(a-\sigma+1)\sup_{\theta\le0}|[\Psi^n(\sigma,t)-\Psi^m(\sigma,t)](\theta)|_X.$$

Moreover, if $-(t-\sigma)\le\theta\le0$, then

$$
\begin{aligned}
|[\Psi^n(\sigma,t)-\Psi^m(\sigma,t)](\theta)|_X \le{}& \left|\int_{t+\theta}^{\min\{t,t+\theta+\frac{1}{n}\}} (n(t+\theta-s)+1)B_nf(s)\mathrm{d}s\right|_X \\
&+\left|\int_{\sigma}^{t+\theta} T(t+\theta-s)(B_nf(s)-B_mf(s))\mathrm{d}s\right|_X \\
&+\left|\lim_{\lambda\to\infty}\int_{\sigma}^{t+\theta} T(t+\theta-s)B_\lambda L(s)(\Psi^n(\sigma,s)-\Psi^m(\sigma,s))\mathrm{d}s\right|_X \\
&+\left|\int_{t+\theta}^{\min\{t,t+\theta+\frac{1}{m}\}} (m(t+\theta-s)+1)B_mf(s)\mathrm{d}s\right|_X .
\end{aligned}
$$

Thus,

$$
\begin{aligned}
|[\Psi^n(\sigma,t)-\Psi^m(\sigma,t)](\theta)|_X \le{}& \sup_{\sigma\le\eta\le a}\left|\int_{\sigma}^{\eta} T(\eta-s)B_nf(s)\mathrm{d}s-\int_{\sigma}^{\eta} T(\eta-s)B_mf(s)\mathrm{d}s\right|_X \\
&+\int_{\sigma}^{t} M_0e^{\omega_0(t+\theta-s)}M_0|L(s)|_{\mathcal{L}(\mathbb{B},X)}|\Psi^n(\sigma,s)-\Psi^m(\sigma,s)|_{\mathbb{B}}\mathrm{d}s \\
&+\sup_{\sigma\le\eta\le a}|f(\eta)|_X\int_{t+\theta}^{t+\theta+\frac{1}{n}} |n(t+\theta-s)+1|\frac{nM_0}{n-\omega_0}\mathrm{d}s \\
&+\sup_{\sigma\le\eta\le a}|f(\eta)|_X\int_{t+\theta}^{t+\theta+\frac{1}{m}} |m(t+\theta-s)+1|\frac{mM_0}{m-\omega_0}\mathrm{d}s.
\end{aligned}
$$

Therefore,

$$
\begin{aligned}
|[\Psi^n(\sigma,t)-\Psi^m(\sigma,t)](\theta)|_X \le{}& M_0\sup_{\sigma\le\eta\le a}|f(\eta)|_X\left(\frac{1}{n-\omega_0}+\frac{1}{m-\omega_0}\right) \\
&+\sup_{\sigma\le\eta\le a}\left|\int_{\sigma}^{\eta} T(\eta-s)B_nf(s)\mathrm{d}s-\int_{\sigma}^{\eta} T(\eta-s)B_mf(s)\mathrm{d}s\right|_X \\
&+M_0^2\sup_{\sigma\le\eta\le a}|L(\eta)|_{\mathcal{L}(\mathbb{B},X)}\int_{\sigma}^{t} e^{\omega_0(t-s)}|\Psi^n(\sigma,s)-\Psi^m(\sigma,s)|_{\mathbb{B}}\mathrm{d}s.
\end{aligned}
$$

If $-(t-\sigma)-\frac{1}{m}\le\theta\le-(t-\sigma)$, then

$$\left|[\Psi^n(\sigma,t)-\Psi^m(\sigma,t)](\theta)\right|_X \le \left|\int_{t+\theta}^{\min\{t,t+\theta+\frac{1}{n}\}}(n(t+\theta-s)+1)B_n f(s)\mathrm{d}s\right|_X + \left|\int_{t+\theta}^{\min\{t,t+\theta+\frac{1}{m}\}}(m(t+\theta-s)+1)B_m f(s)\mathrm{d}s\right|_X .$$

As a result,

$$\begin{aligned}\left|[\Psi^n(\sigma,t)-\Psi^m(\sigma,t)](\theta)\right|_X &\le M_0 \sup_{\sigma\le\eta\le a}|f(\eta)|_X\left(\frac{1}{n-\omega_0}+\frac{1}{m-\omega_0}\right)\\ &\quad+\sup_{\sigma\le\eta\le a}\left|\int_\sigma^\eta T(\eta-s)B_n f(s)\mathrm{d}s-\int_\sigma^\eta T(\eta-s)B_m f(s)\mathrm{d}s\right|_X\\ &\quad+M_0^2\sup_{\sigma\le\eta\le a}|L(\eta)|_{\mathcal{L}(\mathbb{B},X)}\int_\sigma^t e^{\omega_0(t-s)}|\Psi^n(\sigma,s)-\Psi^m(\sigma,s)|_{\mathbb{B}}\mathrm{d}s.\end{aligned}$$

If $-(t-\sigma)-\frac{1}{n}\le\theta\le-(t-\sigma)-\frac{1}{m}$, we have that

$$\begin{aligned}\left|[\Psi^n(\sigma,t)-\Psi^m(\sigma,t)](\theta)\right|_X &\le \left|\int_{t+\theta}^{\min\{t,t+\theta+\frac{1}{n}\}}(n(t+\theta-s)+1)B_n f(s)\mathrm{d}s\right|_X\\ &\le M_0 \sup_{\sigma\le\eta\le a}|f(\eta)|_X\left(\frac{1}{n-\omega_0}+\frac{1}{m-\omega_0}\right)\\ &\quad+\sup_{\sigma\le\eta\le a}\left|\int_\sigma^\eta T(\eta-s)B_n f(s)\mathrm{d}s-\int_\sigma^\eta T(\eta-s)B_m f(s)\mathrm{d}s\right|_X\\ &\quad+M_0^2\sup_{\sigma\le\eta\le a}|L(\eta)|_{\mathcal{L}(\mathbb{B},X)}\int_\sigma^t e^{\omega_0(t-s)}|\Psi^n(\sigma,s)-\Psi^m(\sigma,s)|_{\mathbb{B}}\mathrm{d}s.\end{aligned}$$

Consequently,

$$\begin{aligned}\sup_{\theta\le 0}\left|[\Psi^n(\sigma,t)-\Psi^m(\sigma,t)](\theta)\right|_X &\le M_0 \sup_{\sigma\le\eta\le a}|f(\eta)|_X\left(\frac{1}{n-\omega_0}+\frac{1}{m-\omega_0}\right)\\ &\quad+\sup_{\sigma\le\eta\le a}\left|\int_\sigma^\eta T(\eta-s)B_n f(s)\mathrm{d}s-\int_\sigma^\eta T(\eta-s)B_m f(s)\mathrm{d}s\right|_X\\ &\quad+M_0^2\sup_{\sigma\le\eta\le a}|L(\eta)|_{\mathcal{L}(\mathbb{B},X)}\int_\sigma^t e^{\omega_0(t-s)}|\chi^{n,m}(\sigma,s)|_{\mathbb{B}}\mathrm{d}s.\end{aligned}$$

Therefore,

$$
\begin{aligned}
&\left|\Psi^n(\sigma,t)-\Psi^m(\sigma,t)\right|_{\mathbb{B}}\\
&\quad\le K(a-\sigma+1)M_0\sup_{\sigma\le\eta\le a}|f(\eta)|_X\left(\frac{1}{n-\omega_0}+\frac{1}{m-\omega_0}\right)\\
&\qquad+K(a-\sigma+1)\sup_{\sigma\le\eta\le a}\left|\int_\sigma^\eta T(\eta-s)B_nf(s)ds-\int_\sigma^\eta T(\eta-s)B_mf(s)ds\right|_X\\
&\qquad+K(a-\sigma+1)M_0^2\sup_{\sigma\le\eta\le a}|L(\eta)|_{\mathcal{L}(\mathbb{B},X)}e^{\omega_0(a-\sigma)}\int_\sigma^t\left|\Psi^n(\sigma,s)-\Psi^m(\sigma,s)\right|_{\mathbb{B}}ds.
\end{aligned}
$$

As a consequence of Gronwall's lemma, we conclude, for all $t\ge a$, that

$$
\left|\Psi^n(\sigma,t)-\Psi^m(\sigma,t)\right|_{\mathbb{B}}\le\Theta^{n,m}(a)e^{(K(a-\sigma+1)M_0^2\sup_{\sigma\le\eta\le a}|L(\eta)|_{\mathcal{L}(\mathbb{B},X)}e^{\omega_0(a-\sigma)})(t-\sigma)},
$$

where

$$
\begin{aligned}
\Theta^{n,m}(a)&=K(a-\sigma+1)M_0\sup_{\sigma\le\eta\le a}|f(\eta)|_X\left(\frac{1}{n-\omega_0}+\frac{1}{m-\omega_0}\right)\\
&\quad+K(a-\sigma+1)\sup_{\sigma\le\eta\le a}\left|\int_\sigma^\eta T(\eta-s)B_nf(s)ds-\int_\sigma^\eta T(\eta-s)B_mf(s)ds\right|_X.
\end{aligned}
$$

Furthermore,

$$
\sup_{t\in[\sigma,a]}\left|\Psi^n(\sigma,t)-\Psi^m(\sigma,t)\right|_{\mathbb{B}}\le\Theta^{n,m}(a)e^{(K(a-\sigma+1)M_0^2\sup_{\sigma\le\eta\le a}|L(\eta)|_{\mathcal{L}(\mathbb{B},X)}e^{\omega_0(a-\sigma)})(a-\sigma)}.
$$

From the fact that $(\int_\sigma^\eta T(\eta-s)B_nf(s)ds)_{n\ge0}$ converges uniformly on $[\sigma,a]$ as $n\to+\infty$, we deduce that

$$
\lim_{n,m\to+\infty}\sup_{t\in[\sigma,a]}\left|\Psi^n(\sigma,t)-\Psi^m(\sigma,t)\right|_{\mathbb{B}}=0.
$$

Consequently, $(\Psi^n(\sigma,\cdot))_{n\in\mathbb{N}}$ is a Cauchy sequence in $C([\sigma,a];\mathbb{B})$. Then, there exists $\Phi\in C([\sigma,a];\mathbb{B})$ such that

$$
\lim_{n\to+\infty}\sup_{t\in[\sigma,a]}\left|\Psi^n(\sigma,t)-\Phi(t)\right|_{\mathbb{B}}=0.
$$

Consider the function $\xi:]-\infty,a]\to X$ defined by

$$
\xi(t)=\begin{cases}\lim_{\lambda\to+\infty}\int_\sigma^t T(t-s)B_\lambda(L(s)\Phi(s)+f(s))ds & \text{for }\sigma\le t\le a,\\ 0 & \text{for }t\le\sigma.\end{cases}
$$

Then, for $t\in[\sigma,a]$,

$$\xi_t(\theta) = \begin{cases} \lim_{\lambda\to+\infty} \int_\sigma^{t+\theta} T(t+\theta-s)B_\lambda(L(s)\Phi(s)+f(s))ds & \text{for } \sigma \le t+\theta \le a, \\ 0 & \text{for } t+\theta \le \sigma. \end{cases}$$

By Lemma 5.6.1 and the definition of ξ and $\Psi^n(\sigma,t)$, we can affirm that

$$|\Psi^n(\sigma,t) - \xi_t|_{\mathbb{B}} \le K(a-\sigma+1)\sup_{\theta\le 0}|\Psi^n(\sigma,t)(\theta) - \xi_t(\theta)|_X.$$

Let $\theta \in [-(t-\sigma), 0]$. Then,

$$\begin{aligned} |\Psi^n(\sigma,t)(\theta) - \xi_t(\theta)|_X \le & \left| \int_{t+\theta}^{\min\{t,t+\theta+\frac{1}{n}\}} (n(t+\theta-s)+1)B_n f(s)ds \right|_X \\ & + \left| \int_\sigma^{t+\theta} T(t+\theta-s)(B_n f(s))ds - \lim_{\lambda\to\infty} \int_\sigma^{t+\theta} T(t+\theta-s)B_\lambda f(s)ds \right|_X \\ & + \left| \lim_{\lambda\to\infty} \int_\sigma^{t+\theta} T(t+\theta-s)B_\lambda L(s)(\Psi^n(\sigma,s) - \Phi(s))ds \right|_X . \end{aligned}$$

It follows that

$$\begin{aligned} |\Psi^n(\sigma,t)(\theta) - \xi_t(\theta)|_X \le & \frac{M_0}{n-\omega_0} \sup_{s\in[\sigma,a]} |f(s)|_X \\ & + M_0^2 \sup_{s\in[\sigma,a]} |L(s)|_{\mathcal{L}(\mathbb{B},X)} \int_\sigma^a e^{\omega_0(a-\eta)} |\Psi^n(\sigma,\eta) - \Phi(\eta))|_{\mathbb{B}} d\eta \\ & + \sup_{s\in[\sigma,a]} \left| \int_\sigma^s T(s-\eta)B_n f(\eta)d\eta - \lim_{\lambda\to\infty} \int_\sigma^s T(s-\eta)B_\lambda f(\eta)d\eta \right|_X . \end{aligned}$$

If $-(t-\sigma) - \frac{1}{n} \le \theta \le -(t-\sigma)$, we obtain that

$$\begin{aligned} |\Psi^n(\sigma,t)(\theta) - \xi_t(\theta)|_X \le & \left| \int_\sigma^{\min\{t,t+\theta+\frac{1}{n}\}} (n(t+\theta-s)+1)B_n f(s)ds \right|_X \\ \le & \frac{M_0}{n-\omega_0} \sup_{s\in[\sigma,a]} |f(s)|_X \\ & + M_0^2 \sup_{s\in[\sigma,a]} |L(s)|_{\mathcal{L}(\mathbb{B},X)} \int_\sigma^a e^{\omega_0(a-\eta)} |\Psi^n(\sigma,\eta) - \Phi(\eta))|_{\mathbb{B}} d\eta \\ & + \sup_{s\in[\sigma,a]} \left| \int_\sigma^s T(s-\eta)B_n f(\eta)d\eta - \lim_{\lambda\to\infty} \int_\sigma^s T(s-\eta)B_\lambda f(\eta)d\eta \right|_X . \end{aligned}$$

Consequently,

$$|\Psi^n(\sigma,t)-\xi_t|_{\mathbb{B}} \le K(a-\sigma+1)\Bigg\{\frac{M_0}{n-\omega_0}\sup_{s\in[\sigma,a]}|f(s)|_X + M_0^2\sup_{s\in[\sigma,a]}|L(s)|_{\mathcal{L}(\mathbb{B},X)}\int_\sigma^a e^{\omega_0(a-\eta)}|\Psi^n(\sigma,\eta)-\Phi(\eta))|_{\mathbb{B}}d\eta + \sup_{s\in[\sigma,a]}\left|\int_\sigma^s T(s-\eta)B_nf(\eta)d\eta - \lim_{\lambda\to\infty}\int_\sigma^s T(s-\eta)B_\lambda f(\eta)d\eta\right|_X\Bigg\}.$$

Therefore,

$$\lim_{n\to+\infty}\sup_{t\in[\sigma,a]}|\Psi^n(\sigma,t)-\xi_t|_{\mathbb{B}} = 0.$$

Thus, $\xi_t = \Phi(t) = u_t(\cdot,\sigma,0,L,f)$, for any $t\in[\sigma,a]$. This completes the proof of the theorem. □

As a direct consequence of the previous theorem, we get the following result.

Theorem 5.6.10. *Let assumptions* $(\mathbf{H}_0)$ *and* $(\mathbf{H}_1)$ *be satisfied. Then, for any* $\sigma\in\mathbb{R}$ *and* $\phi\in\mathbb{B}_A$*, the unique integral solution* $u(\cdot,\sigma,\phi,L,f)$ *of* (5.109) *satisfies the following variation of constants formula in* $\mathbb{B}_A$*:*

$$u_t(\cdot,\sigma,\phi,L,f) = \mathbb{U}(t,\sigma)\phi + \lim_{n\to+\infty}\int_\sigma^t \mathbb{U}(t,s)\Upsilon^n f(s)ds \quad \text{for } t\ge\sigma. \tag{5.110}$$

Proof. We can observe, for $t\in[\sigma,+\infty[$, that

$$\begin{aligned}u_t(\cdot,\sigma,\phi,L,f) &= u_t(\cdot,\sigma,\phi,L,0)+u_t(\cdot,\sigma,0,L,f)\\ &= \mathbb{U}(t,\sigma)\phi + u_t(\cdot,\sigma,0,L,f).\end{aligned}$$

Then, the proof is finished by Theorem 5.6.9. □

5.6.4 Almost periodic and almost automorphic solutions for equation (5.109)

Consider the following assumption:
$(\mathbf{H}_2)$ L is 1-periodic on t.

Let $\mathbb{U}(1,0)$ be the monodromy operator associated to equation (5.109) and consider the following linear difference equation:

$$y_{n+1} = \mathbb{U}(1,0)y_n + v_n \quad \text{for } n\in\mathbb{Z}, \tag{5.111}$$

where

$$v_n = \lim_{k\to+\infty} \int_n^{n+1} \mathbb{U}(n+1,s)\Upsilon^k f(s)ds \quad \text{for } n \in \mathbb{Z}.$$

The aim of this section is to study the link between the existence of an almost periodic (resp. almost automorphic) solution of the difference equation (5.111) and the existence of an almost periodic (resp. almost automorphic) integral solution of (5.109). As a result, we show that equation (5.109) has the Bohr–Neugebauer property.

Lemma 5.6.11. *Suppose that* $(\mathbf{H}_0)$, $(\mathbf{H}_1)$, *and* $(\mathbf{H}_2)$ *be satisfied. Let* $(y_n)_{n\in\mathbb{Z}}$ *be a solution of equation* (5.111) *in* $\mathbb{Z}$. *Then, for any* $m \geq n$,

$$y_m = \mathbb{U}(m,n)y_n + \lim_{k\to+\infty} \int_n^m \mathbb{U}(m,s)\Upsilon^k f(s)ds.$$

Proof. The proof of this result is achieved by showing, for all $n \in \mathbb{Z}$, that

$$y_{n+j} = \mathbb{U}(n+j,n)y_n + \lim_{k\to+\infty} \int_n^{n+j} \mathbb{U}(n+j,s)\Upsilon^k f(s)ds \quad \text{for all } j \geq 1. \tag{5.112}$$

When $j = 1$, we have

$$\begin{aligned} y_{n+1} &= \mathbb{U}(1,0)y_n + \lim_{k\to+\infty} \int_n^{n+1} \mathbb{U}(n+1,s)\Upsilon^k f(s)ds \\ &= \mathbb{U}(n+1,n)y_n + \lim_{k\to+\infty} \int_n^{n+1} \mathbb{U}(n+1,s)\Upsilon^k f(s)ds. \end{aligned}$$

So (5.112) is true for $j = 1$. Let $j \geq 1$ and suppose that (5.112) is valid for all $i = 1, \dots, j$. Then,

$$y_{n+j+1} = \mathbb{U}(1,0)y_{n+j} + v_{n+j}.$$

Therefore, by the induction hypothesis, we obtain that

$$\begin{aligned} y_{n+j+1} = \mathbb{U}(1,0)&\left[\mathbb{U}(n+j,n)y_n + \lim_{k\to+\infty} \int_n^{n+j} \mathbb{U}(n+j,s)\Upsilon^k f(s)ds\right] \\ &+ \lim_{k\to+\infty} \int_{n+j}^{n+j+1} \mathbb{U}(n+j+1,s)\Upsilon^k f(s)ds \\ = {}& \mathbb{U}(n+j+1,n)y_n + \int_n^{n+j} \mathbb{U}(n+j+1,s)\Upsilon^k f(s)ds \end{aligned}$$

$$+\int_{n+j}^{n+j+1} \mathbb{U}(n+j+1,s)\Upsilon^k f(s)ds$$

$$= \mathbb{U}(n+j+1,n)y_n + \int_{n}^{n+j+1} \mathbb{U}(n+j+1,s)\Upsilon^k f(s)ds.$$

By the principle of mathematical induction, it follows that (5.112) is true for all $j \geq 1$. □

Lemma 5.6.12. *Assume that* $(\mathbf{H}_0)$, $(\mathbf{H}_1)$, *and* $(\mathbf{H}_2)$ *hold. Let* $(y_n)_{n\in\mathbb{Z}}$ *be a solution of equation* (5.111) *on* $\mathbb{Z}$ *and* $\chi : \mathbb{R} \to \mathbb{B}_A$ *be the function given, for* $t \in \mathbb{R}$, *by*

$$\chi(t) = \mathbb{U}(t,[t])y_{[t]} + \lim_{k\to+\infty} \int_{[t]}^{t} \mathbb{U}(t,s)\Upsilon^k f(s)ds. \tag{5.113}$$

Then, the function $u(t) = \chi(t)(0)$ *is an integral solution of* (5.109) *on* $\mathbb{R}$.

Proof. Let $\sigma \in \mathbb{R}$ and $t \geq \sigma$. Then,

$$\chi(t) = \mathbb{U}(t,[t])y_{[t]} + \lim_{k\to+\infty} \int_{[t]}^{t} \mathbb{U}(t,s)\Upsilon^k f(s)ds.$$

By Lemma 5.6.11, we have that

$$y_{[t]} = \mathbb{U}([t],[\sigma])y_{[\sigma]} + \lim_{k\to+\infty} \int_{[\sigma]}^{[t]} \mathbb{U}([t],s)\Upsilon^k f(s)ds.$$

Therefore,

$$\begin{aligned}
\chi(t) &= \mathbb{U}(t,[\sigma])y_{[\sigma]} + \lim_{k\to+\infty} \int_{[\sigma]}^{t} \mathbb{U}(t,s)\Upsilon^k f(s)ds \\
&= \mathbb{U}(t,\sigma)\left\{\mathbb{U}(\sigma,[\sigma])y_{[\sigma]} + \lim_{k\to+\infty} \int_{[\sigma]}^{\sigma} \mathbb{U}(\sigma,s)\Upsilon^k f(s)ds\right\} + \lim_{k\to+\infty} \int_{\sigma}^{t} \mathbb{U}(t,s)\Upsilon^k f(s)ds \\
&= \mathbb{U}(t,\sigma)\chi(\sigma) + \lim_{k\to+\infty} \int_{\sigma}^{t} \mathbb{U}(t,s)\Upsilon^k f(s)ds.
\end{aligned}$$

As a result of Theorem 5.6.10, we deduce that u is an integral solution of equation (5.109) on $\mathbb{R}$. □

Lemma 5.6.13. *Let the hypotheses* $(\mathbf{H}_0)$, $(\mathbf{H}_1)$, *and* $(\mathbf{H}_2)$ *be satisfied. Then, the statements below are true:*

(1) *If f is an almost periodic function, then $(v_n)_{n\in\mathbb{Z}}$ is almost periodic.*
(2) *If f is almost automorphic, then $(v_n)_{n\in\mathbb{Z}}$ is almost automorphic.*

Proof. Since $(\mathbb{U}(t,s))_{t\geq s}$ is 1-periodic, we can see, for $n \in \mathbb{Z}$, that

$$v_n = \lim_{k\to+\infty} \int_0^1 \mathbb{U}(1,s)\Upsilon^k f(s+n)ds.$$

Let us show (1). Consider $\{m'_j\}_{j\in\mathbb{N}} \subset \mathbb{Z}$. Because f is an almost periodic function, we can extract a subsequence $\{m_j\}_{j\in\mathbb{N}}$ of $\{m'_j\}_{j\in\mathbb{N}}$ such that

$$\lim_{j\to+\infty} f(t+m_j) := \tilde{f}(t),$$

uniformly on $\in \mathbb{R}$. We define the following sequence:

$$\overline{v}_n := \lim_{k\to+\infty} \int_0^1 \mathbb{U}(1,s)\Upsilon^k \tilde{f}(s+n)ds \quad \text{for } n \in \mathbb{Z}.$$

Then, for $n \in \mathbb{Z}$ and $j \in \mathbb{N}$, we have

$$|v_{n+m_j} - \overline{v}_n|_{\mathbb{B}} \leq \tilde{M}K(-1) \sup_{s\in[0,1]} |\mathbb{U}(1,s)|_{\mathcal{L}(\mathbb{B}_A)} \sup_{t\in\mathbb{R}} |f(t+m_j) - \tilde{f}(t)|_X.$$

Hence,

$$\lim_{j\to+\infty} \sup_{n\in\mathbb{Z}} |v_{n+m_j} - \overline{v}_n|_{\mathbb{B}} = 0,$$

which establishes (1).

Now we are going to prove (2). Let $\{m'_j\}_{j\in\mathbb{N}} \subset \mathbb{Z}$. As f is almost automorphic, there exist a subsequence $\{m_j\}_{j\in\mathbb{N}}$ of $\{m'_j\}_{j\in\mathbb{N}}$ and a function $\overline{f} : \mathbb{R} \to X$ such that, for any $t \in \mathbb{R}$,

$$\lim_{j\to+\infty} f(t+m_j) = \overline{f}(t) \quad \text{and} \quad \lim_{j\to+\infty} \overline{f}(t-m_j) = f(t).$$

Let $\tilde{v} : \mathbb{Z} \to \mathbb{B}_A$ be the sequence defined, for $n \in \mathbb{Z}$, by

$$\tilde{v}_n = \lim_{k\to+\infty} \int_0^1 \mathbb{U}(1,s)\Upsilon^k \overline{f}(s+n)ds.$$

For $n \in \mathbb{Z}$ and $j \in \mathbb{N}$, we have

$$|v_{n+m_j} - \tilde{v}_n|_{\mathbb{B}} \leq \tilde{M}K(-1) \sup_{s\in[0,1]} |\mathbb{U}(1,s)|_{\mathcal{L}(\mathbb{B}_A)} \int_0^1 |f(s+n+m_j) - \overline{f}(s+n)|_X ds$$

and

$$|\tilde{v}_{n-m_j} - v_n|_{\mathbb{B}} \le \tilde{M}K(-1) \sup_{s\in[0,1]} |\mathbb{U}(1,s)|_{\mathcal{L}(\mathbb{B}_A)} \int_0^1 |\bar{f}(s+n-m_j) - f(s+n)|_X ds.$$

Lebesgue's dominated convergence theorem yields

$$\lim_{j\to+\infty} v(n+m_j) = \tilde{v}(n) \quad \text{and} \quad \lim_{j\to+\infty} \tilde{v}(n-m_j) = v(n).$$

Therefore, v is almost automorphic. □

Now, we will use the spectral properties of the monodromy operator $\mathbb{U}(1,0)$ to study Massera-type theorem and the Bohr–Neugebauer property for the difference equation (5.111). To do so, we have to introduce the notion of a uniform fading memory space.

Consider the following axiom:

(C) If a uniformly bounded sequence $(\phi_n)_{n\in\mathbb{N}}$ in $C_{00}(\mathbb{R}^-;X)$ converging to a function ϕ compactly on $]-\infty,0]$, then $\phi \in \mathbb{B}$ and $|\phi_n - \phi|_{\mathbb{B}} \to 0$ as $n \to +\infty$.

Proposition 5.6.14 ([181]). *Assume that $\mathbb{B}$ satisfies axiom* **(C)**. *Then,* $BC(\mathbb{R}^-;X) \subset \mathbb{B}$ *and there exists $C_1 > 0$ such that*

$$|\phi|_{\mathbb{B}} \le C_1 \sup_{\theta\le 0} |\phi(\theta)|_X \quad \textit{for all } \phi \in BC(\mathbb{R}^-;X).$$

Moreover, if $\vartheta :]-\infty,a] \to X$, $a \in \mathbb{R}$, is a function such that $\vartheta_t \in \mathbb{B}$ for all $t \le a$, then the mapping $t \to \vartheta_t$ is in $BC(]-\infty,a];\mathbb{B})$ *if and only if $\vartheta \in$* $BC(]-\infty,a];X)$.

For $t \ge 0$, we define the operator $\mathbb{V}(t)$ on $\mathbb{B}_0 := \{\phi \in \mathbb{B}; \phi(0) = 0\}$ by

$$\mathbb{V}(t)\phi(\theta) = \begin{cases} \phi(t+\theta) & \text{for } t+\theta \le 0, \\ 0 & \text{for } t+\theta \ge 0, \end{cases}$$

for $\phi \in \mathbb{B}_0$. Note that the family $(\mathbb{V}(t))_{t\ge 0}$ is the solution semigroup associated to the trivial equation:

$$\begin{cases} \frac{d}{dt}u(t) = 0 & \text{for } t \ge 0, \\ u_0 = \phi \in \mathbb{B}_0 = \{\phi \in \mathbb{B}; \phi(0) = 0\}. \end{cases}$$

Definition 5.6.15 ([181]). Space $\mathbb{B}$ is said to be a fading memory space if $\mathbb{B}$ satisfies axioms **(A)**, **(B)**, **(C)** and

$$\lim_{t\to+\infty} |\mathbb{V}(t)\phi|_{\mathbb{B}} = 0 \quad \text{for } \phi \in \mathbb{B}_0.$$

Moreover, if $\lim_{t\to+\infty} |\mathbb{V}(t)|_{\mathcal{L}(\mathbb{B}_0)} = 0$, then $\mathbb{B}$ is called a uniform fading memory space.

Proposition 5.6.16 ([181]). *The following statements hold:*
(i) *If $\mathbb{B}$ is a fading memory space, then the functions $K(\cdot)$ and $M(\cdot)$ in axiom* (**A**) *can be chosen to be constants.*
(ii) *If $\mathbb{B}$ is a uniform fading memory space, then the functions $K(\cdot)$ and $M(\cdot)$ can be chosen such that $K(\cdot)$ is constant and $M(t) \to 0$ as $t \to +\infty$.*

The following hypotheses will be necessary in the rest of this work:
($\mathbf{H}_3$) Space $\mathbb{B}$ is a uniform fading memory space.
($\mathbf{H}_4$) For all $t > 0$, $T(t)$ is a compact operator.

Lemma 5.6.17. *Suppose that* ($\mathbf{H}_0$)–($\mathbf{H}_4$) *hold and f is an almost periodic (resp. almost automorphic) function. Then, every bounded solution of equation* (5.111) *on $\mathbb{Z}$ is almost periodic (resp. almost automorphic) and, if equation* (5.111) *has a bounded solution on $\mathbb{N}$, then it must have an almost periodic (resp. almost automorphic) solution on $\mathbb{Z}$.*

Proof. As a result of Theorems 5.5.8–5.5.10, the proof of this lemma is completed by proving that

$$r_{\text{ess}}(\mathbb{U}(1,0)) < 1,$$

where $r_{\text{ess}}(\mathbb{U}(1,0))$ is the essential spectral radius of $\mathbb{U}(1,0)$. In fact, from Proposition 5 in [56], we have that

$$\mathbb{U}(1,0) = \mathbb{Y}(1) + \mathbb{W}(1),$$

where $(\mathbb{Y}(t))_{t\geq 0}$ is an exponentially stable semigroup on $\mathbb{B}_A$ and $\mathbb{W}(1)$ is a compact operator. Then,

$$\begin{aligned} r_{\text{ess}}(\mathbb{U}(1,0)) &= r_{\text{ess}}(\mathbb{Y}(1) + \mathbb{W}(1)) \\ &= r_{\text{ess}}(\mathbb{Y}(1)) \\ &\leq r(\mathbb{Y}(1)). \end{aligned}$$

On the other hand, since $(\mathbb{Y}(t))_{t\geq 0}$ is exponentially stable, by [136, Proposition 1.7, p. 299],

$$\omega(\mathbb{Y}) < 0,$$

where $\omega(\mathbb{Y})$ is the growth bound of $(\mathbb{Y}(t))_{t\geq 0}$ (see [136, p. 299]) of the semigroup $(\mathcal{Y}(t))_{t\geq 0}$. From [270, Proposition 4.13, pp. 170–171], we have that

$$r(\mathbb{Y}(1)) = e^{\omega(\mathbb{Y})} < 1.$$

Hence, we conclude that

$$r_{\text{ess}}(\mathbb{U}(1,0)) < 1,$$

completing the proof. □

5.6.5 Almost periodic solution of equation (5.109)

We assume that
($\mathbf{H}_5$) f is almost periodic.

Theorem 5.6.18. *Assume that* ($\mathbf{H}_0$)–($\mathbf{H}_5$) *hold. Then, equation* (5.111) *has an almost periodic solution on* $\mathbb{Z}$ *if and only if equation* (5.109) *has an almost periodic integral solution on* $\mathbb{R}$.

Proof. Let $(y_n)_{n\in\mathbb{Z}}$ be an almost periodic solution of equation on $\mathbb{Z}$. By Lemma 5.6.12, the function $u(t) = \chi(t)(0)$, where χ is the function given by (5.113), is an integral solution of (5.109) on $\mathbb{R}$. Let us show that χ is almost periodic. Let $\tilde{y}$ be the almost periodic extension of $(y_n)_{n\in\mathbb{Z}}$ on $\mathbb{R}$. Then, the function $\mathcal{F} : \mathbb{R} \to X \times \mathbb{B}_A$ defined, for $t \in \mathbb{R}$, by $\mathcal{F}(t) = (f(t), \tilde{y}(t))$, is almost periodic.

Let $\varepsilon > 0$. Then, $\mathbb{Z} \cap \mathbb{T}(\mathcal{F}, \varepsilon)$ is relatively dense. So there exists $l(\varepsilon) > 0$ such that

$$(\mathbb{Z} \cap \mathbb{T}(\mathcal{F}, \varepsilon)) \cap [a, a + l(\varepsilon)] \neq \emptyset \quad \text{for all } a \in \mathbb{R}.$$

For $a \in \mathbb{R}$ and $\tau_\varepsilon \in (\mathbb{Z} \cap \mathbb{T}(\mathcal{F}, \varepsilon)) \cap [a, a + l(\varepsilon)]$, we have

$$\begin{aligned}|\chi(t+\tau_\varepsilon) - \chi(t)|_{\mathbb{B}} \le{}& |\mathbb{U}(t+\tau_\varepsilon, [t+\tau_\varepsilon])y_{[t+\tau_\varepsilon]} - \mathbb{U}(t,[t])y_{[t]}|_{\mathbb{B}} \\ &+ \left|\lim_{k\to+\infty}\int_{[t+\tau_\varepsilon]}^{t+\tau_\varepsilon} \mathbb{U}(t+\tau_\varepsilon, s)\Upsilon^k f(s)ds - \lim_{k\to+\infty}\int_{[t]}^{t} \mathbb{U}(t,s)\Upsilon^k f(s)ds\right|_{\mathbb{B}}\end{aligned}$$

for all $t \in \mathbb{R}$. Therefore,

$$\begin{aligned}|\chi(t+\tau_\varepsilon) - \chi(t)|_{\mathbb{B}} \le{}& |\mathbb{U}(t+\tau_\varepsilon, [t]+\tau_\varepsilon)y_{[t]+\tau_\varepsilon} - \mathbb{U}(t,[t])y_{[t]}|_{\mathbb{B}} \\ &+ \left|\lim_{k\to+\infty}\int_{[t]+\tau_\varepsilon}^{t+\tau_\varepsilon} \mathbb{U}(t+\tau_\varepsilon, s)\Upsilon^k f(s)ds - \lim_{k\to+\infty}\int_{[t]}^{t} \mathbb{U}(t,s)\Upsilon^k f(s)ds\right|_{\mathbb{B}} \\ \le{}& |\mathbb{U}(t,[t])(y([t]+\tau_\varepsilon) - y([t]))|_{\mathbb{B}} \\ &+ \left|\lim_{k\to+\infty}\int_{[t]}^{t} \mathbb{U}(t,s)\Upsilon^k (f(s+\tau_\varepsilon) - f(s))ds\right|_{\mathbb{B}} \\ <{}& \Big(\sup_{s\in[[t],t]} |\mathbb{U}(t,s)|_{\mathcal{L}(\mathbb{B}_A)} + \tilde{M}K(-1)\sup_{s\in[[t],t]} |\mathbb{U}(t,s)|_{\mathcal{L}(\mathbb{B}_A)}\Big)\varepsilon\end{aligned}$$

for all $t \in \mathbb{R}$. By (iii) in Theorem 5.6.7, we get, for $t \in \mathbb{R}$, that

$$|\mathbb{U}(t,s)|_{\mathcal{L}(\mathbb{B}_A)} \le m_1(t,s)e^{m_2(t,s)(t-s)} \quad \text{for all } s \in [[t],t].$$

Since $\mathbb{B}$ is a uniform fading memory space, we have that K is constant and $M(t) \to 0$ as $t \to +\infty$, also, since L is 1-periodic, one obtains $\sup_{s\in\mathbb{R}}|L(s)|_{\mathcal{L}(\mathbb{B},X)} < \infty$. Hence we deduce

$$\tilde{m}_1 := \sup_{t\in\mathbb{R}} \sup_{s\in[[t],t]} m_1(t,s) < \infty \quad \text{and} \quad \tilde{m}_2 := \sup_{t\in\mathbb{R}} \sup_{s\in[[t],t]} m_2(t,s) < \infty.$$

Thus,

$$\sup_{t\in\mathbb{R}}|\chi(t+\tau_\varepsilon) - \chi(t)|_{\mathbb{B}} \le (\tilde{m}_1 e^{\tilde{m}_2} + K\tilde{M}\tilde{m}_1 e^{\tilde{m}_2})\varepsilon.$$

Then, $\tau_\varepsilon \in \mathbb{T}(\chi, (\tilde{m}_1 e^{\tilde{m}_2} + K\tilde{M}\tilde{m}_1 e^{\tilde{m}_2})\varepsilon) \cap [a, a + l(\varepsilon)]$. So $\mathbb{T}(\chi, (\tilde{m}_1 e^{\tilde{m}_2} + K\tilde{M}\tilde{m}_1 e^{\tilde{m}_2})\varepsilon)$ is relatively dense for all $\varepsilon > 0$. As a result, we deduce that χ is almost periodic. Hence, u is also almost periodic.

Let now u be an almost periodic solution of (5.109) on $\mathbb{R}$. Then, u is bounded on $\mathbb{R}$. Therefore,

$$\sup_{t\in\mathbb{R}}|u_t|_{\mathbb{B}} < \infty.$$

In fact, for $t \ge 0$, we have

$$|u_t|_{\mathbb{B}} \le K(t) \sup_{s\in[0,t]} |x(s)|_X + M(t)|u_0|_{\mathbb{B}}.$$

Since $\mathbb{B}$ is a uniform fading memory space, the function K is constant and $M(t) \to 0$ as $t \to +\infty$. Thus,

$$K(\cdot) \equiv K \quad \text{and} \quad \hat{M} := \sup_{t\ge 0} M(t) < \infty.$$

Therefore,

$$|u_t|_{\mathbb{B}} \le K \sup_{s\in\mathbb{R}}|x(s)|_X + \hat{M}|u_0|_{\mathbb{B}} \quad \text{for all } t \ge 0.$$

In the other hand, we have

$$\sup_{t\le 0}|u(t)|_X < \infty.$$

Then, by Proposition 5.6.14, we deduce

$$\sup_{t\leq 0}|u_t|_{\mathbb{B}} < \infty.$$

Thus,

$$\sup_{t\in\mathbb{R}}|u_t|_{\mathbb{B}} < \infty.$$

By Theorem 5.6.10, we can see, for $n \in \mathbb{Z}$, that

$$\begin{aligned} u_{n+1} &= \mathbb{U}(n+1,n)u_n + \lim_{k\to+\infty}\int_n^{n+1} \mathbb{U}(n+1,s)\Upsilon^k f(s)ds \\ &= \mathbb{U}(1,0)u_n + v_n. \end{aligned}$$

Consequently, the sequence $(u_n)_{n\in\mathbb{Z}}$ is a bounded solution of equation (5.111) on $\mathbb{Z}$. By Lemma 5.6.17, we get the result. □

As a result of the previous theorem, we deduce that equation (5.109) has the following Bohr–Neugebauer property.

Theorem 5.6.19. *Suppose that* $(\mathbf{H}_0)$–$(\mathbf{H}_5)$ *are satisfied. Then, the following properties are verified:*
(1) *If equation* (5.109) *admits a bounded integral solution on* $\mathbb{R}^+$, *then it must have an almost periodic integral solution on* $\mathbb{R}$.
(2) *Let* u *be a bounded integral solution of equation* (5.109) *on* $\mathbb{R}$. *Then,* u *and* $t \to u_t$ *are almost periodic.*

Proof. (1) Suppose u is a bounded solution of equation (5.109) on $\mathbb{R}^+$. Then,

$$\sup_{t\in\mathbb{R}^+}|u_t|_{\mathbb{B}} < \infty.$$

By Theorem 5.6.10, we have that $(u_n)_{n\in\mathbb{N}}$ is a bounded solution of (5.111) on $\mathbb{N}$. From Lemma 5.6.17, we deduce that equation (5.111) has an almost periodic solution on $\mathbb{Z}$. By Theorem 5.6.18, we get claim (1).

Let us show (2). First, we observe that $(u_n)_{n\in\mathbb{Z}}$ is a bounded solution of (5.111) on $\mathbb{Z}$. Then, from Lemma 5.6.17 and Theorem 5.6.18, we obtain that $(u_n)_{n\in\mathbb{Z}}$ and the function χ given by (5.113) are almost periodic. On the other hand, we can see that

$$\chi(t) = \mathbb{U}(t,[t])u_{[t]} + \lim_{k\to+\infty}\int_{[t]}^{t} \mathbb{U}(t,s)\Upsilon^k f(s)ds = u_t$$

for all $t \in \mathbb{R}$, completing the proof. □

5.6.6 Almost automorphic solutions

Suppose that
($\mathbf{H}_6$) f is almost automorphic.

Theorem 5.6.20. *Assume that* ($\mathbf{H}_0$)–($\mathbf{H}_4$) *and* ($\mathbf{H}_6$) *hold. If* $(y_n)_{n\in\mathbb{Z}}$ *is an almost automorphic solution of* (5.111) *on* $\mathbb{Z}$, *then the function* $\chi:\mathbb{R}\to\mathbb{B}_A$ *given, for* $t\in\mathbb{R}$, *by*

$$\chi(t)=\mathbb{U}(t,[t])y_{[t]}+\lim_{k\to+\infty}\int_{[t]}^{t}\mathbb{U}(t,s)\mathrm{Y}^k f(s)$$

is almost automorphic. Hence, the function $u(t)=\chi(t)(0)$ *is an almost automorphic integral solution of* (5.109) *on* $\mathbb{R}$.

Proof. As a conclusion of Lemma 5.6.12, we have that u is an integral solution of (5.109) on $\mathbb{R}$. Then, the proof of this theorem is finished by showing that χ is almost automorphic. Indeed, let $(t_n)_{n\in\mathbb{N}}$ be any sequence in $\mathbb{R}$. Set $\varsigma_n=t_n-[t_n]$ for $n\in\mathbb{N}$. Since $(\varsigma_n)_{n\in\mathbb{N}}\subset[0,1]$ and $(y_n)_{n\in\mathbb{Z}}, f$ are almost automorphic, there exist a nondecreasing map $\tau:\mathbb{N}\to\mathbb{N}$, $\varsigma_0\in[0,1]$, a sequence $(\tilde{y}_n)_{n\in\mathbb{N}}\subset\mathbb{B}_A$, and a measurable function $\tilde{f}:\mathbb{R}\to X$ such that

$$\lim_{n\to+\infty}f(t+[t_{\tau(n)}])=\tilde{f}(t)\quad\text{and}\quad\lim_{n\to+\infty}\tilde{f}(t-[t_{\tau(n)}])=f(t)\quad\text{for all } t\in\mathbb{R},$$
$$\lim_{n\to+\infty}y_{m+[t_{\tau(n)}]}=\tilde{y}_m\quad\text{and}\quad\lim_{n\to+\infty}\tilde{y}_{m-[t_{\tau(n)}]}=y_m\quad\text{for all } m\in\mathbb{Z},$$

and

$$\lim_{n\to+\infty}\varsigma_{\tau(n)}=\varsigma_0.$$

We can observe, for $n\in\mathbb{N}$, that

$$\begin{aligned}\chi(t_{\tau(n)}+t)&=\mathbb{U}(t_{\tau(n)}+t,[t_{\tau(n)}]+[t])\chi([t_{\tau(n)}]+[t])\\&\quad+\lim_{k\to+\infty}\int_{[t_{\tau(n)}]+[t]}^{t+t_{\tau(n)}}\mathbb{U}(t+t_{\tau(n)},s)\mathrm{Y}^k f(s)ds\\&=\mathbb{U}(\varsigma_{\tau(n)}+t-[t],0)y_{[t_{\tau(n)}]+[t]}\\&\quad+\lim_{k\to+\infty}\int_{0}^{\varsigma_{\tau(n)}+t-[t]}\mathbb{U}(\varsigma_{\tau(n)}+t-[t],s)\mathrm{Y}^k f([t_{\tau(n)}]+[t]+s)ds.\end{aligned}$$

Consider the function $\overline{\chi}:\mathbb{R}\to\mathbb{B}_A$ defined, for $t\in\mathbb{R}$, by

$$\overline{\chi}(t)=\mathbb{U}(\varsigma_0+t-[t],0)\tilde{y}_{[t]}+u_{\varsigma_0+t-[t]}(\cdot,0,0,L,\tilde{f}([t]+\cdot)).$$

Then,

$$
\begin{aligned}
&|\chi(t_{\tau(n)}+t)-\tilde{\chi}(t)|_{\mathbb{B}}\\
&\quad\le |\mathbb{U}(\varsigma_{\tau(n)}+t-[t],0)y_{[t_{\tau(n)}]+[t]}-\mathbb{U}(\varsigma_0+t-[t],0)\tilde{y}_{[t]}|_{\mathbb{B}}\\
&\qquad+|u_{\varsigma_{\tau(n)}+t-[t]}(\cdot,0,0,L,f([t_{\tau(n)}]+[t]+\cdot))-u_{\varsigma_0+t-[t]}(\cdot,0,0,L,\tilde{f}([t]+\cdot))|_{\mathbb{B}}.
\end{aligned}
$$

Therefore,

$$
\begin{aligned}
&|\chi(t_{\tau(n)}+t)-\tilde{\chi}(t)|_{\mathbb{B}}\\
&\quad\le |\mathbb{U}(\varsigma_{\tau(n)}+t-[t],0)\{y_{[t_{\tau(n)}]+[t]}-\tilde{y}_{[t]}\}|_{\mathbb{B}}\\
&\qquad+|\{\mathbb{U}(\varsigma_{\tau(n)}+t-[t],0)-\mathbb{U}(\varsigma_0+t-[t],0)\}\tilde{y}_{[t]}|_{\mathbb{B}}\\
&\qquad+2|u_{\varsigma_0+t-[t]}(\cdot,0,0,L,f([t_{\tau(n)}]+[t]+\cdot))-u_{\varsigma_0+t-[t]}(\cdot,0,0,L,\tilde{f}([t]+\cdot))|_{\mathbb{B}}\\
&\qquad+|u_{\varsigma_{\tau(n)}+t-[t]}(\cdot,0,0,L,f([t_{\tau(n)}]+[t]+\cdot))-u_{\varsigma_{\tau(n)}+t-[t]}(\cdot,0,0,L,\tilde{f}([t]+\cdot))|_{\mathbb{B}}\\
&\qquad+|u_{\varsigma_{\tau(n)}+t-[t]}(\cdot,0,0,L,\tilde{f}([t]+\cdot))-u_{\varsigma_0+t-[t]}(\cdot,0,0,L,\tilde{f}([t]+\cdot))|_{\mathbb{B}}.
\end{aligned}
$$

By Lemma 5.6.8, we obtain

$$\lim_{n\to+\infty} 2|u_{\varsigma_0+t-[t]}(\cdot,0,0,L,f([t_{\tau(n)}]+[t]+\cdot))-u_{\varsigma_0+t-[t]}(\cdot,0,0,L,\tilde{f}([t]+\cdot))|_{\mathbb{B}}=0.$$

Based on the continuity of the functions $t\mapsto u_t$ and $(t,s)\mapsto \mathbb{U}(t,s)\phi$ for $\phi\in\mathbb{B}_A$, we deduce that

$$\lim_{n\to+\infty}|\chi(t_{\tau(n)}+t)-\tilde{\chi}(t)|_{\mathbb{B}}=0 \quad \text{for all } t\in\mathbb{R}. \tag{5.114}$$

Now, we will show that

$$\lim_{n\to+\infty}|\tilde{\chi}(t-t_{\tau(n)})-\chi(t)|_{\mathbb{B}}=0 \quad \text{for all } t\in\mathbb{R}.$$

First, consider the case when $t-[t]>\varsigma_0$. Then, there exists $N(t)\in\mathbb{N}^*$ such that $\varsigma_{\tau(n)}\le t-[t]$, for all $n\ge N(t)$. In this situation, we get $[t-t_{\tau(n)}]=[t]-[t_{\tau(n)}]$. Therefore, for any $n\ge N(t)$,

$$\tilde{\chi}(t-t_{\tau(n)})=\mathbb{U}(\varsigma_0-\varsigma_{\tau(n)}+t-[t],0)\tilde{y}_{[t]-[t_{\tau(n)}]}+u_{\varsigma_0-\varsigma_{\tau(n)}+t-[t]}(\cdot,0,0,L,\tilde{f}([t]-[t_{\tau(n)}]+\cdot)).$$

By the same arguments used to prove (5.114), we can affirm that

$$\lim_{n\to+\infty}|\tilde{\chi}(t-t_{\tau(n)})-\chi(t)|_{\mathbb{B}}=0.$$

Second, we investigate the case $t-[t]<\varsigma_0$. Now, there is $N(t)\in\mathbb{N}^*$ such that $\varsigma_{\tau(n)}\ge t-[t]$, for all $n\ge N(t)$. In this case, we have that $[t-t_{\tau(n)}]=[t]-[t_{\tau(n)}]-1$, for any $n\ge N(t)$. Thus, for each $n\ge N(t)$,

$$\tilde{\chi}(t - t_{\tau(n)}) = \mathbb{U}(\varsigma_0 - \varsigma_{\tau(n)} + t - [t] + 1, 0)\tilde{y}_{[t]-[t_{\tau(n)}]-1}$$
$$+ u_{\varsigma_0-\varsigma_{\tau(n)}+t-[t]}(\cdot, 0, 0, L, \tilde{f}([t] - [t_{\tau(n)}] - 1 + \cdot)).$$

As a result of Lemma 5.6.8, the strong continuity of the process $(\mathbb{U}(t,s))_{t\geq s}$, and the continuity of the function $t \to u_t$, we conclude that

$$\lim_{n\to+\infty} \tilde{\chi}(t - t_{\tau(n)}) = \mathbb{U}(t - [t] + 1, 0)y_{[t]-1} + \lim_{k\to+\infty} \int_0^{t-[t]+1} \mathbb{U}(t - [t] + 1, s)\Upsilon^k f([t] - 1 + s)ds$$
$$= \mathbb{U}(t, [t] - 1)y_{[t]-1} + \lim_{k\to+\infty} \int_{[t]-1}^{t} \mathbb{U}(t,s)\Upsilon^k f(s)ds$$
$$= \mathbb{U}(t, [t])\left\{\mathbb{U}([t], [t] - 1)y_{[t]-1} + \lim_{k\to+\infty} \int_{[t]-1}^{[t]} \mathbb{U}([t], s)\Upsilon^k f(s)ds\right\}$$
$$+ \lim_{k\to+\infty} \int_{[t]}^{t} \mathbb{U}(t,s)\Upsilon^k f(s)ds$$
$$= \mathbb{U}(t, [t])\chi([t]) + \lim_{k\to+\infty} \int_{[t]}^{t} \mathbb{U}(t,s)\Upsilon^k f(s)ds$$
$$= \chi(t).$$

Finally, if $t - [t] = \varsigma_0$, there exists a subsequence $(\overline{\varsigma_{\tau(n)}})_{n\in\mathbb{N}}$ of $(\varsigma_{\tau(n)})_{n\in\mathbb{N}}$ such that, for all $n \in \mathbb{N}$,

$$\overline{\varsigma_{\tau(n)}} \leq t - [t] \quad \text{or} \quad \overline{\varsigma_{\tau(n)}} \geq t - [t].$$

With no restriction of generality, we can assume, for all $n \in \mathbb{N}$, that

$$\varsigma_{\tau(n)} \leq t - [t] \quad \text{or} \quad \varsigma_{\tau(n)} \geq t - [t].$$

In all cases, we have one of the above situations. Consequently,

$$\lim_{n\to+\infty} \tilde{\chi}(t - t_{\tau(n)}) = \chi(t) \quad \text{for all } t \in \mathbb{R}.$$

Hence, the function $\chi(\cdot)$ is almost automorphic. This ends the proof. □

Theorem 5.6.21. *Assume that* $(\mathbf{H}_0)$, $(\mathbf{H}_1)$, $(\mathbf{H}_3)$, *and* $(\mathbf{H}_6)$ *hold. Then, every almost automorphic integral solution of equation* (5.109) *is compact almost automorphic.*

Proof. Let u be an almost automorphic integral solution of (5.109) on $\mathbb{R}$ and $(t_n, s_n)_{n\in\mathbb{N}}$ be a sequence in $\mathbb{R}^2$ such that $|t_n - s_s| \to 0$ as $n \to +\infty$. Let $n \in \mathbb{N}$ be such that $t_n \geq s_n$. Then,

$$u(t_n) - u(s_n) = T(t_n - s_n)u(s_n) - u(s_n) + \lim_{\lambda \to +\infty} \int_{s_n}^{t_n} T(t - s_n)B_\lambda[L(r)u_s + f(r)]\mathrm{d}r.$$

Thus,

$$|u(t_n) - u(s_n)|_X \le \sup_{\xi \in \mathcal{Q}} |T(t_n - s_n)\xi - \xi|_X + \int_{s_n}^{t_s} M_0^2 e^{(t_n - r)\omega_0}(|L(r)|_{\mathcal{L}(\mathbb{B},X)}|u_r|_{\mathbb{B}} + |f(r)|_X)\mathrm{d}r,$$

where $\mathcal{Q} := \{u(r), r \in \mathbb{R}\}$. By axiom **(A)**, we have that

$$\sup_{t \in \mathbb{R}} |u_t|_{\mathbb{B}} < \infty.$$

Also, since $(t_n)_{n\in\mathbb{N}}$ and $(s_n)_{n\in\mathbb{N}}$ are bounded, we can find a constant $\tilde{C} > 0$ such that

$$\int_{s_n}^{t_s} M_0^2 e^{(t_n - r)\omega_0}\mathrm{d}r \le \tilde{C}(t_n - s_n).$$

Consequently,

$$\begin{aligned}|u(t_n) - u(s_n)|_X &\le \sup_{\xi \in \mathcal{Q}} |T(|t_n - s_n|)\xi - \xi|_X \\ &\quad + M_0^2\tilde{C}\Big(\tilde{L}\sup_{r\in\mathbb{R}}|u_r|_{\mathbb{B}} + \sup_{r\in\mathbb{R}}|f(r)|_X\Big)|t_n - s_n| \quad \text{for all } n \in \mathbb{N},\end{aligned}$$

where

$$\tilde{L} := \sup_{r \in [\min\{\{s_n, t_n; n\in\mathbb{N}\}, \max\{\{s_n, t_n; n\in\mathbb{N}\}]} |L(r)|_{\mathcal{L}(\mathbb{B},X)}.$$

As u is almost automorphic on $\mathbb{R}$, we have that $\mathcal{Q}$ is relatively compact in X. Consequently,

$$\lim_{n\to+\infty} \sup_{\xi\in\mathcal{Q}} |T(|t_n - s_n|)\xi - \xi|_X = 0.$$

Furthermore,

$$\lim_{n\to+\infty} |u(t_n) - u(s_n)|_X = 0.$$

Hence, u is uniformly continuous. Hence, we deduce the result of this theorem. □

Now, we are in a position to establish the Bohr–Neugebauer property for equation (5.109) in the case of almost automorphy.

Theorem 5.6.22. *Suppose that* $(\mathbf{H}_0)$–$(\mathbf{H}_4)$ *and* $(\mathbf{H}_6)$ *are satisfied. If equation* (5.109) *has a bounded integral solution on* $\mathbb{R}^+$, *then it has a compact almost automorphic integral solution on* $\mathbb{R}$. *In addition, if* u *is a bounded integral solution of equation* (5.109) *on* $\mathbb{R}$, *then the history function* $t \mapsto u_t$ *is almost automorphic and* u *is compact almost automorphic.*

Proof. By Theorem 5.6.10 and axiom **(A)**, we can observe that if u is a bounded integral solution of (5.109) on $\mathbb{R}^+$, $(u_n)_{n\in\mathbb{N}}$ is a bounded solution of (5.111) on $\mathbb{N}$. Therefore, by Lemma 5.6.17, equation (5.111) has an almost automorphic solution on $\mathbb{Z}$. As a result of Theorems 5.6.20 and 5.6.21, we conclude that equation (5.109) has a compact almost automorphic integral solution on $\mathbb{R}$.

Let us prove the second part of this theorem. Let u be a bounded integral solution of (5.109) on $\mathbb{R}$. Employing similar arguments used to show (2) in Theorem 5.6.19 and by Lemma 5.6.17, we can assert that the sequence $(u_n)_{n\in\mathbb{Z}}$ is an almost automorphic solution of (5.111). Consequently, the function χ given in Theorem 5.6.20 is almost automorphic and such that $\chi(t) = u_t$. Hence, $t \to u_t$ and u are almost automorphic. Using Theorem 5.6.21, we conclude that u is compact almost automorphic, ending the proof. □

5.6.7 Application

In this section, we apply our main results to establish the existence of an almost periodic and almost automorphic solution for the following model of reaction–diffusion system:

$$\begin{cases} \frac{\partial}{\partial t} w(t,x) = \Delta w(t,x) + \int_{-\infty}^{0} \xi_1(t,\theta) w(t+\theta,x)d\theta & \\ \qquad + \Gamma_1(t)(x) & \text{for } t \in \mathbb{R}^+ \text{ and } x \in \Omega, \\ w(t,x) = 0 & \text{for } x \in \partial\Omega \text{ and } t \in \mathbb{R}^+, \\ w(\theta,x) = \psi(\theta)(x) & \text{for } \theta \leq 0 \text{ and } x \in \overline{\Omega}, \end{cases} \tag{5.115}$$

where Ω is a bounded domain of $\mathbb{R}^n$ with a smooth boundary, Δ is the Laplacian operator in $\mathbb{R}^n$, $\Gamma_1 : \mathbb{R} \to C(\overline{\Omega}; \mathbb{R})$ is a continuous functions, $\psi :]-\infty, 0] \to C(\overline{\Omega}; \mathbb{R})$ is a bounded continuous function, and $\xi_1 : \mathbb{R} \times]-\infty, 0] \to \mathbb{R}^+$ is a function satisfying the following conditions:

$(\mathbf{C}_1)$ there exits $\gamma > 0$ such that, for all $t \in \mathbb{R}$,

$$e^{-\gamma \cdot} \xi_1(t, \cdot) \in L^1(]-\infty, 0]; \mathbb{R}^+),$$

$(\mathbf{C}_2)$ there exists $\xi_0 \in L^1(]-\infty, 0]; \mathbb{R}^+)$ such that

$$e^{\gamma\theta} \xi_1(t,\theta) \leq \xi_0(\theta) \quad \text{for a. e. } \theta \in]-\infty, 0]$$

for all $t \in \mathbb{R}$,

$(\mathbf{C}_3)$ for almost everywhere $\theta \in]-\infty, 0]$, the mapping $\xi_1(\cdot, \theta) : \mathbb{R} \to \mathbb{R}^+$ is a 1-periodic continuous function.

The aim of the following is to write the model (5.115) in an abstract form. Consider $X = C(\overline{\Omega}; \mathbb{R})$ which is endowed with sup norm. Let $(A, D(A))$ be the linear operator defined by

$$\begin{cases} Ay = \Delta y, \\ D(A) = \{y \in X \cap H_0^1(\Omega) : \Delta y \in X\}. \end{cases}$$

Let $\delta_0 > 0$ be the smallest eigenvalue of $(-A, D(A))$ in $H_0^1(\Omega)$. Then, form [139], we have that $]-\delta_0, +\infty[\subset \rho(A)$ and

$$|\mathcal{R}(\lambda, A)^m|_X \leq \frac{\exp\left(\delta_0 |\Omega|^{\frac{2}{n}} (2\pi)^{-1}\right)}{\lambda + \delta_0} \quad \text{for } m \in \mathbb{N} \text{ and } \lambda > -\delta_0.$$

Consequently, $(A, D(A))$ satisfies $(\mathbf{H}_0)$ and $\overline{D(A)} = \{y \in X : y|_{\partial\Omega} = 0\} \neq X$. We denote by $(A_0, D(A_0))$ the part of A in $\overline{D(A)}$, which is given by

$$\begin{cases} D(A_0) = \{y \in D(A) : Ay|_{\partial\Omega} = 0\}, \\ A_0 y = Ay, \quad \forall y \in D(A_0). \end{cases}$$

Lemma 5.6.23 ([139]). *The operator $(A_0, D(A_0))$ generates a compact semigroup $(T(t))_{t\geq 0}$ in $\overline{D(A)}$. Consequently, condition $(\mathbf{H}_4)$ holds.*

We introduce the following phase space:

$$\mathbb{B} = \left\{\phi \in C(]-\infty, 0]; X) : \lim_{\theta \to -\infty} e^{\gamma\theta} \phi(\theta) = 0\right\},$$

equipped with the norm $|\phi|_{\mathbb{B}} = \sup_{\theta\leq 0} e^{\gamma\theta} |\phi(\theta)|_X$, where γ is the same constant given in condition $(\mathbf{C}_1)$. From [181, Theorem 7.3, p. 15], we have that the space $\mathbb{B}$ is a uniform fading memory space. That is, $\mathbb{B}$ satisfies condition $(\mathbf{H}_3)$, namely, the functions in axiom **(A)** are given by

$$N = 1; \quad K(t) = 1 \quad \text{and} \quad M(t) = e^{-\gamma t} \quad \text{for } t \geq 0.$$

Let $L : \mathbb{R} \times \mathbb{B} \to X$ be the mapping defined by

$$L(t)\phi(x) = \int_{-\infty}^{0} \xi_1(t, \theta)\phi(\theta)(x) d\theta \quad \text{for } x \in \overline{\Omega}, t \in \mathbb{R}, \text{and } \phi \in \mathbb{B}.$$

From the conditions $(\mathbf{C}_1)$, $(\mathbf{C}_2)$, and $(\mathbf{C}_3)$, we have that $(L(t))_{t\in\mathbb{R}}$ is well defined and satisfies the assumptions $(\mathbf{H}_1)$ and $(\mathbf{H}_2)$. Consider the function $u : \mathbb{R} \to X$, defined by

$$u(t)(x) = w(t, x) \quad \text{for } x \in \overline{\Omega} \text{ and } t \in \mathbb{R}.$$

Consequently, the model (5.115) takes the following abstract form:

$$\begin{cases} \frac{\mathrm{d}}{\mathrm{d}t}u(t) = Au(t) + L(t)u_t + \Gamma_1(t) & \text{for } t \geq 0, \\ u_0 = \psi. \end{cases} \tag{5.116}$$

Theorem 5.6.24. *Assume that the conditions stated above hold. We suppose further that:*
(1) Γ_1 *is a bounded function,*
(2) $\sup_{t\in\mathbb{R}} \int_{-\infty}^{0} e^{-\gamma\theta}\xi_1(t,\theta)\mathrm{d}\theta < \frac{\delta_0(1-\mu_0)}{\exp(2\delta_0|\Omega|^{\frac{2}{n}}(2\pi)^{-1})}$ *for some* $\mu_0 \in]0,1[$,
(3) $\sup_{\theta\leq 0}|\psi(\theta)|_X \leq \varpi$, *where*

$$\varpi := 1 + \frac{\exp(2\delta_0|\Omega|^{\frac{2}{n}}(2\pi)^{-1})\sup_{t\in\mathbb{R}}|\Gamma_1(t)|_X}{\delta_0\mu_0}.$$

Then, equation (5.116) *has a bounded integral solution on* $\mathbb{R}^+$. *In addition, if* Γ_1 *is almost periodic (resp. almost automorphic), then equation* (5.116) *has an almost periodic (resp. compact almost automorphic).*

Proof. Let us prove by contradiction that

$$|u(t)| \leq \varpi \quad \text{for all } t \geq 0.$$

Suppose that

$$\Theta := \{t \in \mathbb{R}^+, |u(t)|_X > \varpi\} \neq \emptyset.$$

Let $t_0 := \inf \Theta$. Then,

$$u(t_0) = T(t_0)\psi(0) + \lim_{\lambda\to\infty} \int_0^{t_0} T(t_0 - s)B_\lambda\{L(s)u_s + \Gamma_1(s)\}\mathrm{d}s.$$

Therefore,

$$\begin{aligned} |u(t_0)|_X &\leq \exp(\delta_0|\Omega|^{\frac{2}{n}}(2\pi)^{-1})e^{-\delta_0 t_0}|\psi(0)|_X \\ &\quad + \int_0^{t_0} \exp(2\delta_0|\Omega|^{\frac{2}{n}}(2\pi)^{-1})e^{-\delta_0(t_0-s)}\{|L(s)u_s(\cdot,\phi)|_X + |\Gamma_1(s)|_X\}\mathrm{d}s. \end{aligned}$$

Observe that $-\infty < s + \theta \leq \tilde{t}$, for all $\theta \leq 0$ and $s \in [0, t_0]$, and then

$$\sup_{\theta\leq 0}|u(s+\theta)|_X \leq \varpi.$$

Therefore,

$$|u_s|_{\mathbb{B}} = \sup_{\theta\le 0} e^{\gamma\theta}|u(s+\theta)|_X \le \varpi \quad \text{for all } s\in[0,t_0].$$

Thus,

$$\left|L(s)u_s(\cdot)\right|_X \le \varpi \sup_{t\in\mathbb{R}} \int_{-\infty}^{0} e^{-\gamma\theta}\xi_1(t,\theta)d\theta \le \frac{\varpi\delta_0(1-\mu_0)}{\exp\left(2\delta_0|\Omega|^{\frac{2}{n}}(2\pi)^{-1}\right)} \quad \text{for all } s\in[0,t_0].$$

It follows that

$$\begin{aligned}\left|u(t_0)\right|_X &\le e^{-\delta_0 t_0}\varpi + \varpi\delta_0(1-\mu_0)\int_0^{t_0} e^{-\delta_0(t_0-s)}ds \\ &\quad + \exp\left(2\delta_0|\Omega|^{\frac{2}{n}}(2\pi)^{-1}\right)\sup_{t\in\mathbb{R}}\left|\Gamma_1(t)\right|_X \int_0^{t_0} e^{-\delta_0(t_0-s)}ds \\ &\le e^{-\delta_0 t_0}\varpi + \left\{\varpi - \mu_0\varpi + \frac{\exp\left(2\delta_0|\Omega|^{\frac{2}{n}}(2\pi)^{-1}\right)\sup_{t\in\mathbb{R}}|\Gamma_1(t)|_X}{\delta_0}\right\}(1-e^{-\delta_0 t_0}).\end{aligned}$$

By the definition of ϖ, we have that

$$\frac{\exp\left(2\delta_0|\Omega|^{\frac{2}{n}}(2\pi)^{-1}\right)\sup_{t\in\mathbb{R}}|\Gamma_1(t)|_X}{\delta_0} = \mu_0\varpi - \mu_0.$$

Then,

$$\begin{aligned}\left|u(t_0)\right|_X &\le e^{-\delta_0 t_0}\varpi + (\varpi-\mu_0)(1-e^{-\delta_0 t_0}) \\ &\le e^{-\delta_0 t_0}\varpi + \varpi - e^{-\delta_0 t_0}\varpi - \mu_0(1-e^{-\delta_0 t_0}).\end{aligned}$$

Consequently,

$$\left|u(t_0)\right|_X < \varpi. \tag{5.117}$$

Since $t_0 = \inf\Theta$, there exists a sequence $(t_n)_{n\in\mathbb{N}^*}\subset\Theta$ such that $t_n\to t_0$ as $n\to+\infty$. Due to

$$\left|u(t_n)\right|_X > \varpi \quad \text{for all } n\in\mathbb{N}^*,$$

we obtain

$$\left|u(t_0)\right|_X = \lim_{n\to+\infty}\left|u(t_n)\right|_X \ge \varpi,$$

which contradicts (5.117). Consequently, u is a bounded integral solution of equation (5.116) on $\mathbb{R}^+$. The rest of the proof is a consequence of Theorems 5.6.19 and 5.6.22. □

Theorem 5.6.25. *Let the conditions* (1)–(3) *in Theorem* 5.6.24 *be satisfied. Assume further that $\gamma \geq \delta_0$ and Γ_1 is almost periodic (resp. almost automorphic). Then, equation* (5.116) *has a unique almost periodic (resp. compact almost automorphic) solution which is globally attractive.*

Proof. Applying Theorem 5.6.24, we can conclude that equation (5.116) has an almost periodic (resp. a compact almost automorphic) u. Before starting to show the uniqueness, we observe, for $t \geq \sigma$, that

$$|\mathbb{U}(t,\sigma)\phi(0)|_X \leq \exp(\delta_0|\Omega|^{\frac{2}{n}}(2\pi)^{-1})e^{-\delta_0(t-\sigma)}|\phi|_{\mathbb{B}} + \int_\sigma^t \exp(2\delta_0|\Omega|^{\frac{2}{n}}(2\pi)^{-1})e^{-\delta_0(t-s)}|L(s)|_{\mathcal{L}(\mathbb{B},X)}|\mathbb{U}(s,\sigma)\phi|_{\mathbb{B}_A}ds.$$

Then,

$$e^{\delta_0 t}|\mathbb{U}(t,\sigma)\phi(0)|_X \leq \exp(\delta_0|\Omega|^{\frac{2}{n}}(2\pi)^{-1})e^{\delta_0\sigma}|\phi|_{\mathbb{B}} + \exp(2\delta_0|\Omega|^{\frac{2}{n}}(2\pi)^{-1})\sup_{t\in\mathbb{R}}\int_{-\infty}^0 e^{-\gamma\theta}\xi_1(t,\theta)d\theta\int_\sigma^t e^{\delta_0 s}|\mathbb{U}(s,\sigma)\phi|_{\mathbb{B}_A}ds.$$

So, for $\theta \leq 0$ such that $t+\theta-\sigma \geq 0$, we have

$$\begin{aligned} &e^{\delta_0(t+\theta)}|\mathbb{U}(t+\theta,\sigma)\phi(0)|_X \\ &\leq \exp(\delta_0|\Omega|^{\frac{2}{n}}(2\pi)^{-1})e^{\delta_0\sigma}|\phi|_{\mathbb{B}} \\ &\quad + \exp(2\delta_0|\Omega|^{\frac{2}{n}}(2\pi)^{-1})\sup_{t\in\mathbb{R}}\int_{-\infty}^0 e^{-\gamma\theta}\xi_1(t,\theta)d\theta\int_\sigma^{t+\theta} e^{\delta_0 s}|\mathbb{U}(s,\sigma)\phi|_{\mathbb{B}_A}ds. \end{aligned}$$

Thus,

$$\begin{aligned} &e^{\delta_0 t}e^{\gamma\theta}|\mathbb{U}(t,\sigma)\phi(\theta)|_X \\ &\leq \exp(\delta_0|\Omega|^{\frac{2}{n}}(2\pi)^{-1})e^{(-\delta_0+\gamma)\theta}e^{\delta_0\sigma}|\phi|_{\mathbb{B}} \\ &\quad + e^{(-\delta_0+\gamma)\theta}\exp(2\delta_0|\Omega|^{\frac{2}{n}}(2\pi)^{-1})\sup_{t\in\mathbb{R}}\int_{-\infty}^0 e^{-\gamma\theta}\xi_1(t,\theta)d\theta\int_\sigma^t e^{\delta_0 s}|\mathbb{U}(s,\sigma)\phi|_{\mathbb{B}_A}ds. \end{aligned}$$

Since $\gamma \geq \delta_0$, one obtains

$$\begin{aligned} &e^{\delta_0 t}e^{\gamma\theta}|\mathbb{U}(t,\sigma)\phi(\theta)|_X \\ &\leq \exp(\delta_0|\Omega|^{\frac{2}{n}}(2\pi)^{-1})e^{\delta_0\sigma}|\phi|_{\mathbb{B}} \\ &\quad + \exp(2\delta_0|\Omega|^{\frac{2}{n}}(2\pi)^{-1})\sup_{t\in\mathbb{R}}\int_{-\infty}^0 e^{-\gamma\theta}\xi_1(t,\theta)d\theta\int_\sigma^t e^{\delta_0 s}|\mathbb{U}(s,\sigma)\phi|_{\mathbb{B}_A}ds. \end{aligned}$$

On the other hand, for $\theta \le 0$ such that $t + \theta - \sigma \le 0$, we get

$$\begin{aligned} e^{\delta_0 t} e^{\gamma\theta} |\mathbb{U}(t,\sigma)\phi(\theta)|_X &= e^{\delta_0 t} e^{\gamma\theta} |\phi(t+\theta-\sigma)|_X \\ &= e^{(t-\sigma)(\delta_0-\gamma)} e^{\delta_0\sigma} e^{\gamma(t+\theta-\sigma)} |\phi(t+\theta-\sigma)|_X. \end{aligned}$$

Once again, we use the fact that $\gamma \ge \delta_0$ and obtain

$$\begin{aligned} e^{\delta_0 t} e^{\gamma\theta} |\mathbb{U}(t,\sigma)\phi(\theta)|_X \le{}& \exp\big(\delta_0|\Omega|^{\frac{2}{n}}(2\pi)^{-1}\big) e^{\delta_0\sigma}|\phi|_{\mathbb{B}} \\ &+ \exp\big(2\delta_0|\Omega|^{\frac{2}{n}}(2\pi)^{-1}\big) \sup_{t\in\mathbb{R}} \int_{-\infty}^{0} e^{-\gamma\theta}\xi_1(t,\theta)\mathrm{d}\theta \int_{\sigma}^{t} e^{\delta_0 s}|\mathbb{U}(s,\sigma)\phi|_{\mathbb{B}_A}\mathrm{d}s. \end{aligned}$$

Consequently,

$$\begin{aligned} e^{\delta_0 t}|\mathbb{U}(t,\sigma)\phi|_{\mathbb{B}_A} &= e^{\delta_0 t}\sup_{\theta\le 0} e^{\gamma\theta}|\mathbb{U}(t,\sigma)\phi(\theta)|_X \\ &\le \exp\big(\delta_0|\Omega|^{\frac{2}{n}}(2\pi)^{-1}\big) e^{\delta_0\sigma}|\phi|_{\mathbb{B}} \\ &\quad + \exp\big(2\delta_0|\Omega|^{\frac{2}{n}}(2\pi)^{-1}\big) \sup_{t\in\mathbb{R}} \int_{-\infty}^{0} e^{-\gamma\theta}\xi_1(t,\theta)\mathrm{d}\theta \int_{\sigma}^{t} e^{\delta_0 s}|\mathbb{U}(s,\sigma)\phi|_{\mathbb{B}_A}\mathrm{d}s. \end{aligned}$$

Thanks to Gronwall's lemma, we deduce

$$e^{\delta_0 t}|\mathbb{U}(t,\sigma)\phi|_{\mathbb{B}_A} \le \exp\big(\delta_0|\Omega|^{\frac{2}{n}}(2\pi)^{-1}\big) e^{\delta_0\sigma} e^{\exp(2\delta_0|\Omega|^{\frac{2}{n}}(2\pi)^{-1})\sup_{t\in\mathbb{R}}\int_{-\infty}^{0} e^{-\gamma\theta}\xi_1(t,\theta)\mathrm{d}\theta(t-\sigma)}|\phi|_{\mathbb{B}}.$$

Thus, for all $t \ge \sigma$,

$$|\mathbb{U}(t,\sigma)\phi|_{\mathbb{B}_A} \le \exp\big(\delta_0|\Omega|^{\frac{2}{n}}(2\pi)^{-1}\big) e^{(\exp(2\delta_0|\Omega|^{\frac{2}{n}}(2\pi)^{-1})\sup_{t\in\mathbb{R}}\int_{-\infty}^{0} e^{-\gamma\theta}\xi_1(t,\theta)\mathrm{d}\theta-\delta_0)(t-\sigma)}|\phi|_{\mathbb{B}}. \tag{5.118}$$

Now, let v be another almost periodic (resp. compact almost automorphic) solution of equation (5.116). Then, from Theorem 5.6.10, we have, for $t \ge \sigma$, that

$$u_t - v_t = \mathbb{U}(t,\sigma)(u_\sigma - v_\sigma).$$

Using (5.118), we can assert

$$|u_t - v_t|_{\mathbb{B}} \le \exp\big(\delta_0|\Omega|^{\frac{2}{n}}(2\pi)^{-1}\big) e^{(\exp(2\delta_0|\Omega|^{\frac{2}{n}}(2\pi)^{-1})\sup_{t\in\mathbb{R}}\int_{-\infty}^{0} e^{-\gamma\theta}\xi_1(t,\theta)\mathrm{d}\theta-\delta_0)(t-\sigma)} \sup_{\eta\in\mathbb{R}}|u_\eta - v_\eta|_{\mathbb{B}}.$$

Letting $\sigma \to -\infty$, we get $u_t = v_t$. Thus, $u = v$, that is, u is unique.

Let us prove the last part. Let v be any solution of equation (5.116) on $\mathbb{R}^+$. Then, by Theorem 5.6.10, we obtain, for any $t \ge 0$, that

$$\begin{aligned}|u_t - v_t|_{\mathbb{B}} &= \left|\mathbb{U}(t,0)(u_0 - v_0)\right|_{\mathbb{B}} \\ &\leq \exp\left(\delta_0|\Omega|^{\frac{2}{n}}(2\pi)^{-1}\right)e^{(\exp(2\delta_0|\Omega|^{\frac{2}{n}}(2\pi)^{-1})\sup_{t\in\mathbb{R}}\int_{-\infty}^{0}e^{-\gamma\theta}\xi_1(t,\theta)\mathrm{d}\theta-\delta_0)t}|u_0 - v_0|_{\mathbb{B}}.\end{aligned}$$

Therefore,

$$\lim_{t\to+\infty}|u_t - \tilde{u}_t|_{\mathbb{B}} = 0.$$

That is, u is globally attractive, completing the proof. □

Bibliography

[1] M. Adimy, H. Bouzahir, and K. Ezzinbi. Local existence and stability for some partial functional differential equations with infinite delay. *Nonlinear Analysis: Theory, Methods & Applications*, 48(3):323–348, 2002.

[2] M. Adimy, A. Elazzouzi, and K. Ezzinbi. Bohr–Neugebauer type theorem for some partial neutral functional differential equations. *Nonlinear Analysis: Theory, Methods & Applications*, 66(5):1145–1160, 2007.

[3] M. Adimy, A. Elazzouzi, and K. Ezzinbi. Reduction principle and dynamic behaviors for a class of partial functional differential equations. *Nonlinear Analysis: Theory, Methods & Applications*, 71(5):1709–1727, 2009.

[4] M. Adimy and K. Ezzinbi. Equation de type neutre et semigroupes intégrés. *Comptes Rendus de L'Académie des Sciences Paris, Série I*, 318:529–534, 1994.

[5] M. Adimy and K. Ezzinbi. Existence and linearized stability for partial neutral functional differential equations with nondense domains. *Differential Equations and Dynamical Systems*, 7:371–417, 1999.

[6] M. Adimy and K. Ezzinbi. Existence and stability of solutions for a class of partial neutral functional differential equations. *Hiroshima Mathematical Journal*, 34(3):251–294, 2004.

[7] M. Adimy, K. Ezzinbi, and M. Laklach. Spectral decomposition for partial neutral functional differential equations. *Canadian Applied Mathematics Quarterly*, 9(1):1–34, 2001.

[8] M. Adimy, K. Ezzinbi, and A. Ouhinou. Variation of constants formula and almost periodic solutions for some partial functional differential equations with infinite delay. *Journal of Mathematical Analysis and Applications*, 317(2):668–689, 2006.

[9] M. Adimy, K. Ezzinbi, and A. Ouhinou. Behavior near hyperbolic stationary solutions for partial functional differential equations with infinite delay. *Nonlinear Analysis: Theory, Methods & Applications*, 68(8):2280–2302, 2008.

[10] R. Agarwal, M. Belmekki, and M. Benchohra. A survey on semilinear differential equations and inclusions involving Riemann–Liouville fractional derivative. *Advances in Difference Equations*, 2009:981728, 2009.

[11] R. Agarwal, M. Benchohra, and S. Hamani. A survey on existence results for boundary value problems of nonlinear fractional differential equations and inclusions. *Acta Applicandae Mathematicae*, 109:973–1033, 2010.

[12] R. Agarwal, C. Cuevas, and H. Soto. Pseudo-almost periodic solutions of a class of semilinear fractional differential equations. *Journal of Applied Mathematics and Computation*, 37:625–634, 2011.

[13] R. Agarwal, B. de Andrade, and C. Cuevas. On type of periodicity and ergodicity to a class of fractional order differential equations. *Advances in Difference Equations*, 2010:179750, 2010.

[14] R. Agarwal, V. Lakshmikantham, and J. Nieto. On the concept of solution for fractional differential equations with uncertainty. *Nonlinear Analysis: Theory, Methods & Applications*, 72:2859–2862, 2010.

[15] R. P. Agarwal, B. de Andrade, and C. Cuevas. Weighted pseudo-almost periodic solutions of a class of semilinear fractional differential equations. *Nonlinear Analysis: Real World Applications*, 11(5):3532–3554, 2010.

[16] V. Ahn and R. McVinisch. Fractional differential equations driven by Lévy noise. *Journal of Applied Mathematics and Stochastic Analysis*, 16:97–119, 2003.

[17] E. Ait Dads. *Contribution à l'existence de solutions pseudo presque périodiques d'une classe d'équations fonctionnelles non linéaires*. PhD thesis, Cadi Ayyad University, Marrakesh, 1994.

[18] E. Ait Dads and O. Arino. Exponential dichotomy and existence of pseudo almost-periodic solutions of some differential equations. *Nonlinear Analysis: Theory, Methods & Applications*, 27(4):369–386, 1996.

[19] E. Ait Dads, O. Arino, and K. Ezzinbi. Pseudo almost periodic solutions for some differential equations in a Banach space. *Nonlinear Analysis: Theory, Methods & Applications*, 28(7):1141–1155, 1997.

https://doi.org/10.1515/9783111684710-006

[20] E. Ait Dads, P. Cieutat, and K. Ezzinbi. The existence of pseudo-almost periodic solutions for some nonlinear differential equations in Banach spaces. *Nonlinear Analysis: Theory, Methods & Applications*, 69(4):1325–1342, 2008.

[21] E. Ait Dads, B. Es-Sebbar, and L. Lhachimi. On massera and Bohr–Neugebauer type theorems for some almost automorphic differential equations. *Journal of Mathematical Analysis and Applications*, 518(2):126761, 2023.

[22] E. Ait Dads and K. Ezzinbi. Existence of positive pseudo almost periodic solution for a class of functional equations arising in epidemic problems. *Cybernetics and Systems Analysis*, 30(6):900–910, 1994.

[23] E. Ait Dads and K. Ezzinbi. Pseudo almost periodic solutions of some delay differential equations. *Journal of Mathematical Analysis and Applications*, 201(3):840–850, 1996.

[24] E. Ait Dads and K. Ezzinbi. Existence of positive pseudo-almost-periodic solution for some nonlinear infinite delay integral equations arising in epidemic problems. *Nonlinear Analysis: Theory, Methods & Applications*, 41(1):1–13, 2000.

[25] E. Ait Dads and L. Lhachimi. Exponential trichotomy and qualitative properties for ordinary differential equation. *International Journal of Evolution Equations*, 1:57–67, 2010.

[26] E. H. Ait Dads, B. Es-sebbar, K. Ezzinbi, and M. Ziat. Behavior of bounded solutions for some almost periodic neutral partial functional differential equations. *Mathematical Methods in the Applied Sciences*, 40(7):2377–2397, 2017.

[27] P. G. T. Alarcón and Y. Nakata. Stability analysis of a renewal equation for cell population dynamics with quiescence. *SIAM Journal on Applied Mathematics*, 74:1266–1297, 2014.

[28] L. Amerio. Soluzioni quasi-periodiche, o limitate, di sistemi differenziali non lineari quasi-periodici, o limitati. *Annali di Matematica Pura ed Applicata*, 39(1):97–119, 1955.

[29] L. Amerio. Problema misto e quasi-periodicità per l'equazione delle onde non omogenea. *Annali di Matematica Pura ed Applicata*, 49(1):393–417, 1960.

[30] L. Amerio. Problema misto e soluzioni quasiperiodiche dell'equazione delle onde. *Rendiconti del Seminario Matematico e Fisico di Milano*, 30(1):197–222, 1960.

[31] L. Amerio. Almost-periodic solutions of equation of schrodinger type. *Atti della Accademia Nazionale dei Lincei. Classe di Scienze Fisiche, Matematiche e Naturali*, 43(3–4):147–153, 1967.

[32] L. Amerio. Almost-periodic functions in Banach-spaces. *Matematisk-fysiske Meddelelser Kongelige Danske Videnskabernes Selskab*, 42(3):25–33, 1989.

[33] L. Amerio and G. Prouse. On the nonlinear wave equation, I, II. *Rendiconti della Accademia Nazionale dei Lincei*, 44, 1968.

[34] L. Amerio and G. Prouse. Uniqueness and almost periodicity for a nonlinear wave equation. *Rendiconti Lincei-Matematicae Applicazioni*, 56(1), 1969.

[35] L. Amerio and G. Prouse. *Almost-Periodic Functions and Functional Equations. University Series in Higher Mathematics*. Van Nostrand Reinhold Company, 1971.

[36] B. Amir and L. Maniar. Composition of pseudo almost periodic functions and Cauchy problems with operator of non dense domain. *Annales Mathématiques Blaise Pascal*, 6(1):1–11, 1999.

[37] D. Araya, R. Castro, and C. Lizama. Almost automorphic solutions of difference equations. *Advances in Difference Equations*, 2009:591380, 2009.

[38] D. Araya and C. Lizama. Almost automorphic mild solutions to fractional differential equations. *Nonlinear Analysis: Theory, Methods & Applications*, 69:3692–3705, 2008.

[39] W. Arendt, C. J. Batty, M. Hieber, and F. Neubrander. *Vector-valued Laplace Transforms and Cauchy Problems*, volume 96 of *Monographs in Mathematics*. Springer Basel, 2011.

[40] O. Arino and E. Hanebaly. Solutions presque périodiques de: $(dx/dt) + \|x\|^{\alpha}x = h(t)$ $(\alpha \geq 0)$ sur les espaces de Banach. *Comptes Rendus Mathématiques de L'Académie des Sciences Paris, Série I Mathématique*, 306(16):707–710, 1988.

[41] J. P. Aubin. *Applied Functional Analysis*. John Wiley & Sons, 2011.

[42] M. Ayachi, J. Blot, and P. Cieutat. Almost periodic solutions of monotone second-order differential equations. *Advanced Nonlinear Studies*, 11(3):541–554, 2011.

[43] M. Bai and S. Xu. On a size-structured population model with infinite states-at-birth and distributed delay in birth process. *Applicable Analysis*, 92(9):1916–1927, 2013.

[44] J. B. Baillon and A. Haraux. Comportement à l'infini pour les équations d'évolution avec forcing périodique. *Archive for Rational Mechanics and Analysis*, 67(1):101–109, 1977.

[45] M. Baroun, S. Boulite, T. Diagana, and L. Maniar. Almost periodic solutions to some semilinear non-autonomous thermoelastic plate equations. *Journal of Mathematical Analysis and Applications*, 349(1):74–84, 2009.

[46] M. Baroun, L. Maniar, and R. Schnaubelt. Almost periodicity of parabolic evolution equations with inhomogeneous boundary values. *Integral Equations and Operator Theory*, 65(2):169–193, 2009.

[47] O. D. C. Barril, A. Calsina and J. Z. Farkas. On competition through growth reduction. *arXiv preprint*, arXiv:2303.02981, Mar 2023.

[48] B. Basit. Generalization of two theorems of M. I. Kadets concerning the indefinite integral of abstract almost periodic functions. *Matematičeskie Zametki*, 9:311–321, 1971 (in Russian).

[49] B. Basit and H. Günzler. Spectral criteria for solutions of evolution equations and comments on reduced spectra. *arXiv preprint*, arXiv:1006.2169, 2010.

[50] B. Basit and A. J. Pryde. Ergodicity and stability of orbits of unbounded semigroup representations. *Journal of the Australian Mathematical Society*, 77:209–232, 2004.

[51] A. Bátkai, M. Kramar-Fijavž, and A. Rhandi. *Positive Operator Semigroups: From Finite to Infinite Dimensions*, volume 257. Birkhäuser, 2017.

[52] A. Bátkai and S. Piazzera. Semigroups and linear partial differential equations with delay. *Journal of Mathematical Analysis and Applications*, 264(1):1–20, 2001.

[53] A. Bátkai and S. Piazzera. *Semigroups for Delay Equations*. AK Peters Wellesley, 2005.

[54] E. Bazhlekova. *Fractional Evolution Equations in Banach Spaces*. PhD thesis, Eindhoven University of Technology, 2001.

[55] M. Benchohra, J. Henderson, S. Ntouyas, and A. Ouahab. Existence results for fractional order functional differential equations with infinite delay. *Journal of Mathematical Analysis and Applications*, 338:1340–1350, 2008.

[56] R. Benkhalti, H. Bouzahir, and K. Ezzinbi. Existence of a periodic solution for some partial functional differential equations with infinite delay. *Journal of Mathematical Analysis and Applications*, 256(1):257–280, 2001.

[57] R. Benkhalti, B. Es-sebbar, and K. Ezzinbi. On a Bohr–Neugebauer property for some almost automorphic abstract delay equations. *Journal of Integral Equations and Applications*, 30(3):313–345, 2018.

[58] S. Benzoni-Gavage. *Calcul différentiel et équations différentielles-2e éd.: Cours et exercices corrigés*. Dunod, 2021.

[59] A. Besicovitch. On generalized almost periodic functions. *Proceedings of the London Mathematical Society*, 2(1):495–512, 1926.

[60] A. Besicovitch. *Almost Periodic Functions*. Dover, New York, 1954.

[61] P. H. Bezandry and T. Diagana. Existence of almost periodic solutions to some stochastic differential equations. *Applicable Analysis*, 86(7):819–827, 2007.

[62] P. H. Bezandry and T. Diagana. *Almost Periodic Stochastic Processes*. Springer Science & Business Media, 2011.

[63] M. Biroli. Sur les solutions bornées et presque périodiques des équations et inéquations d'évolution. *Annali di Matematica Pura ed Applicata*, 93(1):1–79, 1972.

[64] M. Biroli. On the almost periodic solution to some parabolic quasi-variational inequalities. *Rivista di Matematica della Universitá di Parma*, 4:295–303, 1979.

[65] M. Biroli and F. Dal Fabbro. Bounded or almost periodic solutions of a wave equation with nonlinear viscosity. *Differential Equations*, 127:47–52, 1989.

[66] M. Biroli and A. Haraux. Asymptotic behavior for an almost periodic, strongly dissipative wave equation. *Journal of Differential Equations*, 38(3):422–440, 1980.
[67] J. Blot. Almost periodically forced pendulum. *Funkcialaj Ekvacioj*, 36:235–250, 1993.
[68] J. Blot, P. Cieutat, and K. Ezzinbi. New approach for weighted pseudo-almost periodic functions under the light of measure theory, basic results and applications. *Applicable Analysis*, 92(3):493–526, 2013.
[69] J. Blot, P. Cieutat, and J. Mawhin. Almost-periodic oscillations of monotone second-order systems. *Advances in Differential Equations*, 2(5):693–714, 1997.
[70] J. Blot, G. Mophou, G. M. N'Guérékata, and D. Pennequin. Weighted pseudo almost automorphic functions and applications to abstract differential equations. *Nonlinear Analysis: Theory, Methods & Applications*, 71(3):903–909, 2009.
[71] S. Bochner. Abstrakte fastperiodische funktionen. *Acta Mathematica*, 61(1):149–184, 1933.
[72] S. Bochner. Uniform convergence of monotone sequences of functions. *Proceedings of the National Academy of Sciences of the United States of America*, 47(4):582, 1961.
[73] S. Bochner. A new approach to almost periodicity. *Proceedings of the National Academy of Sciences of the United States of America*, 48(12):2039–2043, 1962.
[74] S. Bochner. Continuous mappings of almost automorphic and almost periodic functions. *Proceedings of the National Academy of Sciences of the United States of America*, 52(4):907–910, 1964.
[75] S. Bochner and J. Von Neumann. Almost periodic functions in groups, II. *Transactions of the American Mathematical Society*, 37(1):21–50, 1935.
[76] S. Bochner and J. Von Neumann. On compact solutions of operational-differential equations. I. *Annals of Mathematics*, 1:255–291, 1935.
[77] P. Bohl. *Uber die Darstellung von Funktionen einer Variablen durch trigonometrische Reihen mit mehreren einer Variablen proportionalen Argumenten*. Master's thesis, 1893.
[78] H. Bohr. Zur theorie der fastperiodischen funktionen. *Acta Mathematica*, 46(1–2):101–214, 1925.
[79] H. Bohr and O. Neugebauer. Über lineare Differentialgleichungen mit konstanten Koeffizienten und fastperiodischer rechter Seite. *Nachrichten von der Gesellschaft der Wissenschaften zu Göttingen, Mathematisch-Physikalische Klasse*, 1926:8–22, 1926.
[80] S. Boulite, L. Maniar, and G. M. N'Guérékata. Almost automorphic solutions for hyperbolic semilinear evolution equations. *Semigroup Forum*, 71(2):231–240, 2005.
[81] S. Boulite, L. Maniar, and G. M. N'Guérékata. Almost automorphic solutions for semilinear boundary differential equations. *Proceedings of the American Mathematical Society*, 134(12):3613–3624, 2006.
[82] D. Breda, O. Diekmann, W. De Graaf, A. Pugliese, and R. Vermiglio. On the formulation of epidemic models (an appraisal of Kermack and McKendrick). *Journal of Biological Dynamics*, 6(sup2):103–117, 2012.
[83] H. Brézis. *Opérateurs Maximaux Monotones et Semi-Groupes de Contractions dans les Espaces de Hilbert*. North-Holland, Amsterdam–London, 1973.
[84] S. H. Bruse. Generation of analytic semigroups by strongly elliptic operators under general boundary conditions. *Transactions of the American Mathematical Society*, 259:299–310, 1980.
[85] D. Bugajewski and G. M. N'Guérékata. On the topological structure of almost automorphic and asymptotically almost automorphic solutions of differential and integral equations in abstract spaces. *Nonlinear Analysis: Theory, Methods & Applications*, 59:1333–1345, 2004.
[86] C. Buse, P. Cerone, S. S. Dragomir, and A. Sofo. Uniform stability of periodic discrete systems in Banach spaces. *Journal of Difference Equations and Applications*, 11(12):1081–1088, 2005.
[87] J. Campos and M. Tarallo. Almost automorphic linear dynamics by Favard theory. *Journal of Differential Equations*, 256(4):1350–1367, 2014.
[88] J. Cao, Q. Yang, and Z. Huang. Existence of anti-periodic mild solutions for a class of semilinear fractional differential equations. *Communications in Nonlinear Science and Numerical Simulation*, 17:277–283, 2012.

[89] T. Caraballo and D. Cheban. Almost periodic and almost automorphic solutions of linear differential/difference equations without Favard's separation condition. I. *Journal of Differential Equations*, 246(1):108–128, 2009.

[90] T. Cazenave and A. Haraux. *Introduction aux Problèmes d'Evolution Semi-linéaires*. Ellipses-Edition, Paris, 1990.

[91] Y. Chang and X. Luo. Pseudo almost automorphic behavior of solutions to a semi-linear fractional differential equation. *Mathematical Communications*, 20:53–68, 2015.

[92] Y. Chang and C. Tang. Asymptotically almost automorphic solutions to stochastic differential equations driven by a Lévy process. *Stochastics. An International Journal of Probability and Stochastic Processes*, 88:980–1011, 2016.

[93] Y. Chang, R. Zhang, and G. M. N'Guérékata. Weighted pseudo almost automorphic mild solutions to semilinear fractional differential equations. *Computers & Mathematics with Applications*, 64:3160–3170, 2012.

[94] R. Chill and J. Prüss. Asymptotic behaviour of linear evolutionary integral equations. *Integral Equations and Operator Theory*, 39:193–213, 2001.

[95] P. Cieutat. Almost periodic solutions of forced vectorial Liénard equations. *Journal of Differential Equations*, 209(2):302–328, 2005.

[96] P. Cieutat and K. Ezzinbi. Existence, uniqueness and attractiveness of a pseudo almost automorphic solutions for some dissipative differential equations in Banach spaces. *Journal of Mathematical Analysis and Applications*, 354(2):494–506, 2009.

[97] P. Cieutat and K. Ezzinbi. Almost automorphic solutions for some evolution equations through the minimizing for some subvariant functional, applications to heat and wave equations with nonlinearities. *Journal of Functional Analysis*, 260(9):2598–2634, 2011.

[98] P. Cieutat, S. Fatajou, and G. M. N'Guérékata. Bounded and almost automorphic solutions of some nonlinear differential equation in Banach spaces. *Nonlinear Analysis: Theory, Methods & Applications*, 71:674–684, 2009.

[99] P. Cieutat, S. Fatajou, and G. M. N'Guérékata. Composition of pseudo almost periodic and pseudo almost automorphic functions and applications to evolution equations. *Applicable Analysis*, 89:11–27, 2010.

[100] P. Cieutat and A. Haraux. Exponential decay and existence of almost periodic solutions for some linear forced differential equations. *Portugaliae Mathematica*, 59(2):141–160, 2002.

[101] R. Cooke. Almost periodicity of bounded and compact solutions of differential equations. *Duke Mathematical Journal*, 36:273–276, 1969.

[102] C. Corduneanu. Almost periodic discrete processes. *Libertas Mathematica (vol. I–XXXI)*, 2:159–170, 1982.

[103] C. Corduneanu. *Almost Periodic Functions*. Chelsea Publishing Company, 1989.

[104] C. Corduneanu. *Principles of Differential and Integral Equations*. American Mathematical Society, 2008.

[105] C. Corduneanu. *Almost Periodic Oscillations and Waves*. Springer Science & Business Media, 2009.

[106] C. D. Coster and P. Habets. *Two-Point Boundary Value Problems: Lower and Upper Solutions*. Elsevier, Amsterdam, 2006.

[107] R. Courant and D. Gilbert. *Methods of Mathematical Physics, V. II. Partial Differential Equations*. Wiley, New York, 1962.

[108] M. G. Crandall and T. M. Liggett. Generation of semi-groups of nonlinear transformations on general Banach spaces. *American Journal of Mathematics*, 1:265–298, 1971.

[109] E. Cuesta. Asymptotic behaviour of the solutions of fractional integro-differential equations and some time discretizations. *Discrete and Continuous Dynamical Systems*, 2007:277–285, 2007.

[110] C. Cuevas and C. Lizama. Almost automorphic solutions to a class of semilinear fractional differential equations. *Applied Mathematics Letters*, 21:1315–1319, 2008.

[111] C. Cuevas and M. Pinto. Existence and uniqueness of pseudo almost periodic solutions of semilinear Cauchy problems with non dense domain. *Nonlinear Analysis: Theory, Methods & Applications*, 45(1):73–83, 2001.

[112] C. Cuevas, A. Sepúlveda, and H. Soto. Almost periodic and pseudo-almost periodic solutions to fractional differential and integro-differential equations. *Applied Mathematics and Computation*, 218:1735–1745, 2011.

[113] G. Da Prato and E. Sinestrari. Differential operators with non dense domain. *Annali della Scuola Normale Superiore di Pisa-Classe di Scienze*, 14(2):285–344, 1987.

[114] C. M. Dafermos. Almost periodic processes and almost periodic solutions of evolution equations. In *Dynamical Systems (Proc. Internat. Sympos., Univ. Florida, Gainesville, Fla., 1976)*, pages 43–57, 1977.

[115] E. B. Davies. *Linear Operators and Their Spectra*, volume 106. Cambridge University Press, 2007.

[116] T. Diagana. Pseudo almost periodic solutions to some differential equations. *Nonlinear Analysis: Theory, Methods & Applications*, 60(7):1277–1286, 2005.

[117] T. Diagana. Weighted pseudo almost periodic functions and applications. *Comptes Rendus. Mathématique*, 343(10):643–646, 2006.

[118] T. Diagana. Existence of weighted pseudo almost periodic solutions to some classes of hyperbolic evolution equations. *Journal of Mathematical Analysis and Applications*, 350(1):18–28, 2009.

[119] T. Diagana. *Almost Automorphic Type and Almost Periodic Type Functions in Abstract Spaces*. Springer, 2013.

[120] T. Diagana, H. R. Henriquez, and E. M. Hernández. Almost automorphic mild solutions to some partial neutral functional-differential equations and applications. *Nonlinear Analysis: Theory, Methods & Applications*, 69(5–6):1485–1493, 2008.

[121] T. Diagana, E. Hernández, and J. dos Santos. Existence of asymptotically almost automorphic solutions to some abstract partial neutral integro-differential equations. *Nonlinear Analysis: Theory, Methods & Applications*, 71:248–257, 2009.

[122] T. Diagana and E. M. Hernández. Existence and uniqueness of pseudo almost periodic solutions to some abstract partial neutral functional–differential equations and applications. *Journal of Mathematical Analysis and Applications*, 327(2):776–791, 2007.

[123] T. Diagana and G. M. N'Guérékata. Stepanov-like almost automorphic functions and applications to some semilinear equations. *Applicable Analysis*, 86(6):723–733, 2007.

[124] T. Diagana, G. M. N'Guérékata, and N. Van Minh. Almost automorphic solutions of evolution equations. *Proceedings of the American Mathematical Society*, 132(11):3289–3298, 2004.

[125] O. Diekmann and M. Gyllenberg. Equations with infinite delay: blending the abstract and the concrete. *Journal of Differential Equations*, 252(2):819–851, 2012.

[126] K. Diethelm. *The Analysis of Fractional Differential Equations*, volume 2004 of *Lecture Notes in Mathematics*. Springer Verlag, Berlin, Heidelberg, 2010.

[127] H. Ding, J. Liang, and T. Xiao. Almost automorphic solutions to abstract fractional differential equations. *Advances in Difference Equations*, 2010:508374, 2010.

[128] H. Ding, T. Xiao, and J. Liang. Asymptotically almost automorphic solutions for some integrodifferential equations with nonlocal initial conditions. *Journal of Mathematical Analysis and Applications*, 338:141–151, 2008.

[129] H.-S. Ding, W. Long, and G. M. N'Guérékata. A composition theorem for weighted pseudo-almost automorphic functions and applications. *Nonlinear Analysis: Theory, Methods & Applications*, 73(8):2644–2650, 2010.

[130] N. Drisi and B. Es-Sebbar. Almost automorphic solutions to logistic equations with discrete and continuous delay. *Comptes Rendus. Mathématique*, 355(12):1208–1214, 2017.

[131] N. Drisi, B. Es-sebbar, and K. Ezzinbi. Compact almost automorphic solutions for some nonlinear dissipative differential equations in Banach spaces. *Numerical Functional Analysis and Optimization*, 39(7):825–841, 2018.

[132] N. Dunford and J. Schwartz. *Linear Operators; Part I: General Theory*. Wiley, New York, 1957.

[133] N. Dunford, J. T. Schwartz, W. G. Bade, and R. G. Bartle. *Linear Operators*. Wiley-Interscience, New York, 1971.

[134] S. Eidelman and A. Kochubei. Cauchy problem for fractional diffusion equations. *Journal of Differential Equations*, 199:211–255, 2004.
[135] M. El-Borai. Some probability densities and fundamental solutions of fractional evolution equations. *Chaos, Solitons and Fractals*, 14:433–440, 2002.
[136] K. J. Engel and R. Nagel. *One-parameter Semigroups for Linear Evolution Equations*, volume 194. Springer Science & Business Media, 2000.
[137] B. Es-sebbar and K. Ezzinbi. Stepanov ergodic perturbations for some neutral partial functional differential equations. *Mathematical Methods in the Applied Sciences*, 39(8):1945–1963, 2016.
[138] B. Es-sebbar and K. Ezzinbi. Almost periodicity and almost automorphy for some evolution equations using Favard's theory in uniformly convex Banach spaces. *Semigroup Forum*, 94(2):229–259, 2017.
[139] B. Es-sebbar, K. Ezzinbi, and G. M. N'Guérékata. Bohr–Neugebauer property for almost automorphic partial functional differential equations. *Applicable Analysis*, 98(1–2):381–407, 2019.
[140] E. Esclangon. Sur une extension de la notion de périodicité. *Comptes Rendus de l'Académie des Sciences*, 135:891–894, 1902.
[141] K. Ezzinbi, S. Fatajou, and G. M. N'Guérékata. C^n-almost automorphic solutions for partial neutral functional differential equations. *Applicable Analysis*, 86(9):1127–1146, 2007.
[142] K. Ezzinbi, S. Fatajou, and G. M. N'Guerekata. Almost automorphic solutions for dissipative ordinary and functional differential equations in Banach spaces. *Communications in Mathematical Analysis*, 4(2):8–18, 2008.
[143] K. Ezzinbi, S. Fatajou, and G. M. N'Guérékata. Pseudo almost automorphic solutions for dissipative differential equations in Banach spaces. *Journal of Mathematical Analysis and Applications*, 351:765–772, 2009.
[144] K. Ezzinbi, S. Fatajou, and G. M. N'Guérékata. Weighted pseudo-almost periodic solutions for some neutral partial functional differential equations. *Electronic Journal of Differential Equations*, 2010(128):1–14, 2010.
[145] K. Ezzinbi and G. M. N'Guérékata. Almost automorphic solutions for partial functional differential equations with infinite delay. *Semigroup Forum*, 75(1):95–115, 2007.
[146] K. Ezzinbi and G. M. N'Guérékata. Almost automorphic solutions for some partial functional differential equations. *Journal of Mathematical Analysis and Applications*, 328(1):344–358, 2007.
[147] J. Favard. Sur les équations différentielles linéaires à coefficients presque-périodiques. *Acta Mathematica*, 51(1):31–81, 1928.
[148] J. Favard. Sur certains systèmes différentiels scalaires linéaires et homogènes à coefficients presque-périodiques. *Annali di Matematica Pura ed Applicata, Series 4*, 62(1):297–316, 1963.
[149] C. Feng. On the existence and uniqueness of almost periodic solutions for delay logistic equations. *Applied Mathematics and Computation*, 136(2):487–494, 2003.
[150] A. Fink. Almost automorphic and almost periodic solutions which minimize functionals. *Tohoku Mathematical Journal, Second Series*, 20(3):323–332, 1968.
[151] A. Fink. Extensions of almost automorphic sequences. *Journal of Mathematical Analysis and Applications*, 27(3):519–523, 1969.
[152] A. Fink. *Almost Periodic Differential Equations*, volume 377 of *Lecture Notes in Mathematics*. Springer-Verlag, Berlin–New York, 1974.
[153] A. Fink. Almost periodic functions. *CUBO, A Mathematical Journal*, 3(1):184–195, 2001.
[154] A. M. Fink. *Almost Periodic Differential Equations*, volume 377 of *Lecture Notes in Mathematics*. Springer-Verlag, Berlin–New York, 1974.
[155] G. Fournier, A. Szulkin, and M. Willem. Semilinear elliptic equations in $\mathbb{R}^n$ with almost periodic or unbounded forcing term. *SIAM Journal on Mathematical Analysis*, 27:1653–1660, 1996.
[156] D. Gilbarg and N. Trudinger. *Elliptic Partial Differential Equations of Second Order*. Springer, Berlin, 1998.
[157] J. A. Goldstein. Convexity, boundedness, and almost periodicity for differential equations in Hillbert space. *International Journal of Mathematics and Mathematical Sciences*, 2(1):1–13, 1979.

[158] J. A. Goldstein and G. M. N'Guérékata. Almost automorphic solutions of semilinear evolution equations. *Proceedings of the American Mathematical Society*, 133(8):2401–2408, 2005.

[159] K. Gopalsamy. *Equations of Mathematical Ecology*. Kluwer A. P., Dordrecht, 1992.

[160] R. Gorenflo and F. Mainardi. Fractional calculus: Integral and differential equations of fractional order. In A. Carpinteri and F. Mainardi, editors, *Fractals and Fractional Calculus in Continuum Mechanics*, pages 223–276. Springer-Verlag, Vienna, New York, 1997.

[161] G. Greiner. A typical Perron-Frobenius theorem with applications to an age-dependent population equation. In *Infinite-Dimensional Systems*, pages 86–100. Springer, Berlin, Heidelberg, 1984.

[162] M. Haase. The functional calculus for sectorial operators. In *Operator Theory: Advances and Applications*, volume 169. Birkhuser Verlag, Basel, 2006.

[163] J. K. Hale. Functional differential equations with infinite delays. *Journal of Mathematical Analysis and Applications*, 48(1):276–283, 1974.

[164] J. K. Hale. Coupled oscillators on a circle. *Resenhas do Instituto de Matemática e Estatística da Universidade de São Paulo*, 1(4):441–457, 1994.

[165] J. K. Hale and J. Kato. Phase space for retarded equations with infinite delay. *Funkcialaj Ekvacioj*, 21(1):11–41, 1978.

[166] J. K. Hale and S. M. V. Lunel. *Introduction to Functional Differential Equations*, volume 99. Springer Science & Business Media, 2013.

[167] E. Hanebaly. Solutions presque-périodiques d'équations différentielles monotones. *Comptes Rendus de L'Académie des Sciences Paris, Série I*, 296:263–265, 1983.

[168] E. Hanebaly. *Contribution à l'étude des solutions périodiques et presque-périodiques d'équations différentielles non-linéaires sur les espaces de Banach*. PhD thesis, Pau, 1988.

[169] A. Haraux. *Opérateurs maximaux monotones et oscillations forcées non linéaires*. Thèse, Université Paris VI, 1978.

[170] A. Haraux. Comportement à l'infini pour certains systèmes dissipatifs non linéaires. *Proceedings of the Royal Society of Edinburgh. Section A. Mathematics*, 84(3–4):213–234, 1979.

[171] A. Haraux. *Nonlinear Evolution Equations-Global Behavior of Solutions*, volume 841 of *Lecture Notes in Mathematics*. Springer-Verlag, 1981.

[172] A. Haraux. Almost-periodic forcing for a wave equation with a nonlinear, local damping term. *Proceedings of the Royal Society of Edinburgh. Section A. Mathematics*, 94(3–4):195–212, 1983.

[173] A. Haraux. Asymptotic behavior of trajectories for some nonautonomous, almost periodic processes. *Journal of Differential Equations*, 49(3):473–483, 1983.

[174] A. Haraux. Asymptotic behavior for two-dimensional, quasi-autonomous, almost-periodic evolution equations. *Journal of Differential Equations*, 66(1):62–70, 1987.

[175] B. He, J. Cao, and B. Yang. Weighted Stepanov-like pseudo-almost automorphic mild solutions for semilinear fractional differential equations. *Advances in Difference Equations*, 2015:74, 2015.

[176] D. Henry. *Geometric Theory of Semilinear Parabolic Equations. Lecture Notes in Mathematics*. Springer-Verlag, Berlin–New York, 1981.

[177] E. Hewitt and K. Ross. *Abstract Harmonic Analysis*, volume 1. Springer, Berlin, 1979.

[178] M. Hideaki, M. Satoru, and N. Van Minh. Decomposition of the phase space for integral equations and variation-of-constant formula in the phase space. *Funkcialaj Ekvacioj*, 55(3):479–520, 2012.

[179] R. Hilfer. *Applications of Fractional Calculus in Physics*. World Scientific, Singapore, 2000.

[180] Y. Hino and S. Murakami. Almost automorphic solutions for abstract functional differential equations. *Journal of Mathematical Analysis and Applications*, 286(2):741–752, 2003.

[181] Y. Hino, T. Naito, and S. Murakami. *Functional Differential Equations with Infinite Delay*. Springer, 1991.

[182] Y. Hino, T. Naito, N. VanMinh, and J. S. Shin. *Almost periodic solutions of differential equations in Banach spaces*. CRC Press, 2001.

[183] M. Iannelli. *Mathematical Theory of Age-Structured Population Dynamics*. Giardini Editori e Stampatori, Pisa, 1994.

[184] H. Ishii. On the existence of almost periodic complete trajectories for contractive almost periodic processes. *Journal of Differential Equations*, 43(1):66–72, 1982.

[185] R. A. Johnson. A linear, almost periodic equation with an almost automorphic solution. *Proceedings of the American Mathematical Society*, 82(2):199–205, 1981.

[186] V. Kavitha, P. Wang, and R. Murugesu. Existence of weighted pseudo almost automorphic mild solutions of fractional integro-differential equations. *Journal of Fractional Calculus and Applications*, 4:1–19, 2013.

[187] W. O. Kermack and A. G. McKendrick. A contribution to the mathematical theory of epidemics. *Proceedings of the Royal Society of London. Series A*, 115:700–721, 1927.

[188] A. Kilbas, H. Srivastava, and J. Trujillo. *Theory and Applications of Fractional Differential Equations. North-Holland Mathematics Studies.*

[189] M. A. Krasnosel'skii, V. S. Burd, and Y. S. Kolesov. *Nonlinear Almost Periodic Oscillations.* Wiley, New York, 1973 (English edition).

[190] N. V. Krylov. *Lectures on Elliptic and Parabolic Equations in Hölder Spaces.* Amer. Math. Soc., Providence, RI, 1996.

[191] K. Kuratowski. *Topology*, volume 1. Academic Press, New York, 1996 (English edition).

[192] V. Lakshmikantham. Theory of fractional differential equations. *Nonlinear Analysis: Theory, Methods & Applications*, 60:3337–3343, 2008.

[193] V. Lakshmikantham and J. Devi. Theory of fractional differential equations in Banach spaces. *European Journal of Pure and Applied Mathematics*, 1:38–45, 2008.

[194] V. Lakshmikantham and S. Leela. *Nonlinear Differential Equations in Abstract Spaces*, volume 2 of International Series in Nonlinear Mathematics: Theory, Methods and Applications. Pergamon Press, Oxford, New York, 1981.

[195] V. Lakshmikantham and A. Vatsala. Theory of fractional differential inequalities and applications. *Communications in Applied Analysis*, 11:395–402, 2007.

[196] V. Lakshmikantham and A. Vatsala. Basic theory of fractional differential equations. *Nonlinear Analysis: Theory, Methods & Applications*, 69:2677–2682, 2008.

[197] A. Leung. *Nonlinear Systems of Partial Differential Equations. Applications to Life and Physical Sciences.* World Sci. Publ., Nackensack, NJ, 2009.

[198] B. Levitan. A new generalization of the almost periodic functions of H. Bohr. *Zapiski Mekhaniko-Matematicheskogo Fakulteta Khar'kovskogo*, 15:334, 1938.

[199] B. M. Levitan and V. V. Zhikov. *Almost Periodic Functions and Differential Equations.* Cambridge University Press, 1982.

[200] H. X. Li, F. L. Huang, and J. Y. Li. Composition of pseudo almost-periodic functions and semilinear differential equations. *Journal of Mathematical Analysis and Applications*, 255(2):436–446, 2001.

[201] J. Liang, L. Maniar, G. M. N'Guérékata, and T. J. Xiao. Existence and uniqueness of $C^{(n)}$-almost periodic solutions to some ordinary differential equations. *Nonlinear Analysis: Theory, Methods & Applications*, 66(9):1899–1910, 2007.

[202] J. Liang, T. J. Xiao, and J. Zhang. Decomposition of weighted pseudo-almost periodic functions. *Nonlinear Analysis: Theory, Methods & Applications*, 73(10):3456–3461, 2010.

[203] J. Liang, J. Zhang, and T. J. Xiao. Composition of pseudo almost automorphic and asymptotically almost automorphic functions. *Journal of Mathematical Analysis and Applications*, 340(2):1493–1499, 2008.

[204] J. Liu, G. M. N'Guérékata, and N. Van Minh. A Massera type theorem for almost automorphic solutions of differential equations. *Journal of Mathematical Analysis and Applications*, 299(2):587–599, 2004.

[205] J. H. Liu, G. M. N'Guérékata, and N. Van Minh. *Topics on stability and periodicity in abstract differential equations*, volume 6. World Scientific, 2008.

[206] J. H. Liu and X. Q. Song. Almost automorphic and weighted pseudo almost automorphic solutions of semilinear evolution equations. *Journal of Functional Analysis*, 258(1):196–207, 2010.

[207] C. Lizama and F. Poblete. Regularity of mild solutions for a class of fractional order differential equations. *Applied Mathematics and Computation*, 224:803–816, 2013.
[208] A. J. Lotka. Relation between birth rates and death rates. *Science*, 26:21–22, 1907.
[209] P. Magal and S. Ruan. *Theory and applications of Abstract Semilinear Cauchy Problems*. Springer, 2018.
[210] M. Maqbul and D. Bahuguna. On the Stepanov-like almost automorphic solutions of abstract differential equations. *Differential Equations and Dynamical Systems*, 20(4):377–394, 2012.
[211] M. Maqbul and D. Bahuguna. Almost periodic solutions for Stepanov-almost periodic differential equations. *Differential Equations and Dynamical Systems*, 22:251–264, 2014.
[212] R. H. Martin. *Nonlinear Operators and Differential Equations in Banach Spaces*. John Wiley & Sons, New York, 1976.
[213] J. L. Massera. The existence of periodic solutions of systems of differential equations. *Duke Mathematical Journal*, 17(4):457–475, 1950.
[214] H. Matsunaga, S. Murakami, and Y. Nagabuchi. Formal adjoint operators and asymptotic formula for solutions of autonomous linear integral equations. *Journal of Mathematical Analysis and Applications*, 410(2):807–826, 2014.
[215] H. Matsunaga, S. Murakami, Y. Nagabuchi, and N. Van Minh. Center manifold theorem and stability for integral equations with infinite delay. *Funkcialaj Ekvacioj*, 58(1):87–134, 2015.
[216] J. Mawhin. Global results for the forced pendulum equation. In A. Cañada, P. Drábek, and A. Fonda, editors, *Handbook of Differential Equations, Ordinary Differential Equations*, volume 1, pages 533–589. Elsevier–North Holland, Amsterdam, 2004.
[217] L. Mawhin. Remarques sur les solutions bornées ou quasi-périodiques de l'équation du pendule forcé. *Annales des Sciences Mathématiques du Québec*, 22:213–224, 1998.
[218] J. A. Metz and O. Diekmann. *The Dynamics of Physiologically Structured Populations*. Springer-Verlag, Berlin, Heidelberg, 1986.
[219] K. Miller and B. Ross. *An Introduction to the Fractional Calculus and Fractional Differential Equations*. John Wiley & Sons, New York, NY, USA, 1993.
[220] G. Mophou. Existence and uniqueness of mild solutions to impulsive fractional differential equations. *Nonlinear Analysis: Theory, Methods & Applications*, 72:1604–1615, 2010.
[221] G. Mophou. Weighted pseudo almost automorphic mild solutions to semilinear fractional differential equations. *Applied Mathematics and Computation*, 217:7579–7587, 2011.
[222] G. Mophou, O. Nakoulima, and G. M. N'Guérékata. Existence results for some fractional differential equations with nonlocal conditions. *Nonlinear Studies*, 17:15–22, 2010.
[223] G. Mophou and G. M. N'Guérékata. Existence of mild solution for some fractional differential equations with nonlocal conditions. *Semigroup Forum*, 79:315–322, 2009.
[224] G. Mophou and G. M. N'Guérékata. On solutions of some nonlocal fractional differential equations with nondense domain. *Nonlinear Analysis: Theory, Methods & Applications*, 71:4668–4675, 2009.
[225] S. Murakami and T. Naito. Fading memory spaces and stability properties for functional differential equations with infinite delay. *Funkcialaj Ekvacioj*, 32:91–105, 1989.
[226] S. Murakami, T. Naito, and N. Van Minh. Massera's theorem for almost periodic solutions of functional differential equations dedicated to professor Yoshiyuki Hino on his sixtieth birthday. *Journal of the Mathematical Society of Japan*, 56(1):247–268, 2004.
[227] R. Nagel. *One-parameter Semigroups of Postive Operators*. Springer-Verlag, New York, 1986.
[228] T. Naito, N. Minh, R. Miyazaki, and Y. Hamaya. Boundedness and almost periodicity in dynamical systems. *Journal of Difference Equations and Applications*, 7(4):507–527, 2001.
[229] Y. Nakata. Note on stability conditions for structured population dynamics models. *Electronic Journal on the Qualitative Theory of Differential Equations*, 2016(78):1–14, 2016.
[230] Y. Nakata, Y. Enatsu, H. Inaba, T. Kuniya, Y. Muroya, and Y. Takeuchi. Stability of epidemic models with waning immunity. *Electronic SUT Journal of Mathematics*, 50(2):205–245, 2014.
[231] Y. Nakata and R. Omori. Delay equation formulation for an epidemic model with waning immunity: an application to mycoplasma pneumoniae. *Electronic SIFAC-PapersOnLine*, 48(18):132–135, 2015.

[232] G. M. N'Guérékata. Sur les solutions presque automorphes d'équations différentielles abstraites. *Annales des Sciences Mathématiques du Québec*, 5(1):69–79, 1981.

[233] G. M. N'Guérékata. Some remarks on asymptotically almost automorphic functions. *Rivista di Matematica della Universitá di Parma*, 13(4):301–303, 1987.

[234] G. M. N'Guérékata. *Almost Automorphic and Almost Periodic Functions in Abstract Spaces*. Springer, 2001.

[235] G. M. N'Guérékata. Existence and uniqueness of almost automorphic mild solutions to some semilinear abstract differential equations. *Semigroup Forum*, 69(1):80–86, 2004.

[236] G. M. N'Guérékata. Comments on almost automorphic and almost periodic functions in Banach spaces. *Far East Journal of Mathematical Sciences*, 17:337–344, 2005.

[237] G. M. N'Guérékata. *Topics in Almost Automorphy*. Springer, New York, Boston, Dordrecht, London, Moscow, 2005.

[238] G. M. N'Guérékata. A Cauchy problem for some fractional abstract differential equation with nonlocal conditions. *Nonlinear Analysis: Theory, Methods & Applications*, 70:1873–1876, 2009.

[239] G. M. N'Guérékata. *Spectral Theory for Bounded Functions and Applications to Evolution Equations*. Nova Science Publishers, New York, 2017.

[240] G. M. N'Guérékata et al.*Almost Periodic and Almost Automorphic Functions in Abstract Spaces*. Springer, 2021.

[241] G. M. N'Guérékata and A. Pankov. Stepanov-like almost automorphic functions and monotone evolution equations. *Nonlinear Analysis: Theory, Methods & Applications*, 68(9):2658–2667, 2008.

[242] V. M. Nguyen. A spectral theory of continuous functions and the Loomis–Arendt–Batty–Vu theory on the asymptotic behavior of solutions of evolution equations. *Journal of Differential Equations*, 247(4):1249–1274, 2009.

[243] D. Ortega and M. Tarallo. Almost periodic upper and lower solutions. *Journal of Differential Equations*, 193:343–358, 2003.

[244] R. Ortega and M. Tarallo. Almost periodic linear differential equations with non-separated solutions. *Journal of Functional Analysis*, 237(2):402–426, 2006.

[245] A. Pankov. Nonlinear second order elliptic equations with almost periodic coefficients. *Ukrainskij Matematičeskij žurnal*, 35(5):649–652, 1983. English translation: Ukr. Math. J., 35(5):564–567, 1983.

[246] A. A. Pankov. *Bounded and Almost Periodic Solutions of Nonlinear Operator Differential Equations*. Springer, 1990.

[247] A. Pazy. *Semigroups of Linear Operators and Applications to Partial Differential Equations*. Springer-Verlag, New York, 1983.

[248] I. Podlubny. *Fractional Differential Equations*. Academic Press, San Diego, 1999.

[249] G. Prouse. Soluzioni quasi-periodiche dell'equazione differenziale di Navier–Stokes in due dimensioni. *Rendiconti del Seminario Matematico della Università di Padova*, 33:186–212, 1963.

[250] G. Prouse. Soluzioni quasi-periodiche dell'equazione non omogenea della membrana vibrante, con termine dissipativo quadratico. *Atti della Accademia Nazionale dei Lincei. Classe di Scienze Fisiche, Matematiche e Naturali*, 37:364–370, 1964.

[251] V. Radŭlescu. *Qualitative Analysis of Nonlinear Partial Differential Equations: Monotonicity, Analytic, and Variational Methods*. Hindawi Publ. Corp., New York, 2008.

[252] A. Rhandi. Positivity and stability for a population equation with diffusion on L^1. *Positivity*, 2(2):101–113, 1998.

[253] A. Rhandi and H. Schnaubelt. Asymptotic behaviour of a non-autonomous population equation with diffusion in L^1. *Discrete and Continuous Dynamical Systems*, 5:663–684, 1999.

[254] W. Ruess and W. Summers. Compactness in spaces of vector valued continuous functions and asymptotic almost periodicity. *Mathematische Nachrichten*, 135:7–33, 1988.

[255] W. Ruess and Q. Vu. Asymptotically almost periodic solutions of evolution equations in Banach spaces. *Journal of Differential Equations*, 122:282–301, 1995.

[256] S. Samko, A. Kilbas, and O. Marichev. *Fractional Integral and Derivatives: Theory and Applications*. Gordon and Breach Science Publishers, Switzerland, 1993.

[257] K. Schmitt and J. R. Ward. Almost periodic solutions of nonlinear second order differential equations. *Results in Mathematics*, 21:190–199, 1992.

[258] L. Schwartz. *Analyse: topologie générale et analyse fonctionnelle*. Hermann, Paris, 1976.

[259] W. Shen and Y. Yi. Almost automorphic and almost periodic dynamics in skew-product semiflows. *Memoirs of the American Mathematical Society*, 136(651):647–647, 1998.

[260] M. A. Shubin. Almost periodic functions and differential equations. *Uspehi Matematičeskih Nauk*, 33(2):3–47, 1978. English translation: Russian Math. Surv., 33(2):1–52, 1978.

[261] D. Smart. *Fixed Point Theorems*. Cambridge University Press, 1980.

[262] J. Smoller. *Shock Waves and Reaction–Diffusion Equations*. Springer, New York, 1994.

[263] W. Stepanoff. Über einige Verallgemeinerungen der fast periodischen Funktionen. *Mathematische Annalen*, 95(1):473–498, 1926.

[264] M. Tarallo. A Stepanov version for Favard theory. *Archiv der Mathematik*, 90(1):53–59, 2008.

[265] H. R. Thieme. Semiflows generated by Lipschitz perturbations of non-densely defined operators. *Differential and Integral Equations*, 3(6):1035–1066, 1990.

[266] N. Van Minh, T. Naito, and G. M. N'Guérékata. A spectral countability condition for almost automorphy of solutions of differential equations. *Proceedings of the American Mathematical Society*, 134(11):3257–3266, 2006.

[267] N. Van Minh, G. M. N'Guérékata, and R. Yuan. *Lectures on the Asymptotic Behavior of Solutions of Differential Equations*. Nova Science Publishers, 2008.

[268] W. A. Veech. Almost automorphic functions on groups. *American Journal of Mathematics*, 87(3):719–751, 1965.

[269] W. A. Veech. On a theorem of Bochner. *Annals of Mathematics*, 86(1):117–137, 1967.

[270] G. F. Webb. *Theory of Nonlinear Age-Dependent Population Dynamics*. CRC Press, 1985.

[271] D. Wexler. Nonlinear passive evolution equations. *Journal of Differential Equations*, 23(3):414–435, 1977.

[272] J. Wu. *Theory and Applications of Partial Functional Differential Equations*, volume 119. Springer Science & Business Media, 2012.

[273] J. Wu and H. Xia. Self-sustained oscillations in a ring array of coupled lossless transmission lines. *Journal of Differential Equations*, 124(1):247–278, 1996.

[274] Z. Xia, M. Fan, and R. Agarwal. Pseudo almost automorphy of semilinear fractional differential equations in Banach spaces. *Fractional Calculus & Applied Analysis*, 19:741–764, 2016.

[275] T. J. Xiao, J. Liang, and J. Zhang. Pseudo almost automorphic solutions to semilinear differential equations in Banach spaces. *Semigroup Forum*, 76(3):518–524, 2008.

[276] T. J. Xiao, X. X. Zhu, and J. Liang. Pseudo-almost automorphic mild solutions to nonautonomous differential equations and applications. *Nonlinear Analysis: Theory, Methods & Applications*, 70(11):4079–4085, 2009.

[277] K. Yosida. *Functional Analysis*. Springer-Verlag, Berlin, Heidelberg, New York, 1980.

[278] R. Yuan. Pseudo-almost periodic solutions of second-order neutral delay differential equations with piecewise constant argument. *Nonlinear Analysis: Theory, Methods & Applications*, 41(7):871–890, 2000.

[279] R. Yuan. On Favard's theorems. *Journal of Differential Equations*, 249(8):1884–1916, 2010.

[280] S. Zaidman. Quasi-periodicità per una equazione operazionale del primo ordine. *Rendiconti della Accademia Nazionale dei Lincei*, 35:152–157, 1963.

[281] S. Zaidman. Remarks on differential equations with Bohr–Neugebauer property. *Journal of Mathematical Analysis and Applications*, 38(1):167–173, 1972.

[282] S. Zaidman. Bohr–Neugebauer theorem for operators of finite rank in Hilbert spaces. *Notices of the American Mathematical Society*, 21(7):A594–A594, 1974.

[283] S. Zaidman. Almost automorphic solutions of some abstract evolutions equations. *Istituto Lombardo-Accademia Di Scienze E Lettre, Estrato dai Rendiconti, Classe di Scienze (A)*, 110:578–588, 1976.

[284] S. Zaidman. *Almost-Periodic Functions in Abstract Spaces*, volume 126. Pitman Advanced Publishing Program, 1985.

[285] S. Zaidman. *Topics in Abstract Differential Equations*, volume 304. Longman Scientific & Technical, Harlow, UK, 1994.

[286] S. Zaidman. On the Bohr transform of almost-periodic solutions for some differential equations in abstract spaces. *International Journal of Mathematics and Mathematical Sciences*, 27:521–534, 2001.

[287] M. Zaki. Almost automorphic integrals of almost automorphic functions. *Canadian Mathematical Bulletin*, 15:433–436, 1972.

[288] M. Zaki. Almost automorphic solutions of certain abstract differential equations. *Annali di Matematica Pura ed Applicata*, 101(1):91–114, 1974.

[289] E. Zeidler and P. R. Wadsack. *Nonlinear Functional Analysis and Its Applications: Fixed-Point Theorems/Transl. by Peter R. Wadsack*. Springer-Verlag, 1993.

[290] C. Y. Zhang. Integration of vector-valued pseudo-almost periodic functions. *Proceedings of the American Mathematical Society*, 121(1):167–174, 1994.

[291] C. Y. Zhang. Pseudo almost periodic solutions of some differential equations. *Journal of Mathematical Analysis and Applications*, 181(1):62–76, 1994.

[292] C. Y. Zhang. Pseudo almost periodic solutions of some differential equations, II. *Journal of Mathematical Analysis and Applications*, 192(2):543–561, 1995.

[293] J. Zhao, Y. Chang, and G. M. N'Guérékata. Existence of asymptotically almost automorphic solutions to nonlinear delay integral equations. *Dynamic Systems and Applications*, 21:339–350, 2012.

[294] Z. Zhao, Y. Chang, and J. Nieto. Square-mean asymptotically almost automorphic process and its application to stochastic integro-differential equations. *Dynamic Systems and Applications*, 22:269–284, 2013.

[295] Z. M. Zheng and H. S. Ding. On completeness of the space of weighted pseudo almost automorphic functions. *Journal of Functional Analysis*, 268(10):3211–3218, 2015.

[296] Y. Zhou. *Basic Theory of Fractional Differential Equations*. World Scientific, Singapore, 2014.

[297] Y. Zhou and L. Peng. On the time-fractional Navier–Stokes equations. *Computers & Mathematics with Applications*, 73:874–891, 2017.

[298] Y. Zhou and L. Peng. Weak solutions of the time-fractional Navier–Stokes equations and optimal control. *Computers & Mathematics with Applications*, 73:1016–1027, 2017.

[299] Y. Zhou, V. Vijayakumar, and R. Murugesu. Controllability for fractional evolution inclusions without compactness. *Evolution Equations and Control Theory*, 4:507–524, 2015.

[300] Y. Zhou and L. Zhang. Existence and multiplicity results of homoclinic solutions for fractional Hamiltonian systems. *Computers & Mathematics with Applications*, 73:1325–1345, 2017.

Index

https://doi.org/10.1515/9783111684710-007

www.ingramcontent.com/pod-product-compliance
Lightning Source LLC
Chambersburg PA
CBHW082103220326
41598CB00066BA/5019
* 9 7 8 3 1 1 1 6 8 3 6 1 4 *